Lecture Notes in Computer Scie

Edited by G. Goos, J. Hartmanis, and J. van Leeuwen

Springer
Berlin
Heidelberg
New York
Hong Kong
London
Milan
Paris
Tokyo

Kenneth G. Paterson (Ed.)

Cryptography and Coding

9th IMA International Conference
Cirencester, UK, December 16-18, 2003
Proceedings

 Springer

Series Editors

Gerhard Goos, Karlsruhe University, Germany
Juris Hartmanis, Cornell University, NY, USA
Jan van Leeuwen, Utrecht University, The Netherlands

Volume Editor

Kenneth G. Paterson
Information Security Group
Royal Holloway, University of London
Egham, Surrey TW20 0EX, UK
E-mail: kenny.paterson@rhul.ac.uk

Cataloging-in-Publication Data applied for

A catalog record for this book is available from the Library of Congress.

Bibliographic information published by Die Deutsche Bibliothek
Die Deutsche Bibliothek lists this publication in the Deutsche Nationalbibliografie;
detailed bibliographic data is available in the Internet at <http://dnb.ddb.de>.

CR Subject Classification (1998): E.3-4, G.2.1, C.2, J.1

ISSN 0302-9743
ISBN 3-540-20663-9 Springer-Verlag Berlin Heidelberg New York

Springer-Verlag is a part of Springer Science+Business Media

springeronline.com

© Springer-Verlag Berlin Heidelberg 2003
Printed in Germany

Typesetting: Camera-ready by author, data conversion by PTP-Berlin, Protago-TeX-Production GmbH
Printed on acid-free paper   SPIN: 10966013   06/3142   5 4 3 2 1 0

Preface

The ninth in the series of IMA Conferences on Cryptography and Coding was held (as ever) at the Royal Agricultural College, Cirencester, from 16–18 December 2003. The conference's varied programme of 4 invited and 25 contributed papers is represented in this volume.

The contributed papers were selected from the 49 submissions using a careful refereeing process. The contributed and invited papers are grouped into 5 topics: coding and applications; applications of coding in cryptography; cryptography; cryptanalysis; and network security and protocols. These topic headings represent the breadth of activity in the areas of coding, cryptography and communications, and the rich interplay between these areas.

Assembling the conference programme and this proceedings required the help of many individuals. I would like to record my appreciation of them here.

Firstly, I would like to thank the programme committee who aided me immensely by evaluating the submissions, providing detailed written feedback for the authors of many of the papers, and advising me at many critical points during the process. Their help and cooperation was essential, especially in view of the short amount of time available to conduct the reviewing task. The committee this year consisted of Mike Darnell, Mick Ganley, Bahram Honary, Chris Mitchell, Matthew Parker, Nigel Smart and Mike Walker.

I would also like to thank those people who assisted the programme committee by acting as "secondary reviewers": Simon Blackburn, Colin Boyd, Alex Dent, Steven Galbraith, Keith Martin, James McKee, Sean Murphy, Dan Page, Matt Robshaw and Frederik Vercauteren. My apologies to any individuals missing from this list.

I am indebted to our four invited speakers for their contributions to the conference and this volume. The best candidates for invited speakers are always the most in-demand, and therefore busiest, people. This year's were no exception. Their contributions provided a valuable framing for the contributed papers.

My thanks too to the many authors who submitted papers to the conference. We were blessed this year with a strong set of submissions, and some good papers had to be rejected. I appreciate the understanding and good grace of those authors who were not successful with their submissions. I trust that they found any feedback from the reviewing process useful in helping to improve their work.

I am also grateful to the authors of accepted papers for their cooperation in compiling this volume: almost all of them met the various tight deadlines imposed by the production schedule. I would like to thank the staff at Springer-Verlag for their help with the production of this volume, especially Alfred Hofmann who answered many questions.

Much assistance was provided by Pamela Bye and Lucy Nye at the IMA. Their help took away much of the administrative burden, allowing the programme committee to focus on the scientific issues.

Valuable sponsorship for the conference was received from Hewlett-Packard Laboratories, Vodafone and the IEEE UKRI Communications Chapter.

Finally, I would like to thank my partner Liz for all her support during what was a very busy professional period for us both.

I Liz, Diolch o galon a llawer o gariad.

October 2003   Kenneth G. Paterson

Table of Contents

Coding and Applications

Recent Developments in Array Error-Control Codes 1
Patrick Guy Farrell

High Rate Convolutional Codes with Optimal Cycle Weights 4
Eirik Rosnes and Øyvind Ytrehus

A Multifunctional Turbo-Based Receiver Using Partial Unit
Memory Codes 24
Lina Fagoonee and Bahram Honary

Commitment Capacity of Discrete Memoryless Channels 35
Andreas Winter, Anderson C.A. Nascimento, and Hideki Imai

Separating and Intersecting Properties of BCH and Kasami Codes 52
Hans Georg Schaathun and Tor Helleseth

Applications of Coding in Cryptography

Analysis and Design of Modern Stream Ciphers...................... 66
Thomas Johansson

Improved Fast Correlation Attack Using Low Rate Codes 67
Håvard Molland, John Erik Mathiassen, and Tor Helleseth

On the Covering Radius of Second Order Binary Reed-Muller Code
in the Set of Resilient Boolean Functions........................... 82
Yuri Borissov, An Braeken, Svetla Nikova, and Bart Preneel

Degree Optimized Resilient Boolean Functions from
Maiorana-McFarland Class 93
Enes Pasalic

Differential Uniformity for Arrays 115
K.J. Horadam

Cryptography

Uses and Abuses of Cryptography.................................. 125
Richard Walton

A Designer's Guide to KEMs 133
Alexander W. Dent

A General Construction of IND-CCA2 Secure Public Key Encryption.... 152
Eike Kiltz and John Malone-Lee

Efficient Key Updating Signature Schemes Based on IBS 167
Dae Hyun Yum and Pil Joong Lee

Periodic Sequences with Maximal Linear Complexity and Almost
Maximal k-Error Linear Complexity 183
Harald Niederreiter and Igor E. Shparlinski

Cryptanalysis

Estimates for Discrete Logarithm Computations in Finite Fields of
Small Characteristic 190
Robert Granger

Resolving Large Prime(s) Variants for Discrete
Logarithm Computation 207
A.J. Holt and J.H. Davenport

Computing the $M = UU^t$ Integer Matrix Decomposition 223
Katharina Geißler and Nigel P. Smart

Cryptanalysis of the Public Key Cryptosystem Based on the Word
Problem on the Grigorchuk Groups 234
George Petrides

More Detail for a Combined Timing and Power Attack against
Implementations of RSA 245
Werner Schindler and Colin D. Walter

Predicting the Inversive Generator 264
*Simon R. Blackburn, Domingo Gomez-Perez, Jaime Gutierrez, and
Igor E. Shparlinski*

A Stochastical Model and Its Analysis for a Physical Random
Number Generator Presented At CHES 2002 276
Werner Schindler

Analysis of Double Block Length Hash Functions 290
Mitsuhiro Hattori, Shoichi Hirose, and Susumu Yoshida

Network Security and Protocols

Cryptography in Wireless Standards (Invited Paper) 303
Valtteri Niemi

On the Correctness of Security Proofs for the 3GPP Confidentiality and
Integrity Algorithms 306
Tetsu Iwata and Kaoru Kurosawa

A General Attack Model on Hash-Based Client Puzzles 319
    Geraint Price

Tripartite Authenticated Key Agreement Protocols from Pairings 332
    Sattam S. Al-Riyami and Kenneth G. Paterson

Remote User Authentication Using Public Information................. 360
    Chris J. Mitchell

Mental Poker Revisited 370
    Adam Barnett and Nigel P. Smart

Author Index 385

Recent Developments in Array Error-Control Codes

(Invited Paper)

Patrick Guy Farrell

Department of Communication Systems Lancaster University,
Lancaster, LA1 4YR, UK
P.G.Farrell@lancaster.ac.uk
Paddy.Farrell@virgin.net

Array error-control codes are linear block or convolutional codes, with code-words or coded sequences constructed by attaching check symbols to arrays of information symbols arranged in two or more dimensions. The check symbols are calculated by taking sums of the the information symbols lying along rows, columns, diagonals or other directions or paths in the information array. The simplest array code is the binary block code obtained by taking single parity checks across the rows and columns of a rectangular array of information bits. Array codes can be constructed with symbols from a field, ring or group, can have a wide range of parameters (block or constraint length, rate, distance, etc), and can be designed to detect and correct random and/or bursts or clusters of errors. The motivation for investigating and applying array codes (apart from their interesting mathematical aspects) is that they often provide a good trade-off between error-control power and complexity of decoding. The rate of a random error-control block array code, such as a product code, for example (classical product codes form a sub-class of array codes), is usually less than that of the best available alternative code with the same distance and length, but in exchange the array code will be much easier to decode [1]. However, in many cases array codes designed to correct burst error patterns can be both optimal (maximum distance separable (MDS), for example) and simpler to decode than other equivalent codes [1].

The aim of this presentation is to highlight the most important developments in array codes which have taken place since my survey paper [1] was published in 1992. The favourable trade-offs described above have in many ways been improved significantly since then, thus further widening the considerable range of practical applications of array codes in information transmission and storage systems. Perhaps the most exciting development comes from the realisation that array codes, and in particular various modified forms of classical product codes, are ideally suited for turbo decoding. Here, for example, iterative soft decoding of the row and column component codes in a two-dimensional array code gives a performance close to the Shannon limit with feasible complexity [2]. This very important development has also motivated much investigation of the weight dis-

K.G. Paterson (Ed.): Cryptography and Coding 2003, LNCS 2898, pp. 1–3, 2003.
© Springer-Verlag Berlin Heidelberg 2003

tributions of the code words of array and product codes, in order to optimise the performance of turbo and other decoding algorithms [3]. A number of novel alternative soft-decision decoding methods for array codes have been devised since 1992, which improve on previous methods [1], but fall short of the performance of turbo decoding.

Another important development arises from the concept of the generalised array code (GAC) and its associated coset trellis. Most well known optimal block codes (eg, Hamming, BCH, RS, etc) can be realised as generalised array codes, and their corresponding sectionalised coset trellises are feasible to decode using the Viterbi algorithm [4]. The GAC construction is a form of the Plotkin or u/u+v construction (augmentation or superimposition). The extent to which an array code can be generalised (augmented) is related to its covering radius, and for certain parameters the GAC construction is equivalent to the generalised concatenated code (GCC) construction.

Several new array code constructions for correcting linear or multidimensional error or erasure bursts have been proposed recently, including the projection, diamond, phased-burst-correcting, multi-track, crisscross-error-correcting, burst identification, EVENODD and X-Code constructions. Several of these constructions generate MDS array codes, and in many cases their encoding algorithms can be optimised to minimise the encoding delay. In addition they are very simple to decode; less complicated, for example, than equivalent RS-based codes [5,6]. Efficient new array code constructions combining error control with channel and system constraints have also emerged, including the run-length-limited, hard square model, checkerboard, balanced and conservative constructions. When encoding burst-error-correcting array codes the positions of the check symbols in the encoded array, and the order in which the symbols of the complete code word or sequence are read out of the encoder, are crucial in determining the error-control performance of the code. The latter process is equivalent to an interleaving operation, and this has motivated investigations of array codes based on explicit two-dimensional interleaving of simple component codes [7].

The range, power and feasibility of array codes has improved dramatically over the last decade: my presentation will attempt to bring this out, together with some of the interesting properties of array codes, and the many fascinating open problems that remain to be investigated.

References

1. P.G. Farrell: A survey of array error control codes; European Transactions on Telecommunications, Vol 3, No 5, pp 441–54, Sept-Oct, 1992.
2. R. Pyndiah: Near-optimum decoding of product codes: block turbo codes; IEEE Trans Commun, Vol 46, No 8, pp 1003–10, Aug, 1998.
3. D.M. Rankin and T.A. Gulliver: Single parity check product codes; IEEE Trans Commun, Vol 49, No 8, pp 1354–62, Aug, 2001.
4. B. Honary and G. Markarian: Trellis Decoding of Block Codes; Kluwer Academic Publishers, 1997.

5. M. Blaum, J. Bruck and A. Vardy: MDS array codes with independent parity checks; IEEEtrans Info Theory, Vol 42, No 2, pp 529–42, March, 1996.

6. Lihao Xu and J. Bruck: X-Code: MDS array codes with optimal encoding; IEEE Trans Info Theory, Vol 45, No 1, pp 272–6, Jan, 1999.

7. M. Blaum, J. Bruck and A. Vardy: Interleaving schemes for multidimensional cluster errors; IEEE Trans Info Theory, Vol 44, No 2, pp 730–43, March, 1998.

High Rate Convolutional Codes with Optimal Cycle Weights

Eirik Rosnes and Øyvind Ytrehus*

The Selmer Center, Dept. of Informatics,
University of Bergen,
P.B. 7800, N-5020 Bergen, Norway
{eirik,oyvind}@ii.uib.no

Abstract. Consider a cycle in the state diagram of a rate $(n-r)/n$ convolutional code. The total Hamming weight of all labels on the edges of the cycle divided by the number of edges is the average cycle weight per edge. Let w_0 denote the *minimum average cycle weight per edge* over all cycles in a minimal state diagram of a convolutional code, excluding the all-zero cycle around the all-zero state. For comparison between codes of different parameters, let $w \triangleq w_0/(n-r)$. This work investigates high rate convolutional codes with large w. We present an explicit construction technique of free distance $d_{\text{free}} = 4$ convolutional codes with limited bit-oriented trellis state complexity, high rate, and large w. The construction produces optimal codes, in the sense of maximizing w, within the classes of rate $(\nu(2^{\nu-1}+1)+2^{\nu}-1)/(\nu(2^{\nu-1}+2)+2^{\nu})$, $\nu \geq 2$, codes with $d_{\text{free}} \geq 3$ and any code degree. An efficient exhaustive search algorithm is outlined as well. A computer search was carried out, and several codes having larger w than codes in the literature were found.

1 Introduction

Let w_0 denote the *minimum average cycle weight per edge* over all cycles in a minimal state diagram of a convolutional code, excluding the all-zero cycle around the all-zero state. Codes with low w_0 contain long codewords of low weight. These codes are susceptible to long error events when used with either maximum likelihood (Viterbi) or sequential decoding [1], [2]. The *active distances* for convolutional codes [3], describing which error patterns are guaranteed to be corrected under a maximum likelihood decoding assumption, are lower bounded by a linearly increasing function with *slope* w_0. In this sense, w_0 determines the code's error correcting capability.

When working with concatenated codes, e.g., serial concatenated convolutional codes, decoded using iterative decoding schemes, simulation results indicate that outer codes with large w_0 and small degree compare favorably with other choices for the outer code [4]. Further, convolutional codes with large w_0 and free distance d_{free} can be used to obtain good tailbiting codes [5].

* This work was supported by Nera Research and the Norwegian Research Council (NFR) Grants 137783/221 and 146874/420.

K.G. Paterson (Ed.): Cryptography and Coding 2003, LNCS 2898, pp. 4–23, 2003.

Huth and Weber [6] derived a general upper bound on w_0 for rate $1/n$ convolutional codes. Recently, Jordan et al. [7] generalized the bound to rate $(n-r)/n$, $r \geq 1$, codes. The new upper bound applies to codes having a canonical generator matrix containing a delay-free $(n-r) \times (n-r)$ minor of degree equal to the code degree. A general lower bound on w_0 can be found in Hole and Hole [2].

In [1], Hemmati and Costello derived a tight upper bound on w_0 for a special class of rate $1/2$ convolutional codes. By generalizing the approach in [1], the authors of [2] showed that the bound in [1] applies to a particular class of rate $(n-1)/n$, $n \geq 2$, codes as well. As shown in [8], the argument can further be generalized to the most general case of rate $(n-r)/n$, $r \geq 1$, codes.

In [9], several convolutional code classes that are not asymptotically catastrophic were presented. In particular, a construction of rate $(2^{2\nu-1} - \nu)/2^{2\nu-1}$, $\nu \geq 2$, codes with code degree ν, $d_{\text{free}} = 4$, and $w_0 = 1$ was given. We will compare this construction with our proposed construction, and show that the new construction compares favorably.

For comparison between codes of different parameters, it is convenient to define instead the *minimum average cycle weight per information bit*, $w \triangleq w_0/(n - r)$ of a rate $(n - r)/n$ convolutional code. In this work we present rate $(n - r)/n$, $r \geq 1$, convolutional codes with larger w compared to existing codes found by Hole and Hole [2], and by Jordan et al. through puncturing of rate $1/2$ mother codes [4]. In [10], Hole presented a general method to obtain an upper bound on w. We will use this method to limit the search space.

The concept of a minimal trellis of a linear block code is well established. Recently, the theory has been generalized to convolutional codes by Sidorenko and Zyablov in [11], and by McEliece and Lin in [12]. A parity check matrix approach was considered in [11], generalizing the BCJR construction [13] to convolutional codes. In fact, this construction is known to be minimal in the sense that it gives the unique trellis which minimizes the number of states in each depth of any trellis representing the code. In [12], a generator matrix approach was outlined. The trellis state complexity profile of a linear block or convolutional code is a vector consisting of the number of trellis states in each depth of a minimal trellis. The bit-oriented trellis state complexity is informally defined here as the base-2 logarithm of the maximum entry in the trellis state complexity profile. A more precise definition is provided in Section 2.

This paper is organized as follows: Section 2 introduces a convenient matrix notation for convolutional codes. An explicit convolutional code construction technique of $d_{\text{free}} = 4$ codes with limited bit-oriented trellis state complexity, high rate, and large w is included in Section 3. In Section 4, an efficient exhaustive search algorithm is described. The results of a computer search are given in Section 5. Finally, some conclusions are drawn in Section 6.

2 Convolutional Code Preliminaries

We use some of the notation and definitions introduced in [14], as well as a convenient matrix notation established in [15], [16], and [9].

A linear $(n, n-r, \nu)$ convolutional code $\mathcal{C}$ is an $(n-r)$-dimensional subspace in an n-dimensional vector space $F(D)^n$, where $F(D)$ is the field of rational functions in the indeterminate D over the field F [14]. The code degree, or the overall constraint length, is denoted by ν. In this work the convolutional code symbols are taken from the binary field $F = GF(2)$. A convolutional code can be defined by an $r \times n$ polynomial parity check matrix $\mathbf{H}(D)$. We assume in general a canonical parity check matrix [14]. Let the jth polynomial in the ith row of $\mathbf{H}(D)$ be denoted by $h_j^{(i)}(D) = h_{j,0}^{(i)} + h_{j,1}^{(i)} D + \cdots + h_{j,\nu_i}^{(i)} D^{\nu_i} \in F[D]$, where $F[D]$ is the ring of all polynomials in D with coefficients in F. The maximum degree of the polynomials in the ith row is the ith row degree, denoted by ν_i. Every canonical parity check matrix of a given convolutional code have the same set of row degrees with $\nu = \sum_{i=1}^{r} \nu_i$ [14].

The coefficients of $h_j^{(i)}(D)$ define a column vector $\mathbf{h}_j^{(i)}$ with $h_{j,0}^{(i)}$ as its topmost element. The n polynomials in the ith row of the parity check matrix $\mathbf{H}(D)$ give rise to a $(\nu_i + 1) \times n$ matrix $\mathbf{H}^{(i)} = (\mathbf{h}_1^{(i)}, \ldots, \mathbf{h}_n^{(i)})$ over the field F. Furthermore, let $\mathbf{H}$, referred to as a *combined parity check matrix*, be defined as

$$\mathbf{H} = \begin{pmatrix} \mathbf{H}^{(1)} \\ \mathbf{H}^{(2)} \\ \vdots \\ \mathbf{H}^{(r)} \end{pmatrix} = (\mathbf{h}_1, \ldots, \mathbf{h}_n). \tag{1}$$

Note that the combined parity check matrix is a $(\nu + r) \times n$ matrix over the field F. The D-transform of $\mathbf{H}$ is $\mathbf{H}(D)$, where the polynomial in the ith row and jth column is the previously defined polynomial $h_j^{(i)}(D)$.

Let $\mathbf{x} = (x_a x_{a+1} \cdots x_b)^T$ be a finite dimensional column vector, where $(\cdot)^T$ denotes the transpose of its argument. The lth *left shift* of $\mathbf{x}$, denoted by $\mathbf{x}^{\llcorner}$, is defined as $\mathbf{x}^{\llcorner} = (x_{a+l} \cdots x_b 0 \cdots 0)^T$ where $l \geq 0$. The last l coordinates in $\mathbf{x}^{\llcorner}$ are equal to zero. Furthermore, the lth *right shift* of $\mathbf{x}$, denoted by $\mathbf{x}^{\lrcorner}$, is defined as $\mathbf{x}^{\lrcorner} = (0 \cdots 0 x_a \cdots x_{b-l})^T$. The first l coordinates in $\mathbf{x}^{\lrcorner}$ are equal to zero.

A codeword in $\mathcal{C}$ is a semi-infinite sequence of n-tuples, or column vectors. An arbitrary codeword sequence is denoted as $\mathbf{v} = (\mathbf{v}_0, \mathbf{v}_1, \ldots)$, where $\mathbf{v}_t$, $t \geq 0$, is an n-tuple (or *label*) given as $\mathbf{v}_t = (v_t^1 \cdots v_t^n)^T$. We now define the ith *syndrome vector* $\mathbf{s}_t^{(i)} = (s_{t,0}^{(i)} \cdots s_{t,\nu_i}^{(i)})^T$ of dimension $(\nu_i + 1)$ at time t, recursively as follows $(t > 0, 1 \leq i \leq r)$:

$$\mathbf{s}_t^{(i)} = (\mathbf{s}_{t-1}^{(i)})^{\llcorner} + \mathbf{H}^{(i)} \mathbf{v}_t, \tag{2}$$

with $\mathbf{s}_0^{(i)}$ equal to a fixed vector, e.g., the all-zero vector. Furthermore, we define the ith *syndrome sequence* as $(s_{0,0}^{(i)} s_{1,0}^{(i)} \cdots)$ where $s_{t,0}^{(i)}$, $t \geq 0$, is the zeroth element in the ith syndrome vector $\mathbf{s}_t^{(i)}$. The code consists of all semi-infinite sequences $\mathbf{v}$ such that all the corresponding syndrome sequences are equal to the all-zero sequence.

A compact form of (2) is obtained using the combined parity check matrix. The *combined syndrome vector* at time t, $\mathbf{s}_t = ((\mathbf{s}_t^{(1)})^T, \ldots, (\mathbf{s}_t^{(r)})^T)^T = (s_{t,0}^{(1)} \cdots s_{t,\nu_1}^{(1)}, \ldots, s_{t,0}^{(r)} \cdots s_{t,\nu_r}^{(r)})^T$ is given by

$$\mathbf{s}_t = (\mathbf{s}_{t-1})^{\mathbf{\downarrow}} + \mathbf{H}\mathbf{v}_t, \tag{3}$$

where $(\mathbf{s}_{t-1})^{\mathbf{\downarrow}} = (((\mathbf{s}_{t-1}^{(1)})^{\mathbf{\downarrow}})^T, \ldots, ((\mathbf{s}_{t-1}^{(r)})^{\mathbf{\downarrow}})^T)^T$, i.e., the shift operator should be applied to each component individually. The combined syndrome vectors in (3) have dimension $\nu + r$. Assuming that the syndrome vectors are computed from codewords, the set of possible combined syndrome vectors $\mathcal{V}_t = \{\mathbf{s}_t : (\mathbf{v}_0, \mathbf{v}_1, \ldots) \in \mathcal{C}\}$ is a vector space of dimension ν (after an initial transient).

Example 1. Consider a $(6, 4, 4)$ binary convolutional code with free distance 5 defined by the polynomial canonical parity check matrix

$$\mathbf{H}(D) = \begin{pmatrix} 3 & 2 & 5 & 3 & 2 & 3 \\ 0 & 1 & 1 & 5 & 7 & 7 \end{pmatrix}, \tag{4}$$

where the entries are given in octal notation in the sense that $3 = 011 = 1 + D$. The corresponding binary combined parity check matrix is

$$\mathbf{H} = \begin{pmatrix} \mathbf{H}^{(1)} \\ \mathbf{H}^{(2)} \end{pmatrix} = \begin{pmatrix} 1 & 0 & 1 & 1 & 0 & 1 \\ 1 & 1 & 0 & 1 & 1 & 1 \\ 0 & 0 & 1 & 0 & 0 & 0 \\ \hline 0 & 1 & 1 & 1 & 1 & 1 \\ 0 & 0 & 0 & 0 & 1 & 1 \\ 0 & 0 & 0 & 1 & 1 & 1 \end{pmatrix}, \tag{5}$$

where the solid line separates the two component matrices $\mathbf{H}^{(1)}$ and $\mathbf{H}^{(2)}$. Further, let $\mathbf{s}_{t-1} = (011, 011)^T$, $\mathbf{v}_t = (000001)^T$, and $\mathbf{v}_{t+1} = (000000)^T$. From (3) we have $\mathbf{s}_t = (000, 001)^T$ and $\mathbf{s}_{t+1} = (000, 010)^T$. The binary column vectors $\mathbf{v}_t$ and $\mathbf{v}_{t+1}$ are contained in a codeword since $s_{t,0}^{(1)} = s_{t,0}^{(2)} = s_{t+1,0}^{(1)} = s_{t+1,0}^{(2)} = 0$. Note that when the label is the all-zero label, the combined syndrome vector at the next time instant is obtained by simply shifting each component of the previous combined syndrome vector individually.

A code $\mathcal{C}$ may be represented by a state diagram, where each state represents a combined syndrome vector. Obviously, the number of states in the state diagram is equal to 2^ν. The set of possible transitions between states is determined by the equation in (3). An important observation is that a transition from state $\mathbf{s}$ on an edge with weight zero will lead to the state $\mathbf{s}^{\mathbf{\downarrow}}$. (The weight of an edge is the Hamming weight of the label on the edge.)

A path of length p in a state diagram consists of p consecutive edges. A *cycle* is a path returning back to the state where it started, in which the intermediate states are distinct and different from the starting state.

The combinatorial decoding complexity of a convolutional code $\mathcal{C}$ under *bit-oriented* maximum likelihood decoding depends on the trellis state complexity profile of $\mathcal{C}$. The minimal trellis of a convolutional code is periodic, since it can be written as an infinite composition of a basic building block called the *trellis module*. Further, the trellis is bit-oriented in the sense that there is only a single bit on each edge in the trellis. As indicated in [16], there are r levels within the trellis module, in which every state have only a single outgoing edge. By a blocking procedure it is possible to reduce the *length* of the trellis module from n to $n - r$ [16]. When we speak of the trellis state complexity profile of a code $\mathcal{C}$ we mean the trellis state complexity profile of this reduced trellis module.

Let $\mathbf{b} = (b_0 \cdots b_n)$ be a row vector with $b_0 = 0$ and $b_j = \mathrm{rank}[\mathbf{H}^L_{n-j+1,n}]$, $1 \le j \le n$, where $\mathbf{H}^L_{n-j+1,n}$ is the submatrix of $\mathbf{H}^L$ consisting of the last j columns, and $\mathbf{H}^L$ is the zero-degree indicator matrix of $\mathbf{H}(D)$. Here, $\mathrm{rank}(\cdot)$ denotes the rank of its matrix argument, where the rank of a matrix is the dimension of its column or row space. The zero-degree indicator matrix of $\mathbf{H}(D)$ is $\mathbf{H}(0)$. Further, let $\mathbf{f} = (f_0 \cdots f_n)$ be a row vector with $f_0 = 0$ and $f_j = \mathrm{rank}[\mathbf{H}^H_{1,j}]$, $1 \le j \le n$, where $\mathbf{H}^H_{1,j}$ is the submatrix of $\mathbf{H}^H$ consisting of the first j columns, and $\mathbf{H}^H$ is the highest-degree indicator matrix of $\mathbf{H}(D)$. The highest-degree indicator matrix of $\mathbf{H}(D)$ is the matrix

$$\begin{pmatrix} D^{\nu_1} & & & \\ & D^{\nu_2} & & \\ & & \ddots & \\ & & & D^{\nu_r} \end{pmatrix} \mathbf{H}(D^{-1}) \tag{6}$$

with $D = 0$, where the first matrix in (6) is a diagonal matrix. A sorted integer set $J = \{j_1, j_2, \ldots, j_{n-r}\}$ is determined by the row vector $\mathbf{b}$. In general,

$$J = \{j : j \in \{0, \ldots, n-1\} \text{ and } b_{n-j} = b_{n-j-1}\}. \tag{7}$$

Then, the ith component of the trellis state complexity profile c_i, $1 \le i \le n - r$, is[1]

$$c_i = 2^{\nu - r + f_{j_i} + b_{n-j_i}}. \tag{8}$$

Definition 1. *The base-2 logarithm of the maximum number of states in the trellis state complexity profile*

$$c_{\max} = c_{\max}(\mathbf{H}) = \log_2 \left[\max_{1 \le i \le n-r} c_i \right] \tag{9}$$

will be called the bit-oriented trellis state complexity.

Example 2. For the code defined by the canonical parity check matrix in (4), the zero-degree indicator matrix and the highest-degree indicator matrix are

$$\mathbf{H}^L = \begin{pmatrix} 1 & 0 & 1 & 1 & 0 & 1 \\ 0 & 1 & 1 & 1 & 1 & 1 \end{pmatrix} \quad \text{and} \quad \mathbf{H}^H = \begin{pmatrix} 0 & 0 & 1 & 0 & 0 & 0 \\ 0 & 0 & 0 & 1 & 1 & 1 \end{pmatrix}, \tag{10}$$

respectively. Here, $J = \{0, 1, 2, 3\}$, $(16, 16, 16, 32)$ is the trellis state complexity profile, and $c_{\max} = 5$.

---

[1] The result is a generalization of a similar result for linear block codes [17].

2.1   Bounds on w

The following theorem, giving an upper bound on w, was proved in [7].

Theorem 1. *For an $(n, n-r)$, $r \geq 1$, convolutional code C with code degree ν and a canonical generator matrix containing a delay-free $(n-r) \times (n-r)$ minor of degree ν,*

$$w \leq \frac{r}{n-r} \frac{2^{\nu-1}}{2^\nu - 1}. \tag{11}$$

Let $d_{\max}(N, K)$ be the largest minimum distance of all (N, K) linear block codes. A new upper bound on w depending on the free distance on the convolutional code was presented and proved in [5]. For convenience, we will restate the theorem below.

Theorem 2. *For an $(n, n-r)$, $r \geq 1$, convolutional code C with free distance d_{free},*

$$w \leq \frac{1}{n-r} \min \left\{ \frac{d_{\max}(ln, l(n-r))}{l} \right\}, \tag{12}$$

where the minimum is taken over all integers $l = 1, 2, \ldots$ such that $d_{\max}(ln, l(n-r)) < d_{\text{free}}$.

Assume that the rows of a canonical parity check matrix $\mathbf{H}(D)$ are ordered such that the row degrees satisfy $\nu_1 = \nu_2 = \cdots = \nu_\gamma = 0$ for some γ, $0 \leq \gamma < r$. If $\gamma > 0$, then the first γ rows of the parity check matrix define an $(n, n-\gamma)$ linear block code $\tilde{C}$ over F, which is the *embedded block code* [18]. For $\gamma = 0$, we define $\tilde{C}$ to be the trivial (n, n) block code consisting of all n-tuples over F. The parity check matrix of $\tilde{C}$ is denoted by $\mathbf{H}_\gamma$, and the minimum distance by $d_{\tilde{C}}$. For $\gamma = 0$, $d_{\tilde{C}} = 1$. In general, the labels on the edges in a state diagram for C, constructed from $\mathbf{H}$, are codewords in $\tilde{C}$. The following general lower bound on w can be found in [2].

Theorem 3. *For an $(n, n-r)$, $r \geq 1$, convolutional code C defined by a canonical combined parity check matrix $\mathbf{H}$ with row degrees ν_i, $1 \leq i \leq r$,*

$$w = w(\mathbf{H}) \geq \frac{1}{n-r} \frac{d_{\tilde{C}}}{\nu_{\max}}, \tag{13}$$

where $\nu_{\max}$ is the maximum row degree.

3   Explicit Constructions

We will first consider a construction of codes with $d_{\text{free}} = 4$ and $w_0 = 1$ presented in [9]. The construction is a special case of a more general construction which appeared in [15]. Secondly, we will present a new construction of $d_{\text{free}} = 4$ codes that compares favorably with the first construction.

3.1   Construction from [9]

Given a code degree $\nu \geq 2$, choose $\gamma = 1$, $\nu_2 = \cdots = \nu_{r-1} = 1$, and $\nu_r = 2$, from which it follows that $r = \nu$. Further, let the construction technique for any ν be denoted by $\tilde{\mathcal{M}}_\nu$, and choose $n = 2^{2\nu-1}$.

Let $\mathbf{Q}(D)$ denote a $1 \times n$ polynomial matrix consisting of every polynomial with a constant term and degree at most $\nu + r - 1 = 2\nu - 1$. Further, construct the parity check matrix of a convolutional code $\mathcal{C}$ as follows:

$$\mathbf{H}_{\text{comp}}(D) = \mathbf{Q}(D). \tag{14}$$

Remark 1. The parity check matrix in (14) is given in *compressed form*, i.e., a combined parity check matrix is obtained by converting each polynomial to a column vector with the coefficient of the constant term and the coefficient of $D^{2\nu-1}$ as its topmost and last element, respectively.

Codes from the construction $\tilde{\mathcal{M}}_\nu$, $\nu \geq 2$, have $d_{\text{free}} = 4$ and $w_0 = 1$ [9], [19]. The codes from $\tilde{\mathcal{M}}_\nu$ are *length optimal* [15], meaning that there do not exist codes with code degree ν, $d_{\text{free}} = 4$, and strictly larger code length within the class of $(n, n-\nu)$ codes. Thus, not surprisingly it turns out that $c_{\text{max}} = \nu + r = 2\nu$ for these codes.

3.2   New Construction of Codes with $d_{\text{free}} = 4$

In this subsection we consider an explicit construction of codes with code degree $\nu \geq 2$, $d_{\text{free}} = 4$, $r \geq 2$, and $c_{\text{max}} = \nu + 1$.

Choose $\gamma = 1$. Let the construction technique for any given $r \geq 2$ and row degrees $1 \leq \nu_2 \leq \cdots \leq \nu_r$ satisfying $\nu = \sum_{i=1}^{r} \nu_i \geq 2$, be denoted by $\mathcal{M}_r(\nu_2, \ldots, \nu_r)$, and choose $n = (r - 1)(2^{\nu-1} + 2) + 2^\nu$. Define

$$I_{\text{const}}[i] = \sum_{j=0}^{i-1} \nu_j + i - 1 \quad \text{and} \quad I_{\text{high}}[i] = \sum_{j=1}^{i} \nu_j + i - 1 \tag{15}$$

for all i, $1 \leq i \leq r$. In the first definition above, ν_0 is defined to be zero.

Let $\mathbf{P}^{(j)}(D)$, $2 \leq j \leq r$, denote a $1 \times (2^{\nu-1} + 2)$ polynomial matrix consisting of 1) every polynomial of degree at most $\nu + r - 1$ with the coefficients of $D^{I_{\text{const}}[1]}$ and $D^{I_{\text{const}}[j]}$ equal to one, the coefficients of $D^{I_{\text{const}}[i]}$, $2 \leq i \leq j - 1$, all equal to zero, and the coefficients of $D^{I_{\text{high}}[i]}$, $j \leq i \leq r$, all equal to zero. Further, 2) $\mathbf{P}^{(j)}(D)$ should contain 2 polynomials of degree at most $\nu + r - 1$ with the coefficients of $D^{I_{\text{const}}[1]}$, $D^{I_{\text{const}}[j]}$, and $D^{I_{\text{high}}[j]}$ all equal to one, the coefficients of $D^{I_{\text{const}}[i]}$, $2 \leq i \leq j - 1$, all equal to zero, and the coefficients of $D^{I_{\text{high}}[i]}$, $j < i \leq r$, all equal to zero. The polynomials in $\mathbf{P}^{(j)}(D)$ are sorted in increasing order with respect to the coefficient of $D^{I_{\text{high}}[j]}$. Let $\mathbf{P}^{(r+1)}(D)$ denote a $1 \times 2^\nu$ polynomial matrix consisting of every polynomial with degree at most $\nu + r - 1$ with the coefficients of $D^{I_{\text{const}}[i]}$, $2 \leq i \leq r$, all equal to zero, and the coefficient of $D^{I_{\text{const}}[1]}$ equal to one. We construct the parity check matrix of a convolutional code $\mathcal{C}$ as follows:

$$\mathbf{H}_{\text{comp}}(D) = \left(\mathbf{P}^{(2)}(D)\,\mathbf{P}^{(3)}(D)\,\cdots\,\mathbf{P}^{(r+1)}(D)\right). \tag{16}$$

The parity check matrix in (16) is given in compressed form, see Remark 1 for details.

Example 3. Choose $r = 3$, $\nu_2 = 1$, and $\nu_3 = 2$. In this case, $n = 20$ and

$$\mathbf{H}(D) = \begin{pmatrix} 1\,1\,1\,1\,1\,1\,1\,1\,1\,1\,1\,1\,1\,1\,1\,1\,1\,1\,1\,1 \\ 1\,1\,1\,1\,3\,3\,0\,0\,2\,2\,2\,2\,0\,2\,0\,2\,0\,2\,0\,2 \\ 0\,2\,1\,3\,2\,1\,1\,3\,1\,3\,5\,7\,0\,0\,2\,2\,4\,4\,6\,6 \end{pmatrix}, \tag{17}$$

where the matrix entries are given in octal notation in the sense that $6 = 110 = D + D^2$. The rate of the code is $17/20$.

Example 4. Choose $r = 4$ and $\nu_2 = \nu_3 = \nu_4 = 1$. In this case, $n = 26$ and

$$\mathbf{H}(D) = \begin{pmatrix} 1\,1\,1\,1\,1\,1\,1\,1\,1\,1\,1\,1\,1\,1\,1\,1\,1\,1\,1\,1\,1\,1\,1\,1\,1\,1 \\ 1\,1\,1\,1\,3\,3\,0\,0\,2\,2\,2\,2\,0\,0\,2\,2\,2\,2\,0\,2\,0\,2\,0\,2\,0\,2 \\ 0\,0\,1\,1\,0\,1\,1\,1\,1\,3\,3\,0\,2\,0\,2\,0\,2\,0\,0\,2\,2\,0\,0\,2\,2 \\ 0\,1\,0\,1\,0\,1\,0\,1\,0\,1\,0\,1\,1\,1\,1\,1\,3\,3\,0\,0\,0\,0\,2\,2\,2\,2 \end{pmatrix}, \tag{18}$$

where the matrix entries are given in octal notation. The rate of the code is $22/26$.

The first row of the parity check matrix in (16), in binary combined form, consists of only 1's, and all columns, viewed as binary vectors, are distinct. This, and the fact that $\nu_1 = 0$, guarantee that the free distance is at least 4 [15].

Lemma 1. *The matrix in (16) is canonical for any $r \geq 2$ and $1 \leq \nu_2 \leq \cdots \leq \nu_r$, $\nu = \sum_{i=1}^{r} \nu_i \geq 2$.*

To prove Lemma 1, we use the fact that an $r \times n$ polynomial parity check matrix is canonical if the greatest common divisor (gcd) of the $r \times r$ minors is equal to 1 and their greatest degree is equal to the code degree ν [14]. It is sufficient to prove that the matrix in (16) contains one $r \times r$ submatrix with determinant 1 and another submatrix with determinant of degree ν. The details are omitted for brevity. The proof of Lemma 2 below is given in the Appendix.

Lemma 2. *The bit-oriented trellis state complexity of codes from the construction $\mathcal{M}_r(\nu_2, \ldots, \nu_r)$ is $\nu + 1$ for any $r \geq 2$ and $1 \leq \nu_2 \leq \cdots \leq \nu_r$, $\nu = \sum_{i=1}^{r} \nu_i \geq 2$.*

Theorem 4. *For a code $\mathcal{C}$ from the construction $\mathcal{M}_r(\nu_2, \ldots, \nu_r)$, $w_0 = 2/\nu_r$ for any $r \geq 2$ and $1 \leq \nu_2 \leq \cdots \leq \nu_r$, $\nu = \sum_{i=1}^{r} \nu_i \geq 2$.*

Proof. The state diagram of a code $\mathcal{C}$ from the construction $\mathcal{M}_r(\nu_2, \ldots, \nu_r)$ contains a cycle consisting of an all-zero path of length $\nu_r - 1$ and an edge of weight 2. An all-zero path from the state $\mathbf{s} = (0, 0 \cdots 0, \ldots, 0 \cdots 0, 0 \cdots 01)^T$ to $\mathbf{s}^{\downarrow_{r-1}}$ exists, from which there is an edge of weight 2 back to $\mathbf{s}$. This is the case since there exist two combined columns $\mathbf{h}$ and $\tilde{\mathbf{h}}$ in any combined parity check matrix from $\mathcal{M}_r(\nu_2, \ldots, \nu_r)$ such that $(\mathbf{s}^{\downarrow_{r-1}})^{\downarrow_1} + \mathbf{h} + \tilde{\mathbf{h}} = \mathbf{s}$. Consequently, there exists a cycle of length ν_r with Hamming weight of 2, which implies that $w_0 \leq 2/\nu_r$. Using Theorem 3, $w_0 \geq 2/\nu_r$ since $d_{\tilde{c}} = 2$ and $\nu_{\max} = \nu_r$. $\qquad\square$

Remark 2. The choice $\nu_2 = \cdots = \nu_r = 1$ will maximize w when ν is given. In this case $r = \nu + 1$ and $w = 2/(\nu(2^{\nu-1} + 1) + 2^\nu - 1)$. The maximum possible rate is obtained for $r = 2$.

Theorem 5. *The construction $\mathcal{M}_r(\nu_2, \ldots, \nu_r)$ with $\nu_2 = \cdots = \nu_r = 1$ and $r \geq 3$ is optimal in the sense of giving codes with maximum possible w within the classes of $(\nu(2^{\nu-1} + 2) + 2^\nu, \nu(2^{\nu-1} + 1) + 2^\nu - 1)$, $\nu \geq 2$, codes with $d_{\text{free}} \geq 3$ and any code degree.*

Proof. We use Theorem 2 with $3 \leq d_{\text{free}} \leq 4$, $r = \nu + 1$, $n = \nu(2^{\nu-1} + 2) + 2^\nu$, and $n - r = \nu(2^{\nu-1} + 1) + 2^\nu - 1$. The proof is 3-fold. We first prove that $d_{\max}(n, n - r) = 2$. Secondly, we prove that $d_{\max}(2n, 2(n - r)) \geq 4$. In part III of the proof, we use Theorem 2 to derive an upper bound on w.

Part I: Since $2 \leq \nu \Rightarrow 4 \leq \nu + 2 \Rightarrow 2^{\nu+1} \leq \nu 2^{\nu-1} + 2^\nu$, it follows that $2^{\nu+1} < \nu(2^{\nu-1} + 2) + 2^\nu = n$, from which we can conclude that it is not possible to construct a parity check matrix with $r = \nu + 1$ rows and n distinct columns different from the all-zero column. Thus, $d_{\max}(n, n - r) = 2$.

Part II: Any linear block code defined by a parity check matrix with distinct, odd weight columns has minimum distance at least 4. If we can show that $2^{2r-1} = 2^{2\nu+1} \geq 2n = 2\nu(2^{\nu-1} + 2) + 2^{\nu+1}$, then it is always possible to construct a $2r \times 2n$ parity check matrix with $2n$ distinct, odd weight columns, and it follows that $d_{\max}(2n, 2(n - r)) \geq 4$.

By insertion, for $\nu = 2$, it holds that $2^{2\nu+1} > 2\nu(2^{\nu-1} + 2) + 2^{\nu+1}$. Thus, in the following we will assume that $\nu \geq 3$. Clearly,

$$\nu < 2^\nu \Rightarrow 2\nu(2^{\nu-1} + 2) + 2^{\nu+1} < 2^{\nu+1}(2^{\nu-1} + 3) \qquad (19)$$

and

$$2^{\nu-1} + 3 < 2^\nu \Rightarrow 2^{\nu+1}(2^{\nu-1} + 3) < 2^{2\nu+1}, \qquad (20)$$

from which it follows that $2\nu(2^{\nu-1} + 2) + 2^{\nu+1} < 2^{2\nu+1}$ since both $\nu < 2^\nu$ and $2^{\nu-1} + 3 < 2^\nu$ are true for all $\nu \geq 3$. Thus, it holds that $d_{\max}(2n, 2(n - r)) \geq 4$ for all $\nu \geq 2$.

Part III: Using Theorem 2,

$$w \leq \frac{1}{n - r} \frac{d_{\max}(n, n - r)}{1} = \frac{2}{n - r}. \qquad (21)$$

The upper bound in (21) applies to free distance ≥ 3 codes with any code degree, and the bound is equal to the w provided by Theorem 4 with $\nu_{\max} = 1$, proving the optimality of the construction $\mathcal{M}_r(\nu_2, \ldots, \nu_r)$ with $\nu_2 = \cdots = \nu_r = 1$ and $r \geq 3$.   □

Note that the optimal codes from the construction $\mathcal{M}_r(\nu_2, \ldots, \nu_r)$ with $\nu_2 = \cdots = \nu_r = 1$ and $r \geq 3$ have limited bit-oriented trellis state complexity. In particular, it is not possible to increase w by relaxing the complexity constraint or the distance constraint from 4 to 3, or by altering the code degree.

3.3 Comparisons

In Fig. 1, the negative of the natural logarithm of the code rate $- \log[(n - r)/n]$, and the minimum average cycle weight per information bit w from the two constructions $\tilde{\mathcal{M}}_\nu$ and $\mathcal{M}_r(1, \ldots, 1)$, $r \geq 3$, are plotted versus bit-oriented trellis state complexity $c_{\max}$. Observe that our proposed construction outperforms the construction $\tilde{\mathcal{M}}_\nu$. This is illustrated in Fig. 1 by the vertical solid line originating at the horizontal axis for $c_{\max} = 8$. Follow this line until the intersection with the curve marked with ∇. The value on the vertical axis gives the obtainable rate from the construction $\mathcal{M}_r(1, \ldots, 1)$. Further, follow the solid line to the right until the curve marked with $+$ is reached. The value on the horizontal axis gives the corresponding $c_{\max}$ from the construction $\tilde{\mathcal{M}}_\nu$. Finally, follow the line downward until the curve marked with $*$ is intersected. The w from the construction $\tilde{\mathcal{M}}_\nu$ is established from the value on the vertical axis, which conveniently can be compared with the w from the construction $\mathcal{M}_r(1, \ldots, 1)$. To summarize, for the same code rate, both a larger w and a smaller $c_{\max}$ is obtained from the construction $\mathcal{M}_r(1, \ldots, 1)$ compared to what is possible with the construction $\tilde{\mathcal{M}}_\nu$.

4 Exhaustive Computer Search

In this section we will formulate several lemmas and one theorem that are important for the exhaustive search algorithm to be described later in this section.

Lemma 3. *Let* $\mathbf{H}$ *denote a combined parity check matrix defining a code* $\mathcal{C}$ *with free distance* d_{free}. *Then the following holds:*

1. $d_{\mathrm{free}} = 1$ *if and only if* $\mathbf{H}$ *contains the all-zero combined column.*
2. *If* $\mathbf{H}$ *contains two equal combined columns, then* $d_{\mathrm{free}} = 2$.

The proof of Lemma 3 is trivial, and is omitted for brevity. In this work only codes with $d_{\mathrm{free}} \geq 3$ are considered. Thus, using Lemma 3, we will assume without loss of generality that every combined parity check matrix have distinct combined columns different from the all-zero combined column.

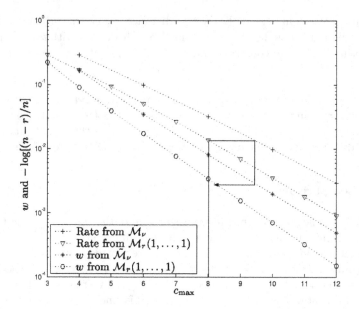

Fig. 1. w and $-\log[(n-r)/n]$ versus $c_{\max}$ from the two constructions $\tilde{\mathcal{M}}_\nu$ and $\mathcal{M}_r(1,\ldots,1)$, $r \geq 3$

Definition 2. *Let* **H** *denote any combined parity check matrix with row degrees* ν_i, $1 \leq i \leq r$, *such that the sum* $\sum_{i=1}^{r} \nu_i = \nu$. *Then define a mapping* φ *as*

$$\varphi : F^{\nu+r} \to F^\infty$$

$$\left(x_0^{(1)} \cdots x_{\nu_1}^{(1)}, \ldots, x_0^{(r)} \cdots x_{\nu_r}^{(r)}\right)^T \mapsto \left(x_0^{(1)} \cdots x_{\nu_1}^{(1)} 0 \cdots, \ldots, x_0^{(r)} \cdots x_{\nu_r}^{(r)} 0 \cdots\right)^T.$$
(22)

Further, let $\mathbf{H}_\infty$ *denote the semi-infinite matrix*

$$\mathbf{H}_\infty = \left(\mathbf{H}\varphi \; (\mathbf{H}\varphi)^{\downarrow_1} \; (\mathbf{H}\varphi)^{\downarrow_2} \cdots\right)$$
(23)

over the field F, *where the shift operator should be applied to each matrix column individually. Furthermore, the shift operator should be applied to each component of a combined column vector individually.*

Definition 3. *Two combined parity check matrices* **H** *and* $\tilde{\mathbf{H}}$ *are said to be equivalent if the set of columns from* $\mathbf{H}_\infty$ *is equal to the set of columns from* $\tilde{\mathbf{H}}_\infty$.

Lemma 4. *If* **H** *and* $\tilde{\mathbf{H}}$ *are two equivalent combined parity check matrices, then the corresponding codes have the same* d_{free} *and* w.

The proof of Lemma 4 is given in the Appendix. The following lemma is a slight generalization of a result that appeared in [15].

Lemma 5. *Every combined parity check matrix* $\mathbf{H}$, *of constraint length* e, *is equivalent to some combined parity check matrix* $\tilde{\mathbf{H}}$, *of constraint length* $\leq e$, *in which, for every column index* j, $1 \leq j \leq n$, *there is some row index* i, $1 \leq i \leq r$, *such that* $\tilde{h}_{j,0}^{(i)} = 1$.

Proof. Consider a combined parity check matrix $\mathbf{H}$. Suppose that there exists a j, $1 \leq j \leq n$, such that $h_{j,0}^{(i)} = 0$ for all i, $1 \leq i \leq r$. Define $\tilde{\mathbf{H}}$ as follows:

$$\tilde{\mathbf{H}} = ((\mathbf{h}_j)^{\perp}, \mathbf{h}_1, \ldots, \mathbf{h}_{j-1}, \mathbf{h}_{j+1}, \ldots, \mathbf{h}_n). \tag{24}$$

After an initial transient, $\mathbf{H}_{\infty}$ and $\tilde{\mathbf{H}}_{\infty}$ contain the same set of column vectors, and the lemma follows (repeating the argument if necessary). □

Lemmas 3 and 5 imply the following.

Corollary 1. *Without loss of generality, only combined parity check matrices* $\mathbf{H}$ *with distinct combined columns containing no combined column* $\mathbf{h} = (h_0^{(1)} \cdots h_{\nu_1}^{(1)}, h_0^{(2)} \cdots h_{\nu_2}^{(2)}, \ldots, h_0^{(r)} \cdots h_{\nu_r}^{(r)})^T$ *of the form*

$$h_0^{(1)} = h_0^{(2)} = \cdots = h_0^{(r)} = 0 \tag{25}$$

need to be considered when searching for codes with free distance $d_{\text{free}} \geq 3$, *where the ith row degree of* $\mathbf{H}$ *is* ν_i, $1 \leq i \leq r$.

Let X denote the set of possible combined columns in accordance with Corollary 1.

Lemma 6. *Given a combined parity check matrix* $\mathbf{H}$, *let* $\tilde{\mathbf{H}}$ *denote a matrix obtained from* $\mathbf{H}$ *by a single transposition of any two columns from* $\mathbf{H}$. *Then it follows that* $\mathbf{H}$ *and* $\tilde{\mathbf{H}}$ *are equivalent.*

Proof. The result follows from the fact that a column transposition in $\mathbf{H}$ corresponds to a column permutation in $\mathbf{H}_{\infty}$. □

Lemma 7 below, whose proof is omitted, appeared in [15].

Lemma 7. *A combined parity check matrix* $\mathbf{H}$, *with row degrees* $(\nu_1, \ldots, \nu_r)$, *is equivalent to the reverse combined parity check matrix* $\tilde{\mathbf{H}}$, *with row degrees* $(\nu_r, \ldots, \nu_1)$, *in which each combined column* $\tilde{\mathbf{h}}_j$ *is the reverse of* $\mathbf{h}_j$.

Lemma 8. *A combined parity check matrix* $\mathbf{H}$ *is given. Let* $\tilde{\mathbf{H}}$ *be a combined parity check matrix obtained from* $\mathbf{H}$ *by extending* $\mathbf{H}$ *with any combined column* $\mathbf{h} \in X$, *i.e.,* $\tilde{\mathbf{H}} = [\mathbf{H}, \mathbf{h}]$. *Then it follows that* $w_0(\mathbf{H}) \geq w_0(\tilde{\mathbf{H}})$.

Remark 3. Using Lemma 8, it follows that $w(\mathbf{H}) > w(\tilde{\mathbf{H}})$, where $\tilde{\mathbf{H}}$ is an extended version of $\mathbf{H}$ as in Lemma 8. This is true since $n - r$ grows with n.

In the following theorem, the combined columns from X are treated as integers with the topmost element being the least significant bit.

Theorem 6. *In an exhaustive search only ordered combined parity check matrices with columns from X need to be considered, i.e., combined parity check matrices satisfying $\mathbf{h}_j \leq \mathbf{h}_{j+1}$ for all j, $1 \leq j \leq n-1$, where $\leq$ denotes comparison of integers.*

The proofs of Lemma 8 and Theorem 6 are given in the Appendix.

4.1   Algorithm

In the following, an $(n, n-r)$ convolutional code represented by a canonical parity check matrix with row degrees ν_i, $1 \leq i \leq r$, will be called an $(n, n-r, (\nu_1, \ldots, \nu_r))$ convolutional code. A w-optimal $(n, n-r, (\nu_1, \ldots, \nu_r))$ code is an $(n, n-r, (\nu_1, \ldots, \nu_r))$ code with the maximum possible w. From Theorem 6, only ordered combined parity check matrices need to be considered in an exhaustive search, from which we get the following algorithm.

Exhaustive Search$(N, r, (\nu_1, \ldots, \nu_r), d_{\text{free,bound}})$:

```
/* Find a w-optimal (N_max, N_max − r, (ν₁,...,νᵣ)) convolutional
code with d_free ≥ d_free,bound ≥ 3, where N_max is the largest code
length ≤ N such that the code exists. */
Start with an empty combined parity check matrix H and let
X₀ = X, N_max = 0, and w_best = −1.
while (|X₀| > 0)
          Set l = |H|, and
(∗)       obtain and remove a random column h ∈ X_l,
          such that h ≥ the last element in H.
(∗∗)      If ((N_max < N or w(H ∪ h) ≥ w_best), d_free(H ∪ h) ≥ d_free,bound,
              and |H| < N) then:
              Set H = H ∪ h and X_{l+1} = X_l.
          Otherwise,
              If (|H| > N_max and H is canonical) then:
                  Make a record.
                  N_max = |H|.
                  w_best = w(H).
              Otherwise if (|H| = N_max and H is canonical) then:
                  Make a record if a better code is found.
          Remove the last column from H.
return (w_best, N_max)
```

In the line marked $(\ast\ast)$, we have incorporated Lemma 8 into the algorithm to limit the size of the search space. Further, Lemma 7 can be incorporated into the algorithm in the line marked $(\ast)$ to further limit the size of the search space.

We may also incorporate limits on the bit-oriented trellis state complexity in the line marked $(\ast\ast)$ [20]. These codes are interesting due to reduced decoding complexity with maximum likelihood decoding [16], [20].

4.2   Determination of Free Distance

An efficient algorithm to compute the free distance of a high rate convolutional code is outlined in [15], [20]. The algorithm, which is uni-directional, can be extended into a bidirectional algorithm. The bidirectional version of the algorithm is similar to the BEAST algorithm in [21]. To keep the description short, we omit the details.

4.3   Cycle Weight Computation

To compute $w_0(\mathbf{H})$ for any given combined parity check matrix $\mathbf{H}$, where the columns are taken from the set X constructed from given values of r and $(\nu_1, \ldots, \nu_r)$, we use Karp's algorithm [22]. As a remark, note that there do exist faster algorithms to compute the minimum average cycle weight per edge in a graph [23]. Karp's algorithm has good performance on smaller graphs, mostly due to its simplicity. However, as the number of nodes in the graph grows, its performance degrades rapidly compared to other algorithms [23].

The asymptotic complexity of Karp's algorithm is $\Theta(2^{2\nu} \cdot \min(2^{n-r}, 2^\nu))$, making it hard to search for codes with large dimension and code degree.

4.4   Speed-up of Processing Using Preprocessing

The exact value of $w(\mathbf{H} \cup \mathbf{h})$ in the line marked (∗∗) can be replaced by any upper bound, $w_{\text{bound}}(\mathbf{H} \cup \mathbf{h})$ on $w(\mathbf{H} \cup \mathbf{h})$. We use the method presented in [10].

Let $\mathbf{H}$ denote a canonical combined parity check matrix defining a code $\mathcal{C}$. A valid state in the state diagram for $\mathcal{C}$ constructed from $\mathbf{H}$ is of the form

$$\mathbf{s} = (0, \ldots, 0, 0s_1^{(\gamma+1)} \cdots s_{\nu_{\gamma+1}}^{(\gamma+1)}, \ldots, 0s_1^{(r)} \cdots s_{\nu_r}^{(r)})^T.$$

If $s_1^{(\gamma+1)} = \cdots = s_1^{(r)} = 0$, then there exists an all-zero path of length p, from $\mathbf{s}$ to $\mathbf{s}^\ell$, where p is a positive integer such that $s_1^{(i)} = \cdots = s_p^{(i)} = 0$ for all i, $\gamma < i \le r$. If $\mathbf{H}$ contains the column $\mathbf{s}^{\ell+1} + \mathbf{s}$, then the state diagram contains a length $p+1$ cycle of weight 1. More generally, if $\mathbf{H}$ contains k columns, denoted as $\mathbf{h}_j$, $1 \le j \le k$, such that their sum is equal to $\mathbf{s}^{\ell+1} + \mathbf{s}$, then the state diagram contains a length $p+1$ cycle of weight k.

Assume that there exists an i, $\gamma < i \le r$, such that $s_1^{(i)} = 1$. Further, assume that $\mathbf{H}$ contains k columns, denoted as $\mathbf{h}_j$, $1 \le j \le k$, such that their sum is equal to $\mathbf{s}^\ell + \mathbf{s}$, then the state diagram contains a length 1 cycle of weight k. These cycles of length ≥ 1 are referred to as *self-cycles generated* by the columns $\mathbf{h}_j$, $1 \le j \le k$. We can preprocess the self-cycles, and more generally cycles containing several all-zero paths connected by edges with nonzero labels.

Example 5. Choose $r = 2$, $\nu_1 = 3$, and $\nu_2 = 3$. Further, assume that a combined parity check matrix $\mathbf{H}$ contains the column $\mathbf{h} = (0000, 1001)^T$. The state diagram of the code defined by $\mathbf{H}$, constructed from $\mathbf{H}$, contains a length 3 self-cycle of weight 1 generated by $\mathbf{h}$ ($\mathbf{s} + \mathbf{s}^3 = \mathbf{h}$ with $\mathbf{s} = (0000, 0001)^T$), and the minimum average cycle weight per edge is upper bounded by $1/3$, which is equal to the general lower bound from Theorem 3 with $d_{\tilde{c}} = 1$. Further, there does not exist a self-cycle generated by the column $(1001, 1101)^T$, illustrating the fact that there can exist columns in X which cannot be decomposed into the sum of a nonzero state and its maximum shift.

5   Codes from Computer Search

The upper bound in Theorem 1 decreases when the code degree grows. This indicates that a w-optimal code with a large code degree has lower w than a w-optimal code with a smaller code degree. The tables in [4] support this claim as well. Consequently, we propose the following search strategy:

1. Choose a convolutional code class in which the search is to be performed, i.e., specify r.
2. Specify the rate and a lower bound on the free distance.
3. Choose the smallest code degree such that there exist codes with the specified parameters.
4. Perform the search.

5.1 w-Optimal Codes from Exhaustive Search

In Tables 1 and 2 the notation $(n, n - r, (\nu_1, \ldots, \nu_r), d_{\text{free,bound}})$ refers to an $(n, n - r, (\nu_1, \ldots, \nu_r))$ code with $d_{\text{free}} \geq d_{\text{free,bound}}$. The new codes are defined by a canonical parity check matrix in *decimal compressed form*. As an example, the 7th entry in Table 2 is a $(6, 4, (2, 2))$ code with free distance 5. The canonical parity check matrix is given in equation (4). From the tables we can observe that better codes are obtained than what was previously known, e.g., the w-optimal $(8, 6)$ code with $d_{\text{free,bound}} = 4$ has $w = 1/6$ for $\nu = 2$. Note that an equivalent code (in the sense of a column permutation) is obtained from the construction $\mathcal{M}_r(\nu_2, \ldots, \nu_r)$ with $r = 2$ and $\nu_2 = 2$, i.e., the construction is in some sense optimal in this case. The w-optimal $(4, 3)$ code with $d_{\text{free,bound}} = 4$ has $w = 2/15$ for $\nu = 3$, which is strictly less than $1/6$. Hence, we get both a larger w and a smaller code degree by looking into the class of $r = 2$ codes.

Within the class of $(n, n - 1)$, $n \geq 9$, codes with $d_{\text{free}} \geq 5$, a code degree ≥ 8 is required for the code to exist [15]. Searching for w-optimal codes is quite time consuming in this case due to the exponential complexity of Karp's algorithm. On the other hand, searching for codes when $d_{\text{free,bound}} = 3$ or 4 is fast. To keep the tables short, these results are omitted.

5.2   Tightness of Upper Bound in Theorem 1

We have looked into the tightness of the upper bound in Theorem 1. Note that the w-optimal codes below are not tabulated in the tables, because the codes have larger code degrees than what is necessary to obtain the given rates for the given values on $d_{\text{free,bound}}$.

Table 1. Canonical parity check matrices in decimal compressed form of w-optimal $(n, n-1, \nu)$ convolutional codes with free distance $d_{\text{free}} \geq d_{\text{free,bound}}$, where $d_{\text{free,bound}}$ is a given lower bound on the free distance. The 4th column contains the minimum average cycle weight per information bit of existing codes. [1] Code found by Hole and Hole [2]. [2] Code found by Jordan *et al.* through puncturing of rate 1/2 mother codes [4]. The 5th column contains upper bounds based on Theorem 1. The codes are found by exhaustive search. The label of the first column, Param. stands for $(n, n-r, \nu, d_{\text{free,bound}})$ (see Section 5.1 for details)

Param.	Parity check matrix	w	w_{lit}	w_{bound}	$\frac{w_{\text{bound}} - w}{w_{\text{bound}}}$
(3,2,2,3)	(1,3,7)	1/3	$1/4^2$	1/3	0
(3,2,3,4)	(7,11,15)	1/4	$1/4^2$	2/7	0.1250
(3,2,4,5)	(11,15,19)	3/14		4/15	0.1964
(3,2,5,6)	(11,19,55)	3/16	$1/6^2$	8/31	0.2734
(3,2,6,7)	(69,101,115)	1/7		16/63	0.4375
(3,2,7,8)	(69,93,251)	1/7		32/127	0.4330
(3,2,9,9)	(613,683,989)	1/8		128/511	0.5010
(3,2,10,10)	(1119,1759,1869)	1/8		256/1023	0.5005
(4,3,2,3)	(1,3,5,7)	1/6	$1/6^2$	2/9	0.2500
(4,3,3,4)	(7,11,13,15)	2/15	$2/15^{1,2}$	4/21	0.3000
(4,3,5,5)	(31,37,45,59)	1/9		16/93	0.3542
(4,3,6,6)	(19,37,91,103)	4/39	$2/21^2$	32/189	0.3942
(4,3,8,7)	(349,399,471,503)	1/12		128/765	0.5020
(4,3,9,8)	(331,565,923,1009)	10/123		256/1533	0.5131
(5,4,3,3)	(3,5,7,11,15)	1/8	$1/8^2$	1/7	0.1250
(5,4,4,4)	(7,11,13,15,31)	1/10	$1/12^2$	2/15	0.2500
(5,4,6,5)	(23,45,67,83,123)	3/40		8/63	0.4094
(5,4,7,6)	(69,103,107,151,251)	1/14		16/127	0.4330
(6,5,3,3)	(1,3,5,7,11,15)	1/10	$1/10^2$	4/35	0.1250
(6,5,4,4)	(15,19,21,23,27,29)	1/15		8/75	0.3750
(6,5,6,5)	(45,67,95,101,105,127)	1/20		32/315	0.5078
(6,5,7,6)	(35,81,91,205,223,239)	1/20		64/635	0.5039
(7,6,3,3)	(1,3,5,7,11,13,15)	1/15	$1/15^1$	2/21	0.3000
(7,6,4,4)	(15,19,21,23,27,29,31)	1/18		4/45	0.3750
(7,6,7,5)	(83,139,157,213,215,217,255)	1/24		32/381	0.5039
(7,6,8,6)	(55,247,313,357,425,465,475)	1/25		64/765	0.5219
(8,7,3,3)	(1,3,5,7,9,11,13,15)	1/21		4/49	0.4167
(8,7,4,4)	(15,19,21,23,25,27,29,31)	2/49		8/105	0.4643
(8,7,7,5)	(87,109,143,159,187,201,213,229)	2/63		64/889	0.5590
(8,7,8,6)	(133,191,289,361,391,403,491,493)	6/217		128/1785	0.6144

Table 2. Canonical parity check matrices in decimal compressed form of w-optimal $(n, n-2, \nu)$ convolutional codes with free distance $d_{\text{free}} \geq d_{\text{free,bound}}$, where $d_{\text{free,bound}}$ is a given lower bound on the free distance. The 4th column contains upper bounds based on Theorem 1. The codes are found by exhaustive search. The label of the first column, Param. stands for $(n, n-r, (\nu_1, \nu_2), d_{\text{free,bound}})$ (see Section 5.1 for details)

Param.	Parity check matrix	w	w_{bound}	$\frac{w_{\text{bound}} - w}{w_{\text{bound}}}$
(5,3,(0,1),3)	(1,2,3,5,7)	1/2	2/3	0.2500
(5,3,(1,1),4)	(1,6,11,12,15)	1/3	4/9	0.2500
(5,3,(1,2),5)	(3,6,11,23,30)	1/3	8/21	0.1250
(5,3,(1,3),6)	(3,7,31,43,46)	4/15	16/45	0.2500
(6,4,(0,1),3)	(1,2,3,5,6,7)	1/4	1/2	0.5000
(6,4,(1,1),4)	(3,6,7,9,12,13)	1/4	1/3	0.2500
(6,4,(2,2),5)	(3,10,13,43,58,59)	3/14	4/15	0.1964
(6,4,(2,2),6)	(7,13,44,49,61,62)	1/6	4/15	0.3750
(7,5,(0,2),3)	(1,3,5,7,9,11,13)	1/5	4/15	0.2500
(7,5,(0,2),4)	(1,3,5,7,9,11,13)	1/5	4/15	0.2500
(7,5,(2,2),5)	(12,15,31,35,46,58,59)	1/7	16/75	0.3304
(7,5,(2,3),6)	(7,21,62,79,97,109,110)	2/15	32/155	0.3542
(8,6,(0,2),3)	(1,3,5,7,9,11,13,15)	1/6	2/9	0.2500
(8,6,(0,2),4)	(1,3,5,7,9,11,13,15)	1/6	2/9	0.2500
(8,6,(2,2),5)	(13,15,30,39,46,49,55,59)	1/10	8/45	0.4375
(8,6,(2,3),6)	(7,21,62,79,97,109,110,121)	4/39	16/93	0.4038
(9,7,(1,1),3)	(3,6,7,9,11,12,13,14,15)	1/7	4/21	0.2500
(9,7,(1,2),4)	(3,6,7,11,14,22,23,30,31)	1/7	8/49	0.1250
(9,7,(2,3),5)	(7,21,26,39,56,75,95,110,117)	2/21	32/217	0.3542
(10,8,(1,1),3)	(1,3,6,7,9,11,12,13,14,15)	1/8	1/6	0.2500
(10,8,(1,2),4)	(3,6,7,11,14,19,22,23,30,31)	1/8	1/7	0.1250
(10,8,(2,3),5)	(13,15,30,61,67,74,87,99,124,127)	1/14	4/31	0.4464
(11,9,(1,1),3)	(1,3,4,6,7,9,11,12,13,14,15)	1/9	4/27	0.2500
(11,9,(1,2),4)	(3,6,7,11,14,15,19,22,23,27,30)	1/9	8/63	0.1250

A w-optimal $(5, 3, (1, 2))$ code with $d_{\text{free}} \geq 3$ has $w = 2/5$. The code, defined by the canonical parity check matrix $(1, 4, 6, 13, 30)$ in decimal compressed form, was found by exhaustive search. The trellis state complexity profile is $(8, 8, 8)$. The upper bound in Theorem 1 is equal to $8/21$, which is strictly less than $2/5$. This is not an inconsistency because there does not exist a canonical generator matrix for the code with a delay-free 3×3 minor of degree $\nu = 3$. This is closely related to the fact that any systematic generator matrix for the code is non-recursive.

A w-optimal $(6, 4, (1, 2))$ code with $d_{\text{free}} \geq 3$ has $w = 8/28$. The code, defined by the canonical parity check matrix $(1, 4, 5, 6, 13, 31)$ in decimal compressed form, was found by exhaustive search. The trellis state complexity profile is

$(8, 8, 8, 8)$. The upper bound in Theorem 1 is equal to $8/28$, which shows that the bound is tight in this case.

6 Conclusions

We have investigated high rate convolutional codes with optimal minimum average cycle weight per information bit w. Furthermore, we have presented an explicit construction technique of $d_{\text{free}} = 4$ codes with high rate, large w, and bit-oriented trellis state complexity equal to $\nu + 1$, where ν is the code degree. The construction produces optimal codes, in the sense of maximizing w, within the classes of $(\nu(2^{\nu-1} + 2) + 2^{\nu}, \nu(2^{\nu-1} + 1) + 2^{\nu} - 1)$, $\nu \geq 2$, codes with $d_{\text{free}} \geq 3$ and any code degree. An efficient exhaustive search algorithm was outlined as well. A computer search was carried out, and codes having larger w than codes in the literature were found.

References

1. Hemmati, F., Costello, Jr., D.J.: Asymptotically catastrophic convolutional codes. IEEE Trans. Inform. Theory **IT-26** (1980) 298–304
2. Hole, M.F., Hole, K.J.: Tight bounds on the minimum average weight per branch for rate $(n - 1)/n$ convolutional codes. IEEE Trans. Inform. Theory **43** (1997) 1301–1305
3. Höst, S., Johannesson, R., Zigangirov, K.S., Zyablov, V.V.: Active distances for convolutional codes. IEEE Trans. Inform. Theory **44** (1999) 658–669
4. Jordan, R., Pavlushkov, V., Zyablov, V.V.: Maximum slope convolutional codes. IEEE Trans. Inform. Theory, submitted for publication (2001)
5. Bocharova, I.E., Handlery, M., Johannesson, R., Kudryashov, B.D.: Tailbiting codes obtained via convolutional codes with large active distance-slopes. IEEE Trans. Inform. Theory **48** (2002) 2577–2587
6. Huth, G.K., Weber, C.L.: Minimum weight convolutional codewords of finite length. IEEE Trans. Inform. Theory **22** (1976) 243–246
7. Jordan, R., Pavlushkov, V., Zyablov, V.V.: An upper bound on the slope of convolutional codes. In: Proc. IEEE Int. Symp. on Inform. Theory (ISIT), Lausanne, Switzerland (2002) 424
8. Rosnes, E.: Upper bound on the minimum average cycle weight of classes of convolutional codes. In: Proc. IEEE Int. Symp. on Inform. Theory (ISIT), Yokohama, Japan (2003) 280
9. Hole, K.J.: On classes of convolutional codes that are not asymptotically catastrophic. IEEE Trans. Inform. Theory **46** (2000) 663–669
10. Hole, K.J.: A note on asymptotically catastrophic convolutional codes of rate $(n - 1)/n$. IEEE Trans. Commun. **45** (1997) 1014–1016
11. Sidorenko, V., Zyablov, V.: Decoding of convolutional codes using a syndrome trellis. IEEE Trans. Inform. Theory **40** (1994) 1663–1666
12. McEliece, R., Lin, W.: The trellis complexity of convolutional codes. IEEE Trans. Inform. Theory **42** (1996) 1855–1864
13. Bahl, L.R., Cocke, J., Jelinek, F., Raviv, J.: Optimal decoding of linear codes for minimizing symbol error rate. IEEE Trans. Inform. Theory **IT-20** (1974) 284–287

14. McEliece, R.J.: The algebraic theory of convolutional codes. In Pless, V.S., Huffman, W.C., eds.: Handbook of Coding Theory. Elsevier, North-Holland, Amsterdam (1998) 1065–1138

15. Ytrehus, Ø.: A note on high rate binary convolutional codes. Tech. Report no. 68, Department of Informatics, University of Bergen, Norway (1992)

16. CharnKeitKong, P., Imai, H., Yamaguchi, K.: On classes of rate $k/(k+1)$ convolutional codes and their decoding techniques. IEEE Trans. Inform. Theory **42** (1996) 2181–2193

17. McEliece, R.J.: On the BCJR trellis for linear block codes. IEEE Trans. Inform. Theory **42** (1996) 1072–1092

18. Hole, K.J., Ytrehus, Ø.: Cosets of convolutional codes with least possible maximum zero- and one-run lengths. IEEE Trans. Inform. Theory **44** (1997) 423–431

19. Hole, K.J.: On the growth rate of convolutional codes. Tech. Report no. 172, Department of Informatics, University of Bergen, Norway (1999)

20. Rosnes, E., Ytrehus, Ø.: On maximum length convolutional codes under a trellis complexity constraint. Journal of Complexity, accepted for publication (2003)

21. Bocharova, I.E., Kudryashov, B.D., Handlery, M., Johannesson, R.: How to efficiently find the minimum distance of tailbiting codes. In: Proc. IEEE Int. Symp. on Inform. Theory (ISIT), Lausanne, Switzerland (2002) 259

22. Karp, R.M.: A characterization of the minimum cycle mean in a digraph. SIAM J. Discr. Math. **23** (1978) 309–311

23. Dasdan, A., Irani, S.S., Gupta, R.K.: Efficient algorithms for optimum cycle mean and optimum cost to time ratio problems. In: Proc. 36th Design Automation Conf. (DAC), New Orleans, LA (1999) 37–42

Appendix

Proof (of Lemma 2). From the construction,

$$
\mathbf{b} = \left(0 \underbrace{11\cdots1}_{2^\nu} \underbrace{22\cdots2}_{2^{\nu-1}+2} \cdots \underbrace{rr\cdots r}_{2^{\nu-1}+2} \right) \tag{26}
$$

and

$$
\mathbf{f} = \left(0 \underbrace{11\cdots1}_{2^{\nu-1}} \underbrace{22}_{2} \cdots \underbrace{r^- r^- \cdots r^-}_{2^{\nu-1}} \underbrace{rr}_{2} \underbrace{rr\cdots r}_{2^\nu} \right) \tag{27}
$$

where $r^- \triangleq r - 1$. From (8), the ith component of the trellis state complexity profile c_i, $1 \le i \le n - r$, is $c_i = 2^{\nu - r + f_{j_i} + b_{n-j_i}}$. In our case, using the definition in (7),

$$
J = \{0, 1, \ldots, n-2\} / \{ i(2^{\nu-1} + 2) - 1 : i \in \{1, \ldots, r^-\} \}, \tag{28}
$$

and the base-2 logarithm of the trellis state complexity profile is the length $(r-1)(2^{\nu-1}+1)+2^{\nu}-1$ row vector

$$
\left(\underbrace{\nu\,(\nu+1)\cdots(\nu+1)}_{2^{\nu}-1}\underbrace{\mathbf{a}\cdots\mathbf{a}}_{r-2}\underbrace{(\nu+1)\cdots(\nu+1)}_{2^{\nu}-1} \right), \quad \mathbf{a} = \left(\underbrace{(\nu+1)\cdots(\nu+1)}_{2^{\nu-1}+1} \right)
$$
(29)

from which it follows that the bit-oriented trellis state complexity is $\nu+1$. $\qquad\square$

Proof (of Lemma 4). The set consisting of the columns in $\mathbf{H}_{\infty}$, and the set consisting of the columns in $\tilde{\mathbf{H}}_{\infty}$ are equal. The corresponding codes' weight distributions are equal since the Hamming weight of a codeword is independent of coordinate order, from which it follows that the corresponding codes have the same free distance.

Any $(n, n-r)$ convolutional code $\mathcal{C}$ can be considered as an $(ln, l(n-r))$ convolutional code $\mathcal{C}^l$ for any positive integer l by consecutive blocking of l n-tuples in the codewords from $\mathcal{C}$ into ln-tuples. Clearly, the sets of codewords are equal, and it holds that $w(\mathcal{C}) = w(\mathcal{C}^l)$ for any $l \geq 1$. A minimal state diagram of $\mathcal{C}^l$ contains edges with ln-bit labels of output bits. Any coordinate order within these ln-bit labels gives the same w_0 and w, from which it follows as $l \to \infty$, that $w(\mathbf{H}) = w(\tilde{\mathbf{H}})$. $\qquad\square$

Proof (of Lemma 8). Suppose we have a combined parity check matrix $\mathbf{H}$ with n columns. The cycle in the state diagram constructed from $\mathbf{H}$ minimizing the average cycle weight per information bit is denoted by $\underline{\alpha} = (\mathbf{s}_0, \mathbf{s}_1, \ldots, \mathbf{s}_l)$, where l is the number of edges on the cycle, $\mathbf{s}_i$, $0 \leq i \leq l$, are state diagram vertices, $\mathbf{s}_0 = \mathbf{s}_l$, and $\mathbf{s}_i \neq \mathbf{s}_j$, $0 \leq i < j \leq l-1$. The string of n-tuples representing the output labels along the cycle is denoted by the column vector $(\mathbf{v}_1^T, \ldots, \mathbf{v}_l^T)^T$. Consider the extended combined parity check matrix $\tilde{\mathbf{H}} = [\mathbf{H}, \mathbf{h}]$, where $\mathbf{h}$ is any combined column in X. The cycle $\underline{\alpha}$ will exist in the state diagram constructed from $\tilde{\mathbf{H}}$ as well. The string of n-tuples representing the output labels along the cycle is the column vector $([\mathbf{v}_1^T 0], \ldots, [\mathbf{v}_l^T 0])^T$. The Hamming weight of the cycle and the number of edges on the cycle are the same, and the result follows immediately. $\qquad\square$

Proof (of Theorem 6). An unordered combined parity check matrix $\mathbf{H}$ with columns from X is given. In $\mathbf{H}$ there exists at least one j, $1 \leq j \leq n-1$, such that $\mathbf{h}_j > \mathbf{h}_{j+1}$. By a single transposition of columns j and $j+1$, the two columns are ordered (since either $\mathbf{h}_j > \mathbf{h}_{j+1}$ or $\mathbf{h}_j \leq \mathbf{h}_{j+1}$). From Lemma 6, we know that any column transposition will result in an equivalent combined parity check matrix, and the result follows from Lemma 4 (repeating the argument if necessary). $\qquad\square$

A Multifunctional Turbo-Based Receiver Using Partial Unit Memory Codes

Lina Fagoonee and Bahram Honary

Dept. of Communication Systems, Lancaster University, Lancaster LA1 4YR. U.K.
{l.fagoonee,b.honary}@lancaster.ac.uk

Abstract. A multifunctional system comprising a turbo decoder, low complexity turbo equalizer and data-aided frame synchronizer is developed to compensate for frequency-selective channels with fading. The turbo codes are based on Partial Unit Memory Codes, which may be constructed with higher free distance than equivalent recursive systematic convolutional codes and can achieve better performance than the latter. The purely digital multifunctional receiver is targeted for end-applications such as combat-radio and radio-relay, providing robust communication at low signal-to-noise ratio with performance approaching the Shannon limit. This paper will present the operation of a multifunctional turbo-based receiver, whose structure is reminiscent of a data-aided turbo synchronizer, both of which are described in detail.

1 Introduction

The idea of a multifunctional receiver based on a turbo scheme is that all components of the receiver are capable of exchanging soft information, so that extrinsic information from a previous stage may assist in an improved estimate of the output of the current stage/operation. The multifunction turbo receiver comprises a Partial Unit Memory (PUM) turbo decoder [1, 2], a low complexity turbo equalizer [3] and a data-aided turbo synchronizer, which is described in detail in Section 2.

1.1 Turbo Codes Based on PUM Codes

The original turbo codes [5] were designed with a number of parallel concatenated Recursive Systematic Convolutional (RSC) codes. The performance of turbo codes can be improved by increasing the constraint length of the component convolutional code. However, this results in an increase in complexity of trellis-based decoding algorithms owing to a corresponding increase in the number of memory states.

In order to reduce the decoding complexity of trellis codes, turbo codes with component PUM codes are proposed as a low-complexity alternative to the classical turbo codes based on RSC codes. PUM codes have fewer number of states than RSC codes in their trellis for the same number of encoder inputs because only a fraction of these inputs are shifted to the memory cells. PUM codes have the additional advantage of their excellent distance properties [4] thus ensuring that the performance

K.G. Paterson (Ed.): Cryptography and Coding 2003, LNCS 2898, pp. 24–34, 2003.
© Springer-Verlag Berlin Heidelberg 2003

of the PUM turbo codes will be maintained, if not improved, relative to the classical RSC turbo codes.

The construction of PUM codes for turbo coding applications is described in detail in [1, 2], where it is proved, by simulation, that turbo codes based on PUM codes can be constructed to outperform the equivalent conventional RSC turbo codes.

The structure of a PUM turbo code with r component (n,k) PUM encoders, which may or may not be identical, is shown in Fig. 1. k is the number of message bits contributing to the codeword and n is the size of the codeword in bits. All r component encoders use the same inputs, but in a different sequence owing to the permutation of the input bits by the interleaver.

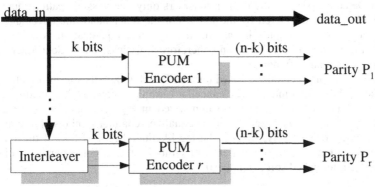

Fig. 1. Block Diagram of a PUM Turbo Code

The encoding structure of turbo codes imposes a constraint on the component encoders. The overall turbo code has a higher throughput if the component codes are systematic instead of non-systematic because in the former case the k input bits are common to all the r parallel encoders and need to be transmitted only once. Thus, for an input sub-block of k bits in the input sequence, there are $(n-k)$ parity bits, generated by each encoder. The rate R_s of the turbo code with systematic component encoders is given by equation 1.

$$R_s = \frac{k}{k + r(n-k)} \qquad (1)$$

It is possible to construct minimal recursive systematic PUM generator encoder structures from equivalent optimal non-systematic generator matrices whilst preserving the maximum free distance of the code [1, 2]. This approach generates codewords whose free distances reach, or are very close to, an upper bound [4] for a given k and n.

The multistage decoding of PUM-characteristic trellises obtained from the systematic encoder structures differs from the classical max-log MAP decoding [6] of turbo codes based on RSC codes by taking into account the contribution from more than 1 information bit during the calculation of the branch transition likelihoods, forward and backward recursive node metrics.

1.2 Low Complexity Turbo Equalizer

Once a high-performance and low-complexity error control code was determined for low signal-to-noise ratios, a suitable equalizer was investigated to be used in conjunction with the chosen codec in an iterative scheme to effectively eliminate the effects of Inter Symbol Interference (ISI) due to multipath. Owing to the complexity and delay of turbo decoding, the low complexity Decision Feedback Equalizer (DFE) is considered the best equalizer to incorporate in a turbo equalization scheme. The DFE uses the minimum Mean Squared Error criterion to optimise the feedback filter tap coefficients. It is less computationally complex than the Maximum Likelihood Symbol Detector, because the former considers only one possible path as opposed to all possible paths considered by the latter. On the other hand, the advantage of the DFE over the least complex linear equalizers is that equalization is accomplished through the feedback of sliced, and therefore noiseless, data. Consequently, this process does not amplify noise.

A sub-optimum turbo equalizer based on a DFE which outputs soft information, via a soft slicer, (SDFE) and interference canceller is implemented, as described in [3]. The operation of the turbo equalizer is summarized in Fig. 2, where R_n is the channel symbol stream at time n, c_n and c_k are the equalized channel symbols before and after de-interleaving respectively, $L_i(c_k)$ is the log likelihood ratio (LLR) of the symbol c_k for the ith iteration, e_k is the equivalent soft symbol estimate corresponding to its LLR and e_n is the interleaved soft symbol estimate.

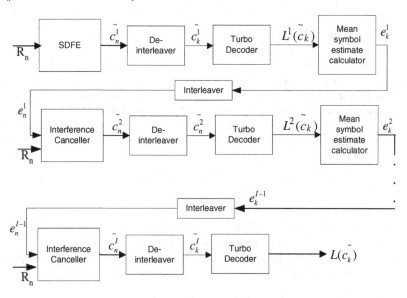

Fig. 2. Operation of Turbo Equalizer for I iterations

The soft output equalizer block represents a SDFE for the first iteration and an interference canceller [7] for all remaining iterations to remove any remaining ISI.

The flow of information between iterations (serial operation) of the combined equalization and decoding process is shown in Fig. 2.

The tap coefficients of the interference canceller filters are updated at each iteration with extrinsic information from the previous decoding operation. Both equalizer and decoder are SISO (Soft Input Soft Output) blocks, that is, they both accept soft information and produce soft information at their output.

1.3 Turbo Synchronization and Multifunctional Receiver

Synchronization is the first process carried out at the receiver so that all later stages such as equalization and coding can distinguish between symbols. Frame synchronization is especially relevant to turbo codes because data bits are encoded in frames whose size is governed by the interleaver size and code rate. Turbo decoding has the added advantage of providing additional information to the frame synchronizer so that it can make a more accurate estimate on the location of frame boundaries in a received symbol stream, assuming that carrier and symbol synchronization have been achieved.

Frame synchronization can be performed with and without the aid of a known set of pilot symbols, that is, synchronization can be data-assisted or intrinsic. The more popular data-assisted methods include the correlation and maximum likelihood approaches. Trellis Extracted Synchronization Techniques (TEST) [8] and decoder-assisted synchronizers [9] are examples of intrinsic synchronization schemes that have been suggested for non-iterative decoding schemes. The advantage of intrinsic methods over data-assisted methods is the absence of the overhead caused by the insertion of pilot symbols into every frame, although pilot symbols provide valuable positioning information especially where the coded symbols can be seriously degraded by the channel.

A data-aided turbo synchronization scheme is developed for time-varying fading channels and is described in detail in Section II. In Section III, the multifunctional system which culminates from a frame synchronizer, low-complexity equalizer and PUM-based decoder is described. The exchange of soft information between these blocks to improve the final decoder performance is also described. Results of the overall receiver are presented in Section IV, using fading frequency-selective mobile channel conditions.

2 Data Assisted Turbo Synchronization

The general operation of the turbo synchronizer in terms of sliding windows and decoder operation, with emphasis on the metric calculation, is described below. The variation of soft information with increasing number of iterations is exploited as a metric. It is proposed that pilot sequences are introduced periodically in the transmitted stream, e.g. after every few frames.

Fig. 3 illustrates the principle of the proposed system in terms of sliding windows. The buffer size B is greater than the transmitted frame size N. An ideal value for B is twice the transmitted frame size. Each sliding window of size N symbols is decoded and the turbo decoder returns a metric associated with that frame. There are N windows, which represent N serial or parallel decoding operations. Decoding is performed similarly to standard turbo decoding, except for the additional calculation of the variance of the decoder soft outputs for the first and last decoding stages.

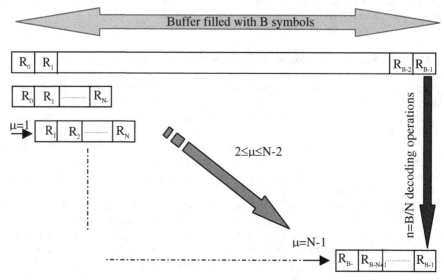

Fig. 3. Sliding windows in ML detection

In [10], the variance of a meta-channel, σ^2 is suggested as a suitable measure for monitoring the soft output after a number of decoding stages. Here the variance σ^2, as shown in equation 2, is the total energy of the soft likelihoods of the soft outputs, $L_i(u)$ of a decoder at the ith decoding stage.

$$\sigma_i^2 = \sum_{j=0}^{N-1} L_i^2(u_j) \qquad 1 \le i \le 2I$$

(2)

There are 2I decoding stages during I turbo decoding iterations. If a frame is out of sync, σ_i^2 hardly changes with variations in i. However, if frame synchronization is achieved, σ_i^2 increases with i. Hence, a good measure of the reliability of a frame from the synchronizer's point of view is the difference between the last and first decoding stages.

The data-assisted frame synchronizer reduces the complexity of going through N ML decoding operations by allowing μ to be chosen from a list of M possible frame delays. This list will be provided by the detection of a preamble of size P, which is inserted at varying frame periods into the transmitted stream.

The operation of the data-aided intrinsic synchronizer at the receiver is shown in Fig. 4. The decoder metrics $L_T(\mu)$ as a function of the decoder output variance and normalized by the interleaver size π, is given by equation 3. Soft information from the pilot detector, $L_p(\mu)$ as given in equation 4, uses an ML-based correlation method and is included in the final synchronizer metric, $L_s(\mu)$. This method estimates M possible μ's. It provides a low complexity ML estimate of μ by determining the minimum error probability of a received frame for a given μ.

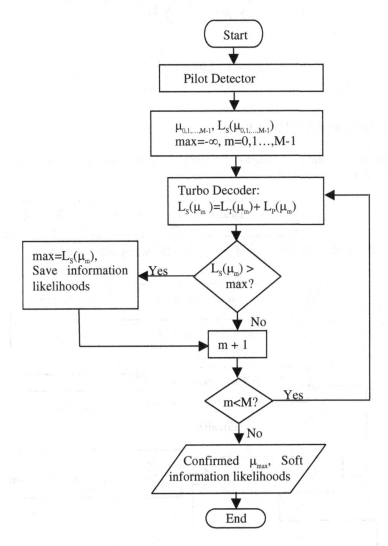

Fig. 4. Operation of data-aided ML synchronizer

$$L_T(\mu) = \frac{1}{\pi}\left(\sigma_{2I}^2 - \sigma_1^2\right) \qquad (3)$$
$$0 \le \mu < M$$

$$L_P(\mu) = \sum_{i=0}^{P-1} R_{\mu+i} s_i - \sum_{i=0}^{P-1} \left|R_{\mu+i}\right| \qquad (4)$$

$$L_S(\mu) = L_T(\mu) + L_P(\mu) \qquad (5)$$

3 Multifunctional Receiver

The operation of the multifunctional system including frame synchronizer, equalizer and decoder is illustrated in Fig. 5.

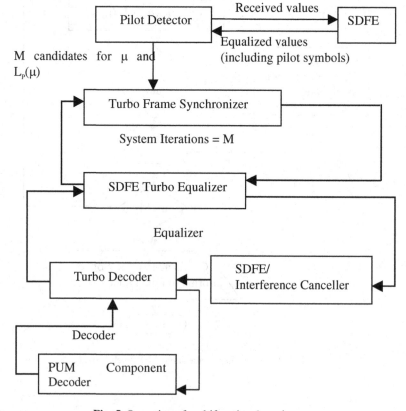

Fig. 5. Operation of multifunctional receiver

The pilot detector first equalizes the received values, which include the pilot symbols. The location μ of M (equal to the number of multifunctional system iterations) pilot symbols are then determined, using a ML-based correlation method, in descending order of the detector metric, $L_p(\mu)$.

The frame synchronizer first strips out the pilot symbols from the original received values (prior to equalization) before calling the turbo equalizer for as many times as there are possible delays μ as determined by the pilot detector. The number of iterations of the multifunctional system is equivalent to the number of potential delays, M, estimated by the pilot detector prior to confirmation by the frame synchronizer.

The turbo equalizer, in turn, calls the SDFE (1st equalizer iteration) or interference canceller (2nd equalizer iteration, etc.) and the turbo decoder. The turbo decoder returns the decoder metric (normalized energy of the soft decoded information) and the soft decoded values to the turbo equalizer, which in turn passes them to the frame synchronizer after the pre-set equalizer iterations. The number of decoder operations is also pre-set.

The frame synchronizer of the multifunctional system operates in a similar way as the data-aided ML frame synchronizer, except that the turbo decoder block is replaced by the SDFE turbo equalizer block.

4   Simulation Results of Multifunctional System

The multifunctional system is simulated with the COST 207 RA channel impulse response and for the case of frequency-selective channels with fading a Doppler frequency of 22 Hz. All simulations assume perfect channel estimation. An interleaver of size 1000 is used.

Figs. 6 and 7 illustrate the Bit Error Rate (BER) and Frame Location Error Rate (FLER) performance, respectively, of the multifunctional system in a frequency selective channel with fading for varying M. The equalizer and decoder iterations are fixed to 2 and 1, respectively. The preamble period is 5 frames. The performance of the turbo receiver is, as expected, proportional to the number of estimates for the sync point, M. Hence, there is a trade-off to make between added complexity and memory requirements compared to a fast, not so accurate, determination of possible sync points.

Figs. 8 and 9 illustrate the performance of the multifunctional turbo receiver for varying preamble periods ranging from 2 frames to 50 frames in frequency selective channels with negligible fading. It is interesting to note that the performance is not significantly degraded by increasing the preamble period. Except for higher signal-to-noise ratios where there is a slight gain in performance by the use of smaller preamble insertion periods, it can be safely be concluded that for periods up to 50 frames there is no reason to use smaller periods that will reduce system throughput.

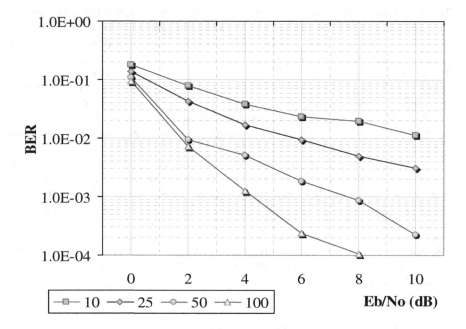

Fig. 6. BER with varying M

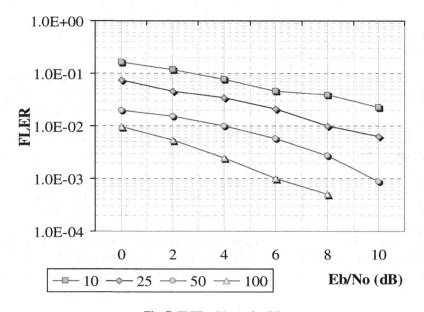

Fig. 7. FLER with varying M

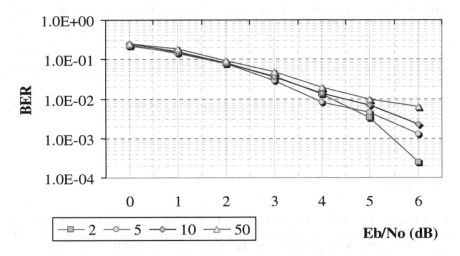

Fig. 8. BER with varying preamble periods

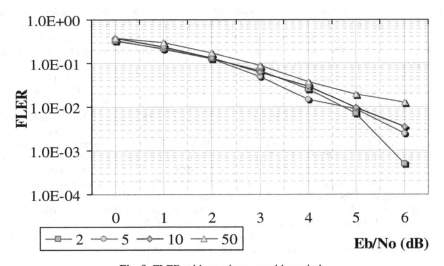

Fig. 9. FLER with varying preamble periods

5 Conclusion

A robust data-aided turbo synchronizer is presented where pilot sequences are included periodically (period > n*frame_size) in the symbol stream. The high SNR method or other correlation methods are used to provide a list of potential delays, one of which (the largest defined metric) is then confirmed by the decoder-based synchronizer as the frame timing offset. This data-aided ML synchronizer is cost-

efficient because even for fast fading channels the preamble insertion period is large enough to result in minimal increase in throughput.
A practical configuration for a turbo-based multifunctional system comprising a turbo decoder for PUM turbo codes, a low complexity turbo equalizer and the data-aided turbo frame synchronizer has been presented that can operate in frequency-selective channels with fading.

Acknowledgement. This work was funded under the Ministry of Defence Applied Research Program, QinetiQ contract number CU009-0055.

References

[1] Fagoonee, L., Honary, B. and Williams, C. PUM-based Turbo Codes. In: Proc. 6th International Symposium on Digital Signal Processing for Communication Systems (DSPCS'2002), Sydney-Manly, Australia, 28–31 January 2002. 191–194.

[2] Fagoonee, L., Honary B. and Williams, C. 2002. PUM-Based Turbo Codes. In: Advanced Digital Signal Processing for Communication Systems. Kluwer Academic Publishers, ISBN: 1-4020-72023, 2002.

[3] Fagoonee, L., Clarke, P. and Honary, B. 2003. Combined Turbo Equalization and Decoding for Frequency Selective Fading Channels. In: Proc. 9th International Conference on HF Radio Systems and Techniques, University of Bath, U.K., 23–26 June 2003.

[4] Lauer, G.S. 1979. Some Optimal Partial-Unit-Memory Codes. IEEE Transactions Information Theory, IT-25(2), 240–243.

[5] Berrou, C., Glavieux, A. and Thitimajshima, P. 1993. Near Shannon Limit Error Correcting Coding and Decoding: Turbo Codes (1). In: Proc. IEEE International Conference in Communications., Geneva, Switzerland, May 1993. 1064–1071.

[6] Hagenauer J., Offer E. and Papke L. 1996. Iterative Decoding of Binary Block and Convolutional Codes. IEEE Transactions on Information Theory, 42(2), 429–445.

[7] Fijalkow, I., Roumy, A., Ronger, S., Pirez, D. and Vila, P. 2000. Improved Interference Cancellation for Turbo-Equalization. In: Proc. IEEE International Conference on Acoustics, Speech and Signal Processing (ICASSP2000), Vol. 1, 2000. 416–419.

[8] Honary, B., Darnell, M. and Markarian, G. 1994. Trellis Extracted Synchronisation Techniques (TEST). UK Patent Application No. 9414275-7, 14th July 1994.

[9] Howlader, M.M.K. and Woerner, B.D. 2001. Decoder-assisted frame synchronization for turbo-coded systems. IEE Electronics Letters, 37(6), 362–363.

[10] Wysocki, T., Razavi, H. and Honary, B. 1997. Digital Signal Processing for Communication Systems. Kluwer Academic Publishers, 81–90.

Commitment Capacity of Discrete Memoryless Channels

Andreas Winter[1], Anderson C.A. Nascimento[2], and Hideki Imai[2]

[1] Department of Computer Science, University of Bristol,
Merchant Venturers Building, Woodland Road, Bristol BS8 1UB, United Kingdom
winter@cs.bris.ac.uk
[2] Imai Laboratory, Information and Systems, Institute of Industrial Science,
University of Tokyo, 4-6-1 Komaba, Meguro–ku, Tokyo 153–8505, Japan.
{anderson,imai}@imailab.iis.u-tokyo.ac.jp

Abstract. In extension of the bit commitment task and following work initiated by Crépeau, we introduce and solve the problem of characterising the optimal rate at which a discrete memoryless channel can be used to for bit commitment. It turns out that the answer is very intuitive: it is the maximum equivocation of the channel (after removing trivial redundancy), even when unlimited noiseless bidirectional side communication is allowed. By a well–known reduction, this result provides a lower bound on the channel's capacity for implementing coin tossing.

The method of proving this relates the problem to Wyner's wire–tap channel in an amusing way. There is also an extension to quantum channels.

1 Introduction

Noise is a powerful resource for the implementation of cryptographic primitives: it allows for the construction of *information theoretically secure* cryptographic protocols — a task typically impossible without the noise, and in practice done by relaxing to *computational security*, assuming conjectures from complexity theory.

In his famous paper [26], Wyner was the first to exploit noise in order to establish a secure channel in the presence of an eavesdropper. These results were extended in studies of secret key distillation by Maurer [19], Ahlswede and Csiszár [1] and followers. The noise in these studies is assumed to affect the eavesdropper: thus, to work in practice, it has to be guaranteed or certified somehow. This might be due to some — trusted — third party who controls the channel (and thus prevents the cryptographic parties from cheating), or due to physical limitations, as in quantum key distribution [4,5]. Crépeau and Kilian [13] showed how information theoretically secure bit commitment can be implemented using a binary symmetric channel, their results being improved in [12] and [15].

The object of the present study is to optimise the use of the noisy channel, much as in Shannon's theory of channel capacities: while the previous studies

K.G. Paterson (Ed.): Cryptography and Coding 2003, LNCS 2898, pp. 35–51, 2003.
© Springer-Verlag Berlin Heidelberg 2003

have concentrated on the *possibility* of bit commitment using noisy channels, here we look at committing to one out of a larger message set, e.g. a bit string. We are able, for a general discrete memoryless channel, to characterise the commitment capacity by a simple (single–letter) formula (theorem 2), stated in section 2, and proved in two parts in sections 3 and 4. A few specific examples are discussed in section 5 to illustrate the main result. In section 6 we close with a discussion. An appendix collects some facts abut typical sequences used in the main proof.

2   Definitions and Main Result

In the commitment of a message there are two parties, called *Alice* and *Bob*, the first one given the message a from a certain set $\mathcal{A}$. The whole procedure consists of two stages: first the *commit phase*, in which Alice (based on a) and Bob exchange messages, according to a protocol. This will leave Bob with a record (usually called *view*), to be used in the second stage, the *reveal phase*. This consists of Alice disclosing a and other relevant information to Bob. Bob performs a test on all his recorded data which accepts if Alice followed the rules and disclosed the correct information in the second stage, and rejects if a violation of the rules is discovered.

To be useful, such a scheme has to fulfill two requirements: it must be "concealing" as well as "sound" and "binding": the first property means that after the commit phase Bob has no or almost no information about a (i.e., even though Alice has "committed" herself to something by the communications to Bob, this commitment remains secret), and this has to hold even if Bob does not follow the protocol, while Alice does. Soundness means that if both parties behave according to the protocol, Bob's test will accept (with high probability) after the reveal phase. The protocol to be binding means that Bob's test is such that whatever Alice did in the commit phase (with Bob following the rules) there is only at most one a she can "reveal" which passes Bob's test.

In our present consideration there is an unlimited bidirectional noiseless channel available between Alice and Bob, and in addition a discrete memoryless noisy channel $W : \mathcal{X} \longrightarrow \mathcal{Z}$ from Alice to Bob, which may be used n times: on input $x^n = x_1 \ldots x_n$, the output distribution on $\mathcal{Z}^n$ is $W_{x^n}^n = W_{x_1} \otimes \cdots \otimes W_{x_n}$.

Definition 1 *The channel W is called* non–redundant, *if none of its output distributions is a convex combination of its other output distributions:*

$$\forall y \forall P \text{ s.t. } P(y) = 0 \quad W_y \neq \sum_x P(x) W_x.$$

In geometric terms this means that all distributions W_x are distinct extremal points of the polytope $\mathcal{W} = \text{conv}\{W_x : x \in \mathcal{X}\}$, the convex hull of the output distributions within the probability simplex over $\mathcal{Z}$. Clearly, we can make W into a non–redundant channel $\widetilde{W}$ by removing all input symbols x whose output distribution W_x is not extremal. The old channel can be simulated by the new *one, because by feeding it distributions over input symbols one can generate the output distributions of the removed symbols.*

The channel W is called trivial, *if after making it non–redundant its output distributions have mutually disjoint support. This means that from the output one can infer the input with certainty.*

With this we can pass to a formal definition of a protocol: this, consisting of the named two stages, involves creation on Alice's side of either messages intended for the noiseless channel, or inputs to the noisy channel, based on previous messages receive from Bob via the noiseless channel, which themselves are based on data received before, etc. Both agents may employ probabilistic choices, which we model by Alice and Bob each using a random variable, M and N, respectively. This allows them to use *deterministic* functions in the protocol. Note that this makes all messages sent and received into well-defined random variables, *dependent on a*.

Commit Phase: The protocol goes for r rounds of Alice–to–Bob and Bob–to–Alice noiseless communications U_j and V_j. After round r_i ($r_1 \leq \ldots \leq r_n \leq r$) Alice will also send a symbol X_i down the noisy channel W, which Bob receives as Z_i. Setting $r_0 = 0$ and $r_{n+1} = r$:
Round r_i+k ($1 \leq k \leq r_{i+1}-r_i$): Alice sends $U^{r_i+k} = f_{r_i+k}(a, M, V^{r_i+k-1})$ noiselessly. Bob answers $V_{r_i+k} = g_{r_i+k}(Z^i, N, U^{r_i+k})$, also noiselessly. After round r_i and before round $r_i + 1$ ($1 \leq i \leq n$), Alice sends $X_i = F_i(a, M, V^{r_i})$, which Bob receives as $Z_i = W(X_i)$.

Reveal Phase: A similar procedure as the Commit Phase, but without the noisy channel uses, including Alice's sending a to Bob. At the end of the exchange Bob performs a test as to whether to accept Alice's behaviour or not. It is easily seen that this procedure can be simulated by Alice simply telling Bob a and M, after which Bob performs his test $\beta(Z^n, N, U^r; a, M) \in \{\text{ACC}, \text{REJ}\}$. I.e., requiring Alice to reveal M and a makes cheating for her only more difficult.

We shall, for technical reasons, impose the condition that the range of the variable U^r is bounded:

$$|U^r| \leq \exp(Bn), \tag{1}$$

with a constant B. Note that exp and log in this paper are always to basis 2, unless otherwise stated.

Now, the mathematical form of the conditions for concealing as well as for soundness and binding is this: we call the above protocol ϵ–*concealing* if for any two messages $a, a' \in \mathcal{A}$ and any behaviour of Bob during the commit phase,

$$\frac{1}{2}\left\|\text{Distr}_a(Z^n NU^r) - \text{Distr}_{a'}(Z^n NU^r)\right\|_1 \leq \epsilon, \tag{A}$$

where $\text{Distr}_a(Z^n NU^r)$ is the distribution of the random variables $Z^n NU^r$ after completion of the commit phase which Alice entered with the message a and the randomness M, and with the ℓ_1–norm $\|\cdot\|_1$; the above expression is identical to the total variational distance of the distributions. This is certainly the strongest requirement one could wish for: it says that no statistical test of Bob immediately after the commit phase can distinguish between a and a' with probability larger than ϵ. Note that V^r is a function of $Z^n N U^r$, and hence could be left out in eq. (A). Assuming any probability distribution on the messages, a is the value of a random variable A, and it is jointly distributed with all other variables of the protocol. Then, whatever Bob's strategy,

$$I(A \wedge Z^n N U^r) \le \epsilon' = H(2\epsilon, 1 - 2\epsilon) + 2n\epsilon(\log B + \log |\mathcal{Z}|), \qquad \text{(A')}$$

where

$$I(X \wedge Y) = H(X) + H(Y) - H(XY)$$

is the (Shannon) mutual information between X and Y, and

$$H(X) = -\sum_x \Pr\{X = x\} \log \Pr\{X = x\}$$

is the (Shannon) entropy of X [25].

We call the protocol δ-*sound and* -*binding* (δ-*binding* for short), if for Alice and Bob following the protocol, for all $a \in \mathcal{A}$,

$$\Pr\{\beta(Z^n N U^r; aM) = \text{ACC}\} \ge 1 - \delta, \qquad \text{(B1)}$$

and, whatever Alice does during the commit phase, governed by a random variable S with values σ (which determines the distribution of $Z^n N U^r$), for all $A = a(S, V^r)$, $A' = a'(S, V^r)$, $\widetilde{M} = \mu(S, V^r)$ and $\widetilde{M}' = \mu'(S, V^r)$ such that $A \ne A'$ with probability 1,

$$\Pr\left\{\beta(Z^n N U^r; A\widetilde{M}) = \text{ACC} \ \& \ \beta(Z^n N U^r; A'\widetilde{M}') = \text{ACC}\right\} \le \delta. \qquad \text{(B2)}$$

Note that by convexity the cheating attempt of Alice is w.l.o.g. *deterministic*, which is to say that S takes on only one value σ with non-zero probability, hence $\Pr\{S = \sigma\} = 1$.

We call $\frac{1}{n} \log |\mathcal{A}|$ the *(commitment) rate* of the protocol. A rate R is said to be *achievable* if there exist commitment protocols for every n with rates converging to R, which are ϵ-concealing and δ-binding with $\epsilon, \delta \to 0$ as $n \to \infty$. The *commitment capacity* $C_{\text{com}}(W)$ of W is the supremum of all achievable rates.

The main result of this paper is the following theorem:

Theorem 2 *The commitment capacity of the discrete channel W (assumed to be non-redundant) is*

$$C_{\text{com}}(W) = \max\{H(X|Z) : X, Z \text{ RVs}, \text{Distr}(Z|X) = W\},$$

i.e., the maximal equivocation *of the channel over all possible input distributions.*

Corollary 3 *Every non-trivial discrete memoryless channel can be used to perform bit commitment.* □

By invoking the well–known reduction of coin tossing to bit commitment [7] we obtain:

Corollary 4 *The channel W can be used for secure two–party coin tossing at rate at least $C_{\text{com}}(W)$. I.e., for the naturally defined coin tossing capacity $C_{\text{c.t.}}(W)$, one has $C_{\text{c.t.}}(W) \geq C_{\text{com}}(W)$.* □

This theorem will be proved in the following two sections (propositions 7 and 8): first we construct a protocol achieving the equivocation bound, showing that exponential decrease of ϵ and δ is possible, and without using the noiseless side channels at all during the commit phase. Then we show the optimality of the bound.

To justify our allowing small errors both in the concealing and the binding property of a protocol, we close this section by showing that demanding too much trivialises the problem:

Theorem 5 *There is no bit–commitment via W which is ϵ–concealing and 0–binding with $\epsilon < 1$. I.e., not even two distinct messages can be committed: $|\mathcal{A}| = 1$.*

Proof. If the protocol is 0–sound, this means that for *every* value μ attained by M with positive probability, Bob will accept the reveal phase if Alice behaved according to the protocol. On the other hand, that the protocol is 0–binding means that for $a \neq a'$ and arbitrary μ', Bob will never accept if Alice behaves according to the protocol in the commit phase, with values $a\mu$ but tries to "reveal" $a'\mu'$. This opens the possibility of a decoding method for a based on $Z^n N U^r$: Bob simply tries out all possible $a\mu$ with his test β — the single a which is accepted must be the one used by Alice. Hence the scheme cannot be ϵ–concealing with $\epsilon < 1$. □

3   A Scheme Meeting the Equivocation Bound

Here we describe and prove security bounds of a scheme which is very simple compared to the generality we allowed in section 4: in the commit phase it consists only of a single block use of the noisy channel W^n, with no public discussion at all, where the input X^n is a random one–to–many function of a (in particular, a is a function of the x^n chosen). In the reveal phase Alice simply announces X^n to Bob.

Proposition 6 *Given $\sigma, \tau > 0$, and a distribution P of $X \in \mathcal{X}$, with the output $Z = W(X)$, $Q = \text{Distr}(Z)$. Then there exists a collection of codewords*

$$\left(\xi_{a\mu} \in \mathcal{X}^n : a = 1, \dots, K, \ \mu = 1, \dots, L \right)$$

with the following properties:

1. For all $(a, \mu) \neq (a', \mu')$, $d_{\text{H}}(\xi_{a\mu}, \xi_{a'\mu'}) \geq 2\sigma n$.

2. *For every a:*

$$\frac{1}{2}\left\|\frac{1}{L}\sum_{\mu=1}^{L}W_{\xi_{a\mu}}^{n}-Q^{\otimes n}\right\|_{1}\leq 25|\mathcal{X}||\mathcal{Z}|\exp(-n\tau).$$

3. *There are constants G, G' and a continuous function G'' vanishing at 0 such that*

$$K\geq\frac{1}{2n}(3+\log|\mathcal{X}|+\log|\mathcal{Z}|)^{-1}\exp(nH(X|Z)-n\sqrt{2\tau}G'-nG''(\sigma)),$$

$$L\leq n(3+\log|\mathcal{X}|+\log|\mathcal{Z}|)\exp(nI(X\wedge Z)+n\sqrt{2\tau}G).$$

Proof. To get the idea, imagine a wiretap channel [26] with W as the stochastic matrix of the eavesdropper and a symmetric channel $S_\sigma : \mathcal{X} \longrightarrow \mathcal{Y} = \mathcal{X}$ for the legal user:

$$S_\sigma(y|x)=\begin{cases}1-\sigma & \text{if } x=y,\\ \frac{1}{|\mathcal{X}|}\sigma & \text{if } x\neq y.\end{cases}$$

The random coding strategy for such a channel, according to Wyner's solution [26] (but see also [14] and [10]) will produce a code with the properties 2 and 3. Because the code for the legal user must fight the noise of the symmetric channel S_σ, we can expect its codewords to be of large mutual Hamming distance, i.e., we should get property 1.

In detail: pick the $\xi_{a\mu}$ i.i.d. according to the distribution $\widetilde{P}^{n}$, which is 0 outside $\mathcal{T}_{P,\sqrt{2\tau}}^{n}$ (the typical sequences, see appendix A) and $P^{\otimes n}$ within, suitably normalised. Also introduce the subnormalised measures $\widehat{W}_{x^{n}}^{n}$: this is identical to $W_{x^{n}}^{n}$ within $\mathcal{T}_{W,\sqrt{2\tau}}^{n}$ and 0 outside. We will show that with high probability we can select codewords with properties 2 and 3, and only a small proportion of which violate property 1; then by an expurgation argument will we obtain the desired code.

By eqs. (8) and (13) in the appendix we have

$$\frac{1}{2}\left\|\mathbb{E}\widehat{W}_{\xi_{a\mu}}^{n}-Q^{\otimes n}\right\|\leq 3|\mathcal{X}||\mathcal{Z}|\exp(-n\tau),\tag{2}$$

with the expectation referring to the distribution $\widetilde{P}^{n}$ of the $\xi_{a\mu}$. Observe that the support of all $\widehat{W}_{\xi_{a\mu}}^{n}$ is contained in $\mathcal{T}_{Q,2|\mathcal{X}|\sqrt{\tau}}^{n}$, using eq. (17) of the appendix. Now, $\mathcal{S}$ is defined as the set of those z^{n} for which

$$\mathbb{E}\widehat{W}_{\xi_{a\mu}}^{n}(z^{n})\geq T:=\exp(-n\tau)\exp(-nH(Q)-n\sqrt{2\tau}|\mathcal{X}|F),$$

with $F=\sum_{z:Q(z)\neq 0}-\log Q(z)$, and define $\widetilde{W}_{x^{n}}^{n}(z^{n})=\widehat{W}_{x^{n}}^{n}(z^{n})$ if $z^{n}\in\mathcal{S}$ and 0 otherwise. With the cardinality estimate eq. (10) of the appendix and eq. (2) we obtain

$$\frac{1}{2}\left\|\mathbb{E}\widetilde{W}_{\xi_{a\mu}}^{n}-Q^{\otimes n}\right\|\leq 4|\mathcal{X}||\mathcal{Z}|\exp(-n\tau).\tag{3}$$

The Chernoff bound allows us now to efficiently sample the expectation $\widetilde{Q}^n :=$ $\mathbb{E}\widetilde{W}^n_{\xi_{a\mu}}$: observe that all the values of $\widetilde{W}^n_{\xi_{a\mu}}$ are upper bounded by

$$t := \exp\left(-nH(W|P) + n\sqrt{2\tau}|\mathcal{X}|\log|\mathcal{Z}| + n\sqrt{2\tau}E\right),$$

using eq. (14). Thus, rescaling the variables, by lemma 1 and with the union bound, we get

$$\Pr\left\{\forall a \forall z^n \in \mathcal{S} \ \frac{1}{L'}\sum_{\mu=1}^{L'}\widetilde{W}^n_{\xi_{a\mu}}(z^n) \in \left[(1\pm\exp(-n\tau))\widetilde{Q}^n(z^n)\right]\right\}$$

$$\leq 2K|\mathcal{S}|\exp\left(-L\frac{T}{t}\frac{\exp(-2n\tau)}{2\ln 2}\right), \tag{4}$$

which is smaller than $1/2$ if

$$L' > 2 + n(\log|\mathcal{X}| + \log|\mathcal{Z}|)\exp\left(nI(X\wedge Z) + n\sqrt{2\tau}G\right),$$

with $G = 3 + |\mathcal{X}|F + |\mathcal{X}|\log|\mathcal{Z}| + E$. Note that in this case, there exist values for the $\xi_{a\mu}$ such that the averages $\frac{1}{L'}\sum_\mu W^n_{\xi_{a\mu}}$ are close to $Q^{\otimes n}$.

Now we have to enforce property 1: in a random batch of $\xi_{a\mu}$ we call $a\mu$ bad if $\xi_{a\mu}$ has Hamming distance less than $2\sigma n$ from another $\xi_{a'\mu'}$. The probability that $a\mu$ is bad is easily bounded:

$$\Pr\{a\mu \text{ bad}\} \leq 2|\mathcal{X}|\exp(-n\tau) + P^{\otimes n}\left(\bigcup_{a'\mu'\neq a\mu} B_{2n\sigma}(\xi_{a'\mu'})\right)$$

$$\leq 2|\mathcal{X}|\exp(-n\tau) + \max\left\{P^{\otimes n}(\mathcal{A}) : |\mathcal{A}| \leq K'L'\binom{n}{2n\sigma}|\mathcal{X}|^{2n\sigma}\right\}$$

$$\leq 5|\mathcal{X}|\exp(-n\tau),$$

by eq. (12) in the appendix, because we choose

$$K' \leq \frac{1}{n}(3 + \log|\mathcal{X}| + \log|\mathcal{Z}|)^{-1}$$
$$\exp\left(nH(X|Z) - n\sqrt{2\tau}G - 2n\sqrt{2\tau}D - nH(2\sigma, 1-2\sigma) - 2n\sigma\log|\mathcal{X}|\right),$$

hence

$$K'L'\binom{n}{2n\sigma}|\mathcal{X}|^{2n\sigma} < \exp\left(nH(P) - 2n\sqrt{2\tau}D\right).$$

Thus, with probability at least $1/2$, only a fraction of $10|\mathcal{X}|\exp(-n\tau)$ of the $a\mu$ are bad. Putting this together with eq. (4), we obtain a selection of $\xi_{a\mu}$ such that

$$\forall a \ \frac{1}{2}\left\|\frac{1}{L'}\sum_\mu W^n_{\xi_{a\mu}} - Q^{\otimes n}\right\|_1 \leq 5|\mathcal{X}||\mathcal{Z}|\exp(-n\tau) \tag{5}$$

and only a fraction of $10|\mathcal{X}|\exp(-n\tau)$ of the $a\mu$ are bad.

This means that for at least half of the a, w.l.o.g. $a = 1, \ldots, K = K'/2$, only a fraction $20|\mathcal{X}|\exp(-n\tau)$ of the μ form bad pairs $a\mu$, w.l.o.g. for $\mu = L+1, \ldots, L'$, with $L = (1 - 20|\mathcal{X}|\exp(-n\tau))L'$. Throwing out the remaining a and the bad μ, we are left with a code as desired.   □

Observe that a receiver of Z^n can efficiently check claims about the input $\xi_{a\mu}$ because of property 1, that distinct codewords have "large" Hamming distance. The non–redundancy of W shuns one–sided errors in this checking, as we shall see. The test β is straightforward: it accepts iff $Z^N \in \mathcal{T}^n_{W,\sqrt{2\tau}}(\xi_{a\mu})$, the set of *conditional typical sequences*, see appendix A. This ensures soundness; for the bindingness we refer to the following proof.

We are now in a position to describe a protocol, having chosen codewords according to proposition 6:

Commit phase: To commit to a message a, Alice picks $\mu \in \{1, \ldots, L\}$ uniformly at random and sends $\xi_{a\mu}$ through the channel. Bob obtains a channel output z^n.

Reveal phase: Alice announces a and μ. Bob performs the test β: he accepts if $z^n \in \mathcal{B}_{a\mu} := \mathcal{T}^n_{W,\sqrt{2\tau}}(\xi_{a\mu})$ and rejects otherwise.

Proposition 7 *Assume that for all $x \in \mathcal{X}$ and distributions P with $P(x) = 0$,*

$$\left\| W_x - \sum_y P(y)W_y \right\|_1 \geq \eta.$$

Let $\tau = \frac{\sigma^4\eta^2}{8|\mathcal{X}|^4|\mathcal{Z}|^2}$: then the above protocol implements an ϵ–concealing and δ–binding commitment with rate

$$\frac{1}{n}\log K \geq H(X|Z) - \sqrt{2\tau}G' - H(2\sigma, 1 - 2\sigma) - 2\sigma\log|\mathcal{X}| - \frac{\log n}{n} - O\left(\frac{1}{n}\right)$$

and exponentially bounded security parameters:

$$\epsilon = 50|\mathcal{X}||\mathcal{Z}|\exp(-n\tau),$$
$$\delta = 2|\mathcal{X}||\mathcal{Z}|\exp(-2n\tau^2).$$

Proof. That the protocol is ϵ–concealing is obvious from property 2 of the code in proposition 6: Bob's distribution of Z^n is always $\epsilon/2$-close to $Q^{\otimes n}$, whatever a is.

To show δ–bindingness observe first that if Alice is honest, sending $\xi_{a\mu}$ in the commit phase and later revealing $a\mu$, the test β will accept with high probability:

$$\Pr\{Z^n \in \mathcal{B}_{a\mu}\} = W^n_{\xi_{a\mu}}\left(\mathcal{T}^n_{W,\sqrt{2\tau}}(\xi_{a\mu})\right)$$
$$\geq 1 - 2|\mathcal{X}||\mathcal{Z}|\exp(-n\tau) \geq 1 - \delta,$$

by eq. (13) in the appendix.

On the other hand, if Alice cheats, we may — in accordance with our definition — assume her using a deterministic strategy: i.e., she "commits" sending some x^n and later attempts to "reveal" either $a\mu$ or $a'\mu'$, with $a \neq a'$. Because of property 1 of the code in proposition 6, at least one of the codewords $\xi_{a\mu}$, $\xi_{a'\mu'}$ is at Hamming distance at least σn from x^n: w.l.o.g., the former of the two. But then the test β accepts "revelation" of $a\mu$ with small probability:

$$\Pr\{Z^n \in \mathcal{B}_{a\mu}\} = W_{x^n}^n\left(\mathcal{T}_{W,\sqrt{2\tau}}^n(\xi_{a\mu})\right) \leq 2\exp(-2n\tau^2) \leq \delta,$$

by lemma 2 in the appendix.   □

4   Upper Bounding the Achievable Rate

We assume that W is non–redundant. We shall prove the following assertion, assuming a uniformly distributed variable $A \in \mathcal{A}$ of messages.

Proposition 8 *Consider an ϵ–concealing and δ–binding commitment protocol with n uses of W. Then*

$$\begin{aligned}
\log|\mathcal{A}| \leq &n \max\{H(X|Z) : \mathrm{Distr}(Z|X) = W\} \\
&+ n\left(\epsilon(\log B + \log|\mathcal{Z}|) + 5\sqrt[3]{\delta}\log|\mathcal{X}|\right) + 2.
\end{aligned} \tag{6}$$

The key, as it turns out, of its proof, is the insight that in the above protocol, should it be concealing and binding, x^n together with Bob's view of the commit phase (essentially) determine a. In the more general formulation we permitted in section 2, we prove :

$$H(A|Z^n N U^r; X^n) \leq \delta' = H\left(5\sqrt[3]{\delta}, 1 - 5\sqrt[3]{\delta}\right) + 5\sqrt[3]{\delta}\log|\mathcal{A}|. \tag{B'}$$

Intuitively, this means that with the items Alice entered into the commit phase of the protocol and those which are accessible to Bob, not too many values of A should be consistent — otherwise Alice had a way to cheat.

Proof of eq. (B'). For each $a\mu$ the commit protocol (both players being honest) creates a distribution $\Delta_{a\mu}$ over *conversations* $(x^n v^r; z^n u^r)$. We leave out Bob's random variable N here, noting that he can create its correct conditional distribution from $z^n u^r; v^r$, which is his view of the conversation. The only other place where he needs it is to perform the test β. We shall in the following assume that it includes this creation of N, which makes β into a probabilistic test, depending on $(a\mu v^r; z^n u^r)$.

The pair $a\mu$ has a probability $\alpha_{a\mu}$ that its conversation with subsequent revelation of $a\mu$ is accepted. By soundness, we have

$$\sum_{\mu} \Pr\{M = \mu\}\alpha_{a\mu} \geq 1 - \delta,$$

for every a. Hence, by Markov inequality, there exists (for every a) a set of μ of total probability $\geq 1 - \sqrt[3]{\delta^2}$ for which $\alpha_{a\mu} \geq 1 - \sqrt[3]{\delta}$. We call such μ *good for a*.

From this we get a set $\mathcal{C}_{a\mu}$ of "partial" conversations $(x^n v^r; u^r)$, with probability $\Delta_{a\mu}(\mathcal{C}_{a\mu}) \geq 1 - \sqrt[3]{\delta}$, which are accepted with probability at least $1 - \sqrt[3]{\delta}$. (In the test also Z^n enters, which is distributed according to $W_{x^n}^n$.)

Let us now define the set

$$\mathcal{C}_a := \bigcup_{\mu \text{ good for } a} \mathcal{C}_{a\mu},$$

which is a set of "partial conversations" which are accepted with probability at least $1 - \sqrt[3]{\delta}$ and

$$\Delta_a(\mathcal{C}_a) \geq 1 - 2\sqrt[3]{\delta},$$

with the distribution

$$\Delta_a := \sum_\mu \Pr\{M = \mu\}\Delta_{a\mu}$$

over "partial conversations": it is the distribution created by the commit phase give the message a.

We claim that

$$\Delta_a\left(X^n V^r; U^r \in \bigcup_{a' \neq a} \mathcal{C}_{a'}\right) \leq 3\sqrt[3]{\delta}. \tag{7}$$

Indeed, if this were not the case, Alice had the following cheating strategy: in the commit phase she follows the protocol for input message a. In the reveal phase she looks at the "partial conversation" $x^n v^r; u^r$ and tries to "reveal" some $a'\mu'$ for which the partial conversation is in $\mathcal{C}_{a'\mu'}$ (if these do not exist, $a'\mu'$ is arbitrary). This defines random variables A' and $\widetilde{M'}$ for which it is easily checked that

$$\Pr\{\beta(Z^n N\dot{U}^r; aM) = \text{ACC} \ \& \ \beta(Z^n NU^r; A'\widetilde{M'}) = \text{ACC}\} > \delta,$$

contradicting the δ–bindingness condition.

Using eq. (7) we can build a decoder for A from $X^n V^r; U^r$: choose $\widehat{A} = a$ such that $X^n V^r; U^r \in \mathcal{C}_a$ — if there exists none or more than one, let $\widehat{A}$ be arbitrary. Clearly,

$$\Pr\{A \neq \widehat{A}\} \leq 5\sqrt[3]{\delta},$$

and invoking Fano's inequality we are done.   □

Armed with this, we can now proceed to the *Proof of proposition 8*. We can successively estimate,

$$
\begin{aligned}
H(X^n|Z^n) &\geq H(X^n|Z^n NU^r) \\
&= H(AX^n|Z^n NU^r) - H(A|Z^n NU^r; X^n) \\
&\geq H(A|Z^n NU^r) - H(A|Z^n NU^r; X^n) \\
&\geq H(A|Z^n NU^r) - \delta' \\
&= H(A) - I(A \wedge Z^n NU^r) - \delta' \\
&\geq H(A) - \epsilon' - \delta',
\end{aligned}
$$

using eq. (B') in the fourth, eq. (A') in the sixth line. On the other hand, subadditivity and the conditioning inequality imply

$$
H(X^n|Z^n) \leq \sum_{k=1}^{n} H(X_k|Z_k),
$$

yielding the claim, because $H(A) = \log|\mathcal{A}|$.

The application to the proof of the converse of theorem 2 is by observing that $\epsilon', \delta' = o(n)$. $\square$

Note that for the proof of the proposition we considered only a very weak attempt of Alice to cheat: she behaves according to the protocol during the commit phase, and only at the reveal stage she tries to be inconsistent. Similarly, our concealingness condition considered only passive attempts to cheat by Bob, i.e., he follows exactly the protocol, and tries to extract information about A only by looking at his view of the exchange.

Thus, even in the model of *passive cheating*, which is less restrictive than our definition in section 2, we obtain the upper bound of proposition 8

5   Examples

In this section we discuss some particular channels, which we present as stochastic matrices with the rows containing the output distributions.

1. Binary symmetric channel B_p: Let $0 \leq p \leq 1$. Define

$$
B_p := \quad
\begin{array}{c|c|c}
\cdot & 0 & 1 \\
\hline\hline
0 & 1-p & p \\
\hline
1 & p & 1-p
\end{array}
$$

The transmission capacity if this channel is easily computed from Shannon's formula [25]: $C(B_p) = 1 - H(p, 1-p)$, which is non–zero iff $p \neq 1/2$. The optimal input distribution is the uniform distribution $(1/2, 1/2)$ on $\{0, 1\}$. Note that this channel is trivial if $p \in \{0, 1/2, 1\}$, hence $C_{\text{com}}(B_p) = 0$ for these values of p. We may thus, w.l.o.g., assume that $0 < p < 1/2$, for which B_p is non–redundant. It is not hard to compute the optimal input distribution as the uniform distribution, establishing $C_{\text{com}}(B_p) = H(p, 1 - p)$.

The result is in accordance with our intuition: the noisier the channel is, the worse it is for transmission, but the better for commitment.

2. A trivial channel: Consider the channel

$$T := \begin{array}{c|c|c} \cdot & 0 & 1 \\ \hline a & 1/2 & 1/2 \\ \hline b & 1 & 0 \\ \hline c & 0 & 1 \end{array}$$

Clearly, T is trivial, hence $C_{\text{com}}(T) = 0$. Still it is an interesting example in the light of our proof of proposition 7: for assume a wiretap channel for which T is the stochastic matrix of the eavesdropper, while the legal user obtains a noiseless copy of the input. Then clearly the wiretap capacity of this system is 1, with optimal input distribution $(1/2, 1/4, 1/4)$.

3. Transmission and commitment need not be opposites: We show here an example of a channel where the optimising input distributions for transmission and for commitment are very different:

$$V := \begin{array}{c|c|c} \cdot & 0 & 1 \\ \hline 0 & 1/2 & 1/2 \\ \hline 1 & 1 & 0 \end{array}$$

It can be easily checked that the maximum of the mutual information, i.e. the transmission capacity, is attained for the input distribution

$$P(0) = \frac{2}{5} = 0.4, \quad P(1) = \frac{3}{5} = 0.6,$$

from which we obtain $C(V) \approx 0.3219$. On the other hand, the equivocation is maximised for the input distribution

$$P'(0) = 1 - \sqrt{\frac{1}{5}} \approx 0.5528, \quad P'(1) = \sqrt{\frac{1}{5}} \approx 0.4472,$$

from which we get that $C_{\text{com}}(V) \approx 0.6942$. The maximising distributions are so different that the sum $C(V) + C_{\text{com}}(V) > 1$, i.e. it exceeds the maximum input and output entropies of the channel.

6   Discussion

We have considered bit–string commitment by using a noisy channel and have characterised the exact capacity for this task by a single–letter formula. This implies a lower bound on the coin tossing capacity of that channel by the same formula.

Satisfactory as this result is, it has to be noted that we are not able in general to provide an explicit protocol: our proof is based on the random coding technique and shows only existence. What is more, even if one finds a good code it will most likely be inefficient: the codebook is just the list of $\xi_{a\mu}$. In this connection we conjecture that the commitment capacity can be achieved by random linear codes (compare the situation for channel coding!). It is in any case an open problem to find efficient good codes, even for the binary symmetric channel. Note that we only demand efficient *encoding* — there is no decoding of errors in our scheme, only an easily performed test.

Our scheme is a block–coding method: Alice has to know the whole of her message, say a bit string, before she can encode. One might wan to use our result as a building block in other protocols which involve committing to bits at various stages — then the natural question arises whether there is an "online" version which would allow Alice to encode and send bits as she goes along.

In the same direction of better applicability it would be desirable to extend our results to a more robust notion of channel: compare the work of [15] where a cheater is granted partial control over the channel characteristics. Still, the fixed channel is not beyond application: note that it can be simulated by pre–distributed data from a trusted party via a "noisy one–time pad" (compare [3] and [24]).

Another open question of interest is to determine the reliability function, i.e., the optimal asymptotic rate of the error $\epsilon + \delta$ (note that implicit in our proposition 7 is a lower bound): it is especially interesting at $R = 0$, because there the rate tells exactly how secure single–bit commitment can be made.

In the journal version of this paper, we outline that a class of quantum channels also allows bit commitment: they even have a commitment capacity of the same form as the classical result. This opens up the possibility of unconditionally secure bit commitment for other noisy quantum channels.

We hope that our work will stimulate the search for optimal rates of other cryptographic primitives, some of which are possible based on noise, e.g. oblivious transfer.

Acknowledgements. We thank Ning Cai and Raymond W. Yeung for sharing their ideas on the quantum wiretap channel, and making available to us their manuscript [10]. We also acknowledge interesting discussions with J. Müller–Quade and P. Tuyls at an early stage of this project.

AW was supported by the U.K. Engineering and Physical Sciences Research Council. ACAN and HI were supported by the project "Research and Development on Quantum Cryptography" of Telecommunications Advancement Organization as part of the programme "Research and Development on Quantum Communication Technology" of the Ministry of Public Management, Home Affairs, Posts and Telecommunications of Japan.

A   Typical Sequences

This appendix collects some facts about typical sequences used in the main body of the text. We follow largely the book of Csiszár and Körner [14].

The fundamental fact we shall use is the following large deviation version of the law of large numbers:

Lemma 1 (Chernoff [11]). *For i.i.d. random variables $X_1, \dots, X_N$, with $0 \le X_n \le 1$ and with expectation $\mathbb{E}X_n = p$:*

$$\Pr\left\{\frac{1}{N}\sum_{n=1}^{N} X_n \ge (1+\eta)p\right\} \le \exp\left(-N\frac{p\eta^2}{2\ln 2}\right),$$

$$\Pr\left\{\frac{1}{N}\sum_{n=1}^{N} X_n \le (1-\eta)p\right\} \le \exp\left(-N\frac{p\eta^2}{2\ln 2}\right).$$

$\square$

For a probability distribution P on $\mathcal{X}$ and $\epsilon > 0$ define the set of ϵ–typical sequences:

$$\mathcal{T}_{P,\epsilon}^n = \left\{x^n : \forall x \; \left|N(x|x^n) - P(x)n\right| \le \epsilon n \; \& \; P(x) = 0 \Rightarrow N(x|x^n) = 0\right\},$$

with the number $N(x|x^n)$ denoting the number of letters x in the word x^n. The probability distribution $P_{x^n}(x) = \frac{1}{n}N(x|x^n)$ is called the *type* of x^n. Note that $x^n \in \mathcal{T}_{P,\epsilon}^n$ is equivalent to $|P_{x^n}(x) - P(x)| \le \epsilon$ for all x.

These are the properties of typical sequences we shall need:

$$P^{\otimes n}\left(\mathcal{T}_{P,\epsilon}^n\right) \ge 1 - 2|\mathcal{X}|\exp\left(-n\epsilon^2/2\right). \tag{8}$$

This is an easy consequence of the Chernoff bound, lemma 1, applied to the indicator variables X_k of the letter x in position k in X^n, with $\eta = \epsilon P(x)^{-1}$.

$$\forall x^n \in \mathcal{T}_{P,\epsilon}^n \quad \begin{cases} P^{\otimes n}(x^n) \le \exp\left(-nH(P) + n\epsilon D\right), \\ P^{\otimes n}(x^n) \ge \exp\left(-nH(P) - n\epsilon D\right), \end{cases} \tag{9}$$

with the constant $D = \sum_{x:P(x)\neq 0} -\log P(x)$. See [14].

$$\left|\mathcal{T}_{P,\epsilon}^n\right| \le \exp\left(nH(P) + n\epsilon D\right), \tag{10}$$

$$\left|\mathcal{T}_{P,\epsilon}^n\right| \ge \left(1 - 2|\mathcal{X}|\exp\left(-n\epsilon^2/2\right)\right)\exp\left(nH(P) - n\epsilon D\right). \tag{11}$$

This follows from eq. (9). These estimates also allow to lower bound the size of sets with large probability: assume $P^{\otimes n}(\mathcal{C}) \ge \eta$, then

$$|\mathcal{C}| \ge \left(\eta - 2|\mathcal{X}|\exp\left(-n\epsilon^2/2\right)\right)\exp\left(nH(P) - n\epsilon D\right). \tag{12}$$

We also use these notions in the "non–stationary" case: consider a channel $W : \mathcal{X} \longrightarrow \mathcal{Z}$, and an input string $x^n \in \mathcal{X}^n$. Then define, with $\epsilon > 0$, the set of conditional ϵ–typical sequences:

$$\mathcal{T}_{W,\epsilon}^n(x^n) = \left\{ z^n : \forall x, z \; \left| N(xz|x^n z^n) - nW(z|x)P_{x^n}(x) \right| \leq \epsilon n \right.$$

$$\left. \& \; W(z|x) = 0 \Rightarrow N(xz|x^n z^n) = 0 \right\}$$

$$= \prod_x \mathcal{T}_{W_x, \epsilon P_{x^n}(x)^{-1}}^{\mathcal{I}_x},$$

with the sets $\mathcal{I}_x$ of positions in the word x^n where $x_k = x$. The latter product representation allows to easily transport all of the above relations for typical sequences to conditional typical sequences:

$$W_{x^n}^n \left(\mathcal{T}_{W,\epsilon}^n(x^n) \right) \geq 1 - 2|\mathcal{X}||\mathcal{Z}| \exp\left(-n\epsilon^2/2 \right). \tag{13}$$

$$\forall x^n \in \mathcal{T}_{W,\epsilon}^n(x^n) \quad \begin{cases} W_{x^n}^n(x^n) \leq \exp\left(-nH(W|P_{x^n}) + n\epsilon E \right), \\ W_{x^n}^n(x^n) \geq \exp\left(-nH(W|P_{x^n}) - n\epsilon E \right), \end{cases} \tag{14}$$

with $E = \max_x \sum_{z : W_x(z) \neq 0} - \log W_x(z)$ and the conditional entropy $H(W|P) = \sum_x P(x)H(W_x)$.

$$\left| \mathcal{T}_{W,\epsilon}^n(x^n) \right| \leq \exp\left(nH(W|P_{x^n}) + n\epsilon E \right), \tag{15}$$

$$\left| \mathcal{T}_{W,\epsilon}^n(x^n) \right| \geq \left(1 - 2|\mathcal{X}||\mathcal{Z}| \exp\left(-n\epsilon^2/2 \right) \right) \exp\left(nH(W|P_{x^n}) - n\epsilon E \right). \tag{16}$$

A last elementary property: for x^n of type P and output distribution Q, with $Q(z) = \sum_x P(x)W_x(z)$,

$$\mathcal{T}_{W,\epsilon}^n(x^n) \subset \mathcal{T}_{Q, \epsilon|\mathcal{X}|}^n. \tag{17}$$

As an application, let us prove the following lemma:

Lemma 2. *For words x^n and y^n with $d_H(x^n, y^n) \geq \sigma n$, and a channel W such that*

$$\forall x \in \mathcal{X}, P \text{ p.d. with } P(x) = 0 \quad \left\| W_x - \sum_y P(y)W_y \right\|_1 \geq \eta,$$

one has, with $\epsilon = \frac{\sigma^2 \eta}{2|\mathcal{X}|^2|\mathcal{Z}|}$,

$$W_{y^n}^n \left(\mathcal{T}_{W,\epsilon}^n(x^n) \right) \leq 2 \exp(-n\epsilon^4/2)$$

Proof. There exists an x such that the word $x^{\mathcal{I}_x}$ (composed of letters x only) has distance at least $\frac{1}{|\mathcal{X}|} \sigma n$ from $y^{\mathcal{I}_x}$. In particular, $N_x := N(x|x^n) = |\mathcal{I}_x| \geq \frac{1}{|\mathcal{X}|} \sigma n$.

This implies also, by assumption on the channel,

$$\left\| \frac{1}{N_x} \sum_{k \in \mathcal{I}_x} W_{y_k} - W_x \right\|_1 \geq \frac{1}{|\mathcal{X}|} \sigma \eta.$$

Hence there must be a $z \in \mathcal{Z}$ with

$$\left| \frac{1}{N_x} \sum_{k \in \mathcal{I}_x} W_{y_k}(z) - W_x(z) \right| \geq \frac{1}{|\mathcal{X}||\mathcal{Z}|} \sigma \eta.$$

By definition, this in turn implies that for all $z^n \in \mathcal{T}_{W,\epsilon}^n(x^n)$,

$$\left| N\left(z | z^{\mathcal{I}_x}\right) - \sum_{k \in \mathcal{I}_x} W_{y_k}(z) \right| \geq \frac{1}{2|\mathcal{X}||\mathcal{Z}|} \sigma \eta N_x.$$

Introducing the sets $\mathcal{J}_{xy} = \{k \in \mathcal{I}_x : y_k = y\}$, with cardinalities $N_{xy} = |\mathcal{I}_{yx}|$, there is a y such that (still for all $z^n \in \mathcal{T}_{W,\epsilon}^n(x^n)$),

$$\left| N\left(z | z^{\mathcal{J}_{xy}}\right) - N_{xy} W_y(z) \right| \geq \frac{1}{2|\mathcal{X}|^2|\mathcal{Z}|} \sigma \eta N_x$$

$$\geq \frac{1}{2|\mathcal{X}|^2|\mathcal{Z}|} \sigma \eta N_{xy}.$$

This implies

$$N_{xy} \geq \frac{1}{4|\mathcal{X}|^2|\mathcal{Z}|} \sigma \eta N_x \geq \frac{1}{4|\mathcal{X}|^3|\mathcal{Z}|} \sigma^2 \eta n,$$

and with lemma 1 we obtain the claim.   □

References

1. R. Ahlswede, I. Csiszár, "Common Randomness in Information Theory and Cryptography – Part I: Secret Sharing", IEEE Trans. Inf. Theory, vol. 39, no. 4, pp. 1121–1132, 1993.
2. R. Ahlswede, A. Winter, "Strong converse for identification via quantum channels", IEEE Trans. Inf. Theory, vol. 48, no. 3, pp. 569–579, 2002. Addendum *ibid.*, vol. 49, no. 1, p. 346, 2003.
3. D. Beaver, "Commodity–Based Cryptography" (Extended Abstract), Proc. 29th Annual ACM Symposium on the Theory of Computing (El Paso, TX, 4–6 May 1997), pp. 446–455, ACM, 1997.
4. C. H. Bennett, G. Brassard, "Quantum Cryptography: Public Key Distribution and Coin Tossing", Proc. IEEE Int. Conf. on Computers Systems and Signal Processing, Bangalore (India), pp. 175–179, 1984.
5. C. H. Bennett, G. Brassard, C. Crépeau, U. Maurer, "Generalized Privacy Amplification", IEEE Trans. Inf. Theory, vol. 41, no. 6, pp. 1915–1923, 1995.
6. C. H. Bennett, P. W. Shor, "Quantum Information Theory", IEEE Trans. Inf. Theory, vol. 44, no. 6, pp. 2724–2742, 1998.

7. M. Blum, "Coin flipping by telephone: a protocol for solving impossible problems", Proc. IEEE Computer Conference, pp. 133–137, 1982.

8. G. Brassard, D. Chaum, C. Crepeau, "Minimum disclosure proofs of knowledge", J. Computer Syst. Sci. , vol. 37, pp. 156–189, 1988.

9. G. Brassard, C. Crepeau, M. Yung, "Constant–round perfect zero–knowledge computationally convincing protocols", Theoretical Computer Science, vol. 84, pp. 23–52, 1991.

10. N. Cai, R. W. Yeung, "Quantum Privacy and Quanum Wiretap Channels", manuscript, 2003.

11. H. Chernoff, "A measure of asymptotic eciency for tests of a hypothesis based on the sum of observations", Ann. Math. Statistics, vol. 23, pp. 493–507, 1952.

12. C. Crépeau, "Efficient Cryptographic Protocols Based on Noisy Channels", Advances in Cryptology: Proc. EUROCRYPT 1997 , pp. 306–317, Springer 1997.

13. C. Crépeau, J. Kilian, "Achieving oblivious transfer using weakened security assumptions", Proc. 29th FOCS, pp. 42–52. IEEE, 1988.

14. I. Csiszár, J. Körner, *Information Theory: Coding Theorems for Discrete Memoryless Channels*, Academic Press, NY 1981.

15. I. B. Damgård, J. Kilian, L. Salvail, "On the (Im)possibility of Basing Oblivious Transfer and Bit Commitment on Weakened Security Assumptions", Advances in Cryptology: EUROCRYPT 1999, pp. 56–73, Springer 1999.

16. S. Halevi, "Efficient commitment schemes with bounded sender and unbounded receiver", Proc. CRYPTO 1995, pp. 84–96. LNCS 963, Springer Verlag, 1995.

17. S. Halevi, S. Micali, "Practical and Provably-Secure Commitment Schemes from Collision Free Hashing", Advances in Cryptology: CRYPTO 1996, pp. 201–215, LNCS 1109, Springer Verlag, 1996.

18. A. S. Holevo, "Bounds for the quantity of information transmitted by a quantum channel", Probl. Inf. Transm., vol. 9, no. 3, pp. 177–183, 1973.

19. U. Maurer, "Protocols for Secret Key Agreement by Public Discussion Based on Common Information", Advances in Cryptology: CRYPTO 1992, pp. 461–470, Springer 1993. "Secret Key Agreement by Public Discussion", IEEE Trans. Inf. Theory, vol. 39, no. 3, pp. 733–742, 1993.

20. D. Mayers, "Unconditionally secure quantum bit commitment is impossible", Phys. Rev. Letters, vol. 78, no. 17, pp. 3414–3417, 1997.

21. M. Naor, "Bit commitment using pseudo–randomness", J. Cryptology, vol. 2, no. 2, pp. 151–158, 1991.

22. M. A. Nielsen, I. L. Chuang, *Quantum Computation and Quantum Information*, Cambridge University Press, 2000.

23. R. Ostrovsky, R. Venkatesan, M. Yung, "Secure commitments against a powerful adversary", Proc. STACS 1992, pp. 439–448, LNCS 577, Springer Verlag, 1992.

24. R. L. Rivest, "Unconditionally Secure Commitment and Oblivious Transfer Schemes Using Private Channels and a Trusted Initializer", unpublished manuscript, 1999.

25. C. E. Shannon, "A mathematical theory of communication", Bell System Tech. Journal, vol. 27, pp. 379–423 and 623–656, 1948.

26. A. Wyner, "The Wire Tap Channel", Bell System Tech. Journal, vol. 54, pp. 1355–1387, 1975.

Separating and Intersecting Properties of BCH and Kasami Codes

Hans Georg Schaathun and Tor Helleseth*

Dept. Informatics,
University of Bergen
Pb. 7800
N-5020 Bergen
Norway

Abstract. Separating codes have recently been applied in the construction of collusion secure fingerprinting schemes. They are related to other combinatorial concepts like intersecting codes, superimposed codes, hashing families, and group testing. In this paper we study some good, binary asymptotic constructions of such codes.

Keywords: separating systems, fingerprinting, BCH codes, Kasami codes

1 Introduction

Copyright violations are of increasing concern to artists and distributors, as production of copies get simpler and cheaper for common people. Digital fingerprinting and traitor tracing is a technique to trace guilty users or pirates when illegal copies are found.

A digital fingerprinting scheme [4] marks every copy sold with an individual mark, such that if one pirate reproduces his copy, the illegal copies may be traced back to him. In traitor tracing schemes [6,7], a similar technique is applied to the decryption keys of a broadcast encryption system. The fingerprint must be hidden such that a user cannot change it by inspecting only his own copy.

If several pirates collude, they can compare their copies, and identify portions where they differ, which must then be part of the fingerprint. Thus having identified parts of the fingerprint, they can also change it, producing a hybrid copy which cannot trivially be traced. It is generally assumed that in each mark or symbol of the fingerprint, the pirate coalition can choose the symbol from either of their copies, but nothing else. Collusion secure fingerprinting schemes are designed to trace at least one pirate when a coalition is guilty.

We view the fingerprints as codewords over some alphabet Q. The fingerprints the pirates are able to forge form the so-called feasible set, defined as

---

* Both authors were supported by the Norwegian Research Council under grant number 146874/420 and by the Aurora programme.

$$F(T) := \{(v_1, \dots, v_n) \in Q^n \mid \forall i, 1 \leq i \leq n, \exists (a_1, \dots, a_n) \in T, a_i = v_i\},$$

where T is the set of fingerprints held by the pirates, Q is the alphabet, and n is the length of a fingerprint. If the code of valid fingerprints still makes it possible to trace at least one guilty pirate out of a coalition of size t or less, we say that the code has the t-identifiable parent property (t-IPP). If the pirates are able to forge the fingerprint of an innocent user, we say that this user is framed. Codes which prevent framing by t pirates are called t-frameproof or $(t, 1)$-separating codes. A code is (t, t)-separating, or t-secure frameproof, if no two disjoint coalitions of size t or less can produce the same fingerprint.

Unfortunately (combinatorial) t-IPP codes are possible only for huge alphabets. Therefore it is interesting to study probabilistic t-IPP, where we permit a small non-zero probability ϵ of incorrect tracing. Recently, (t, t)-separating codes were used to construct probabilistic t-IPP codes [3,2]. In [22] it was proved that the best known asymptotic $(2, 2)$-separating codes is also probabilistic 2-IPP.

The case of $(2, 2)$-separation was introduced by Sagalovich in the context of automata: two such systems transiting simultaneously from state a to a' and from b to b' respectively should be forbidden to pass through a common intermediate state. A state of the system in this case is an n-bit binary string, and the moving from one state to another is obtained by flipping bits one by one. Only shortest paths from the old to the new state are allowed, so moving from a to a' will only involve flipping bits where a and a' differ. The set of valid states Γ forms a $(2, 2)$-separating system, if for any four distinct states, a, a', b, and b' from Γ, the transitions $a \rightarrow a'$ and $b \rightarrow b'$ cannot pass through any common state. Sagalovich's contribution on this topic is substantial and has been surveyed in [21].

The design of self-checking asynchronous networks has been a challenging problem. Friedmann et al. [15] have shown that the unicode single-transition-time asynchronous state assignment corresponds to $(2, 2)$- and $(2, 1)$-separating systems. The coding problem for automata states also motivated research on $(3, 3)$-SS [27]. In [22] it was proved that the best known asymptotic $(2, 2)$-separating codes is also 2-IPP.

Separating codes have also been studied in a set-theoretic framework, e.g. [18], and Körner [19] gives a series of problems equivalent to $(2, 1)$-separation.

In this paper we present new binary, asymptotic constructions of (t, u)-separating codes. We compute the rates for (t, u)-SS when $2 \leq t, u \leq 5$ find that our constructions improve on previous ones. The constructions work for arbitrary t and u, but for $(t, 1)$-SS previous constructions based on designs [9] are still the best.

2   Preliminary Definitions and Bounds

Let Q be an additive group (often a field) called the alphabet, and denote by q its order. Let $\mathbb{V}$ be the set of n-tuples over Q. An $(n, M)_q$ code Γ is an M-subset $\Gamma \subseteq \mathbb{V}$. If Q is a field of q elements and C is a k-dimensional subspace $C \leq \mathbb{V}$, then we say that C is a $[n, k]_q$ (linear) code. We will refer to the elements of $\mathbb{V}$ as words. Let d_1 and m_1 denote respectively the minimum and maximum (Hamming) distance of the code.

Definition 1. *A pair (T, U) of disjoint sets of words is called a (t, u)-configuration if $\#T = t$ and $\#U = u$. The separating weight $\theta(T, U)$ is the number of positions i, where every word of T is different from any word of U on position i.*

The (t, u)-separating weight $\theta_{t,u}(C)$ of a code C, is the least separating weight of any (t, u)-configuration of the code. If $\theta_{t,u}(C) > 0$, then we say that C is a (t, u)-separating code or a (t, u)-SS (separating system).

In earlier works on watermarking and fingerprinting, (t, t)-separating codes have been called t-SFP (secure frameproof) [24,25,23]. The current terminology appears to be older though [21]. It is well known that codes with sufficiently large minimum distance are separating [21].

Lemma 1. *If Γ is a code with minimum distance d_1 and maximum distance m_1, then $2\theta_{2,1} \geq 2d_1 - m_1$.*

Proof. Let $(\mathbf{c}; \mathbf{a}, \mathbf{b})$ be a $(2, 1)$-configuration. Letting the three words be rows of a matrix, we have essentially four types of columns: Type 0 where all the elements are equal, Type I where $\mathbf{a}$ or $\mathbf{b}$ differs from the two others, Type A where $\mathbf{c}$ differs from the two others, and Type B with three different elements. Let v_i denote the number of elements of Type i.

Consider the sum

$$\Sigma := w(\mathbf{c} - \mathbf{a}) + w(\mathbf{c} - \mathbf{b}) \geq 2d_1.$$

Observe that $\Sigma = 2(v_A + v_B) + v_I$. Clearly we have $\theta(\mathbf{c}; \mathbf{a}, \mathbf{b}) = v_A + v_B$, and $w(\mathbf{a} - \mathbf{b}) = v_B + v_I$, so $v_I \leq m_1$, and the theorem follows.

Remark 1. In the binary case, there are no columns of Type B, and therefore

$$2\theta(\mathbf{c}; \mathbf{a}, \mathbf{b}) = w(\mathbf{c} - \mathbf{a}) + w(\mathbf{c} - \mathbf{b}) - w(\mathbf{a} - \mathbf{b}),$$

and consequently we get equality $\theta_{2,1} = d_1 - m_1/2$ if and only if there are three codewords $\mathbf{a}, \mathbf{b}, \mathbf{c}$ such that $w(\mathbf{c} - \mathbf{a}) = w(\mathbf{c} - \mathbf{b}) = d_1$ and $w(\mathbf{b} - \mathbf{a}) = m_1$.

A similar argument also gives the following result [21].

Theorem 1. *Let Γ be a code with minimum distance d_1 and maximum distance m_1. Then $4\theta_{2,2} \geq 4d_1 - 2m_1 - n$. If Γ is linear, then $4\theta_{2,2} \geq 4d_1 - 3m_1$.*

Proposition 1. *Any $(n, M, d_1)_q$ code Γ has $\theta_{t,u} \geq n - tu(n - d_1)$.*

Corollary 1. *An $(n, M, d_1)_q$ code Γ is (t, u)-separating if $d_1/n > 1 - 1/(tu)$.*

Proof. Consider any (t, u)-configuration (T, U) from Γ, and define the sum

$$\Sigma := \sum_{(x,y) \in T \times U} d(x, y).$$

This is the sum of (T, U) distances in the code, so $\Sigma \geq tud_1$. Each coordinate can contribute at most tu to the sum Σ, but if any coordinate does contribute that much, then the configuration is separated on this coordinate. Hence we get that $\Sigma \leq n(tu - 1) + \theta_{t,u}$. The proposition follows by combining the upper and lower bounds and simplifying.

It must be noted that, to get infinite families of separating codes with good rate, the alphabet size q grows extremely rapidly in the t and u, due to the Plotkin bound. On the other hand, for sufficiently large alphabets, we can use the following lemma by Tsfasman [26].

Theorem 2 (The Tsfasman Codes). *For any $\alpha > 0$ there are constructible, infinite families of codes $\mathfrak{A}(N)$ with parameters $[N, NR, N\delta]_q$ for $N \geq N_0(\alpha)$ and*

$$R + \delta \geq 1 - (\sqrt{q} - 1)^{-1} - \alpha.$$

Infinite families of separating codes over small alphabets can be built by concatenation [1]. The outer codes for concatenation will very often be Tsfasman codes.

Definition 2 (Concatenation). *Let C_1 be a $(n_1, Q)_q$ and let C_2 be an $(n_2, M)_Q$ code. Then the concatenated code $C_1 \circ C_2$ is the $(n_1 n_2, M)_q$ code obtained by taking the words of C_2 and mapping every symbol on a word from C_1.*

Proposition 2. *Let Γ_1 be a $(n_1, M)_{M'}$ code with minimum (t, u)-separating weight $\theta_{t,u}^{(1)}$, and let Γ_2 be a $(n_2, M')_q$ code with separating weight $\theta_{t,u}^{(1)}$. Then the concatenated code $\Gamma := \Gamma_2 \circ \Gamma_1$ has minimum separating weight $\theta_{t,u} = \theta_{t,u}^{(1)} \cdot \theta_{t,u}^{(2)}$.*

Note that Γ will usually not satisfy the requirements of Proposition 1.

3   Intersection Gives Separation

The first relationship between intersecting codes and separating codes appeared in [5], and further links have been explored in [11,10] (see also [13]).

Definition 3. *A linear code C of dimension $k \geq t$ is said to be t-wise intersecting if any t linearly independent codewords have intersecting supports. If $t > k$, we say that C is t-wise intersecting if and only if it is k-wise intersecting.*

It is easy to verify that any t-wise intersecting code is also $(t-1)$-wise intersecting. The following relation between intersection and separation is well known [5,11].

Proposition 3. *For a linear, binary code, is*

1. *2-wise intersecting if and only if it is $(2,1)$-separating, and*
2. *3-wise intersecting if and only if it is $(2,2)$-separating.*

Due to this proposition, we can use many bounds on separating codes as bounds on intersecting codes. For instance, by Theorem 1, every code with $4d > 3m$ is 3-wise intersecting.

It was shown in [13], that if C is a (t,u)-SS, then any $\bar\iota(t,u)$ codewords must be linearly independent, where

$$\bar\iota(t,u) := \begin{cases} t+u, & \text{when } t \equiv u \equiv 1 \pmod 2, \\ t+u-1, & \text{when } t \not\equiv u \pmod 2, \\ t+u-2, & \text{when } t \equiv u \equiv 0 \pmod 2. \end{cases}$$

If the codewords are taken as a non-linear subcode of a $(t+u-1)$-wise intersecting code, this condition is also sufficient. The following theorem is from [10], but we include a proof for completeness.

Theorem 3. *Let $i, j \geq 1$ be integers such that $t := i+j-1 \geq 2$. Consider a t-wise intersecting, binary, linear code C, and a non-linear subcode $\Gamma \subseteq C$. The code Γ is (i,j)-separating if any $\bar\iota(i,j)$ non-zero codewords are linearly independent.*

Proof. We start by proving that any $t+1$ codewords being linearly independent is sufficient for Γ to be (i,j)-separating. This holds as the theorem states irrespectively of the parities of i and j. Afterward we will strengthen the result in the cases where i and j are not both odd.

Choose any (two-part) sequence Y' of $t+1$ codewords from Γ,

$$Y' := (\mathbf{a}'_1, \dots, \mathbf{a}'_j; \mathbf{c}'_1, \dots, \mathbf{c}'_{t+1-j}).$$

We have that Y' is $(j, t+1-j)$-separated if and only if $Y := Y' - \mathbf{c}'_{t+1-j}$ is. Hence it suffices to show that

$$Y = (\mathbf{a}_1, \dots, \mathbf{a}_j; \mathbf{c}_1, \dots, \mathbf{c}_{t-j}, \mathbf{0})$$

is $(j, t+1-j)$-separated.

Since the $t+1$ codewords of Y' are linearly independent, so are the t first codewords of Y. Now, consider

$$X := \{\mathbf{a}_1 + \mathbf{c}_1, \dots, \mathbf{a}_1 + \mathbf{c}_{t-j}; \mathbf{a}_1, \dots, \mathbf{a}_j\},$$

which is a set of linearly independent codewords from C, and hence all non-zero on some coordinate i. Since $\mathbf{a}_1 + \mathbf{c}_l$ is non-zero on coordinate i, $\mathbf{c}_l$ must be zero for all l. Hence Y, and consequently Y', is separated on coordinate i.

This completes the first step. In the case where $i \not\equiv j \pmod 2$, we get that t is even, and consequently the t first codewords of Y are linearly independent whenever any t words of Y' are. Therefore it is sufficient that any t codewords of Γ be linearly independent.

Finally, we consider the case where i and j are both even. We shall again show that Y' is separated. If all the $t + 1$ words of Y' are linearly independent, then we are done by the first part of the proof. By assumption, we know that any $t - 1$ words are linearly independent. This gives two cases to consider:

1. $\mathbf{c}'_{t+1-j}$ is the sum of the t first words, which are linearly independent.
2. $\mathbf{c}'_{t-j}$ is the sum of the $t - 1$ first words and $\mathbf{c}'_{t+1-j}$ is independent of the others.

Let Y', Y, and X be defined as before. Consider the first case first. Any $t - 1$ non-zero words of Y are linearly independent, while all the t non-zero words sum to $\mathbf{0}$. Hence, the only linear independence found between the elements of X is that

$$\mathbf{0} = \mathbf{b}_1 + \ldots + \mathbf{b}_{t-j} + \mathbf{a}_2 + \ldots + \mathbf{a}_j, \tag{1}$$

where $\mathbf{b}_i = \mathbf{c}_i + \mathbf{a}_1$. It follows that the $t - 1$ first words of X intersect, since C is t-wise intersecting. Thus there is a position l, where $\mathbf{a}_i$ is 1 for $i = 1, \ldots, j - 1$ and $\mathbf{c}_{i'}$ is zero for $i' = 1, \ldots, t - j$. Furthermore, $\mathbf{a}_j$ is one in position l by (1). Hence Y is separated.

In the second case, we get that the t non-zero words of Y are linearly independent. Thus the result follows like the first part of the proof.

It is perhaps not obvious how these propositions may be used to construct non-linear separating codes with a reasonable rate. The following lemma [11] does the trick.

Lemma 2. *Given an $[n, rm]$ linear, binary code C, we can extract a non-linear subcode Γ of size 2^r such that any $2m$ non-zero codewords are linearly independent.*

Proof. Let C' be the $[2^r - 1, 2^r - 1 - rm, 2m + 1]$ BCH code. The columns of the parity check matrix of C' make a set Γ' of $2^r - 1$ vectors from $\mathsf{GF}(2)^{rm}$, such that any $2m$ of them are linearly independent. Now there is an isomorphism $\phi : \mathsf{GF}(2)^{rm} \to C$, so let $\Gamma = \phi(\Gamma') \cup \{\mathbf{0}\}$.

There is a sufficient condition for intersecting codes, resembling the results we have for separating codes in Proposition 1 and Theorems 1 and 1 [8].

Proposition 4. *Let C be a binary linear code. Any t independent codewords intersect in at least $d_1 - m_1(1 - 2^{1-t})$ coordinate positions.*

Remark 2. The code C has t-wise intersection weight exactly $d_1 - m_1(1 - 2^{1-t})$ if and only if there are subcodes $D_0 \subseteq D_1 \subseteq C$ such that D_0 has dimension $t - 1$ and contains $2^{t-1} - 1$ words of maximum weight, and D_1 has dimension t containing 2^{t-1} words of minimum weight.

4   Kasami Codes

Let T_m denote the Froebenius trace from $\mathsf{GF}(q^m)$ to $\mathsf{GF}(q)$, defined as

$$T_m(x) = \sum_{i=0}^{m-1} x^{q^i}.$$

It is well-known that

$$T_m(x + y) = T_m(x) + T_m(y),$$
$$T_m(x) = T_m(x^q),$$

and if x runs through $\mathsf{GF}(q^m)$, then $T_m(x)$ takes each value in $\mathsf{GF}(q)$ exactly q^{m-1} times. The original Kasami code is a binary code, so let $q = 2$ and write $Q = 2^m$.

Definition 4 (The Kasami Codes). *The* $[2^{2m} - 1, 3m, 2^{2m-1} - 2^{m-1}]$ *Kasami code is the set*

$$\mathcal{K}_m = \{\mathbf{c}(a,b) : a \in \mathsf{GF}(Q^2), b \in \mathsf{GF}(Q)\},$$

where

$$\mathbf{c}(a,b) = \big(T_{2m}(ax) + T_m(bx^{Q+1}) : x \in \mathsf{GF}(Q^2)^*\big).$$

The Kasami codes have three different non-zero weights, given by the following lemma [16].

Lemma 3. *The weight of a codeword* $\mathbf{c}(a,b) \in \mathcal{K}_m$ *is given by*

$$w(\mathbf{c}(a,b)) = \begin{cases} 2^{2m-1} - 2^{m-1}, & \text{if } b \neq 0 \text{ and } T_m(a^{Q+1}/b) = 1, \\ 2^{2m-1} + 2^{m-1}, & \text{if } b \neq 0 \text{ and } T_m(a^{Q+1}/b) = 0, \\ 2^{2m-1}, & \text{if } b = 0 \text{ and } a \neq 0, \\ 0, & \text{if } b = 0 \text{ and } a = 0. \end{cases} \quad (2)$$

Using Proposition 4, we get the following result.

Proposition 5. *The Kasami code* $\mathcal{K}_m$ *is m-wise intersecting, and its t-wise intersection weight is at least*

$$\ell_t(\mathcal{K}_m) \geq 2^m(2^{m-t} - 1) + 2^{m-t}.$$

This implies that the $\mathcal{K}_m$ is a $(2,1)$-SS for $m \geq 2$ and a $(2,2)$-SS for $m \geq 3$. For $t = 2$, the above bound is tight as the following proposition shows. It can be shown by exhaustive search that the bound is tight for $t = m = 3$ as well, but it is an interesting open problem whether the bound is tight in general.

Proposition 6. *The Kasami code $\mathcal{K}_m$ has $(2,1)$-separating weight*

$$\theta_{2,1} = \max\{0, (2^m - 3)2^{m-2}\}.$$

Proof. Recall that $\theta_{2,1} \geq d_1 - m_1/2 = (2^m - 3)2^{m-2}$. This is negative if and only if $m = 1$. Observe that $\mathcal{K}_1$ contains all words of length 3, and thus has $\theta_{2,1} = 0$, as required.

If $m \geq 2$, we get that $\theta_{2,1} > 0$, and by Remark 1 it remains to prove that there are are two codewords **a** and **b** of minimum weight such that $\mathbf{a}+\mathbf{b}$ has maximum weight. This is fulfilled for $\mathbf{a} = \mathbf{c}(\gamma b, b^2)$ and $\mathbf{b} = \mathbf{c}(\gamma b, f^2 b^2)$ if $T_m(\gamma^{Q+1}) = T_m(\gamma^{Q+1}/f^2) = 1$. Such an f exists as long as $m \geq 2$.

5   BCH Codes

Several good (t,u)-separating codes may be constructed from intersecting codes and columns from the parity check matrices of BCH codes. In the tables at the end of this section, we use non-linear subcodes of dual BCH codes as inner codes.

5.1   Finite Constructions of Intersecting Codes

The intersecting properties of the duals of 2-error-correcting BCH codes were first pointed out in [8]. In the sequel, we describe the intersecting properties of arbitrary dual BCH codes.

In MacWilliams and Sloane [20], we find the following lemma.

Lemma 4. *Let C be a BCH code of length $2^m - 1$ and designed distance $d' = 2e + 1$, where $2e - 1 < 2^{\lceil m/2 \rceil} + 1$. For any non-zero words in $C^{\perp}$, the weight w lies in the range*

$$2^{m-1} - (e-1)2^{m/2} \leq w \leq 2^{m-1} + (e-1)2^{m/2}.$$

By using Proposition 4, we get the following result.

Proposition 7. *The dual of a $[2^m - 1, me]$ BCH code with designed distance $d' = 2e + 1$ has t-wise intersection weight*

$$\ell_t \geq 2^{m-t} + (e-1)2^{m/2+1-t} - (e-1)2^{m/2+1}$$
$$= 2^{m/2+1}\left(2^{m/2-t-1} - (e-1)(1 - 2^{-t})\right).$$

Corollary 2. *The dual of the e-error-correcting BCH code with parameters $[2^m - 1, me]$, is t-wise intersecting if*

$$m > 2\left(1 + \log(e-1) + \log(2^t - 1)\right).$$

The bounds in Lemma 4 are not necessarily tight, and for $e = 2, 3$, the exact maximum and minimum weights are known [17].

Lemma 5. *Let C be a 2-error-correcting BCH code of length $2^m - 1$. Then*

$$d_1 = 2^{m-1} - 2^{\lfloor m/2 \rfloor},$$
$$m_1 = 2^{m-1} + 2^{\lfloor m/2 \rfloor}.$$

Proposition 8. *The dual of the 2-error-correcting BCH code with parameters $[2^{2t+1} - 1, 4t + 2, 2^{2t} - 2^t]$, is t-wise intersecting, with intersecting weight $\ell_t \geq 2$.*

This proposition is a direct consequence of the preceding lemma [8].

Lemma 6. *Let C be a 3-error-correcting BCH code of length $2^m - 1$ for $m \geq 4$. Then*

$$d_1 = 2^{m-1} - 2^{\lceil m/2 \rceil},$$
$$m_1 = 2^{m-1} + 2^{\lceil m/2 \rceil}.$$

Proposition 9. *The punctured dual of the 3-error-correcting BCH code with parameters $[2^{2t+2} - 1, 6t + 6]$, is t-wise intersecting, with intersecting weight $\ell_t \geq 4$.*

5.2   Infinite Families of Intersecting Codes

The following lemma was found in [8].

Lemma 7. *Let C_1 be an $[n_1, k_1, d_1]_q$ code with $q = 2^{k_2}$ and minimum distance $d_1 > n_1(1 - 2^{1-t})$. Let C_2 be an $[n_2, k_2, d_2]$ binary t-wise intersecting code. Then the concatenation $C_1 \circ C_2$ is a binary t-wise intersecting $[n_1 n_2, k_1 k_2, d_1 d_2]$ code.*

Lemma 8. *There are constructive infinite sequences of t-wise intersecting binary codes with rates arbitrarily close to*

$$R_t^{(2)} = \left(2^{1-t} - \frac{1}{2^{2t+1} - 1} \right) \frac{2t + 1}{2^{2t} - 1} = 2^{2-3t}(t + o(t)),$$
$$R_t^{(3)} = \left(2^{1-t} - \frac{1}{2^{3t+3} - 1} \right) \frac{3t + 3}{2^{2t+1} - 2} = 2^{-3t}(3t + o(t)).$$

Proof. By concatenating geometric $[N, K, D]_q$ codes from Theorem 2 satisfying $D > N(1 - 2^{1-t})$ with $q = 2^{4t+2}$, and with a rate arbitrarily close to $2^{1-t} - 1/(\sqrt{q} - 1)$, with the $[2^{2t+1} - 2, 4t + 2, 2^{2t} - 2^t - 1]$ code of Proposition 8, we obtain the result.

5.3   Constructions of Separating Codes

There are two basic techniques for constructing asymptotic separating codes from intersecting codes.

Technique I uses a finite intersecting $[n, k]$ code C' as a seed. Then a non-linear subcode is extracted from C' to form an separating inner code C_I. Finally, C_I is concatenated with a separating Tsfasman code C_O. The rate is given by

$$R^{\mathrm{I}} = \frac{\log Q}{n} \left(\frac{1}{uv} - \frac{1}{\sqrt{Q} - 1} \right), \tag{3}$$

where $Q = q^2 \leq 2^{2k/\bar{\iota}(u,v)}$ is as large as possible with q a prime power.

Technique II uses a finite intersecting code C_I as a seed, which is concatenated with a Tsfasman code with minimum distance at least $(1 - 2^{1-t})$ in concordance with Lemma 7, to form an asymptotic intersecting code C'. The asymptotic separating code is a non-linear subcode of C'. The resulting rate is

$$R^{\mathrm{II}} = \frac{k}{n} \left(2^{2-u-v} - \frac{1}{2^{k/2} - 1} \right) \frac{2}{\bar{\iota}(u, v)}, \tag{4}$$

provided k is even. Otherwise 2^k is replaced by $Q = q^2 \leq 2^k$ where q is the largest possible prime power.

Comparing (3) and (4), we see that the difference is in the parenthesised expression. Except when $t \leq 3$, we have $2^{2-u-v} > 1/uv$ which tend to give Technique I a better rate. However Technique II uses a larger alphabet for the outer code, which tends to decrease the penalty factor and hence improve the rate of the outer code.

The following proposition gives the rates obtained when Technique II is applied on the duals of BCH(2) and BCH(3). It is easy to check that $R_2^{II} > R_3^{II}$ (except for the degenerate case $u = v = 1$).

Proposition 10. *There are constructive infinite sequences of binary (u, v)-separating codes of rate*

$$R_2^{II}(u, v) = \frac{4(u+v) - 2}{\bar{\iota}(u, v) \left(2^{2(u+v-1)} - 1 \right)} \left(2^{2-u-v} - \frac{1}{2^{2(u+v)-1} - 1} \right)$$

$$\geq 2^{-3(u+v-2)} (1 + o(1)),$$

$$R_3^{II}(u, v) = \frac{3(u+v)}{\bar{\iota}(u, v) \left(2^{2(u+v-1)} - 1 \right)} \left(2^{2-u-v} - \frac{1}{2^{3(u+v)} - 1} \right).$$

For Technique I, we do not obtain such nice closed form formulæ, because we do not have a nice expression for the alphabet size Q of the outer code.

In Table 1 and 2, we present some good separating codes from duals of BCH(2) and BCH(3) with Technique I. The constructions with BCH(2) are known from [12,13], while the BCH(3)-constructions are new. The symbol K denotes the log-cardinality of the inner code. For big u and v, the inner code resulting from BCH(2) are so small that we do not get a positive rate for the

Table 1. Good (u, v)-SS from BCH(2).

(u,v)	m	$[n,k]$	K	inner rate	outer rate	total rate
$(2,2)$	7	$[126, 14]$	14	$1.111 \cdot 10^{-1}$	$2.421 \cdot 10^{-1}$	$2.690 \cdot 10^{-2}$
$(2,3)$	9	$[510, 18]$	9	$1.666 \cdot 10^{-2}$	$1.111 \cdot 10^{-1}$	$1.851 \cdot 10^{-3}$
$(2,4)$	11	$[2046, 22]$	11	$5.304 \cdot 10^{-3}$	$1.012 \cdot 10^{-1}$	$5.367 \cdot 10^{-4}$
$(2,5)$	13	$[8190, 26]$	8	$9.768 \cdot 10^{-4}$	$3.333 \cdot 10^{-2}$	$3.256 \cdot 10^{-5}$
$(3,3)$	11	$[2046, 22]$	7	$3.382 \cdot 10^{-3}$	$1.111 \cdot 10^{-2}$	$3.757 \cdot 10^{-5}$
$(3,4)$	13	$[8190, 26]$	8	$9.768 \cdot 10^{-4}$	$1.667 \cdot 10^{-2}$	$1.628 \cdot 10^{-5}$
$(4,4)$	15	$[32766, 30]$	10	$3.052 \cdot 10^{-4}$	$3.024 \cdot 10^{-2}$	$9.230 \cdot 10^{-6}$

Table 2. Good (u, v)-SS from BCH(3).

(u,v)	m	$[n,k]$	K	inner rate	outer rate	total rate
$(2,2)$	8	$[252, 24]$	24	$9.524 \cdot 10^{-2}$	$2.498 \cdot 10^{-1}$	$2.379 \cdot 10^{-2}$
$(2,3)$	10	$[1020, 30]$	15	$1.471 \cdot 10^{-2}$	$1.611 \cdot 10^{-1}$	$2.369 \cdot 10^{-3}$
$(2,4)$	12	$[4092, 36]$	18	$4.399 \cdot 10^{-3}$	$1.230 \cdot 10^{-1}$	$5.412 \cdot 10^{-4}$
$(2,5)$	14	$[16380, 42]$	14	$8.547 \cdot 10^{-4}$	$9.213 \cdot 10^{-2}$	$7.874 \cdot 10^{-5}$
$(3,3)$	12	$[4092, 36]$	12	$2.933 \cdot 10^{-3}$	$9.524 \cdot 10^{-2}$	$2.793 \cdot 10^{-4}$
$(3,4)$	14	$[16380, 42]$	14	$8.547 \cdot 10^{-4}$	$7.546 \cdot 10^{-2}$	$6.450 \cdot 10^{-5}$
$(3,5)$	16	$[65532, 48]$	12	$1.831 \cdot 10^{-4}$	$5.079 \cdot 10^{-2}$	$9.301 \cdot 10^{-6}$
$(4,4)$	16	$[65532, 48]$	16	$2.442 \cdot 10^{-4}$	$5.858 \cdot 10^{-2}$	$1.430 \cdot 10^{-5}$
$(4,5)$	18	$[262140, 54]$	13	$4.941 \cdot 10^{-5}$	$3.864 \cdot 10^{-2}$	$1.909 \cdot 10^{-6}$
$(5,5)$	20	$[1048572, 60]$	12	$1.144 \cdot 10^{-5}$	$2.413 \cdot 10^{-2}$	$2.761 \cdot 10^{-7}$

Table 3. Some good (u, v)-separating codes from duals of BCH codes.

	Technique II		Technique I		
(u,v)	BCH(2)	BCH(3)	BCH(2)	BCH(3)	BCH(5)
$(2,2)$	$2.690 \cdot 10^{-2}$	$2.379 \cdot 10^{-2}$	$2.690 \cdot 10^{-2}$	$2.379 \cdot 10^{-2}$	
$(2,3)$	$2.171 \cdot 10^{-3}$	$1.838 \cdot 10^{-3}$	$1.851 \cdot 10^{-3}$	$2.369 \cdot 10^{-3}$	$2.026 \cdot 10^{-4}$
$(2,4)$	$3.334 \cdot 10^{-4}$	$2.749 \cdot 10^{-4}$	$5.367 \cdot 10^{-4}$	$5.412 \cdot 10^{-4}$	$4.045 \cdot 10^{-5}$
$(2,5)$	$3.294 \cdot 10^{-5}$	$2.671 \cdot 10^{-5}$	$3.256 \cdot 10^{-5}$	$7.874 \cdot 10^{-5}$	$6.324 \cdot 10^{-6}$
$(3,3)$	$2.223 \cdot 10^{-4}$	$1.833 \cdot 10^{-4}$	$3.757 \cdot 10^{-5}$	$2.793 \cdot 10^{-4}$	$2.396 \cdot 10^{-5}$
$(3,4)$	$3.294 \cdot 10^{-5}$	$2.671 \cdot 10^{-5}$	$1.628 \cdot 10^{-5}$	$6.450 \cdot 10^{-5}$	$5.270 \cdot 10^{-6}$
$(3,5)$	$3.570 \cdot 10^{-6}$	$2.861 \cdot 10^{-6}$	0	$9.301 \cdot 10^{-6}$	$8.269 \cdot 10^{-7}$
$(4,4)$	$4.759 \cdot 10^{-6}$	$3.815 \cdot 10^{-6}$	$9.230 \cdot 10^{-6}$	$1.430 \cdot 10^{-5}$	$1.105 \cdot 10^{-6}$
$(4,5)$	$5.062 \cdot 10^{-7}$	$4.023 \cdot 10^{-7}$	0	$1.909 \cdot 10^{-6}$	$1.669 \cdot 10^{-7}$
$(5,5)$	$5.660 \cdot 10^{-8}$	$4.470 \cdot 10^{-8}$	0	$2.761 \cdot 10^{-7}$	$2.969 \cdot 10^{-8}$

outer code. This could have been amended by increasing m, but better results are obtained by increasing e. Therefore these codes are omitted from Table 1.

In Table 3, we can compare the constructions for Techniques I and II. Technique I using BCH(3) gives the best rate in all cases studied except for $(2,2)$-SS. However, it is interesting to note that Technique II gives a non-zero rates for some seed codes which give zero rate with Technique I.

We have not presented any results using BCH(4). It is easy to check that when $t \geq 5$, the minimum required value of m, according to Corollary 2, is the same for $e = 4$ and $e = 5$. Consequently, there is no reason for using duals of BCH(4) when $t \geq 5$; using BCH(5) instead can only improve the rate. It can also be checked that BCH(4) is inferior to BCH(5) in the other cases.

The minimum value of m is $2t+1$ for BCH(2). It increases only by 1 to $2t+2$ for BCH(3). Moving to BCH(4), m must make a jump by 3 or more, depending on the value of t. This is of course because the bounds on d_1 and m_1 are much worse for BCH(e) when $e > 3$. It explains why the rates for the inner codes as well as for the outer codes in Tables 1 and 2 are so close together. The big increase needed in m from BCH(3) to BCH(4) is only worthwhile when the rate of the outer code is only small fraction of the ideal rate $1/uv$. For $t, u \leq 5$, BCH(3) performs very well, and BCH(5) cannot improve the overall rate. However, for $(7, 9)$-SS we would not get a positive rate using BCH(3), but BCH(5) does the trick.

6   Conclusion

We have shown that Kasami codes and BCH codes have certain separating properties, and that they can be used to construct record breaking families of separating codes. We only have lower bounds on the t-wise intersection weights for $t > 2$. It would be interesting to find the exact intersection weights, and if the bounds are not tight, the constructed rates may be slightly improved.

The fingerprinting schemes of [2] uses (t, t)-SS as components. The present constructions with improved rates for (t, t)-SS, will thus make it possible to build fingerprinting schemes with better rates as well.

Acknowledgement. The authors wish to thank prof. Gérard Cohen for many a useful conversation on fingerprinting and separation.

References

1. N. Alon. Explicit construction of exponential sized families of k-independent sets. *Discrete Math.*, 58(2):191–193, 1986.
2. A. Barg, G. R. Blakley, and G. A. Kabatiansky. Digital fingerprinting codes: Problem statements, constructions, identification of traitors. *IEEE Trans. Inform. Theory*, 49(4):852–865, 2003.
3. A. Barg, G. Cohen, S. Encheva, G. Kabatiansky, and G. Zémor. A hypergraph approach to the identifying parent property. *SIAM J. Disc. Math.*, 14(3):423–431, 2001.
4. Dan Boneh and James Shaw. Collusion-secure fingerprinting for digital data. *IEEE Trans. Inform. Theory*, 44(5):1897–1905, 1998. Presented in part 1995, see Springer LNCS.
5. Bella Bose and T. R. N. Rao. Separating and completely separating systems and linear codes. *IEEE Trans. Comput.*, 29(7):665–668, 1980.

6. B. Chor, A. Fiat, and M. Naor. Tracing traitors. In *Advances in Cryptology - CRYPTO '94*, volume 839 of *Springer Lecture Notes in Computer Science*, pages 257–270. Springer-Verlag, 1994.
7. B. Chor, A. Fiat, M. Naor, and B. Pinkas. Tracing traitors. *IEEE Trans. Inform. Theory*, 46(3):893–910, May 2000.
8. G rard Cohen and Gilles Zémor. Intersecting codes and independent families. *IEEE Trans. Inform. Theory*, 40:1872–1881, 1994.
9. G rard D. Cohen and Sylvia B. Encheva. Efficient constructions of frameproof codes. *Electronics Letters*, 36(22), 2000.
10. G rard D. Cohen, Sylvia B. Encheva, Simon Litsyn, and Hans Georg Schaathun. Erratum to 'intersecting codes and separating codes'. *Discrete Applied Mathematics*, 2003. Submitted.
11. G rard D. Cohen, Sylvia B. Encheva, Simon Litsyn, and Hans Georg Schaathun. Intersecting codes and separating codes. *Discrete Applied Mathematics*, 128(1):75–83, 2003.
12. G rard D. Cohen, Sylvia B. Encheva, and Hans Georg Schaathun. On separating codes. Technical report, Ecole Nationale Sup rieure des T l communications, 2001.
13. G rard D. Cohen and Hans Georg Schaathun. Asymptotic overview on separating codes. Available at http://www.ii.uib.no/publikasjoner/texrap/index.shtml, May 2003.
14. P. Delsarte and J.-M. Goethals. Tri-weight codes and generalized Hadamard matrices. *Information and Control*, 15:196–206, 1969.
15. A. D. Friedman, R. L. Graham, and J. D. Ullman. Universal single transition time asynchronous state assignments. *IEEE Trans. Comput.*, 18:541–547, 1969.
16. Tor Helleseth and P. Vijay Kumar. The weight hierarchy of the Kasami codes. *Discrete Math.*, 145(1-3):133–143, 1995.
17. Tadao Kasami. Weight distributions of Bose-Chaudhuri-Hocquenghem codes. In *Combinatorial Mathematics and its Applications (Proc. Conf., Univ. North Carolina, Chapel Hill, N.C., 1967)*, pages 335–357. Univ. North Carolina Press, Chapel Hill, N.C., 1969.
18. J. Körner and G. Simonyi. Separating partition systems and locally different sequences. *SIAM J. Discrete Math.*, 1:355–359, 1988.
19. János Körner. On the extremal combinatorics of the Hamming space. *J. Combin. Theory Ser. A*, 71(1):112–126, 1995.
20. F. J. MacWilliams and N. J. A. Sloane. *The Theory of Error-Correcting Codes*. North-Holland, Amsterdam, 1977.
21. Yu. L. Sagalovich. Separating systems. *Problems of Information Transmission*, 30(2):105–123, 1994.
22. Hans Georg Schaathun. Fighting two pirates. In *Applied Algebra, Algebraic Algorithms and Error-Correcting Codes*, volume 2643 of *Springer Lecture Notes in Computer Science*, pages 71–78. Springer-Verlag, May 2003.
23. Jessica N. Staddon, Douglas R. Stinson, and Ruizhong Wei. Combinatorial properties of frameproof and traceability codes. *IEEE Trans. Inform. Theory*, 47(3):1042–1049, 2001.
24. D. R. Stinson and R. Wei. Combinatorial properties and constructions of traceability schemes and frameproof codes. *SIAM J. Discrete Math.*, 11(1):41–53 (electronic), 1998.
25. D.R. Stinson, Tran Van Trung, and R. Wei. Secure frameproof codes, key distribution patterns, group testing algorithms and related structures. *J. Stat. Planning and Inference*, 86(2):595–617, 2000.

26. Michael A. Tsfasman. Algebraic-geometric codes and asymptotic problems. *Discrete Appl. Math.*, 33(1-3):241–256, 1991. Applied algebra, algebraic algorithms, and error-correcting codes (Toulouse, 1989).
27. S. H. Unger. *Asynchronous Sequential Switching Circuits*. Wiley, 1969.

Analysis and Design of Modern Stream Ciphers
(Invited Paper)

Thomas Johansson

Dept. of Information Technology, Lund University,
P.O. Box 118, 221 00 Lund, Sweden.
thomas@it.lth.se

Abstract. When designing symmetric ciphers, security and performance are of utmost importance. When selecting a symmetric encryption algorithm, the first choice is whether to choose a block cipher or a stream cipher. Most modern block ciphers offer a sufficient security and a reasonably good performance. But a block cipher must usually be used in a "stream cipher" mode of operation, which suggests that using a pure stream cipher primitive might be beneficial.

Modern stream ciphers will indeed offer an improved performance compared with block ciphers (typically at least a factor 4-5 if measured in speed). However, the security of modern stream ciphers is not as well understood as for block ciphers. Most stream ciphers that have been widely spread, like RC4, A5/1, have security weaknesses.

It is clear that modern stream cipher designs, represented by proposals like Panama, Mugi, Sober, Snow, Seal, Scream, Turing, Rabbit, Helix, and many more, are very far from classical designs like nonlinear filter generators, nonlinear combination generators, etc. One major difference is that classical designs are bit-oriented, whereas modern designs tend to operate on (e.g. 32 bit) words to provide efficient software implementations. This leads to usage of different operations. Modern stream ciphers use building blocks very similar to those used in block ciphers. Essentially all modern stream cipher designs use S-boxes in one way or the other and combine this with various linear operations, essentially following the old confuse and diffuse paradigm from Shannon.

In this invited talk, we will overview various methods for cryptanalysis of modern stream ciphers. This will include time-memory tradeoff attacks, correlation attacks, distinguishing attacks of different kinds, guess-and-determine type of attacks, and the recent and very interesting algebraic attacks. This will give us lots of useful feedback when considering the design of secure and fast stream ciphers.

K.G. Paterson (Ed.): Cryptography and Coding 2003, LNCS 2898, p. 66, 2003.
© Springer-Verlag Berlin Heidelberg 2003

Improved Fast Correlation Attack Using Low Rate Codes

Håvard Molland, John Erik Mathiassen, and Tor Helleseth

The Selmer Center* Dept. of Informatics, University of Bergen
P.B. 7800 N-5020 BERGEN, Norway
{molland,johnm,tor.helleseth}@ii.uib.no

Abstract. In this paper we present a new and improved correlation attack based on maximum likelihood (ML) decoding. Previously the code rate used for decoding has typically been around $r = 1/2^{14}$. Our algorithm has low computational complexity and is able to use code rates around $r = 1/2^{33}$. This way we get much more information about the key bits. Furthermore, the run time for a successful attack is reduced significantly and we need fewer key stream bits.

1 Introduction

Linear feedback shift registers, *LFSRs*, are popular building blocks for stream ciphers, since they are easy to implement, easy to analyze, and they have nice cryptographic properties. But a linear shift register is not a cryptographic secure function in itself. Assuming we know the connection points in the LFSRs, we just need to know a few bits of the key stream to find the key bits, by using the linear properties in the streams to set up an equation set that is easily solved.

To make such a cipher system more secure, it is possible to combine n LFSRs with a nonlinear function f in such a way that linear complexity becomes very high. Fig 1 describes an example for this model. The key stream $\mathbf{z} = (z_0, z_1, ..., z_t, ..., z_{N-1})$ is generated by $z_t = f(u_t^1, u_t^2, ..., u_t^n)$ and the linearity in the bit streams $\mathbf{u}^i = (u_0^i, u_1^i, ..., u_t^i, ..., u_{N-1}^i)$ from the n LFSRs is destroyed. The plain text $\mathbf{m}$ of length N is then encrypted to cipher text $\mathbf{c}$ by $c_t = z_t \oplus m_t$, $0 \le t < N$.

There exist different types of attacks on systems based on this scheme. The type of attack we describe in this paper is the correlation attack. The attack uses the fact that there often exist some correlations between the bits in some of the shift register streams and the key stream $\mathbf{z}$. This can be formulated as the crossover probability $p = P(u_t \neq z_t)$, where u_t is the bit stream from a LFSR that has a correlation with $\mathbf{z}$. When $p \neq 0.5$, it is possible to do a correlation attack. If $p = 0.5$ there would be no correlation, and a correlation attack could not be done. But it is a well known fact that there always exists a correlation between sums of $\mathbf{u}$ and $\mathbf{z}$ in the model described in Fig. 1. That is

---

* This work was supported by the Norwegian Research Council under Grant 146874/420.

K.G. Paterson (Ed.): Cryptography and Coding 2003, LNCS 2898, pp. 67–81, 2003.

$P(u_{t+j_1} + u_{t+j_2} + ... + u_{t+j_M} \neq z_t) \neq 0.5$ for given M and $(j_1, j_2, ..., j_M)$. When we add a LFSR bit stream with a shift of itself, we always get a new LFSR bit stream. Thus, the model is to decode a LFSR stream that has been sent through a binary symmetric channel (BSC) with crossover probability p.

The simplest correlation attack[6] chooses the shift register $LFSR_i$ that has a correlation to the key stream bit z. Then the initialization bits $\hat{\mathbf{u}}^I$ for the LFSR are guessed and the bit stream $\hat{\mathbf{u}} = (\hat{u}_0, \hat{u}_1, ..., \hat{u}_{N-1})$ is generated. If for a chosen threshold p_{tr} there exists a correlation between the guessed bit stream $\hat{\mathbf{u}}$ and $\mathbf{z}$ such that $P(u_t \neq z_t) < p_{tr} < 0.5$ for $0 \leq t < N$, it is assumed that the correct initialization bits are found. This attack has a complexity of $O(2^{l_i} \cdot N)$ which is much better than $O(2^{l_1+l_2+...+l_n})$, the complexity for guessing the initialization bits for all the LFSRs.

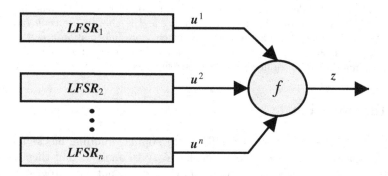

Fig. 1. An example of a stream cipher we are are able to attack using fast correlation attacks. The linear feedback shift registers $LFSR_i$ of length l_i, for $1 \leq i \leq N$, are sent through a nonlinear function f to generate the key stream $\mathbf{z}$

The complexity for guessing all the bits in a given $LFSR_i$ can be too high. To get around this, the fast correlation attack was developed[7,8] by Meier and Staffelbach. This attack uses parity check equations and reconstructs $\mathbf{u}$ from $\mathbf{z}$ using an iterative decoding algorithm. The attack works well when the polynomial that defines the $LFSR_i$ has few taps, but fails when the polynomial has many taps.

In [4] Johansson and Jönsson presented a better attack that works for LFSRs with many taps. Using a clever search algorithm, they find parity equations that are suitable for convolutional codes. The decoding is done using the Viterbi algorithm, which is maximum likelihood. This attack is briefly explained in Sect. 2.

In [1] David Wagner found a new algorithm to solve the generalized birthday problem. In this paper we present an algorithm based on the same idea that finds many equations suitable for correlation attacks. The problem with this algorithm is that it finds many but weak equations, and previous attacks would not be very effective since the code rate will be very low.

In this paper we present an improvement on the attacks based on ML decoding. While Johansson and Jönsson use few but strong equations, we go in the opposite direction and use many and weak equations. We present a new algorithm that is capable of performing an efficient ML decoding even when the code rate is very low. This gives us much more information about the secret initialization bits, and the run time complexity goes down considerably. For a crossover probability $p = 0.47$, polynomial of degree $l = 60$ and the number of known key stream bits $N = 100 \cdot 10^6$, our attack has complexity of order 2^{39}, while the previous convolutional code attack[4,5] has complexity of order 2^{48}. See Table 2 in Sect. 5 for more simulation results compared to previous attacks.

The paper will be organized as follows. First we will give a brief description of the systems we try to attack. In Sect. 2 we will describe the basic mathematics and some important previous attacks. In Sect. 3 we describe an efficient method for finding parity check equations, using the generalized birthday problem. In Sect. 4 we present a new algorithm that is capable of using the huge number of equations found by the method in Sect. 3.

2   Definitions and Previous Attacks

First we will define the basic mathematics for the correlation attacks in this paper.

2.1   The Generator Matrix

Let $g(x) = 1 + g_{l-1}x + g_{l-2}x^2 + ... + g_1x^{l-1} + x^l$ be the primitive feedback polynomial over $\mathbb{F}_2$ of degree l for a linear feedback register, LFSR, that generates the sequence $u = (u_0, u_1, ..., u_{N-1})$. The corresponding recurrence is $u_t = g_1u_{t-1} + g_2u_{t-2} + u_{t-l}$. Let α be defined by $g(\alpha) = 0$. From this we get the reduction rule $\alpha^l = g_1\alpha^{l-1} + g_2\alpha^{l-2} + ... + g_{l-1}\alpha + 1$. Then we can define the generator matrix for sequence u_t, $0 < t < N$ by the $l \times N$ matrix

$$G = [\alpha^0\alpha^1\alpha^2...\alpha^{N-1}]. \tag{1}$$

For each $i > l$, using the reduction rule, α^i can be written as $\alpha^i = h_{l-1}^i\alpha^{l-1} + ... + h_2^i\alpha^2 + h_1^i\alpha + h_0^i$. We see that every column $i \geq l$ is a combination of the first l columns. Any column i in G can be represented by

$$\mathbf{g}_i = [h_0^i, h_1^i, ..., h_{l-1}^i]^{\mathbf{T}}. \tag{2}$$

Thus the sequence u with length N and initialization bits $\mathbf{u}^I = (u_0, u_1, ..., u_{l-1})$, can be generated by

$$\mathbf{u} = \mathbf{u}^I G.$$

The shift register is now turned into a (N, l) block code.

Example 1. Let $g(x) = x^4 + x^3 + 1$. Using the reduction rule we get $\alpha^4 = \alpha^3 + 1$, $\alpha^5 = \alpha(\alpha^3 + 1) = \alpha^4 + \alpha = \alpha^3 + \alpha + 1$ and so on. We choose $N = 10$, and set $G = [\alpha^0\alpha^1...\alpha^9]$. The sequence **u** is generated by the 4×10 matrix G like this,

$$\mathbf{u} = \mathbf{u}^I G = [u_0, u_1, u_2, u_3] \begin{bmatrix} 1 & 0 & 0 & 0 & 1 & 1 & 1 & 1 & 0 & 1 \\ 0 & 1 & 0 & 0 & 0 & 1 & 1 & 1 & 1 & 0 \\ 0 & 0 & 1 & 0 & 0 & 0 & 1 & 1 & 1 & 1 \\ 0 & 0 & 0 & 1 & 1 & 1 & 1 & 0 & 1 & 0 \end{bmatrix}. \tag{3}$$

The reason that we use a generator matrix, is that we easily can see from G which initialization bits $(u_0, u_1, ..., u_{l-1})$ sum to u_i for every $0 \leq i < N$ by looking at column i. For example the bit u_9 (last column) in the example above is calculated by $u_9 = u_0 + u_2$, and it is independent of the initialization bits u_1 and u_3.

2.2 Equations

In [4] Johansson and Jönsson presented the following method for finding equations that are usable for decoding.

Let u be a sequence generated by the generator polynomial $g(x)$ with degree l. If we can find w columns in the generator matrix G that summarize to zero in the $l - B$ last bits,

$$(\mathbf{g}_{i_1} + \mathbf{g}_{i_2} + ... + \mathbf{g}_{i_w})^{\mathbf{T}} = (c_0, c_1, ..., c_{B-1}, \underbrace{0, 0, ..., 0}_{l-B}), \tag{4}$$

for a given B, $0 < B \leq l$, and $l \leq i_1, i_2, ..., i_w < N$, we get an equation of the form

$$c_0 u_0 + c_1 u_1 + ... + c_{B-1} u_{B-1} = u_{i_1} + u_{i_2} + ... + u_{i_w}. \tag{5}$$

This can be seen by noting that column i in G shows which of the initialization bits $\mathbf{u}^I = (u_0, u_1, ..., u_{l-1})$ that summarize to the bit u_i in the sequence **u**. When two columns i and j in G sum to zero in the last $l - B$ entries $(u_B, u_{B+1}, ..., u_{l-1})$, the sum $u_i + u_j$ is independent of those bits. Then we can concentrate on finding just the B first bit of $\mathbf{u}^I$. The equation (5) is cyclic and can therefore be written as

$$c_0 u_t + c_1 u_{t+1} + ... + c_{B-1} u_{t+B-1} = u_{t+i_1} + u_{t+i_2} + ... + u_{t+i_w}, \tag{6}$$

for $0 \leq t < N - i_w$.

Example 2. Let $w = 2$, and $B = 1$. If we examine the matrix G in equation (3), we see that $(\mathbf{g}_6 + \mathbf{g}_8)^{\mathbf{T}} = (1, 1, 1, 1) + (0, 1, 1, 1) = (1, 0, 0, 0)$. From this we get $c_0 = 1$, $i_1 = 6$ and $i_2 = 8$ and the equation is $u_0 = u_6 + u_8$. Because of the cyclic structure we finally get $u_t = u_{t+6} + u_{t+8}$. This equation will hold for every sequence that is generated with $g(x) = x^4 + x^3 + 1$ as feedback polynomial.

In section 3 we will go further into how the actual search for columns that sum to zero in the last $l - B$ bits can be done efficiently.

2.3   Principle for Decoding

In [2] Chepyzhov, Johansson and Smeets presented a simple maximum likelihood algorithm that uses equations found in Sect. 2.2 for decoding. We will now briefly describe this algorithm. First we take equation (5) and make the right side of the equation point to the corresponding key stream bits z instead of u. From this we get the following equation,

$$c_0 u_0 + c_1 u_1 + \ldots + c_{B-1} u_{B-1} \approx z_{i_1} + z_{i_2} + \ldots + z_{i_w}. \tag{7}$$

Let m be the number of equations found by the method in Sect. 2.2. Then we get the equation set

$$c_{0,0} u_0 + c_{0,1} u_1 + \ldots + c_{0,B-1} u_{B-1} \approx z_{i_{1,1}} + z_{i_{1,2}} + \ldots + z_{i_{1,w}}$$
$$c_{1,0} u_0 + c_{1,1} u_1 + \ldots + c_{2,B-1} u_{B-1} \approx z_{i_{2,1}} + z_{i_{2,2}} + \ldots + z_{i_{2,w}}. \tag{8}$$
$$\vdots$$
$$c_{m,0} u_0 + c_{m,1} u_1 + \ldots + c_{m,B-1} u_{B-1} \approx z_{i_{m,1}} + z_{i_{m,2}} + \ldots + z_{i_{m,w}}$$

We use $'\approx'$ to notify that the equations only hold with a certain probability.

Here $(u_0, u_1, \ldots, u_{B-1})$ are the unknown secret bits we want to find and z is the key stream. Remember that u_t and z_t are equal with a probability $1 - p$ where $p = P(u_t \neq z_t)$. Thus, each equation in (8) will hold with a probability

$$P_w = \frac{1}{2} + 2^{w-1}(\frac{1}{2} - p)^w, \tag{9}$$

using the Piling up lemma[9]. Replace the bits $(u_0, u_1, \ldots, u_{B-1})$ in the set (8) with a guess $\hat{U} = (\hat{u}_0, \hat{u}_1, \ldots, \hat{u}_{B-1})$. If $(\hat{u}_0, \hat{u}_1, \ldots, \hat{u}_{B-1}) \neq (u_0, u_1, \ldots, u_{B-1})$, (that is, if one or more of the guessed bits are wrong) each equation will hold with a probability $P = 0.5$. If the guess is right, each equation will hold with a probability $P_w > 0.5$. We see that the (N, l) block-code is reduced to a (m, B) block-code, and the decoding problem is to decode message blocks of length B that are sent as codewords of length m through a channel with crossover probability $1 - P_w$.

The decoding can be done the following way. For all the 2^B possible guesses for $\hat{U} = (\hat{u}_0, \hat{u}_1, \ldots, \hat{u}_{B-1})$, test $\hat{U}$ with all the equations in the set (8), and give the guess one point for every equation in the set that holds. Afterward, assume that $(u_0, u_1, \ldots, u_B) = \hat{U}$ for the guess of $\hat{U}$ that has the highest score. In this way we get the first B bits of the secret initialization bits $(u_0, u_1, \ldots, u_l)$. The procedure can be repeated to find the rest of the bits $(u_B, u_{B+1}, \ldots, u_{l-1})$.

The complexity for this algorithm is

$$O(2^B \cdot m) \tag{10}$$

since we have to test m equations on the 2^B different guesses of $\hat{U}$.

2.4   Fast Correlation via Convolutional Codes

In [4] Johansson and Jönsson showed how the equation set (8) can be used to decode the key stream z via convolutional codes. The problem is formulated as decoding of a $(m, 1, B)$ convolutional code, and the decoding is done using the Viterbi algorithm. This algorithm is optimal, but has relatively high usage of memory. In convolutional codes the coding is done over T bits. Using the fact that the equations are cyclic, the algorithm in Sect. 2.3 is used for calculating the metrics for each state $\hat{U}$ at time t, $0 \leq t < T$. The algorithm in Sect. 2.3 is actually a special case of the fast correlation attack via convolutional code with $T = 1$. When the metrics are calculated, we try to find the longest possible path through the states $0 \leq t < T$. We see that the problem is transformed into finding the longest path trough a $2^B \times T$ trellis. The Viterbi algorithm is optimal for solving this problem. We refer [4] for details about the convolutional attacks.

2.5   Theoretical Analysis and Complexity

In [5] Johansson and Jönsson, presented a theoretical estimate of the success rate for fast correlation attacks via convolutional codes.

For a given bit stream of length N generated by a shift register with feedback polynomial $g(x)$, the expected number of equations of type (5) is

$$E(m) = \frac{\binom{N-T-l}{w}}{2^{l-B}} \approx \frac{\binom{N}{w}}{2^{l-B}} \tag{11}$$

Let $p_e < l \cdot 2^{-B}$ and $p = P(z_t \neq u_t)$. Then the convolutional attack described in Sect. 2.4 has a success with probability $1 - p_e$ if

$$p \leq \frac{1}{2} - \frac{1}{2} \left(\frac{8 ln 2}{m} \right)^{\frac{1}{2w}}. \tag{12}$$

The probability p is set by the stream cipher. The closer p is to 0.5, the more equations are needed to fulfill (12). One way to find more equations is to increment w, the number of bits on the right hand side of the equations. If we do this, each equation we find gets weaker, see equation (9). But although each equation is weaker, we find so many of them that they together give more information about the unknown key bits u^I. The problem with this is, as shown below, that the complexity of the attack also increases when we use many more equations. In Sect. 4 we will describe a new method to solve this problem.

The complexity of the convolutional attack in [4,5] is $O(2^B \cdot m \cdot T)$, since we decode over T bits. This can be rewritten using equation (11) and noting that $m = 2^B \frac{\binom{N}{w}}{2^l}$. Let $o = \frac{\binom{N}{w}}{2^l}$. In this way we see that the complexity can be formulated as

$$O(2^{2B} \cdot o \cdot T), \tag{13}$$

The complexity for the simple attack in [2] is $O(2^{2B} \cdot o)$. Using this formulation, we see that if we use all the equations for given w, N, B and l, the run time complexity increases with a factor 4, when we increment B by one.

3   Methods for Finding Equations

In this section we will describe a fast method for finding many equations. The method is in some ways similar to the solution of the generalized birthday problem that Wagner presented in [1].

We have an equation of the form (5) if we find w columns in the generator matrix G of length N that sum to zero in the last $l - B$ positions. For $w = 2$ we sort the N columns from the generator matrix. Equal columns will then be located next to each other. The complexity of this method is $O(N \log N)$.

3.1   A Basic Method for Finding Equations with $w > 2$

We will now describe a simple and well known algorithm for finding equations when $w > 2$.

First we sort the columns in the $l \times N$ generator matrix G according to the values in last $l - B$ bits. Then we run through all possible sums of $w - 1$ columns, and search for columns in G that are equal to the sums in the last $l - B$ bits. The sum of these w columns is then zero in the $l - B$ last bits. The complexity of this algorithm will be $O(N^{w-1} \log N)$.

This method is straightforward and seems good since we find all the equations. The problem is when $l - B$ becomes big, since it is less likely that the sum of the combination of $w - 1$ columns matches a single column. The number of possible different values in the $l - B$ last bits are 2^{l-B}. If we pick a random combination of $w - 1$ columns we will have a probability less than $P_m = N/2^{l-B}$ of getting a match from the sorted generator matrix. If $N = 2^{20}$, $B = 15$ and $l = 40$ then $P_m = 2^{-5}$, so on average each 32'th sum combination will give an equation. If we increase the degree of the feedback polynomial to $l = 60$, the probability of finding an equation for given $w - 1$ columns will be reduced to $P_m = 2^{-25}$. Since an equation with $w = 4$ is a very weak equation, we need millions of equations in most cases.

Table 1. The table shows the percentage of the total number of equations we need for a successful convolutional attack when $N = 2^{21}$, $l = 60$, $w = 4$ and $B = 20$.

p	v
0,41	0.00068%
0,43	0,0051%
0,45	0,075%
0,47	4.5%

The method above finds all the equations, but in fact we do not need all the equations for the attack to succeed. From (12) we get the equation

$$m_s = \frac{8 \ln 2}{(1 - 2p)^{2w}},$$

where m_s is the number of equations needed for success for a given crossover probability p. Then $v = \frac{m_s}{m}$ will give us the rate of the total number of equations m needed for a successful attack. Table 1 shows different rates needed for different attacks. The fact that we do not need all the equations indicates that we may use a fast method to find a subset of them.

3.2 A Method for Finding All the Equations with $w = 4$

The method described here works in certain situations where the parameters are small. The algorithm works as follows. In the first step we go through all the possible sums of pairs of columns in G. These sums are stored in a matrix G_2 and the indexes of the two columns in G are also stored. In the second step we sort the matrix G_2 according to the last $l - B$ bits. Then we search for the columns in G_2 that are equal in the last $l - B$ bits. In this way we get weight 4 equations of the form:

$$(\mathbf{f}_{j_1} + \mathbf{f}_{j_2})^{\mathbf{T}} = (\mathbf{g}_{i_1} + \mathbf{g}_{i_2} + \mathbf{g}_{i_3} + \mathbf{g}_{i_4})^{\mathbf{T}} = (c_0, c_1, ..., c_{B-1}, \underbrace{0, ..., 0}_{l-B}) \qquad (14)$$

where the f_j's are columns in G_2 and the g_j's are columns in G.

By this method we will find all the equations 3 times. The reason for this is illustrated by the equation $g_{i_1} + g_{i_2} + g_{i_3} + g_{i_4} = 0 \iff g_{i_1} + g_{i_2} = g_{i_3} + g_{i_4}$. Two other pairs giving the same equation is $g_{i_1} + g_{i_4} = g_{i_2} + g_{i_3}$ and $g_{i_1} + g_{i_3} = g_{i_2} + g_{i_4}$. This collisions are avoided if the pairing in the second step has a restriction. All the indexes on the left side of the equation must all be less or greater than the indexes on the right side. In this way $\frac{2}{3}$ of the equations will be thrown away, but the remaining $\frac{1}{3}$ will represent all the equations. This method will be impractical if N is big, since G_2 will have a length of $N_2 = \binom{N}{2}$.

3.3 A Fast Method for Finding a Subset of the Equations with $w = 4$

Here we will solve the problem concerning memory requirement in the algorithm presented above. Using this algorithm we are able to find all the equations, but the number of possible sums in step one is far too many. If we can reduce the size of G_2, without reducing the number of equations significantly, we have succeeded.

The algorithm is divided into two steps. In step one we find a subset of all the sums, where the pairing in step 2 only involves elements in that subset. The sum of two columns that are unequal in the last bits will never be zero. Therefore we may look for sums of pairs in step 1 where we require a certain value in the last $l - B_2$ positions. Without loss of generality we require zeroes in the last $l - B_2$ positions in G_2.

Let $B_4 < B_2 < l$. First we sort the columns in G according to the last $l - B_2$ positions. Then we go through the matrix and find the columns that sum to zero in the last $l - B_2$ positions and store them in matrix G_2. The original positions

Algorithm 1 Algorithm for finding a subset of all the equations with $w = 4$

Input: G, N, B_2, $B_4 < B_2$, l.
Step 1:
sort the $l \times N$ matrix G according to the last $l - B_2$ bits.
For $0 \leq i_1, i_1 < N$ find all pairs of columns $\mathbf{g}_{j_1}$ and $\mathbf{g}_{j_2}$ that sums to

$$\mathbf{f}_j^T = (\mathbf{g}_{i_1} + \mathbf{g}_{i_2})^T = (d_0, d_1, ..., d_{B_2-1}, \underbrace{0, ..., 0}_{l-B_2})$$

Add $\mathbf{f}_j$ and indexes $i_1 + i_2$ to matrix G_2.
Step 2:
sort $l \times N_2$ matrix G_2 according to the last $l - B_4$ bits.
For $0 \leq j_2, j_4 < N_2$ find all pairs of columns $\mathbf{g}_{j_1}$ and $\mathbf{g}_{j_2}$ that sums to

$$(\mathbf{f}_{j_1} + \mathbf{f}_{j_2})^T = (\mathbf{g}_{i_1} + \mathbf{g}_{i_2} + \mathbf{g}_{i_3} + \mathbf{g}_{i_4})^T = (c_0, c_1, ..., c_{B_4-1}, \underbrace{0, ..., 0}_{l-B_4})$$

Add $c_0, c_1, ..., c_{B-1}$ and the indexes i_1, i_2, i_3, i_4 to F
Return: F

of the columns in the sum are also stored. The size of G_2 is thereby reduced by a factor of 2^{l-B_2}. In the second step we repeat the algorithm using B_4 on G_2. We sort the matrix G_2 according to the last $l - B_4$ bits, in order to find pairs of columns from G_2, where the sum is zero in the last $l - B_4$ bits. In this way we get weight 4 equations of the form (5). The pseudo code for this is shown in Algorithm 1.

Algorithm 1 is a method which may keep the memory requirements sufficiently low. From (11) we get the size N_2 of G_2 ,

$$N_2 = \frac{\binom{N}{2}}{2^{l-B_2}} \approx \frac{N^2}{2^{l-B_2+1}}.$$

It is possible to run this algorithm several times to find even more equations. Instead of keeping the last $l - B_2$ bits zero in the first step, we may repeat this algorithm requiring these bits having the fixed values $(d_{B_2}, d_{B_2+1}, ..., d_l) \neq \mathbf{0}$. We may choose to only vary the first two bits, and run the algorithm 2^2 times. Thus we get four times as many equations compared to running it only once. The cost is that we have to sort the matrixes G and G_2

Using Algorithm 1 some of the equations we get will be equal, called collisions. If we use algorithm 1 repeatedly changing r bits (that is: we repeat 2^r times) this bound is

$$p(collision) < 2^{(B_2+r)-l} + 2^{2(B_2+r-l)} < 2 \cdot 2^{(B_2+r-l)} = 2^{(B_2+r+1-l)}$$

since $B_2 + r - l < 0$. If we do not use repetitions we set $r = 0$. In practical attacks, this probability will be very low, and the simulations show that this has little impact on the decoding.

4   Fast Decoding Using Quick Metric

In Sect. 3 we presented a fast method for finding a huge number of equations. These equations can give us a lot of information about the initialization bits. But since there are so many of them, we get two new problems. It will take too much memory to store all the equations, and the complexity will be too high when we use them to calculate the metrics during decoding. Thus, we need an efficient method for storing the equations, and an efficient method for using them.

The complexity for calculating the metrics by the method in Sect. 2.3, is $O(2^B \cdot m)$, where m is the number of equations and B is the message block size of the code. If m is very high, the decoding problem can be to complex. We reduce the decoding complexity to $O(2^{2B} + m)$ by the following two methods referred to as *Quick Metric*.

4.1   A New and Efficient Method for Storing the Equations

Let $m \gg 2^B$ be the number of equations found using the method described in Sect. 3 with $B = B_4$. We get an equation set like (8). The main observation here is that although there are m different equations, there exist only 2^B different versions of the left side of the equations. This means that many equations will share the same left sides defined by $(c_0, c_1, ..., c_{B-1})$ when $m \gg 2^B$. We can now use *counting sort* to store the equations. Let E be an integer array of size 2^B. When an equation of the form (5) is found, we set

$$e = c_0 + 2c_1 + 2^2 c_2 + ... + 2^{B-1} c_{B-1}. \tag{15}$$

Then we count the equation by setting $E(e) \leftarrow E(e) + 1$.

At this point we have stored the left side of the equation. To store the right side, we use another integer array, $sum()$, of size 2^B. Then we calculate the binary sum $s = (z_{i_1} + z_{i_1} + ... + z_{i_w}) \bmod 2$ for the given $(i_1, i_2, ..., i_w)$. Finally we set $Sum(e) \leftarrow Sum(e) + s$.

When the search for equations is finished, $E(e)$ is the number of the equations of type e that was found, and $Sum(e)$ is the number of equations of type e that sum to 1 on the right hand side for a given key stream **z**.

Algorithm 2 shows a pseudo code for this idea. Here the idea is expanded so that it works with decoding via convolutional codes as presented in [4,5]. We assume that the search methods in Algorithm 1 are used to find the w columns that sum to zero in the last B bits. When decoding is done via convolutional codes, the equations are cycled T times when we decode over T bits. This means that we have to calculate $Sum(e)$ for every $0 \leq t < T$, since the right side of (7) is not cyclic itself. From this we get the 2-dimensional array $Sum(e, t)$. One little detail to make this work with convolutional codes, is that the bit c_B in the sum of the columns has to be 1. But this has no impact on the complexity for the algorithm.

Algorithm 2 Algorithm for storing equations (first step)

Input: G, N, T, B,w and z.
For every $i_1, i_2, ..., i_w$, $T \leq i_1, i_2, ..., i_w < N - T$,
If the columns $\mathbf{g}_{i_1}, \mathbf{g}_{i_2}, ..., \mathbf{g}_{i_w}$ in G summarize to

$$(\mathbf{g}_{i_1} + \mathbf{g}_{i_2} + ... + \mathbf{g}_{i_w})^{\mathbf{T}} = (c_0, c_1, ..., c_{B-1}, \underbrace{1, 0, ..., 0}_{l-B})$$

Let e be the integer value of the bits $(c_0, c_1, ..., c_{B-1})$.
$E(e) \leftarrow E(e) + 1$
For every t, $1 \leq t \leq T$,
$Sum(e, t) \leftarrow Sum(e, t) + (z_{t+i} + z_{t+j} + z_{t+k} \bmod 2)$
Return: The integer arrays Sum and E

4.2   A New and Efficient Method for Calculating the Metrics

Assume we have done the search for equations according to Sect. 3.3 and Algorithm 2. After this preprocessing step, we have the two arrays E and sum. Let $m_e = E(e)$ be the number of equations found of type e. Now we can test m_e equations on a guess $\hat{\mathbf{U}}$ in just one step instead of m_e .

Make a guess $\hat{\mathbf{U}}$ for $\mathbf{u}^I$. For every equation type e, do as follows: If the sum $c_0\hat{u}_0 + c_1\hat{u}_1 + ... + c_{B-1}\hat{u}_{B-1}$ corresponding to the equation type e (see equation 15) is 1, the number of the $m_e = E(e)$ equations that hold is $sum(e)$. The metric for the guess $\hat{\mathbf{U}}$ is incremented with $sum(e)$. If $c_0\hat{u}_0 + c_1\hat{u}_1 + ... + c_{B-1}\hat{u}_{B-1} = 0$, the number of the m_e equations that hold is $m_e - sum(e)$, and the metric is incremented with $m_e - sum(e)$. Algorithm 3 shows the pseudo code for this idea.

Now we have calculated the metric for one guess in just 2^B steps instead of $m > 2^B$ steps. The complexity for this part of the attack is actually independent of the amount of equations that are used, and the complexity for calculating the metrics for all the 2^B guesses is $O(2^{2B})$. The reason that the overall complexity is $O(2^{2B} + m)$, is that we have to go through all m equations once in the preprocessing, each time we want to analyze a new key stream z. Using the search algorithm in Sect. 3, we can do some processing independently from z. But in the end we have to go through m equations and save the $z_{i_1} + z_{i_2} + ... + z_{i_w}$ in array sum for each equation. This part of the search algorithm is almost linear in m.

4.3   Complexity and Properties

When we use Quick Metric, the decoding is done in two steps. The first step is the building of the equation count matrix E. The second step is decoding using the Viterbi algorithm with complexity $O(T \cdot 2^{2B})$, because of Quick Metric. The building of matrix E can be divided into 3 parts. First the sorting of G of length N, then the sorting of G_2 of length N_2. Finally we have to go through the sorted G_2 and save all the equations in E. Thus, the total complexity for the first step

---

Algorithm 3 Quick Metric algorithm (second step)

Input: state $\widehat{\mathbf{U}}$, time t, and the tables *Sum* and E .
$metric_{\widehat{U}} \leftarrow 0$
For every e, $0 \leq e < 2^B$
  If equation e over state $\widehat{\mathbf{U}}$ sums to 1,
$metric_{\widehat{U}} \leftarrow metric_{\widehat{U}} + Sum(e, t)$
  Else
$metric_{\widehat{U}} \leftarrow metric_{\widehat{U}} + (E(e) - Sum(e, t))$
Return: $metric_{\widehat{U}}$

---

is $O(N \cdot logN + N_2 \cdot logN_2 + T \cdot m)$. Since m has to be very high for our attack, the complexity is most often dominated by $T \cdot m$, and the overall complexity for the first step is $O(T \cdot m)$.

It will vary which of the two steps that will dominate the complexity. Thus, the total run time complexity for both step is given by

$$O(T \cdot m + T \cdot 2^{2B}).$$

To guarantee success $(99, 9\%)$, the number of equations m and the convolutional memory B should satisfy equation (12) where p and l is are set by the cipher system. T must be high enough so that the algorithm converge to the right path in the trellis. $T \approx l$ is enough for most cases. The complexity for the attack in [4,5] is $O(2^{2B} \cdot o \cdot T)$, where $o = \frac{\binom{N}{w}}{2^l}$.

The first observation is that when we use Quick Metric, the computational complexity for the Viterbi algorithm is independent from the number of equations m that is used for decoding. The main difference from the attacks in [4,5] is that we just have to go through all m equations once in the first step. In [4, 5] they have to go through all the m equations for every time they test one of the 2^B states. Thus, our algorithm has a big advantage when we choose to use more than 2^B equations.

A drawback for our algorithm is that we have to do the first step every time we want to decode a new stream generated by the same system. In [4,5], they just have to do the preprocessing once for each cipher system. Therefor we have to keep the complexity in the first step as low as possible. There is actual a trade off between the two steps. When the first step takes long time, the second step takes less time, and the other way around. This means that we have to choose the parameters N, B and m carefully to get the best possible attack.

The next observation is that our algorithm is stronger for last B, since we can use many more equations. That means that we can attack a system using a last B than is possible with the attacks in [4,5]. Thus, the run time for given B, w and m goes down considerably since B has a huge impact on the complexity.

Table 2. Our attack compared to previous attacks. The generator polynomial degree l for the LFSR is 60 for all the simulations. We set $T = 60$. The * is a theoretical estimate using the success rate equation (12).

Improved convolutional attack

B	p	N	w	*Total decoding complexity*
14	0.43	$15 \cdot 10^6$	4	2^{35}
10	0.43	$100 \cdot 10^6$	4	2^{31}
16	0.47	$100 \cdot 10^6$	4	2^{39}
11	0.43	$40 \cdot 10^6$	4	2^{30}

Previous convolutional attack[5]

B	p	N	w	*Decoding complexity*
20	0.43	$100 \cdot 10^6$	2	2^{38}
18	0.37	$600 \cdot 10^3$	3	2^{37}
25*	0.47	$100 \cdot 10^6$	2	2^{48}

Previous attack through reconstruction of linear polynomials[3]

B	p	N	w	Rounds n	*Decoding complexity*
25	0.43	$40 \cdot 10^6$	2	4	$2^{41.5}$

5   Simulations

The evaluation of the attacks need some explanation. The interesting parameters of the cipher systems we attack, are the polynomial degree l and the crossover probability p. Finally we are given a key stream of length N. We want for a given high l to be able to decode a key stream z where the crossover probability $p = (u_i \neq z_i)$ is as near 0.5 as possible. Of course we want to use few key stream bits and low run time complexity.

To be able to compare the different attacks, we compute the complexity for decoding as the total number of times we have to test an equation on a guessed state. The complexity for the pre-computation is computed as the number of table lookups that have to be done during the search for equations. When we use Quick Metric we have 2 steps, so the overall complexity is given by the sum of the two steps.

See Table 2 for the simulation results. It is important to notice that we have programmed and tested all the attacks in the table, and the results for p come from these tests, not the theoretical estimate (12). For this purpose we used a 2.26 GHz Pentium IV with 1 gigabyte memory running on Linux. The algorithm is fast in practice and even the biggest attack ($p = 0.47$) was done in just a few hours including the search for equations.

From the table we see that our attack is best when p is close to 0.5. For $p = 0.47$ the run time complexity of our attack is dominated by the pre-computation step which is $m{\cdot}T \approx 2^{39}$. The parameters for this attack is $B_2 = 34$, $B = B_4 = 16$ and $m = 2^{33}$ which gives the code rate $r = 2^{-33}$. If we use the method in [4,5], the estimated run time complexity is 2^{48}.

Another attack from Johansson and Jönsson is the the fast correlation attack through reconstruction of linear polynomials[3]. This attack has lower complexity than fast correlation via convolutional codes and it uses less memory. We can apply Quick Metric on the reconstruction algorithm, but unfortunately this will not give a better result than using it on the convolutional code attack. The reason for this is that in each round in the algorithm we would have to repeat the search for equations. To keep B sufficient low, we would have to use many rounds. Thus, the computational complexity for this would become too high.

But when we use Quick Metric on the convolutional attack, the attack achieves in most cases a much lower run time complexity than the attack in [3]. This is shown by the two attacks in Table 2 using $N = 40 \cdot 10^6$.

6 Conclusion

We have presented a new method for calculating the metrics in fast correlation attacks. This method enable us to handle the huge number of parity check equations we get when we use $w = 4$ and the method in Sect. 3. Earlier it has only been possible to handle convolutional code rates down to around $r = 1/2^{14}$. Using our method we have decoded convolutional codes with rates down to $1/2^{32}$ in just a few hours. Because of this we have done attacks on cipher systems with higher crossover probability p than before.

An open problem is the search for equations with $w = 3$. We use $w = 4$ since there exists a fast method for finding those equations. But the equations with $w = 4$ are weak, and this gives the first step high complexity. A good solution would be to use $w = 3$, with a fast search algorithm.

References

1. David Wagner, "A Generalized Birthday Problem", *CRYPTO* 2002, LNCS 2442, Springer-Verlag, 2002, pp. 288–303.
2. V. Chepyzhov, T. Johansson, and B. Smeets, "A simple algorithm for fast correlation attacks on stream ciphers", *Fast Software Encryption, FSE'2000*, Lecture notes in Computer Science, Springer-Verlag, 2001, pp. 181–195.
3. T. Johansson, F. Jönsson, "Fast correlation attacks through reconstruction of linear polynomials", *Advances in Cryptology*-CRYPTO'2000, Lecture Notes in Computer Science, vol. 1880, Springer- Verlag, 2000, pp. 300–315.
4. T. Johansson, and F. Jönsson, "Fast Correlation Attacks on Stream Ciphers via Convolutional Codes", *Advances in Cryptology*-EUROCRYPT'99, Lecture Notes in Computer Science, Vol. 1592, Springer-Verlag, 1999, pp. 347–362.

5. T. Johansson, and F. Jönsson, "Theoretical Analysis of a Correlation Attack based on Convolutional Codes", *Proceedings of 2000 IEEE International Symposium on Information Theory*, IEEE, 2000, pp. 212.

6. T. Siegenthaler, "Decrypting a class of stream ciphers using ciphertext only", *IEEE Trans. on Computers,* vol. C-34, 1985, pp. 81–85.

7. W. Meier, and O. Staffelbach, "Fast correlation attacks on stream ciphers", *Advances in Cryptology*-EUROCRYPT'88, Lecture Notes in computer Science, vol 330, Springer Verlag, 1988, pp. 301–314.

8. W. Meier, and O. Staffelbach, " Fast correlation attacks on certain stream ciphers", *Journal of Cryptology,* vol. 1, 1989, pp. 159–176.

9. M.Matsui, "Linear Cryptanalysis Method for DES Cipher", *Advances in Cryptology*-EUROCRYPT'93, LNCS 765, 1994, pp. 386–397.

On the Covering Radius of Second Order Binary Reed-Muller Code in the Set of Resilient Boolean Functions

Yuri Borissov[1], An Braeken[2], Svetla Nikova[2], and Bart Preneel[2]

[1] Institute of Mathematics and Informatics,
Bulgarian Academy of Sciences,
8 G.Bonchev, 1113 Sofia, Bulgaria
yborisov@moi.math.bas.bg
[2] Department Electrical Engineering - ESAT/SCD/COSIC,
Katholieke Universiteit Leuven, Kasteelpark Arenberg 10,
B-3001 Leuven, Belgium
{an.braeken,svetla.nikova,bart.preneel}@esat.kuleuven.ac.be

Abstract. Let $\mathcal{R}_{t,n}$ denote the set of t-resilient Boolean functions of n variables. First, we prove that the covering radius of the binary Reed-Muller code $RM(2,6)$ in the sets $\mathcal{R}_{t,6}$, $t = 0, 1, 2$ is **16**. Second, we show that the covering radius of the binary Reed-Muller code $RM(2,7)$ in the set $\mathcal{R}_{3,7}$ is **32**. We derive a new lower bound for the covering radius of the Reed-Muller code $RM(2,n)$ in the set $\mathcal{R}_{n-4,n}$. Finally, we present new lower bounds in the sets $\mathcal{R}_{t,7}$, $t = 0, 1, 2$.

1 Introduction

In an important class of stream ciphers, called combination generators, the key stream is produced by combining the outputs of several independent Linear Feedback Shift Register (LFSR) sequences with a nonlinear Boolean function. Siegenthaler [15] was the first to point out that the combining function should possess certain properties in order to resist divide-and-conquer attacks. A Boolean function to be used in stream ciphers should satisfy several properties. *Balancedness* – the Boolean function has to output zeros and ones with equal probabilities. *High nonlinearity* - the Boolean function has to be at sufficiently high distance from any affine function. *Correlation-immunity* (of order t) - the output of the function should be statistically independent of the combination of any t of its inputs. A balanced correlation-immune function is called *resilient*. Other important factors are high algebraic degree and simple implementation in hardware. It is known that there are certain trade-offs involved among these parameters. In order to achieve the desired trade-offs designers typically fix one or two parameters and try to optimize the others.

Recently, also algebraic attacks [3,4] have been applied successfully to stream ciphers. The central idea in the algebraic attacks is to use a lower degree approximation of the combining Boolean function and then to solve an over-defined system of nonlinear multivariate equations of low degree by efficient methods such

as XL or simple linearization [6]. In order to resist these attacks, the Boolean
function should also have a high distance to lower order degree functions.

Kurosawa *et al.* [8] have introduced a new covering radius, which measures the maximum distance between t-resilient functions and r-th degree
functions or the r-th order Reed-Muller code $RM(r,n)$. That is $\hat{\rho}(t,r,n) =
\max d(f(\overline{x}), RM(r,n))$, where the maximum is taken over the set $\mathcal{R}_{t,n}$ of t-
resilient Boolean functions of n variables. They also provide a table with certain
lower and upper bounds for $\hat{\rho}(t,r,n)$.

In this paper we prove exact values of the covering radius $\hat{\rho}(t,2,6)$, for $t =
0,1,2$ and $\hat{\rho}(3,2,7)$. We also generalize our method and find a new lower bound
for the covering radius of the Reed-Muller code $RM(2,n)$ in the set $\mathcal{R}_{n-4,n}$.

The rest of the paper is organized as follows. In Sect. 2 we give some definitions and known results that will be used later in our investigations. Our main
results are described in Sect. 3 and 4. In Sect. 3 we prove that the covering
radius of the binary Reed-Muller code $RM(2,6)$ in the sets $\mathcal{R}_{t,6}$, $t = 0,1,2$ is **16**
and in Sect. 4 we present a proof that the covering radius of the binary Reed-
Muller code $RM(2,7)$ in the set $\mathcal{R}_{3,7}$ is **32**. In this section we also derive a new
lower bound for the covering radius of the Reed-Muller code $RM(2,n)$ in the set
$\mathcal{R}_{n-4,n}$. Finally, the lower bounds of [8] in the sets $\mathcal{R}_{t,7}$, $t = 0,1,2$ are improved.
Conclusions and open problems are presented in Sect. 5.

2   Background and Related Work

Let $f(\overline{x})$ be a Boolean function on $\mathbb{F}_2^n$. Any Boolean function can be uniquely
expressed in the algebraic normal form (ANF):

$$f(\overline{x}) = \sum_{(a_1,\dots,a_n)\in\mathbb{F}_2^n} h(a_1,\dots,a_n)x_1^{a_1}\cdots x_n^{a_n},$$

with h a function on $\mathbb{F}_2^n$, defined by $h(\overline{a}) = \sum_{\overline{x}\leq\overline{a}} f(\overline{x})$ for any $\overline{a} \in \mathbb{F}_2^n$, where
$\overline{x} \leq \overline{a}$ means that $x_i \leq a_i$ for all $i \in \{1,\dots,n\}$. The *algebraic degree* of f,
denoted by $\deg(f)$, is defined as the number of variables in the highest term
$x_1^{a_1}\cdots x_n^{a_n}$ in the ANF of f, for which $h(a_1,\dots,a_n) \neq 0$. If the highest term
of f that contains x_i has degree at most one, x_i is called a linear variable. If
$\deg(f) \leq 1$ then f is called an affine function. The minimum Hamming distance
between f and the set of all affine functions is called the *nonlinearity* of f and
is denoted by N_f.

Let $\overline{x} = (x_1, x_2, \dots, x_n)$, $\overline{\omega} = (\omega_1, \omega_2 \cdots \omega_n)$ be vectors in $\mathbb{F}_2^n = GF(2)^n$, and
$\overline{x} \cdot \overline{\omega} = x_1\omega_1 + x_2\omega_2 + \cdots + x_n\omega_n$ be their *dot product*. The Walsh transform of
$f(\overline{x})$ is a real-valued function over $\mathbb{F}_2^n$ that is defined as

$$W_f(\overline{\omega}) = \sum_{\overline{x}\in\mathbb{F}_2^n} (-1)^{f(\overline{x})+\overline{x}\cdot\overline{\omega}}.$$

A very important equation related to the values in the Walsh spectrum of a Boolean function $f(\overline{x})$ is the *Parseval equation*:

$$\sum_{\overline{\omega} \in \mathbb{F}_2^n} W_f^2(\overline{\omega}) = 2^{2n}.$$

A Boolean function $f(\overline{x})$ on $\mathbb{F}_2^n$ is said to be a *plateaued* function [2,18] if its Walsh transform W_f takes only three values 0 and $\pm\lambda$, where λ is a positive integer, called amplitude of the plateaued function. Because of the Parseval's relation, λ cannot be zero and must be a power 2^r, where $\frac{n}{2} \le r \le n$.

Correlation-immune Boolean functions can be defined in various ways, but for our purposes it is convenient to use as a definition the following spectral characterization given by Xiao and Massey [17].

Definition 1. *A function $f(\overline{x})$ is t-th order correlation-immune if and only if its Walsh transform W_f satisfies $W_f(\overline{\omega}) = 0$, for $1 \le wt(\overline{\omega}) \le t$, where $wt(\overline{x})$ denotes the Hamming weight of the vector $\overline{x}$, i.e., the number of nonzero components of the vector $\overline{x}$. Balanced t-th order correlation-immune functions are called t-resilient functions, i.e. $W_f(\overline{\omega}) = 0$, for $0 \le wt(\overline{\omega}) \le t$.*

Siegenthaler's Inequality [14] states that if the function f is a correlation-immune function of order t then $\deg(f) \le n - t$. Moreover, if f is t-resilient then $\deg(f) \le n - t - 1$, $t < n - 1$.

If a variable x_i is linear for a function f we can present f in the form

$$f(x_1, \ldots, x_{i-1}, x_i, x_{i+1}, \ldots, x_n) = g(x_1, \ldots, x_{i-1}, x_{i+1}, \ldots, x_n) + x_i.$$

Lemma 1. *[16] Let x_{n+1} be a linear variable, i.e., $f(x_1, \ldots, x_n, x_{n+1}) = g(x_1, \ldots, x_n) + c\,x_{n+1}$, where $c \in \{0, 1\}$ and $g(x_1, \ldots, x_n)$ is t-resilient. Then $f(x_1, \ldots, x_n, x_{n+1})$ is $t + 1$-resilient and $N_f = 2N_g$.*

In 2000 Sarkar and Maitra [12], Tarannikov [16], and Zhang and Zheng [19] have proved independently that when $t > \frac{n-1}{2}$, the nonlinearity N_f of a t-resilient function satisfies the condition $N_f \le 2^{n-1} - 2^{t+1}$.

Let f be a Boolean function on $\mathbb{F}_2^n$ and $\overline{\omega}$ be a vector in $\mathbb{F}_2^n$, such that $wt(\overline{\omega}) = r$. By $f_{\overline{\omega}}$ we denote the Boolean function on $\mathbb{F}_2^{n-r}$, defined as follows. Let $i_1, \ldots, i_r$ be such that $\omega_{i_1} = \cdots = \omega_{i_r} = 1$ and $\omega_j = 0$ for $j \notin \{i_1, \ldots, i_r\}$. Then $f_{\overline{\omega}}$ is formed from f by setting the variable x_j to 0 if and only if $j \in \{i_1, \ldots, i_r\}$.

Theorem 1. *[11] Let $f(x_1, \ldots, x_n)$ be a Boolean function and $\overline{\omega} \in \mathbb{F}_2^n$. Then*

$$\sum_{\overline{\theta} \le \overline{\omega}} W_f(\overline{\theta}) = 2^n - 2^{wt(\overline{\omega})+1} wt(f_{\overline{\omega}}).$$

It is well known that the codewords of the r-th order Reed-Muller code of length 2^n (denoted by $RM(r,n)$) may be presented by the set of Boolean functions of degree $\leq r$ in n variables. The *covering radius* of $RM(r,n)$ is defined as

$$\rho(r,n) = \max d(f(\overline{x}), RM(r,n)),$$

where the maximum is taken over all Boolean functions $f(\overline{x})$, and $d(.,.)$ denotes the Hamming distance between two vectors, i.e., the number of position in which they differ.

A new covering radius of $RM(r,n)$ from a cryptographic point of view was introduced in [8]. It is defined as the maximum distance between t-*resilient* functions $\mathcal{R}_{t,n}$ and the r-th order Reed-Muller code $RM(r,n)$. That is,

$$\hat{\rho}(t,r,n) = \max_{f(\overline{x}) \in \mathcal{R}_{t,n}} d(f(\overline{x}), RM(r,n)).$$

It is clear that $0 \leq \hat{\rho}(t,r,n) \leq \rho(r,n)$. Siegentaler's inequality gives that $\hat{\rho}(t,r,n) \neq 0$, when $n > t + r + 1$. Note that $N_f = d(f, RM(1,n))$. In fact, in this terminology an upper bound on $\hat{\rho}(t,1,n)$ has been derived in [12,16,19].

For the new covering radius $\hat{\rho}(t,r,n)$ the authors in [8] derived some lower and upper bounds which are presented in Table 1. The entry $a - b$ means that $a \leq \hat{\rho}(t,r,n) \leq b$

Table 1. Numerical data of the bounds on $\hat{\rho}(t,r,n)$

	n	1	2	3	4	5	6	7
	$r=1$		0	2	4	12	26	56
	$r=2$			0	2	6	12-18	36-44
$t=0$	$r=3$				0	2	6-8	18-22
	$r=4$					0	2	6-8
	$r=5$						0	2
	$r=6$							0
	n	1	2	3	4	5	6	7
	$r=1$			0	4	12	24	56
	$r=2$				0	6	12-18	28-44
$t=1$	$r=3$					0	4-8	8-22
	$r=4$						0	4-8
	$r=5$							0
	n	1	2	3	4	5	6	7
	$r=1$				0	8	24	48-56
	$r=2$					0	12-16	24-44
$t=2$	$r=3$						0	8-22
	$r=4$							0

3 The Covering Radius of $RM(2,6)$ in the Set $\mathcal{R}_{t,6}$ for $t = 0, 1, 2$

We begin with the following lemma.

Lemma 2. *If a Boolean function $f(\overline{x})$ on $\mathbb{F}_2^6$ is at distance 18 from $RM(2,6)$, then its degree is 3.*

Proof. Suppose f is a Boolean function on $\mathbb{F}_2^6$ which is at distance 18 from $RM(2,6)$. Let $\overline{f} = (\overline{a}, \overline{b})$ be the truth table of $f(\overline{x})$, where $\overline{a}$ and $\overline{b}$ are two binary vectors of length 32. This means that we can represent $f(\overline{x})$ as follows:

$$f(\overline{x}) = a(\overline{x})(x_6 + 1) + b(\overline{x})x_6, \tag{1}$$

First we will prove that:

(i) $\overline{a}$ belongs to a coset of $RM(2,5)$ of minimal weight 6.

(ii) There exists a $\overline{u} \in RM(2,5)$, such that $\overline{b} + \overline{u}$ belongs to a coset of $RM(1,5)$ of minimal weight 12.

Consider the coset $C_{\overline{a}}$ of $RM(2,5)$, which contains the vector $\overline{a}$. Suppose that the minimal weight of $C_{\overline{a}}$ is less than 6 and let $\overline{u} \in RM(2,5)$, such that $d(\overline{a}, \overline{u}) = \min_{w \in RM(2,5)} d(\overline{a}, \overline{w}) < 6$. Let us also consider the coset $C_{\overline{b}+\overline{u}}$ of $RM(1,5)$, which contains the vector $\overline{b}+\overline{u}$. Since the covering radius of $RM(1,5)$ is 12 (see [1]), there exists a vector $\overline{v} \in RM(1,5)$, such that $d(\overline{b} + \overline{u}, \overline{v}) \le 12$. Since $d[(\overline{a}, \overline{b}), (\overline{u}, \overline{u} + \overline{v})] = d(\overline{a}, \overline{u}) + d(\overline{b} + \overline{u}, \overline{v})$, the vector $(\overline{u}, \overline{u} + \overline{v}) \in RM(2,6)$ (see [10]) is at distance less than 18 from $\overline{f} = (\overline{a}, \overline{b})$.

But this contradicts our assumption that $\overline{f}$ is at distance 18 from $RM(2,6)$. Hence $C_{\overline{a}}$ is with maximal possible minimal weight 6 (see [1,9]), i.e., $\min_{w \in RM(2,5)} d(\overline{a}, \overline{w}) = d(\overline{a}, \overline{u}) = 6$. Similarly, we can prove that $C_{\overline{b}+\overline{u}}$ has minimal weight 12. Note that *(ii)* holds for $\overline{a}$ as well, since $(\overline{b}, \overline{a})$ is at distance 18 from $RM(2,6)$.

Table II from [1] shows that by an appropriate affine transformation of the variables, any Boolean function of 5 variables can be reduced to one of the functions with 8 possible parts consisting of terms of degree greater than 2. It is easy to see that only in the following two cases the minimal weight of the corresponding coset of $RM(2,5)$ is 6:

1. $x_2x_3x_4x_5 + x_1x_2x_3 + x_1x_4x_5$;

2. $x_1x_2x_3 + x_1x_4x_5$.

Consulting Table I from [1], we can conclude that the first case is not possible for cosets of $RM(1,5)$ with minimal weight 12. Therefore in the representation (1) of f both $a(\overline{x})$ and $b(\overline{x})$ have degree 3.

Similar representation (with subfunctions having degree 3) holds for any other variable x_j, $j = 1, \ldots, 5$. Therefore all functions $f(\overline{x}|x_j = const)$, $j = 1, \ldots, 6$ are of degree 3 and hence f is of degree equal to 3. $\square$

Remark 1. J. Schatz proves in [13] that the covering radius of $RM(2,6)$ is 18 by constructing a coset which has a minimal weight 18. This coset can be written as $\overline{f} + RM(2,6)$, where $f(\overline{x}) = (x_1x_2x_3 + x_1x_4x_5 + x_2x_3 + x_2x_4 + x_3x_5)x_6 + (x_1x_2x_3 + x_1x_4x_5)(x_6 + 1)$.

Lemma 3. *The Boolean function* $g_1(\overline{x}) = x_1x_2x_3 + x_2x_4x_5 + x_3x_4x_6 + x_1x_2 + x_1x_3 + x_2x_5 + x_2 + x_3 + x_4 + x_5 + x_6$ *is 2-resilient and it is at distance 16 from* $RM(2,6)$.

Proof. By computing the Walsh transform and checking the spectrum, we see that $g_1(\overline{x})$ is 2-resilient. The cubic part of g_1 coincides with the Boolean function $f_5(\overline{x}) = x_1x_2x_3 + x_2x_4x_5 + x_3x_4x_6$ from [7], where Hou shows that the coset $\overline{f}_5 + RM(2,6)$ has minimal weight 16. Therefore the function $g_1(\overline{x})$ is at distance 16 from $RM(2,6)$. □

From $g_1(\overline{x})$, by using the translation $\overline{x} \to \overline{x} + \overline{\alpha}, \overline{\alpha} \in \mathbb{F}_2^6$ and complementing the values, we can obtain 128 functions, which possess the same properties as $g_1(\overline{x})$.

The function $g_1(\overline{x})$ from Lemma 3 achieves maximal possible nonlinearity 24 among the 1-resilient functions of 6 variables, i.e., it is at distance 24 from $RM(1,6)$. This holds since the $g_1(\overline{x})$ is a plateaued function with amplitude 16.

Theorem 2. *The covering radius of* $RM(2,6)$ *in the sets* $\mathcal{R}_{t,6}, t = 0,1,2$ *is 16, i.e.,*

$$\hat{\rho}(t,2,6) = 16, \quad t = 0,1,2.$$

Proof. According to Lemma 2, any Boolean function at distance 18 from $RM(2,6)$ has degree 3. By using the results in [7, p.113] we see that the unique orbit of the general linear group $GL(6,2)$ in $RM(3,6)/RM(2,6)$, which has as a representative a coset of minimal weight 18, does not contain balanced functions. Therefore there exist no resilient functions at distance 18 from $RM(2,6)$. On the other hand by Lemma 3 there exists a 2-resilient function at distance 16 from that code. To complete the proof we only need the obvious inclusion $\mathcal{R}_{t,n} \subset \mathcal{R}_{t-1,n}$. □

4 The Covering Radius of $RM(2,7)$ in the Set $\mathcal{R}_{t,7}$ for $t = 0,1,2,3$

First, we shall prove that the covering radius of $RM(2,7)$ in the set $\mathcal{R}_{3,7}$ is 32. Recall that due to the Siegenthaler's upper bound the degree of any 3-resilient function on $\mathbb{F}_2^7$ must be at most 3. From now on, when we say that a Boolean function f is linearly equivalent to $\tilde{f}$, we actually mean that f can be reduced by an invertible linear transformation of the variables to the Boolean function $\tilde{f}$.

The following lemma summarizes the results from Theorem 8.1 and Theorem 8.3 from [7].

Lemma 4. *Any Boolean function on $\mathbb{F}_2^7$ of degree 3 is linearly equivalent to a function with cubic part among:*

$$f_2 = x_1x_2x_3;$$
$$f_3 = x_1x_2x_3 + x_2x_4x_5;$$
$$f_4 = x_1x_2x_3 + x_4x_5x_6;$$
$$f_5 = x_1x_2x_3 + x_2x_4x_5 + x_3x_4x_6;$$
$$f_6 = x_1x_2x_3 + x_1x_4x_5 + x_2x_4x_6 + x_3x_5x_6 + x_4x_5x_6;$$
$$f_7 = x_1x_2x_7 + x_3x_4x_7 + x_5x_6x_7;$$
$$f_8 = x_1x_2x_3 + x_4x_5x_6 + x_1x_4x_7;$$
$$f_9 = x_1x_2x_3 + x_2x_4x_5 + x_3x_4x_6 + x_1x_4x_7;$$
$$f_{10} = x_1x_2x_3 + x_4x_5x_6 + x_1x_4x_7 + x_2x_5x_7;$$
$$f_{11} = x_1x_2x_3 + x_1x_4x_5 + x_2x_4x_6 + x_3x_5x_6 + x_4x_5x_6 + x_1x_6x_7;$$
$$f_{12} = x_1x_2x_3 + x_1x_4x_5 + x_2x_4x_6 + x_3x_5x_6 + x_4x_5x_6 + x_1x_6x_7 + x_2x_4x_7.$$

Let μ_j be the minimal weight of the coset $\overline{f}_j + RM(2,7)$, $2 \leq j \leq 12$. Then $\mu_2 = 16, \mu_3 = 24, \mu_4 = 28, \mu_5 = 32, \mu_6 = 36, \mu_7 = 28, \mu_8 = 32, \mu_9 = 36, \mu_{10} = 36, \mu_{11} = 40$ and $\mu_{12} = 36$.

Lemma 5. *Let f be a Boolean function on $\mathbb{F}_2^7$ of degree 3, linearly equivalent to a function with cubic part among f_4, f_6, f_8, f_{10}, f_{11} or f_{12}. Then f cannot be 2-resilient.*

Proof. Suppose that f is 2-resilient and let $\widetilde{f}$ be the image of f under an invertible linear transformation, such that the cubic part of $\widetilde{f}$ belongs to one of the classes of $\{f_4, f_6, f_8, f_{10}, f_{11}, f_{12}\}$. By [20, Lemma 2] and [12] the Walsh transform values of $\widetilde{f}$ are divisible by 16. Now applying Theorem 1 we get

$$W_{\widetilde{f}}(0,0,\ldots,1) + W_{\widetilde{f}}(0,0,\ldots,0) = 128 - 4 \cdot wt(\widetilde{f}_{(0,0,\ldots,1)}).$$

Thus, 4 is a divisor of $wt(\widetilde{f}_{(0,0,\ldots,1)})$. If $\widetilde{f} \in \{f_4, f_8, f_{10}\}$ the function $\widetilde{f}_{(0,0,\ldots,1)}$ belongs to the coset $\overline{f}_4 + RM(2,6)$, if $\widetilde{f} \in \{f_6, f_{11}, f_{12}\}$ then $\widetilde{f}_{(0,0,\ldots,1)}$ belongs to the coset $\overline{f}_6 + RM(2,6)$ (recall that the subfunction $\widetilde{f}_{(0,0,\ldots,1)}$ is obtained by setting $x_7 = 0$). But from [7, .p.113] we see that there is no weight divisible by 4 in these cosets, which leads to a contradiction. □

Lemma 6. *Let f be a Boolean function on $\mathbb{F}_2^7$ of degree 3 linearly equivalent to a function with cubic part equal to $f_9 = x_1x_2x_3 + x_2x_4x_5 + x_3x_4x_6 + x_1x_4x_7$. Then f cannot be 3-resilient.*

Proof. We first prove by contradiction that a function $\widetilde{f}$ of the form $\widetilde{f} = f_9 + g(\overline{x})$, where $g(\overline{x}) \in RM(2,7)$, cannot be 3-resilient. Suppose the function $\widetilde{f}$ is 3-resilient. Notice that the weight of $\widetilde{f}_{\overline{w}}$ is even, for each w with Hamming weight at most 3. By our assumption and Theorem 1, $W_{\tilde{f}}(\overline{w}) = 0$ for all $\overline{w}$ with Hamming weight at most 3. Consider the following vectors $\overline{w}_i$, for $i = 1, \ldots, 4$ with Hamming weight 4:

$$(0001111), (1010011), (1100101), (0110110).$$

These vectors are the only ones of Hamming weight 4 such that the corresponding function $\tilde{f}_w$ from Theorem 1 has maximum degree and thus has odd Hamming weight. Applying Theorem 1 and Definition 1, those vectors have Walsh transform values which are in absolute value equal to $32k$ with k an odd integer.

The vectors $\tilde{\omega}$ for which the set $\{\theta | \ \theta < \tilde{\omega}, \ wt(\tilde{\omega}) > 4 \ \text{and} \ |W_{\tilde{f}}(\theta)| = 32k,$ with k odd$\}$ has odd cardinality, will also have absolute value of the Walsh transform equal to $32k$ with k odd, based on the same arguments.

So, we also get the following vectors: 12 vectors of Hamming weight 5, formed by extending each of the previous vectors of Hamming weight 4:

$$(0011111), (0101111), (1001111),$$
$$(1110011), (1011011), (1010111),$$
$$(1110101), (1101101), (1100111),$$
$$(1110110), (0111110), (0110111),$$

and four vectors of Hamming weight 6:

$$(1110111), (1111011), (1111101), (1111110).$$

In total we have 20 vectors which have nonzero Walsh transform values divisible by 32. The Parseval equation $\sum_\omega W_{\tilde{f}}^2(\omega) = 2^{14}$, leads to a contradiction.

However, because resiliency is not a linear invariant property, this proof does not imply that any other function which is linearly equivalent to a function from the class of f_9, cannot be 3-resilient. If there exists a 3-resilient function which is linearly equivalent to a function from the coset of f_9, it should be a plateaued function with amplitude 32. This is explained by the fact that for a 3-resilient function the Walsh transform values should be divisible by 32 and the maximum of their absolute values cannot be greater or equal to 64 (otherwise the nonlinearity would be less or equal to 32, which contradicts the minimal weight of the class f_9). As the frequency distribution of the Walsh transform values is a linear invariant property, it suffices to show that there are no plateaued functions with amplitude 32 in the set $\{f_9 + g(\overline{x}) \mid g(\overline{x}) \in RM(2,7)\}$. By computer search over $RM(2,7)/RM(1,7)$ (with complexity $c\,2^{21}$ for a small constant c, which is comparable with the weight distribution problem for the cosets of $RM(2,6)$), we have not found any plateaued function with amplitude 32.   □

Lemma 7. *The Boolean function* $g_2(\overline{x}) = x_1x_2x_3 + x_2x_4x_5 + x_3x_4x_6 + x_1x_2 + x_1x_3 + x_2x_5 + x_2 + x_3 + x_4 + x_5 + x_6 + x_7$ *is 3-resilient and it is at distance 32 from* $RM(2,7)$.

Proof. Since $g_2(x_1, \dots, x_7) = g_1(x_1, \dots, x_6) + x_7$ and g_1 achieves the covering radius of $RM(2,6)$ in $\mathcal{R}_{2,6}$ by Lemma 3, $g_2(\overline{x})$ is 3-resilient and it is at distance 32 from $RM(2,7)$.   □

The function $g_2(\overline{x})$ is plateaued with amplitude 32. Therefore the distance between g_2 and $RM(1,7)$ is the maximal possible, namely 48.

Theorem 3. *The covering radius of $RM(2,7)$ in the set $\mathcal{R}_{3,7}$ is 32.*

Proof. Lemma 5 and Lemma 6 imply that if a Boolean function is at distance greater than 32 from $RM(2,7)$, it cannot be 3-resilient. On the other hand, by Lemma 7 there exists a 3-resilient function at distance 32 from $RM(2,7)$, which completes the proof.   □

Theorem 4. *The covering radius of the Reed-Muller code $RM(2,n)$ in the set $\mathcal{R}_{n-4,n}$ is bounded from below by 2^{n-2}, when $n \geq 6$, i.e.*

$$2^{n-2} \leq \hat{\rho}(n-4,2,n).$$

Proof. The proof is by induction on n. Lemma 3 is in fact the basis of the induction and then we proceed in a similar way as in Lemma 7.

In the following propositions we improve the lower bounds for the covering radius in the sets $\mathcal{R}_{t,7}$ for $t = 0, 1$.

Proposition 1. *The Boolean function $g_3(\overline{x}) = x_1x_2x_3 + x_1x_4x_5 + x_2x_4x_6 + x_3x_5x_6 + x_4x_5x_6 + x_1x_6x_7 + x_1 + x_2$ is 0-resilient (balanced) and it is at distance 40 from $RM(2,7)$.*

Proposition 2. *The Boolean function $g_4(\overline{x}) = x_1x_2x_3 + x_1x_4x_7 + x_2x_4x_5 + x_3x_4x_6 + x_1x_7 + x_5 + x_6 + x_7$ is 1-resilient and it is at distance 36 from $RM(2,7)$.*

In Table 2 below we present the numerical values of $\hat{\rho}(t,2,n)$ that are obtained from Theorem 2, Theorem 3, Proposition 1 and Proposition 2.

Table 2. Numerical results for $\hat{\rho}(t,2,n)$ from Theorem 2, Theorem 3, Proposition 1 and Proposition 2 (marked by (a), (b), (c) and (d) respectively). The entry $a - b$ means that $a \leq \hat{\rho}(t,r,n) \leq b$

	n	1	2	3	4	5	6	7
$t = 0$	$r = 2$				0	2	6	$16^{(a)}$	$40^{(c)}$-44
$t = 1$	$r = 2$					0	6	$16^{(a)}$	$36^{(d)}$-44
$t = 2$	$r = 2$						0	$16^{(a)}$	$32^{(b)}$-44
$t = 3$	$r = 2$							0	$32^{(b)}$

5   Conclusions and Open Problems

In this paper we study the covering radius in the set of resilient functions, which has been defined by Kurosawa *et al.* in [8]. This new concept is meaningful to cryptography especially in the context of the new class of algebraic attacks on stream ciphers proposed by Courtois and Meier at Eurocrypt 2003 [4] and Courtois at Crypto 2003 [5]. In order to resist such attacks the combining Boolean function should be at high distance from lower order functions.

Using results from coding theory, we establish exact values of the covering radius in dimension 6 and improve the bounds in dimension 7 for the covering radius in the set of t-resilient functions with $t \leq 3$ of the second order Reed-Muller code. We also generalize our methods to find a new lower bound for the covering radius in the set of $(n-4)$-resilient functions of the second order Reed-Muller code. An open problem that still remains, is the improvement of the rest of the lower bounds from Table 1. It would also be interesting to generalize our results for higher dimensions.

Acknowledgements. The joint work on the paper started during the visit of Yuri Borissov at the K.U.Leuven. He would like to thank for the hospitality and the creative working environment at COSIC. Part of the work was done during the visit of Svetla Nikova at the Ruhr University, Bochum. The authors were partially supported by Concerted Research Action GOA-MEFISTO-666 of the Flemish Government and An Braeken is research assistant of the Fund for Scientific research - Flanders (Belgium).

References

1. E. Berlekamp, L. Welch, Weight Distribution of the Cosets of the (32,6) Reed-Muller Code, *IEEE IT*, vol. 18, January 1972, pp. 203–207.
2. C. Carlet, E. Prouff, On Plateaued Functions and their Constructions, *FSE'03*, Lund, Sweden, February 24–26, 2003, pp. 57–78.
3. N. Courtois, Higher Order Correlation Attacks, *XL* Algorithm and Cryptanalysis of *Toyocrypt*, ePrint Archive 2002/087, 2002.
4. N. Courtois, W. Meier, Algebraic Attacks on Stream Ciphers with Linear Feedback, *Eurocrypt'03*, LNCS 2656, Springer-Verlag 2003, pp. 345–359.
5. N. Courtois, Fast Algebraic Attacks on Stream Ciphers with Linear Feedback *Crypto'03*, LNCS 2729, Springer-Verlag 2003, pp. 176–194.
6. N. Courtois, A. Klimov, J. Patarin, A. Shamir, Efficient Algorithms for Solving Overdefined Systems of Multivariate Polynomial Equations, *Eurocrypt'00*, LNCS 1807, Springer-Verlag, 2000, pp. 392–407.
7. X. D. Hou, $GL(m,2)$ Acting on $R(r,m)/R(r-1,m)$, *Discrete Mathematics* vol. 149, 1996, pp. 99–122.
8. K. Kurosawa, T. Iwata, T. Yoshiwara, New Covering Radius of Reed-Muller Codes for t-Resilient Functions, *SAC'01*, LNCS 2259, Springer-Verlag 2001, pp. 75–86.
9. A. McLoughlin, The Covering Radius of the $(m-3)$-rd Order Reed-Muller Codes and a Lower Bound on the $(m-4)$-th Order Reed-Muller Codes, *SIAM J. Appl. Mathematics*, vol. 37, No. 2, October 1979, pp. 419–422.

10. F. J. MacWilliams, N.J.A. Sloane, *The Theory of Error-Correcting Codes*, North-Holland Publishing Company, 1977.
11. C. Carlet, P. Sarkar, Spectral Domain Analysis of Correlation Immune and Resilient Boolean Functions, *Finite Fields and Applications*, vol. 8, No. 1, January 2002, pp. 120–130.
12. P. Sarkar, S. Maitra, Nonlinearity Bounds and Constructions of Resilient Boolean Functions, *Crypto'00*, LNCS 1880, Springer-Verlag, 2000, pp. 515–532.
13. J. Schatz, The Second Order Reed-Muller Code of Length 64 has Covering Radius 18, *IEEE IT*, vol. 27, 1981, pp. 529–530.
14. T. Siegenthaler, Correlation-Immunity of Non-linear Combining Functions for Cryptographic Applications, *IEEE IT*, vol. 30, No. 5, 1984, pp. 776–780.
15. T. Siegenthaler, Decrypting a Class of Stream Ciphers Using Cyphertext Only, *IEEE Trans. Comp.*, vol 34, No. 1, 1985, pp. 81–85.
16. Y. Tarannikov, On Resilient Boolean Functions with Maximal Possible Nonlinearity, *Indocrypt'00*, LNCS 1977, Springer-Verlag 2000, pp. 19–30.
17. X. Guo-Zhen, J. Massey, A Spectral Characterization of Correlation-Immune Combining Functions, *IEEE IT*, vol. 34, No. 3, May 1988, pp. 569–571.
18. Y. Zheng, X. M. Zhang, Plateaued Functions, *ICICS'99*, LNCS 1726, Springer-Verlag 1999, pp. 284–300.
19. Y. Zheng, X. M. Zhang, Improved Upper Bound on the Nonlinearity of High Order Correlation Immune Functions, *SAC'00*, LNCS 2012, Springer-Verlag 2001, pp. 262–274.
20. Y. Zheng, X. M. Zhang, New Results on Correlation Immunity, *The 3rd International Conference on Information Security and Cryptography (ICISC'00)*, LNCS 2015, Springer-Verlag 2001, pp. 49–63.

Degree Optimized Resilient Boolean Functions from Maiorana-McFarland Class

Enes Pasalic

INRIA, Domaine de Voluceau, BP 105 - 78153, Le Chesnay Cedex, FRANCE;
enes.pasalic@inria.fr

Abstract. In this paper we present a construction method of degree optimized resilient Boolean functions with very high nonlinearity. We present a general construction method valid for any $n \geq 4$ and for order of resiliency t satisfying $t \leq n - 3$. The construction is based on the modification of the famous Marioana-McFarland class in a controlled manner such that the resulting functions will contain some extra terms of high algebraic degree in its ANF including one term of highest algebraic degree. Hence, the linear complexity is increased, the functions obtained reach the Siegentheler's bound and furthermore the nonlinearity of such a function in many cases is superior to all previously known construction methods. This construction method is then generalized to the case of vectorial resilient functions, that is $\{F\} : \mathbb{F}_2^n \mapsto \mathbb{F}_2^m$, providing functions of very high algebraic degree almost reaching the Siegenthaler's upper bound.

Keywords: Boolean Function, Resiliency, Nonlinearity, Algebraic Degree.

1 Introduction

Resilient Boolean functions have important applications in a nonlinear combiner model of stream cipher. Construction of resilient Boolean functions, with as high nonlinearity as possible, has been an important research question from mid eighties.

More generally, the main cryptographic criteria of Boolean functions used in either nonlinear filtering generator or in nonlinear combining scenario have more or less been identified. A function used in such an application should posses a high algebraic degree (actually the best option is a function reaching the Siegenthaler's bound) to increase the linear complexity of the keystream sequence. In addition, very recently algebraic attacks based on the low degree approximation/decomposition of Boolean functions has been introduced [9,10]. In order to circumvent correlation attacks the function should have a modest order of resiliency and a high nonlinearity.

Not all of these criteria can be satisfied simultaneously and concerning the resiliency order, denoted by t, Siegenthaler [23] proved that $t \leq n - d - 1$ for

K.G. Paterson (Ed.): Cryptography and Coding 2003, LNCS 2898, pp. 93–114, 2003.
© Springer-Verlag Berlin Heidelberg 2003

balanced functions, where d denotes the algebraic degree. Such a function, reaching this bound, is called *degree optimized*. Recently (since 2000), a lot of new results have been published in a very short time which include nontrivial nonlinearity (upper) bounds [21,24,27,2,4] and construction of resilient functions attaining either those bounds or reaching very close. In such a scenario, getting resilient functions with a nonlinearity, that has not been demonstrated earlier, is becoming harder.

Considering a Boolean function on n variables with order of resiliency $(t > \frac{n}{2} - 2)$ and attaining maximum possible nonlinearity, generalized construction methods have been proposed in [24,18]. Construction of highly nonlinear functions with lower order of resiliency has been discussed in [20,16]. But unfortunately none of these methods is general in the sense that they would be able to generate a function for any input size n and any order of resiliency t. This is only true for the Maiorana-McFarland class, where on the other hand the major drawback of this technique is that generally it does not yield degree optimized functions.

Furthermore, the nonlinearity value for the functions in this class reaches its maximum value for a high resiliency order, whereas in case of low or modest resiliency the nonlinearity is very high but in most of the cases not reaching the upper bound on nonlinearity. However, it is not clear whether there exist classes of functions reaching this bound, especially for large ambient spaces.

In this paper, we derive a new class of degree optimized functions by modifying the Maiorana-McFarland class of resilient functions with respect to the function's algebraic degree. Actually, we prove that this new class always exists exactly in those cases when the standard Maiorana-McFarland technique cannot generate degree optimized functions whatever is the choice of its affine subfunctions. The procedure of obtaining a degree optimized function, starting with a function of relatively low degree from the Maiorana-McFarland class, may be viewed as a simple adding of the terms of high algebraic order in a certain manner. The functions obtained in such a way will in many cases exhibit the best known cryptographic criteria, more precisely for a given input space n, order of resiliency t our functions attain in many cases the highest nonlinearity value for degree optimized functions, that is for $d = n - t - 1$.

The technique proposed for Boolean function is then extended to the case of vectorial Boolean functions, i.e. mappings from $\mathbb{F}_2^n$ to $\mathbb{F}_2^m$. In this scenario the success of our method will partly depend on the existence of suitable linear codes. When using the simplex code our method is proved to generate the functions of very high algebraic degree. Unfortunately, this feature is traded-off against a smaller resiliency order and a small decrease in nonlinearity (that can be neglected) when compared to the Maiorana-McFarland construction extended to vectorial functions.

The rest of the paper is organized as follows. Section 2 introduces basic definitions and cryptographic criteria relevant for Boolean mappings. Here, we also review the most important construction techniques and briefly discuss their cryptographic properties. In Section 3 a deeper background on the Maiorana-

McFarland class is presented. Then we propose a modification of this class, which results in a new class of degree optimized resilient functions of very high nonlinearity. In certain cases, as illustrated by two examples, the functions designed by means of our method exhibit the best known trade-off between design parameters which has not been demonstrated before.

In Section 4 we extend our method presented in Section 3 to construct nonlinear vectorial resilient functions of very high algebraic degree. The algebraic degree of these functions is very close to the Siegenthaler's bound implying that such functions are much more resistant to attacks which exploit a low linear complexity of the keystream sequence. Finally, some concluding remarks are given in Section 5.

2 Preliminaries

A Boolean function on n variables may be viewed as a mapping from $\{0,1\}^n$ into $\{0,1\}$. A Boolean function $f(x_1,\ldots,x_n)$ is also interpreted as the output column of its *truth table* f, i.e., a binary string of length 2^n,

$$f = [f(0,0,\cdots,0), f(1,0,\cdots,0), f(0,1,\cdots,0), \ldots, f(1,1,\cdots,1)].$$

The *Hamming distance* between S_1, S_2 is denoted by $d(S_1, S_2)$, i.e.,

$$d(S_1, S_2) = \#(S_1 \neq S_2).$$

Also the *Hamming weight* or simply the weight of a binary string S is the number of ones in S. This is denoted by $wt(S)$. An n-variable function f is said to be *balanced* if its output column in the truth table contains equal number of 0's and 1's (i.e., $wt(f) = 2^{n-1}$).

Addition operator over $GF(2)$ is denoted by $\oplus$, and if no confusion is to arise we use the usual addition operator $+$. Sometimes, abusing the notation, $''+''$ is also used for a bitwise vector addition and in such cases we emphasize such an ambiguity. The Galois field of order 2^n will be denoted by $\mathbb{F}_{2^n}$ and the corresponding vector space by $\mathbb{F}_2^n$. An n-variable Boolean function $f(x_1,\ldots,x_n)$ can be considered to be a multivariate polynomial over $\mathbb{F}_2$. This polynomial can be expressed as a sum of products representation of all distinct k-th order products $(0 \leq k \leq n)$ of the variables. More precisely, $f(x_1,\ldots,x_n)$ can be written as

$$f(x_1,\ldots,x_n) = \sum_{u \in \mathbb{F}_2^n} \lambda_u \left(\prod_{i=1}^{n} x_i^{u_i} \right), \quad \lambda_u \in \mathbb{F}_2, u = (u_1,\ldots,u_n). \quad (1)$$

This representation of f is called the *algebraic normal form* (ANF) of f. The *algebraic degree* of f, denoted by $deg(f)$ or sometimes simply d, is the maximal value of the Hamming weight of u such that $\lambda_u \neq 0$. There is a one-to-one correspondence between the truth table and the ANF via so called inversion formulae.

The set of all Boolean functions in n variables is denoted by $\mathcal{B}_n$. For any $0 \le b \le n$ an n-variable function is called non degenerate on b variables if its ANF contains exactly b distinct input variables. Functions of degree at most one are called *affine* functions. An affine function with constant term equal to zero is called a *linear* function. The set of all n-variable affine (respectively linear) functions is denoted by $\mathcal{A}_n$ (respectively $\mathcal{L}_n$). The *nonlinearity* of an n-variable function f is

$$\mathcal{N}_f = min_{g \in \mathcal{A}_n}(d(f, g)). \tag{2}$$

That is, the nonlinearity is the distance from the set of all n-variable affine functions.

For $x, \omega \in \mathbb{F}_2^n$, the dot or inner product is defined as $x \cdot \omega = x_1\omega_1 + \ldots + x_n\omega_n$. Let $f(x)$ be a Boolean function on n variables. Then the *Walsh transform* of $f(x)$ is a real valued function over $\mathbb{F}_2^n$ which is defined as

$$W_f(\omega) = \sum_{x \in \mathbb{F}_2^n} (-1)^{f(x) \oplus x \cdot \omega}. \tag{3}$$

In terms of Walsh spectra, the nonlinearity of f is given by

$$\mathcal{N}_f = 2^{n-1} - \frac{1}{2} \max_{\omega \in \mathbb{F}_2^n} |W_f(\omega)|. \tag{4}$$

In [25], an important characterization of resilient functions has been provided. A function $f(x_1, \ldots, x_n)$ is *t-resilient* iff its Walsh transform satisfies

$$W_f(\omega) = 0, \text{ for } 0 \le wt(\omega) \le t.$$

As the notation used in [20,21], by an (n, t, d, σ) function we denote an n-variable, t-resilient function with degree d and nonlinearity σ.

Besides the Maiorana-McFarland class, which is thoroughly treated in the next section, the design of highly nonlinear resilient functions is limited to a few iterative techniques proposed recently. Recursive by its nature, they generate an infinite sequence of degree optimized functions [24,18], where in each iterative step the input variable space is increased by 3 whereas the resiliency order increases by 2. These methods have proved to be very efficient for a large order of resiliency $t > n/2 - 2$ (the parameters of resulting sequence depends also on the properties of input function) but still for a low order of resiliency and for arbitrary n it seems that the only efficient construction technique is the Maiorana-McFarland method.

In certain cases, highly nonlinear functions may be obtained using the theory of finite fields (namely through trace mappings), but this method is not general especially when dealing with resilient functions.

3   A Maiorana-McFarland Class Revisited

We first give a brief background on the functions in Maiorana-McFarland class.

Definition 1. *For any positive integers s, k such that $n = s + k$ a Maiorana-McFarland function is a Boolean function on $\mathbb{F}_2^n$ defined by,*

$$f(y, x) = \phi(y) \cdot x + h(y), \quad x \in \mathbb{F}_2^k, y \in \mathbb{F}_2^s. \tag{5}$$

Here, h is any Boolean function on $\mathbb{F}_2^s$ and ϕ is any mapping from $\mathbb{F}_2^s$ to $\mathbb{F}_2^k$.

Notice that in the definition of Maiorana-McFarland class ϕ is an arbitrary function from $\mathbb{F}_2^s$ to $\mathbb{F}_2^k$. By imposing the restriction that ϕ is injective we must have $k \geq s$. A special case of this method is a construction of bent functions. Taking $n = 2k$, i.e., $s = k$ and any bijective mapping ϕ (ϕ is a permutation on $\mathbb{F}_2^s$) will result in a bent function. It is also easy to verify that requiring ϕ to be such that $wt(\phi(y)) \geq t + 1$ for any $y \in \mathbb{F}_2^s$ correspond to a t-resilient function f.

Construction of resilient functions by concatenating the truth tables of small affine functions was first described in [1]. The concatenation simply means that the truth tables of the functions are merged. For instance, for $f = f_1 \| f_2$ the upper half part of the truth table of f correspond to f_1 and the lower part to f_2. However, the analysis has been made in terms of orthogonal arrays. This construction has been revisited in more details in [22] where the authors considered the algebraic degree and nonlinearity of the functions.

Moreover, in [7], a construction of functions with concatenation of small affine functions, specifying the conditions on ϕ has been discussed. All these constructions use small affine functions exactly once (ϕ is injective) and they belong to the Maiorana-McFarland class.

A major advancement in this area has been done in [20], where each affine function has been used more than once in the form of a composition with nonlinear functions. Two generalized algorithms, called Algorithm A and Algorithm B in [20] outline a framework in this direction. However, the nonlinearity of such functions is decreased in comparison to the concatenation of only affine functions. In [3], the concatenation of affine functions is replaced by concatenation of quadratic functions but this method does not produce the degree optimized functions in general.

Let us investigate the consequences of the condition that $\phi : \mathbb{F}_2^s \mapsto \mathbb{F}_2^k$ is injective of weight greater than t, for $t \geq 0$. As already noticed [7], there is a binomial relationship between the parameters involved (note that $n = s + k$),

$$\binom{k}{t+1} + \binom{k}{t+2} + \cdots + \binom{k}{k} \geq 2^{n-k}. \tag{6}$$

Hence, for $n = s + k$ such that (6) holds, then there will exist injective mappings ϕ and consequently the function f will be a t-resilient functions with nonlinearity $\mathcal{N}_f = 2^{n-1} - 2^{k-1}$ [7] (see also Theorem 1 below). Obviously, the aim is to minimize the parameter k with respect to (6). Therefore, for fixed integers t and $n = s + k$, $0 \leq t < n$ we define

$$\mathbf{k} = \min_{t<k} \left\{ k \mid \sum_{i=0}^{k-(t+1)} \binom{k}{t+1+i} \geq 2^{n-k} \right\}. \tag{7}$$

The ANF of f, as defined by (5), is more easily comprehend when f is represented as a concatenation of linear functions from $\mathcal{L}_{\mathbf{k}}$. Also, for practical implementations and for exact nonlinearity calculation it is of relevance to specify the mapping ϕ [1] (or at least to prove the existence of such mapping). Let for any $0 \leq t < k$, $\mathcal{L}_{\mathbf{k}}^t$ denote the set of all linear functions on $\mathbb{F}_2^{\mathbf{k}}$ non-degenerated on at least $t + 1$ variables, that is,

$$\mathcal{L}_{\mathbf{k}}^t = \{ \varphi_c(x) = c \cdot x \mid c \in \mathbb{F}_2^{\mathbf{k}}, wt(c) > t \}. \tag{8}$$

Then, the following properties have been proved in [7].

Theorem 1. [7] *For any $0 \leq t < n$, let $\mathbf{k}$ be defined by (7) and $\mathcal{L}_{\mathbf{k}}^t$ by (8). Let us choose $2^{n-\mathbf{k}}$ distinct linear functions in $\mathcal{L}_{\mathbf{k}}^t$, each being labeled by an element of $\mathbb{F}_2^{n-\mathbf{k}}$ as follows:*

$$\tau \in \mathbb{F}_2^{n-\mathbf{k}} \longleftrightarrow \ell_{[\tau]} \in \mathcal{L}_{\mathbf{k}}^t, \text{ where } [\tau] = \sum_{i=1}^{n-\mathbf{k}} \tau_i 2^{i-1}.$$

Then the Boolean function defined for all $(y, x) \in \mathbb{F}_2^{n-\mathbf{k}} \times \mathbb{F}_2^{\mathbf{k}}$ by

$$f(y, x) = \sum_{\tau \in \mathbb{F}_2^{n-\mathbf{k}}} (y_1 + \tau_1 + 1) \cdots (y_{n-\mathbf{k}} + \tau_{n-\mathbf{k}} + 1) \ell_{[\tau]}(x), \tag{9}$$

is a t-resilient function with nonlinearity $\mathcal{N}_f = 2^{n-1} - 2^{\mathbf{k}-1}$. In general $\deg(f) \leq n - \mathbf{k} + 1$ with equality if there exists a variable x_i, $i = 1, \ldots, \mathbf{k}$, which occurs an odd number of times in $\ell_{[\tau]}(x)$ when τ runs through $\mathbb{F}_2^{n-\mathbf{k}}$.

Remark 1. *The authors in [7] only consider a concatenation of linear functions. A more general approach is to use affine functions instead, that is to replace $\ell_{[\tau]}(x)$ by $a_{[\tau]}(x)$. However none of the cryptographic parameters is affected by this replacement so one can equally well consider only linear functions. Referring to Definition 1 and the equation (9) one can define f to be a concatenation of affine functions, that is,*

$$f(y, x) = \sum_{\tau \in \mathbb{F}_2^{n-\mathbf{k}}} (y_1 + \tau_1 + 1) \cdots (y_{n-\mathbf{k}} + \tau_{n-\mathbf{k}} + 1) a_{[\tau]}(x), \tag{10}$$

where $a_{[\tau]}(x) = \phi(\tau) \cdot x + h(\tau)$. Then the set $\mathcal{L}_{\mathbf{k}}^t$ above is replaced by $\mathcal{A}_{\mathbf{k}}^t = \{ a_c(x) = c \cdot x + b_c \mid c \in \mathbb{F}_2^{\mathbf{k}}, b_c \in \mathbb{F}_2, wt(c) > t \}$.

---

[1] It turns out that both the autocorrelation properties as well as the algebraic degree of function f will depend on the choice of ϕ.

Moreover it is easy to characterize the zeros of its Walsh-spectrum and its propagation characteristics, which has been studied in details [6].

Thus, according to Theorem 1 the algebraic degree of f is upper bounded by $deg(f) \leq n - \mathbf{k} + 1$. On the other hand the degree of any t-resilient function satisfies $deg(f) \leq n - t - 1$. An obvious consequence is that the functions in the Maiorana-McFarland class are in general not degree optimized. This is always true for any $t < \mathbf{k} - 2$. Especially, since $\mathbf{k} \geq \lceil \frac{n+1}{2} \rceil$ for any $t > 0$, the functions in the Maiorana-McFarland class cannot be degree optimized for $t < \lceil \frac{n+1}{2} \rceil - 2$. Hence, for a relatively low order of resiliency this class never provides degree optimized functions and the linear complexity of the keystream sequence generated by such a function cannot be high.

3.1   Degree Optimization of Maiorana-McFarland Class

In our method discussed below, to construct an n-variable t-resilient function, we use a set of $2^k - 1$ affine functions (each exactly once) in k-variables and exactly one specific nonlinear t-resilient function on k variables. Here k is the design parameter which can be calculated for any n and t through the formula (12) below. Hence the function obtained through our method may be represented as a concatenation of the truth tables as $g = a_1 || \cdots || a_{i-1} || \pi || a_{i+1} || \cdots || a_{2^{n-k}}$, where each a_j, $j \neq i$ is affine function in $\mathcal{A}_k$ and π is a nonlinear function in $\mathcal{B}_k$.

For our purpose we are interested in the restriction of the set $\mathcal{A}_k^t$. It is more convenient to consider the elements of $\mathbb{F}_2^k$ than affine functions. Hence, for $k > t$ and $t \geq 0$, for some fixed $\eta = (0, \ldots, 0, \overset{i_1}{\overbrace{1}}, \ldots, \overset{i_{t+1}}{\overbrace{1}})$ of weight $t + 1$, we define the set $S_k^t(\eta) \subset \mathbb{F}_2^k$ as follows:

$$S_k^t(\eta) = \eta \cup \{ \gamma \mid \gamma \in \mathbb{F}_2^k, wt(\gamma) > t \text{ and } \exists j \in [1, t+1] \text{ s.t. } \gamma_{i_j} = 0 \}. \quad (11)$$

In other words, the vector $\eta \in S_k^t(\eta)$, whose coordinates $\eta_{i_j} = 1$ if and only if $j \in [1, t+1]$, is not covered by any $\beta \in S_k^t(\eta) \setminus \{\eta\}$, where the relation $\eta \preccurlyeq \beta$ means that β covers η, i.e., $\eta_i \leq \beta_i$ for all i in the range $[1, k]$.

Clearly the cardinality of this set is $\#S_k^t(\eta) = \sum_{i=0}^{k-(t+1)} \binom{k}{t+1+i} - 2^{k-t-1} + 1$.

To verify this, w.l.o.g. assume that $\eta = (\overset{t+1}{\overbrace{1, \ldots, 1}}, 0, \ldots, 0)$. Keeping the first $t + 1$ coordinates fixed there will be exactly $2^{k-t-1} - 1$ vectors in $\mathbb{F}_2^k \setminus \{\eta\}$ which cover η. Therefore, for fixed integers t and $n = s + k$, $0 \leq t < n$ we define

$$\mathbf{k} = \min_{t < k} \left\{ k \mid \sum_{i=0}^{k-(t+1)} \binom{k}{t+1+i} - 2^{k-t-1} + 1 \geq 2^{n-k} \right\}. \quad (12)$$

Henceforth, we assume that k is always chosen to be the minimum positive integer satisfying (12), that is, $k = \mathbf{k}$. Furthermore f will always denote the function given in Remark 1, that is f will "refer" to the standard Maiorana-McFarland class.

In the sequel we will prove that degree optimized functions of the same nonlinearity and resiliency order may be obtained for any n and t, t satisfying $t < k - 2$, by adding exactly $\binom{n-k}{i}$ terms of order $n - t - i - 1$ in the ANF of function f, where $i = 0, \ldots, n - k$. It is well known that any function of resiliency order $t > n - 3$ must be linear. Hence, we only consider the cases $n \geq t + 3$.

Construction 1. *Let t be a nonnegative integer, and let $n \geq t + 3$ be the input variable space. For a positive integer k defined by equation (12) and for a fixed $\eta \in \mathbb{F}_2^k$ of weight $t + 1$ (with $\eta_{i_j} = 1$ iff $j \in [1, t + 1]$), let the set $S_k^t(\eta)$ be given by (11). Denote by ϕ any injective mapping from $\mathbb{F}_2^{n-k}$ to $S_k^t(\eta)$ satisfying $\phi(\delta) \cdot x = x_{i_1} + \cdots + x_{i_{t+1}}$ for some $\delta \in \mathbb{F}_2^{n-k}$. Such mappings exist due to the fact that k satisfies (12). Then, for $(y, x) \in \mathbb{F}_2^{n-k} \times \mathbb{F}_2^k$ we construct the function $g : \mathbb{F}_2^n \mapsto \mathbb{F}_2$ as follows,*

$$g(y, x) = \begin{cases} \phi(y) \cdot x + h(y), & y \neq \delta; \\ \phi(\delta) \cdot x + x_{i_{t+2}} x_{i_{t+3}} \cdots x_{i_k} + h(\delta), & y = \delta, \end{cases}$$

where h is any Boolean function on $\mathbb{F}_2^{n-k}$.

To simplify the proofs concerning the main properties of functions proposed by Construction 1 and to emphasize the connection to Maiorana-McFarland class, we first derive a result which interlinks Construction 1 with the pure affine concatenation as given in Remark 1.

Proposition 1. *Let $f(y, x)$ be a function in the standard Maiorana-McFarland class, defined by means of Remark 1, that is,*

$$f(y, x) = \sum_{\tau \in \mathbb{F}_2^{n-k}} (y_1 + \tau_1 + 1) \cdots (y_{n-k} + \tau_{n-k} + 1) a_{[\tau]}(x),$$

where $a_{[\tau]}(x) = \phi(\tau) \cdot x + h(\tau)$ and ϕ, h are the same mappings used to define g in Construction 1. Assume that $t < k - 2$. Then the function $g(y, x)$, as defined in Construction 1, is a degree optimized function whose algebraic normal form is given by,

$$g(y, x) = f(y, x) + x_{i_{t+2}} x_{i_{t+3}} \cdots x_{i_k} \prod_{i=1}^{n-k} (y_i + \delta_i + 1). \tag{13}$$

Proof. We first prove that the algebraic normal form of g is given as above. Note that we can write,

$$f(y, x) = \sum_{\tau \in \mathbb{F}_2^{n-k}} \left(\prod_{i=1}^{n-k} (y_i + \tau_i + 1) \right) a_{[\tau]}(x) = \sum_{\tau \in \mathbb{F}_2^{n-k}/\delta} \left[\left(\prod_{i=1}^{n-k} (y_i + \tau_i + 1) \right) a_{[\tau]}(x) \right] +$$

$$+ \left(\prod_{i=1}^{n-k} (y_i + \delta_i + 1) \right) a_{[\delta]}(x).$$

Then clearly,

$$g(y,x) = f(y,x) + \Big(\overbrace{\prod_{i=1}^{n-k}(y_i + \delta_i + 1)\big(a_{[\delta]}(x) + \phi(\delta)\cdot x + h(\delta)}^{=0} + x_{i_{t+2}}x_{i_{t+3}}\cdots x_{i_k}\big),$$

and we obtain (13) as stated.

To prove that g is degree optimized it suffices to note that any term in the ANF of f can only contain one x_i for $i = 1,\ldots,k$. Hence any term in the expression such as $x_{i_{t+2}}x_{i_{t+3}}\cdots x_{i_k}\prod_{i=1}^{n-k}(y_i + \delta_i + 1)$ will not be present in the ANF of f assuming that $k - t > 2$. Since the term $x_{i_{t+2}}x_{i_{t+3}}\cdots x_{i_k}y_1\cdots y_{n-k}$ is of degree $k - (t+1) + n - k = n - t - 1$, g is degree optimized. □

Remark 2. *Note that the assumption that $t < k - 2$ perfectly matches to those functions in Maiorana-McFarland class which cannot be degree optimized. Furthermore, remark that the function g as described above besides the term of the highest degree order introduces many terms of order $n - t - 2$ down to $k - (t+2)$ none of which is present in the ANF of f. This results in a significantly increased linear complexity of g in comparison to f. Another important observation is that the number of terms present in the ANF of g will depend on the value of δ, and the maximum number is obtained for $\delta = (0,\ldots,0)$.*

Next we prove that the function g is t-resilient having the same nonlinearity as f.

Theorem 2. *The function g proposed by Construction 1 is an $(n, t, n - t - 1, 2^{n-1} - 2^{k-1})$ function. Furthermore, the Walsh spectra of g is seven-valued, and more precisely $W_g(w) \in \{0, \pm 2^{k-t}, \pm 2^{k-t}(2^t - 1), \pm 2^k\}$.*

Proof. By Proposition 1 g is a degree optimized function. Note that $g(y,x)$ is an affine t-resilient functions for any fixed $y \neq \delta$. Hence, to show that g is t-resilient it is enough to show that the function $x_{i_1} + \cdots + x_{i_{t+1}} + x_{i_{t+2}}x_{i_{t+3}}\cdots x_{i_k}$ is a t-resilient function. This is obviously true since this function contains $t + 1$ linear terms from a disjoint variable space than the nonlinear term. Then g can be viewed as a concatenation of t-resilient functions, hence t-resilient itself.

Abusing the notation, we use the addition operator $''+''$ for a componentwise bit addition of vectors, i.e. for $\alpha, \beta \in \mathbb{F}_2^k$ we compute $\alpha + \beta = (\alpha_1 + \beta_1, \ldots, \alpha_k + \beta_k)$, but also for a usual integer addition. It should be clear from the context which operation is performed.

To prove that the nonlinearity value is the same as for f we consider the Walsh transform of g. Then for any $(\beta, \alpha) \in \mathbb{F}_2^{n-k} \times \mathbb{F}_2^k$ we have,

$$W_g((\beta,\alpha)) = \sum_{y\in\mathbb{F}_2^{n-k}}\sum_{x\in\mathbb{F}_2^k}(-1)^{g(y,x)\oplus(y,x)\cdot(\beta,\alpha)} = \sum_{y\in\mathbb{F}_2^{n-k}\setminus\delta}(-1)^{y\cdot\beta\oplus h(y)}\sum_{x\in\mathbb{F}_2^k}(-1)^{(\phi(y)+\alpha)\cdot x}$$

$$+ (-1)^{\delta\cdot\beta\oplus h(\delta)}\sum_{x\in\mathbb{F}_2^k}(-1)^{x_{i_1}\oplus\cdots\oplus x_{i_{t+1}}\oplus\alpha\cdot x\oplus x_{i_{t+2}}x_{i_{t+3}}\cdots x_{i_k}}. \tag{14}$$

There are three cases to consider. The first case arise when $\alpha \in \mathbb{F}_2^k$ is such that $\alpha = \phi(y)$ for some $y \neq \delta$, that is $\alpha \in S_k^t(\eta)$. Then obviously the first sum in (14) is equal to 2^k, where this nonzero contribution is obtained for $y = \delta$.

But $(-1)^{\delta \cdot \beta \oplus h(\delta)} \sum_{x \in \mathbb{F}_2^k} (-1)^{x_{i_1} \oplus \cdots \oplus x_{i_{t+1}} \oplus \alpha \cdot x \oplus x_{i_{t+2}} x_{i_{t+3}} \cdots x_{i_k}} = 0$ since the exponent is a balanced function in x. To verify this, notice that $\alpha = \phi(y)$ for some $y \neq \delta$, and since ϕ is injective it implies that $\phi(\delta) + \alpha \neq \mathbf{0}$ in the exponent of the second sum. Due to the properties of the set $S_k^t(\eta)$ and since α is an element of this set, α cannot cover $\phi(\delta)$, or equivalently $(\phi(\delta) + \alpha) \cdot x$ will contain at least one x_j, $j \in \{i_1, \ldots, i_{t+1}\}$.

The second case to consider is the case when $\alpha = \phi(\delta)$. Clearly the first sum in (14) is zero. The second sum is of the form $(-1)^{\delta \cdot \beta \oplus h(\delta)} \sum_{x \in \mathbb{F}_2^k} (-1)^{x_{i_{t+2}} \cdots x_{i_k}}$ implying that $|W_g((\beta, \alpha))| = 2^k - 2 \cdot 2^{k-(t+1)} = 2^{k-t}(2^t - 1)$.

Finally, the third case arise when $\alpha \notin S_k^t(\eta)$. Then the first sum in (14) is obviously zero, whereas the second sum may take three different values depending on the value of α. Indeed, since $(\phi(\delta) + \alpha) \cdot x \neq \mathbf{0}$ the second sum is either 0 or $\pm 2 \cdot 2^{k-(t+1)} = \pm 2^{k-t}$ depending on whether $(\phi(\delta) + \alpha) \cdot x \oplus x_{i_{t+2}} \cdots x_{i_k}$ is balanced or not. The value 0 corresponds to the case of balancedness, that is $(\phi(\delta) + \alpha) \cdot x$ contains some x_j such that $j \notin \{i_{t+2}, \ldots, i_k\}$. When $(\phi(\delta) + \alpha) \cdot x$ does not contain some x_j such that $j \notin \{i_{t+2}, \ldots, i_k\}$ two cases are possible. Then $(\phi(\delta) + \alpha) \cdot x + x_{i_{t+2}} \cdots x_{i_k}$ is either balanced $(W_g((\beta, \alpha)) = 0)$, or this function is balanced on all $(k - t - 1)$-dimensional flats except of being constant on exactly one flat of dimension $k - t - 1$ corresponding to the nonlinear term $x_{i_{t+2}} \cdots x_{i_k}$. In the latter case $W_g((\beta, \alpha)) = \pm 2^{k-t}$. To summarize, when $\alpha \notin S_k^t(\eta)$, $W_g((\beta, \alpha)) \in \{0, \pm 2^{k-t}\}$.

Hence, $\mathcal{N}_g = 2^{n-1} - \frac{1}{2} \max_{(\beta, \alpha) \in \mathbb{F}_2^{n-k} \times \mathbb{F}_2^k} |W_g((\beta, \alpha))| = 2^{n-1} - 2^{k-1}$. Also, from the details of the proof it is clear that $W_g(w) \in \{0, \pm 2^{k-t}, \pm 2^{k-t}(2^t - 1), \pm 2^k\}$. □

Remark 3. *The concept of using the nonlinear functions may naturally be extended to include even more nonlinear functions on the subspaces of dimension k. Notice that using more such functions will additionally increase the complexity of the keystream sequence but this feature is traded-off against more rigorous conditions on the set $S_k^t(\eta)$. Thus defining the set $T \subseteq P = \{p \in \mathbb{F}_2^k \mid wt(p) = t+1\}$ the problem is transformed to finding such $k = \min_{t<k} \{ k \mid \#S_k^t(T) \geq 2^{n-k} \}$, where $S_k^t(T) = \{c \mid c \in \mathbb{F}_2^k, wt(c) > t \text{ and } p \nleq c \text{ for any } p \in T \}$. Now taking any injective mapping ϕ from $\mathbb{F}_2^{n-k}$ to $S_k^t(T)$ the function g^* can be defined in a similar way as above,*

$$g^*(y, x) = \begin{cases} \phi(y) \cdot x + h(y), & y \mid \phi(y) \in S_k^t(T) \setminus T; \\ \phi(y) \cdot x + \prod_{i=1}^{k} x_i^{\phi(y)_i \oplus 1} + h(y), & y \mid \phi(y) \in T, \end{cases}$$

where h is any Boolean function on $\mathbb{F}_2^{n-k}$, and $\phi(y)_i$ denotes the i-th coordinate of the image of ϕ.

We give two important examples to emphasize the importance of this construction. These examples demonstrate the possibility of constructing degree optimized resilient functions with nonlinearity which has not been achieved previously.

Example 1. *A construction of an* $(11, 2, 8, \mathcal{N}_f)$ *function has been discussed in the literature. Using a recursive procedure called Algorithm B, an* $(11, 2, 8, 984)$ *function has been obtained in [20], which so far gives the highest nonlinearity for fixed* $n = 11, t = 2, d = 8$.

According to the weight divisibility results $W_f(\alpha) \equiv 0 \pmod{2^{t+2+\lfloor \frac{n-t-2}{d} \rfloor}}$ *,* $\forall \; \alpha \in \mathbb{F}_2^n$ *and for any* (n, t, d) *function* f, *see [2,21]. Since the order of resiliency* $t \leq n/2 - 2$, *the upper bound on nonlinearity is obtained by combining the bound of bent functions and the weight divisibility results. It can be verified that for* $n = 11, t = 2, d = 8$, $|\max_{\alpha \in \mathbb{F}_2^n} W_f(\alpha)| = k \cdot 16$, *where* $k \geq 3$. *Note that the standard Maiorana-McFarland technique would require the value* $k = 6$ *to improve upon the nonlinearity of the above result [20] (for* $k = 6$ *we would have* $\mathcal{N}_f = 2^{n-1} - 2^{k-1} = 992$). *But then* $t < k - 2$ *implying that this method cannot generate a degree optimized function (actually the maximum degree through this technique is* $d = n - k + 1 = 6$). *It can be verified that for* $n = 11, t = 2$ *and* $k = 6$,

$$\#S_k^t(\eta) = \sum_{i=0}^{k-(t+1)} \binom{k}{t+1+i} - 2^{k-t-1} + 1 = 35 \geq 2^{n-k} = 32,$$

implying that **for the first time**, *using Construction 1, we can construct an* $(11, 2, 8, 992)$ *function* g.

Unfortunately, the ANF of g discussed in Example 1 is far too large to be given in the explicit form. On the other hand the ANF of g is easily computed using Proposition 1. Note that for the given parameters above $(y, x) \in \mathbb{F}_2^5 \times \mathbb{F}_2^6$. Without loss of generality let us fix $\eta = (1, 1, 1, 0, 0, 0)$. That is, we consider $S_6^2(\eta)$ meaning that none of the elements in this set (except of η itself) covers the element η. Then one can choose any other 31 elements from $S_6^2(\eta) \setminus \eta$ to construct a function f by means of associating these elements to linear functions and then labeling these functions as it was done in Theorem 1. Hence the ANF of the function f (which is a Maiorana-McFarland function) is easily deduced by using equation (9). Now if $f(\delta, x) = x_1 + x_2 + x_3$ for some $\delta \in \mathbb{F}_2^5$, then the degree optimized function g is given as $g(y, x) = f(y, x) + x_4 x_5 x_6 \prod_{i=1}^5 (y_i + \delta_i + 1)$.

Now we discuss the same input parameters but including even more nonlinear terms as remarked above.

Example 2. *In order to preserve the same nonlinearity value we use the same* k *as in the example above. Then for* $n = 11, t = 2$ *and* $k = 6$, *let* $T = \{(1, 1, 1, 0, 0, 0), (1, 1, 0, 1, 0, 0)\}$. *For such a choice of* T *we have* $\#S_k^t(T) = \sum_{i=0}^{k-(t+1)} \binom{k}{t+1+i} - 2^{k-t-1} + 1 - 3 = 32 \geq 2^{n-k} = 32$ *implying that using the extension of Construction 1 as given in the above remark we can construct an* $(11, 2, 8, 992)$ *function* g^* *having much more terms of high algebraic degree than the function* g *in the Example 1.*

The choice of T in the above example is not arbitrary. For some other choices of this set it can happen that $\#S_k^t(T) < 2^{n-k}$ implying a decrease in nonlinearity since a larger k must be used.

Open Problem 1. *Derive a general explicit formula for the cardinality of $S_k^t(T)$ as a function of k, t and $|T|$. In particular, for given n, t and the minimal k satisfying the condition $\sum_{i=0}^{k-(t+1)} \binom{k}{t+1+i} - 2^{k-t-1} + 1 \geq 2^{n-k}$ determine the maximum cardinality of T (where T has more than one element) such that $\#S_k^t(T) \geq 2^{n-k}$.*

4   Resilient Functions of High Algebraic Degree

When constructing multiple output Boolean functions one applies similar cryptographic criteria as in the Boolean case. Since this mapping is defined as $F : \mathbb{F}_2^n \mapsto \mathbb{F}_2^m$ all the criteria of concern is defined with respect to all nonzero linear combinations of the output functions $f_0, \ldots, f_{m-1}$. That is, $F = (f_0, \ldots, f_{m-1})$ is viewed as a set of Boolean functions. Hence, the three main cryptographic criteria are defined as below.

Lemma 1. *[26] A function $F = (f_0, \ldots, f_{m-1})$ is an (n, m, t)-resilient function if and only if all nonzero linear combinations of $f_0, \ldots, f_{m-1}$ are $(n, 1, t)$-resilient functions.*

The definition of nonlinearity follows in a similar manner, taken from [17].

Definition 2. *The nonlinearity of $F = (f_0, \ldots, f_{m-1})$, denoted by $\mathcal{N}_F$, is defined as*

$$\mathcal{N}_F = \min_{\alpha \in \mathbb{F}_2^m \setminus \{0\}} \mathcal{N}_{f_\alpha}, \tag{15}$$

where $f_\alpha = \sum_{i=0}^{m-1} \alpha_i f_i$, $\alpha_i \in \mathbb{F}_2$.

Similarly, the algebraic degree of F is defined as the minimum of degrees of all nonzero linear combinations of the component functions of F, namely,

$$deg(F) = \min_{\alpha \in \mathbb{F}_2^m \setminus \{0\}} deg(\sum_{i=0}^{m-1} \alpha_i f_i). \tag{16}$$

We use the same notation most often found in the literature. An (n, m, t) function will denote an n-input, m-output, t-resilient function. The information about the nonlinearity and degree will be given additionally.

There are several approaches when designing nonlinear resilient functions. The method of Kurosawa uses bent concatenation together with a function composition [14]. Another technique [26] uses a linear resilient function, obtained from an error correcting code, and applies a highly nonlinear permutation on such a function. One particularly important example of this approach is construction of nonlinear resilient functions by applying a nonlinear permutation to the $[2^m - 1, m, 2^{m-1}]$ simplex code. For this particular case the authors obtain $2^m!$ distinct $(2^m - 1, m, 2^{m-1} - 1)$ functions some of which have a nonlinearity of at least $2^{2^m - 2} - 2^{2^m - 1 - \frac{1}{2}m}$ and whose algebraic degree is $m - 1$. These functions achieve the upper bound on resiliency given by Friedman [11]. Note that this construction is not degree optimized since $d = m - 1 < 2^m - 1 - (2^{m-1} - 1) - 1$ for any $m \geq 3$ (here we again use the Siegenthaler's inequality $d \leq n - t - 1$).

An application of linearized polynomials in construction of nonlinear resilient functions has been proposed in [8]. Based on the existence of a linear $[u, m, t+1]$ code it was proved that there exist nonlinear $(u + \triangle + 1, m, t + 1)$ resilient functions with algebraic degree $\triangle$, for any $\triangle \geq 0$.

For the purpose of this paper we mostly confine ourselves to the methods in [8,26] for the purpose of comparison, and since the basic idea of our construction is the use of linear codes we will recall the methods given in [13,19]. The main result in [13] is the following lemma.

Lemma 2. *[13] Let $\Theta_0, \dots, \Theta_{m-1}$ be a basis of a binary $[u, m, t+1]$ linear code C. Let β be a primitive element in $\mathbb{F}_{2^m}$ and $(1, \beta, \dots, \beta^{m-1})$ be a polynomial basis of $\mathbb{F}_{2^m}$. Define a bijection $\rho : \mathbb{F}_{2^m} \mapsto C$ by*

$$\rho(a_0 + a_1\beta + \cdots + a_{m-1}\beta^{m-1}) = a_0\Theta_0 + a_1\Theta_1 + \cdots + a_{m-1}\Theta_{m-1}.$$

Consider the matrix

$$A^* = \begin{pmatrix} \rho(1) & \rho(\beta) & \cdots & \rho(\beta^{m-1}) \\ \rho(\beta) & \rho(\beta^2) & \cdots & \rho(\beta^m) \\ \vdots & \vdots & \ddots & \vdots \\ \rho(\beta^{2^m - 2}) & \rho(1) & \cdots & \rho(\beta^{m-2}) \end{pmatrix},$$

of size $(2^m - 1) \times m$, whose entries are elements of $\mathbb{F}_2^u$ (actually the codewords of C). For any linear combination of columns (not all zero) of the matrix A^*, each nonzero codeword of C will appear exactly once in such a nonzero linear combination.

Remark 4. *Since the elements of A^* are vectors we need a triple index set to refer to a specific coordinate of some entry of A^*. For convenience, we adopt the following notation. $A_{i,j}^*(k)$ will indicate the k-th position of the entry (vector) found in the intersection of the i-th row and j-th column of A^*. Hence, for A^* of size $r \times s$ with elements in $\mathbb{F}_2^u$ we let the indices set run as follows: $i = 0, \dots, r-1; j = 0, \dots, s-1; k = 1, \dots, u$. To refer to the whole vector we simply write $A_{i,j}^*$. We also keep '[ ]' to denote the decimal representation of vectors, i.e. for $\alpha \in \mathbb{F}_2^u$, $[\alpha] = \sum_{l=1}^{u} \alpha_l 2^{l-1}$.*

This lemma actually shows how to use the codewords of a linear code efficiently when constructing a function $F : \mathbb{F}_2^n \mapsto \mathbb{F}_2^m$. Then in case that $u > n - m$ one can use any 2^{n-m} rows of A^* to define $F : \mathbb{F}_2^n \mapsto \mathbb{F}_2^m$ through the columns of A, where A is obtained by deleting some rows of A^*. The component functions of

$F = (f_0, \ldots, f_{m-1})$ are simply defined as follows. For any $y \in \mathbb{F}_2^{n-u}$ we define $f_j(y,x) = A_{[y],j} \cdot x, \ j = 0, \ldots, m-1$.

When the parameters of C are such that $u \leq n - m$, the alternative is to use a set of disjoint $[u, m, t+1]$ linear codes as originally proposed in [13].

Definition 3. *[13] A set of linear $[n', m, t+1]$ codes $\{C_1, C_2, \ldots, C_s\}$ such that*

$$C_i \cap C_j = \{0\}, \quad 1 \leq i < j \leq s$$

is called a set of linear $[n', m, t+1]$ disjoint (nonintersecting) codes.

Then the following result was given in [13], which enables us to construct an (n, m, t) function by using the codewords of several disjoint codes.

Theorem 3. *[13] If there exists a set of linear $[u, m, t+1]$ disjoint codes with cardinality $\lceil 2^{n-u}/(2^m - 1) \rceil$ then there exists a t-resilient function $F : \mathbb{F}_2^n \mapsto \mathbb{F}_2^m$ with nonlinearity*

$$\mathcal{N}_F = 2^{n-1} - 2^{u-1}.$$

Though this approach (which is actually an extension of Maiorana-McFarland class) generates highly nonlinear resilient functions, assuming the existence of the set of disjoint codes, it suffers the same drawback as in the Boolean case, namely a low algebraic order. Another problem is an efficient way of finding a set of disjoint codes. In [13] a computer search has been used and in certain cases such a set could be obtained using the tools of projective geometry.

However, very recently a new method has been proposed, which at least for certain parameter values allows us to find such a set [5]. In particular when the length of the code is of the form $u = 2^m - 1$, the method in [5] gives a certain number of disjoint $[2^m - 1, m, 2^{m-1}]$ simplex-like codes. We call this codes simplex-like since there are all derived from the simplex code. On the other hand the simplex code is considered to be the unique dual code of the Hamming code.

We combine these results with the construction idea discussed above to obtain nonlinear resilient functions of very high algebraic degree. From now on we assume that the cardinality of the set of disjoint $[u, m, t+1]$ linear codes is b for some $b \geq 1$. We denote this set by $\mathcal{C}$, i.e. $\mathcal{C} = \{C_1, \ldots, C_b\}$. It is easily verified that denoting by $e = \lfloor log_2 b(2^m - 1) \rfloor$, the input space n is given as $n = e + u$, where u is the length of the codes. Note that a straightforward application of Theorem 3 would result in functions of degree $d \leq e + 1$, which for small e is far from its optimized value $n - t - 1$.

Thus in order to increase the algebraic order, we replace linear functions at certain positions by nonlinear ones. Then it turns out that there is a trade-off between the algebraic degree and resiliency. Remark that Construction 2 below utilizes a set of disjoint simplex-like codes for which the exact calculation of resiliency order is very simple. In general, the resiliency order will depend on the properties of the code in a rather complicated way.

Construction 2. *Let* $\mathcal{C} = \{C_1, \ldots, C_b\}$ *be a set of disjoint* $[2^m - 1, m, 2^{m-1}]$ *simplex-like codes, and associate to each code a mapping* $\rho_r : \mathbb{F}_{2^{2m-1}} \mapsto C_r$, $1 \le r \le b$, *so that*

$$(a_0, a_1\beta, \ldots, a_{m-1}\beta^{m-1}) \xrightarrow{\rho_r} a_0\Theta_0^r + \cdots + a_{m-1}\Theta_{m-1}^r,$$

where $\Theta_0^r, \ldots, \Theta_{m-1}^r$ *is a basis of* C_r, $a_i \in \mathbb{F}_2$, *and* β *is primitive in* $\mathbb{F}_{2^m}$. *Let* A_r *be the associated matrix of* C_r *as in Lemma 2. Let* $A = (A_1^T | A_2^T | \cdots | \hat{A}_b^T)^T$, *where* $\hat{A}_b$ *denotes that some rows of* A_b *may be deleted to adjust* A *to be of size* $2^e \times m$, *and denote by* $e = \lfloor \log_2 b(2^m - 1) \rfloor$. *Let* $G = (g_0, \ldots, g_{m-1})$ *be a function from* $\mathbb{F}_2^{e+2^m-1}$ *to* $\mathbb{F}_2^m$, *whose component functions are defined for any* $(y, x) \in \mathbb{F}_2^e \times \mathbb{F}_2^{2^m-1}$ *as,*

$$g_j(y, x) = \begin{cases} A_{[y],j} \cdot x, & y \mid [y] \ne j; \\ A_{[y],j} \cdot x + \prod_{k=1}^{2^m-1} x_k^{A_{[y],j}(k)\oplus 1}, & y \mid [y] = j, \end{cases}$$

where $j = 0, \ldots, m - 1$.

To clarify further the structure of G in Construction 2, we associate to G a function matrix A^f of size $2^e \times m$ as below,

$$A^f = \begin{pmatrix} \rho_1(1) \cdot x + x^{\rho_i(\beta^0)} & \rho_1(\beta) \cdot x & \cdots & \rho_1(\beta^{m-1}) \cdot x \\ \rho_1(\beta) \cdot x & \rho_1(\beta^2) \cdot x + x^{\rho_i(\beta^2)} & \cdots & \rho_1(\beta^m) \cdot x \\ \vdots & \vdots & \ddots & \vdots \\ \rho_1(\beta^{m-1}) \cdot x & \rho_1(\beta^m) \cdot x & \cdots & \rho_1(\beta^{2m-2}) \cdot x + x^{\rho_i(\beta^{2m-2})} \\ \rho_1(\beta^m) \cdot x & \rho_1(\beta^{m+1}) \cdot x & \cdots & \rho_1(\beta^{2m-1}) \cdot x \\ \vdots & \vdots & \ddots & \vdots \\ \rho_1(\beta^{2^m-2}) \cdot x & \rho_1(1) \cdot x & \cdots & \rho_1(\beta^{m-2}) \cdot x \\ \vdots & \vdots & \ddots & \vdots \\ \rho_b(\beta^{2^m-2}) \cdot x & \rho_b(1) \cdot x & \cdots & \rho_b(\beta^{m-2}) \cdot x \end{pmatrix},$$

where $x^{\rho_i(\beta^z)} = \prod_{k=1}^{2^m-1} x_k^{\rho_i^k(\beta^2)\oplus 1}$. Note that for any $y \in \mathbb{F}_2^e$, we have $g_j(y, x) = A_{[y],j}^f$, $0 \le j \le m - 1$.

A rather technical and lengthy proof of the theorem below, concerning the properties of Construction 2, is given in the Appendix.

Theorem 4. *Let* $\mathcal{C}$ *be a given set of disjoint* $[2^m - 1, m, 2^{m-1}]$ *simplex-like codes with* $|\mathcal{C}| = b$, *and let* $e = \lfloor \log_2 b(2^m - 1) \rfloor$. *Then the function* $G : \mathbb{F}_2^{2^m-1+e} \mapsto \mathbb{F}_2^m$, *constructed by means of Construction 2, is a* $(2^{m-2} - 1)$-*resilient function, with* $\mathcal{N}_G \ge 2^{2^m-2+e} - 2^{2^m-2} - m2^{2^m-1-1}$ *and* $\deg(G) = e + 2^{m-1} - 1$.

We utilize the approach in [13], based on a set of disjoint codes, just to illustrate a wider framework in which our method may be applied. We can equally well use the results given in [19] or in [12], where a single $[u, m, t + 1]$ linear code has been used to provide nonlinear resilient functions for any $n > u$. When compared to the result given in Theorem 3 we deduce the following. Utilizing the set of simplex-like codes the method in [13] would generate $(2^m - 1 + e, m, 2^{m-1} - 1)$ functions of nonlinearity $\mathcal{N}_F = 2^{2^m - 2 + e} - 2^{2^m - 2}$, and the degree $deg(F) \le e + 1$. Hence, there is a drop of nonlinearity when using Construction 2. It equals to $m 2^{2^{m-1} - 1}$ and can be neglected in comparison to the term $2^{2^m - 2}$. Then we can assume that there is a trade-off between the resiliency and algebraic degree only. Actually, our method gives a higher algebraic degree, i.e. $deg(G) = 2^{m-1} - 1 + e$ compared to $deg(F) \le e + 1$ and the gain is at least $2^{m-1} - 2$. On the other hand the resiliency order is decreased by the value 2^{m-2}.

Suppose that we start with the simplex $[2^m - 1, m, 2^{m-1}]$ code to construct a nonlinear resilient function using the technique in [19]. Then to construct a function $G : \mathbb{F}_2^n \mapsto \mathbb{F}_2^m$ (for $n > 2^m - 1$), the procedure in [19] essentially consists of two steps. First, one constructs a matrix A of size $2^{m-1} \times m$ by choosing any 2^{m-1} rows of A^*, where A^* is obtained through Lemma 2. Then a mapping $F = (f_0, \ldots, f_{m-1})$ may be defined through the columns of A, as already discussed above. Notice that $F : \mathbb{F}_2^{2^m - 1 + m - 1} \mapsto \mathbb{F}_2^m$.

Then the second step is to define another set of functions $(g_0, \ldots, g_{m-1})$ as follows. Let $g_j(y, x) = h_j(y) + f_j(x)$ for $j = 0, \ldots, m - 1$, where the functions h_j are chosen in such a manner that any nonzero linear combination of type $\sum_{j=0}^{m-1} a_j h_j$ ($a_i \in \mathbb{F}_2$) is of very high nonlinearity. Then $G = (g_0, \ldots, g_{m-1})$ is also of high nonlinearity with the resiliency order equal to that of $F = (f_0, \ldots, f_{m-1})$. It is easy to verify that if each h_j is a function in k variables then each g_j is a function on $n = (2^m - 1) + (m - 1) + k = 2^m + m + k - 2$ variables.

Then by simply making this function nonlinear on exactly m flats of dimension $2^m - 1$, in the same manner as it was done in Construction 2, the effect on algebraic degree will be exactly the same as discussed above.

4.1   Comparison

We now give a short comparison to the known construction methods only in terms of algebraic degree and resiliency since a detailed examination involving all cryptographic criteria would be very tedious and dependent on the choice of input parameters n, m and t. On the other hand it has been proved that the construction results proposed in [12,13,19] in most of the cases are superior in terms of nonlinearity compared to other methods. The essence of Construction 2 does not depend on the method utilized, so long these methods are restricted to the use of linear codes. Hence the degradation of nonlinearity for a small portion and the drop of resiliency, when the Construction 2 is incorporated in the methods in [12,13,19], is traded-off against a significant increase of algebraic degree.

In terms of algebraic degree the method of Cheon [8] using linearized polynomials provides functions with highest degree for a sufficiently large input space. Based on the existence of a single $[u, m, t + 1]$ linear codes this method generates nonlinear $(n = u + \triangle + 1, m, t)$-resilient functions with degree $d = \triangle$ for any $\triangle \geq 0$. If one starts with the simplex code then nonlinear $(2^m - 1 + \triangle, m, 2^{m-1} - 1)$-resilient functions can be obtained. By Theorem 4, the degree of G (G is a nonlinear $(2^m - 1 + e, m, 2^{m-2} - 1)$-resilient function) is $2^{m-1} - 1 + e$, which for the same input space is obviously higher than $\triangle$. On the other hand the resiliency order is smaller compared to the method of Cheon.

When compared to the construction of Zhang and Zheng [26], in the case that the simplex code is utilized, we can deduce the following. Clearly, for a small input space $n = 2^m - 1$ the method of Zhang and Zheng is better than our, since our construction needs to use as many as possible codewords not giving any nonlinearity for $n = u$. Thus to make a fair comparison we have to investigate the existence of an $[n = 2^m - 1 + e, m', 2^{m-2}]$ linear code, where m' satisfies $m' > 2^{m-1} + e$. This is because in this case the degree of the Zhang and Zheng method, given as $d = m' - 1$, will be larger than $deg(G) = 2^{m-1} - 1 + e$ in Theorem 4. Recall that $e = \lfloor log_2 b(2^m - 1) \rfloor \geq m - 1$, where b is the cardinality of the set of disjoint codes. It seems that in general the codes of the parameters above do exist for small e close to $m - 1$ but not for $e \gg m - 1$. Hence, assuming that many disjoint codes are available our construction is better, whereas for a small cardinality of a set of disjoint codes the method of Zhang and Zheng seems to be better.

In [12] a simple modification of the method of Zhang and Zheng was proposed. The authors simply apply the method to a code of the same length and a larger dimension. Hence to construct a nonlinear (n, m, t) function of degree $d > m - 1$, in [12] a nonlinear permutation is applied to an $[n, d, t + 1]$, where $d > m$. The reliability of this approach is a bit ambiguous (it is quite natural for given n and t to use the code of highest dimension m, and if such a code is utilized in the method of Zhang and Zheng there will not exist $d > m$!?) and the same conclusion as above applies here.

5   Conclusions

The basic construction idea common to both Construction 1 and Construction 2 is to replace exactly one linear subfunction with its nonlinear counterpart for each constituent function $f_1, \ldots, f_m$, $m \geq 1$. In the case of Boolean functions ($m = 1$) the construction proposed here is very elegant and general. The only requirement is the existence of the set $S(\eta)$ or alternatively $S(T)$ of sufficiently large cardinality. It is very important to solve the open problem posed in Section 3 for several reasons.

Firstly, we can deduce the boundaries on the number of nonlinear subfunctions for our method. Maybe more importantly, designing a function g with its supporting nonlinear set T such that $|T| > 1$ will give a better resistance to certain algebraic attacks. Due to the space limitations a thorough treatment of some topics presented here will be given in the extended version of this paper.

In the case of multiple output functions the exact calculation of resiliency order is rather tedious and it depends on the properties of the codes used in the construction. In the case of the simplex code we could derive the exact resiliency order. Actually, the codes with a sparse weight distribution are more suitable for our construction than those having codewords with many different weights.

Acknowledgments. The author wants to thank Pascale Charpin for pointing out the importance of specifying the Walsh spectra of functions in Theorem 2 and for the proofreading of this work.

References

1. P. Camion, C. Carlet, P. Charpin, and N. Sendrier. On correlation-immune functions. In *Advances in Cryptology—EUROCRYPT'91*, volume LNCS 547, pages 86–100. Springer-Verlag, 1991.
2. C. Carlet. On the coset weight divisibility and nonlinearity of resilient and correlation-immune functions. *Discrete Mathematics and Theoretical Computer Science*, 2001.
3. C. Carlet. A larger class of cryptographic Boolean functions via a study of the Maiorana-McFarland constructions. In *Advances in Cryptology—CRYPTO 2002*, volume LNCS 2442, pages 549–564. Springer-Verlag, 2002.
4. C. Carlet and P. Sarkar. Spectral domain analysis of correlation immune and resilient Boolean functions. *Finite Fields and Their Applications*, vol. 8(1):120–130, 2002.
5. P. Charpin and E. Pasalic. Disjoint linear codes in construction of nonlinear resilient functions. Preprint, to be submitted, 2003.
6. P. Charpin and E. Pasalic. On propagation properties of resilient functions. In *Selected Areas in Cryptography—SAC 2002*, volume LNCS 2595, pages 356–365. Springer-Verlag, 2003.
7. S. Chee, S. Lee, D. Lee, and H. S. Sung. On the correlation immune functions and their nonlinearity. In *Advances in Cryptology—ASIACRYPT'96*, volume LNCS 1163, pages 232–243. Springer-Verlag, 1996.
8. J. H. Cheon. Nonlinear vector resilient functions. In *Advances in Cryptology—CRYPTO 2001*, volume LNCS 2139, pages 181–195. Springer-Verlag, 2001.
9. N. Courtois. Higher order correlation attacks, XL algorithm and cryptanalysis of Toyocrypt. In *ICISC 2002*, volume LNCS 2587, pages 182–199. Springer-Verlag, 2002.
10. N. Courtois and W. Meier. Algebraic attacks on stream ciphers with linear feedback. In *Advances in Cryptology—EUROCRYPT 2003*, volume LNCS 2656, pages 346–359. Springer-Verlag, 2003.
11. J. Friedman. On the bit extraction problem. In *33rd IEEE Symposium on Foundations of Computer Science*, pages 314–319, 1982.
12. K.C. Gupta and P. Sarkar. Improved constructions of nonlinear resilient S-boxes. In *Advances in Cryptology—ASIACRYPT 2002*, volume LNCS 2501, pages 466–483. Springer-Verlag, 2002.
13. T. Johansson and E. Pasalic. A construction of resilient functions with high nonlinearity. *IEEE Trans. on Inform. Theory*, IT-49(2), February 2003.
14. K. Kurosawa, T. Satoh, and K. Yamamoto. Highly nonlinear t-resilient functions. *Journal of Universal Computer Science*, vol. 3 (6):721–729, 1997.

15. F. J. MacWilliams and N. J. A. Sloane. *The Theory of Error-Correcting Codes.* North-Holland, Amsterdam, 1977.
16. S. Maitra and E. Pasalic. Further constructions of resilient Boolean functions with very high nonlinearity. *IEEE Trans. on Inform. Theory*, IT-48(7):1825–1834, 2002.
17. K. Nyberg. On the construction of highly nonlinear permutations. In *Advances in Cryptology—EUROCRYPT'92*, volume LNCS 658, pages 92–98. Springer-Verlag, 1992.
18. E. Pasalic, T. Johansson, S. Maitra, and P. Sarkar. New constructions of resilient and correlation immune Boolean functions achieving upper bounds on nonlinearity. In *Workshop on Coding and Cryptography Proceedings*, volume 6, pages 425–435. Elsevier Science, 2001.
19. E. Pasalic and S. Maitra. Linear codes in generalized construction of resilient functions with very high nonlinearity. *IEEE Trans. on Inform. Theory*, IT-48(8):2182–2191, 2002.
20. P. Sarkar and S. Maitra. Construction of nonlinear Boolean functions with important cryptographic properties. In *Advances in Cryptology—EUROCRYPT 2000*, volume LNCS 1807, pages 485–506. Springer-Verlag, 2000.
21. P. Sarkar and S. Maitra. Nonlinearity bounds and constructions of resilient Boolean functions. In *Advances in Cryptology—CRYPTO 2000*, volume LNCS 1880, pages 515–532. Springer-Verlag, 2000.
22. J. Seberry, X. M. Zhang, and Y. Zheng. On constructions and nonlinearity of correlation immune Boolean functions. In *Advances in Cryptology—EUROCRYPT'93*, volume LNCS 765, pages 181–199. Springer-Verlag, 1993.
23. T. Siegenthaler. Correlation-immunity of nonlinear combining functions for cryptographic applications. *IEEE Trans. on Inform. Theory*, IT-30:776–780, 1984.
24. Y. Tarannikov. On resilient Boolean functions with maximal possible nonlinearity. In *Proceedings of Indocrypt*, volume LNCS 1977, pages 19–30. Springer-Verlag, 2000.
25. G-Z. Xiao and J. L. Massey. A spectral characterization of correlation-immune combining functions. *IEEE Trans. on Inform. Theory*, IT-34:569–571, 1988.
26. X. M. Zhang and Y. Zheng. Cryptographically resilient functions. *IEEE Trans. on Inform. Theory*, IT-43(5):1740–1747, 1997.
27. Y. Zheng and X. M. Zhang. Improving upper bound on nonlinearity of high order correlation immune functions. In *Selected Areas in Cryptography–SAC '2000*, volume LNCS 2012, pages 264–274. Springer-Verlag, 2000.

A   The Proof of Theorem 4

For convenience we recall the claim of Theorem 4.

4 *Let $\mathcal{C}$ be a given set of disjoint $[2^m - 1, m, 2^{m-1}]$ simplex-like codes with $|\mathcal{C}| = b$, and let $e = \lfloor log_2 b(2^m - 1) \rfloor$. Then the function $G : \mathbb{F}_2^{2^m - 1 + e} \mapsto \mathbb{F}_2^m$, constructed by means of Construction 2, is a $(2^{m-2} - 1)$-resilient function, with $\mathcal{N}_G \geq 2^{2^m - 2 + e} - 2^{2^m - 2} - m2^{2^{m-1} - 1}$ and $deg(G) = e + 2^{m-1} - 1$.*

Proof. We again abuse the addition operator $"+"$ to perform componentwise bit addition of vectors, bit addition and usual integer addition. Which operation is performed should be clear from the context.

Combining the result of Proposition 1 and the particular placement of nonlinear functions, the ANF of g_j can be written as

$$g_j(y,x) = f_j(y,x) + \prod_{k=1}^{2^m-1} x_k^{A_{j,j}(k)\oplus 1} \left(\prod_{l=1}^{e}(y_l + \tau_l^j + 1)\right) \text{ for } j = 0,\dots,m-1,$$

where $\tau^j \in \mathbb{F}_2^e$ satisfies $[\tau^j] = j$, and the ANF of f_j is obtained from the j-th column of the matrix A. Hence the ANF of f_j is given by $f_j(y,x) = \sum_{\tau \in \mathbb{F}_2^e} \left(\prod_{l=1}^{e}(y_l + \tau_l + 1)\right) A_{[\tau],j} \cdot x$.

Then for any nonzero $a = (a_0,\dots,a_{m-1}) \in \mathbb{F}_2^m$ we may write,

$$\sum_{j=0}^{m-1} a_j g_j(y,x) = \sum_{j=0}^{m-1} a_j f_j(y,x) + \sum_{j=0}^{m-1} a_j \left[\prod_{k=1}^{2^m-1} x_k^{A_{j,j}(k)\oplus 1}\left(\prod_{l=1}^{e}(y_l + \tau_l^j + 1)\right)\right].$$

Since for any nonzero a the sum $\sum_{j=0}^{m-1} a_j f_j(y,x)$ is of degree $d \leq e+1$ we only have to prove that the terms of the form $\prod_{k=1}^{2^m-1} x_k^{A_{j,j}(k)\oplus 1}\left(\prod_{l=1}^{e}(y_l + \tau_l^j + 1)\right)$ are not canceled in the function $\sum_{j=0}^{m-1} a_j g_j(y,x)$.

This is obviously true as any linear combination of A's columns gives a rise to two-by-two distinct codewords, that is for any $y' \neq y''$ and nonzero a,

$$\sum_{j=0}^{m-1} a_j A_{[y'],j} \neq \sum_{j=0}^{m-1} a_j A_{[y''],j}.$$

Then any term of the form $y_1 \cdots y_e \prod_{k=1}^{2^m-1} x_k^{A_{j,j}(k)\oplus 1}$ is present in $\sum_{j=0}^{m-1} a_j g_j(y,x)$. Note that the number of such terms is exactly the weight of a. Since the disjoint codes are simplex-like codes all the codewords are of the same weight 2^{m-1}. Then $wt(\prod_{k=1}^{2^m-1} x_k^{A_{j,j}(k)\oplus 1}) = 2^{m-1} - 1$. Hence $deg(G) = e + 2^{m-1} - 1$ as stated.

To prove that G is $(2^{m-2} - 1)$-resilient, note that for any fixed y the function G is either linear function of weight 2^{m-1} (hence $(2^{m-1} - 1)$-resilient) or it is of the form

$$\left(\sum_{\substack{j=0 \\ j\neq s}}^{m-1} a_j A_{[y],j}\right) \cdot x + A_{[y],s} \cdot x + \prod_{k=1}^{2^m-1} x_k^{A_{[y],s}(k)\oplus 1},$$

for some $s \in [0, m-1]$. Then the order of resiliency is determined by the nonlinear term above. We use simple coding arguments to prove that $\left(\sum_{j=0|j\neq s}^{m-1} a_j A_{[y],j}\right) \cdot x + A_{[y],s} \cdot x$ has exactly 2^{m-2} variables not contained in $\prod_{k=1}^{2^m-1} x_k^{A_{[y],s}(k)\oplus 1}$. Let $\mathbf{u} = A_{[y],s} + (1,\dots,1)$, and $\mathbf{v} = \sum_{j=0|j\neq s}^{m-1} a_j A_{[y],j} + A_{[y],s}$, where $\mathbf{u}, \mathbf{v} \in \mathbb{F}_2^{2^m-1}$. Note that $wt(\mathbf{u}) = 2^{m-1} - 1$, $wt(\mathbf{v}) = 2^{m-1}$, and $wt(\mathbf{u} + \mathbf{v}) = 2^{m-1} - 1$.

We simply show that $wt(\mathbf{u} * \mathbf{v}) = 2^{m-2}$, where $\mathbf{u} * \mathbf{v} = (u_1 v_1, \dots, u_{2^m-1} v_{2^m-1})$. We know that [15],

$$wt(\mathbf{u} + \mathbf{v}) = wt(\mathbf{u}) + wt(\mathbf{v}) - 2wt(\mathbf{u} * \mathbf{v}).$$

Then $wt(\mathbf{u} * \mathbf{v}) = \frac{1}{2}(wt(\mathbf{u}) + wt(\mathbf{v}) - wt(\mathbf{u} + \mathbf{v}))$, and substituting the weight values $wt(\mathbf{u} * \mathbf{v}) = 2^{m-2}$. Hence G is a $(2^{m-2} - 1)$-resilient function.

To prove the nonlinearity assertion, we first note that for any nonzero $a \in \mathbb{F}_2^m$ such that $wt(a) = s$, $g(y, x) = \sum_{j=0}^{m-1} a_j g_j(y, x)$ is nonlinear on exactly s flats of dimension $2^m - 1$. Let $J = \{j_1, \dots, j_s\}$, $0 \le j_1 \ne \cdots \ne j_s \le m - 1$, be a support set of a, i.e. $a_j = 1$ for all $j \in J$. By the construction $g(y, x)$ is nonlinear for those y satisfying $[y] \in J$, where $[y]$ denotes the decimal representation of y. Then,

$$W_g((\beta, \alpha)) = \sum_{y \in \mathbb{F}_2^e} \sum_{x \in \mathbb{F}_2^{2^m - 1}} (-1)^{g(y,x) + (y,x) \cdot (\beta, \alpha)} =$$

$$\sum_{y \in \mathbb{F}_2^e | [y] \notin J} (-1)^{y \cdot \beta} \sum_{x \in \mathbb{F}_2^k} (-1)^{g(y,x) + x \cdot \alpha} + \sum_{y \in \mathbb{F}_2^e | [y] \in J} (-1)^{y \cdot \beta} \sum_{x \in \mathbb{F}_2^{2^m - 1}} (-1)^{g(y,x) + x \cdot \alpha} (17)$$

Note that for those y, such that $[y] \notin J$, we have $g(y, x) + x \cdot \alpha = (\sum_{j \in J} A_{[y], j} + \alpha) \cdot x$, hence the exponent above is a linear function in x. Also for a given y such that $[y] \in J$, the exponent above can be written as $g(y, x) + x \cdot \alpha = (\sum_{j \in J} A_{[y], j} + \alpha) \cdot x + \prod_{k=1}^{2^m - 1} x_k^{A_{[y], [y]}(k) \oplus 1}$. There are two cases to be considered.

Firstly suppose that $\alpha \in \mathbb{F}_2^{2^m - 1}$ is such that $\alpha \ne \sum_{j \in J} A_{[y], j}$ for any y such that $[y] \notin J$. Then the first sum of equation (17) is obviously zero. Computing the second sum, the worst case arises if $\alpha = \sum_{j \in J} A_{[y], j}$ for some y such that $[y] \in J$. But this y is unique due to the properties of the construction. Denote this y by y', and also for convenience let $\prod_{k=1}^{2^m - 1} x_k^{A_{[y'], [y']}(k) \oplus 1} = x_{u_1} x_{u_2} \cdots x_{u_{2^{m-1} - 1}}$, and $c = \sum_{j \in J} A_{[y], j} + \alpha$. Thus the second sum in (17) can be written as,

$$\left| \sum_{y \in \mathbb{F}_2^e | [y] \in J} (-1)^{y \cdot \beta} \sum_{x \in \mathbb{F}_2^{2^m - 1}} (-1)^{g(y,x) + x \cdot \alpha} \right| \le \left| (-1)^{y' \cdot \beta} \underbrace{\sum_{x \in \mathbb{F}_2^{2^m - 1}} (-1)^{x_{u_1} x_{u_2} \cdots x_{u_{2^{m-1} - 1}}}}_{2^{2^m - 1} - 2^{2^{m-1}}} \right| +$$

$$\left| \sum_{\substack{y \in \mathbb{F}_2^e | [y] \in J \\ y \ne y'}} (-1)^{y \cdot \beta} \underbrace{\sum_{x \in \mathbb{F}_2^{2^m - 1}} (-1)^{\overset{\ne 0}{\overbrace{c \cdot x}} + x_{i_1} x_{i_2} \cdots x_{i_{2^{m-1} - 1}}}}_{0 \text{ or } -2^{2^{m-1}}} \right| \le (2^{2^m - 1} - 2^{2^{m-1}}) + (s - 1) 2^{2^{m-1}}.$$

Hence in this case $|W_g((\beta, \alpha))| \le (2^{2^m - 1} - 2^{2^{m-1}}) + (s - 1) 2^{2^{m-1}} = 2^{2^m - 1} + (s - 2) 2^{2^{m-1}}$. Remark that α can be such that $\alpha \ne \sum_{j \in J} A_{[y], j}$ for any y, either $[y] \in J$ or $[y] \notin J$. Then it is easy to verify that $|W_g((\beta, \alpha))| \le s 2^{2^{m-1}}$, for such an α.

When $\alpha \in \mathbb{F}_2^{2^m - 1}$ is such that $\alpha = \sum_{j \in J} A_{[y], j}$ for some y such that $[y] \notin J$, then the first some above is equal to $2^{2^m - 1}$. Then using the similar calculation as above, the second sum in (17) satisfies

$$\left| \sum_{y \in \mathbb{F}_2^e | [y] \in J} (-1)^{y \cdot \beta} \sum_{x \in \mathbb{F}_2^{2^m - 1}} (-1)^{g(y,x) + x \cdot \alpha} \right| \le s 2^{2^{m-1}}.$$

Thus, for such an α we have $|W_g((\beta, \alpha))| \leq (2^{2^m-1}) + s2^{2^{m-1}}$.

Obviously the maximum value in the Walsh spectra correspond to the latter case and takes the highest value when $wt(a) = m$,

$$\max_{(\beta, \alpha) \in \mathbb{F}_2^e \times \mathbb{F}_2^{2^m-1}} |W_G((\beta, \alpha))| \leq (2^{2^m-1}) + m2^{2^{m-1}},$$

and the statement is proved.   □

Differential Uniformity for Arrays

K.J. Horadam

RMIT University, Melbourne VIC 3001, AUSTRALIA
horadam@rmit.edu.au

Abstract. The susceptibility of iterated block ciphers to differential cryptanalysis is minimised by using S-box functions with low differential uniformity.

We extend the idea of differential uniformity to S-boxes with array inputs, giving a unified perspective from which to approach existence and construction problems for highly nonlinear functions. Properties of 2D differentially m-uniform functions are derived, two constructions are given and relationships with known 1D PN and APN functions are demonstrated.

1 Introduction

The differential attack introduced by Biham and Shamir [1] has had a substantial effect on the design of block encryption algorithms, requiring the establishment of theoretical measures of resistance to differential attacks and the construction of resistant functions, as well as study of the links with other design criteria.

Nyberg [15, p.15] introduced differential m-uniformity as one such measure: she defined a function $f : G \to C$ of abelian groups to be differentially m-uniform if $m = \max |\{h \in G : f(g + h) - f(h) = c : g \in G, g \neq 0, c \in C\}|$, where $|X|$ denotes the cardinality of set X. If f is an S-box function, its susceptibility to differential cryptanalysis is minimised if m is as small as possible.

Nyberg's original *perfect nonlinear* (PN) functions [14, Def. 3.1] have $G = \mathbf{Z}_n^a$ and $C = \mathbf{Z}_n^b$, $a \geq b$, and when $n = 2$ they are precisely the *(vectorial) bent* functions (cf. Chabaud and Vaudenay [4, p.358]). When $a = b$, the concepts of differentially 1-uniform, PN and planar functions all coincide. We know examples of such PN functions exist when n is an odd prime p, but cannot exist when $p = 2$. When $a = b$, a differentially 2-uniform function is also termed *almost perfect nonlinear* (APN) and when, additionally, n is even, it is a *semi-planar* function (Coulter and Henderson [5, §2]). When n is odd, any maximally nonlinear function is APN (cf. [4, Theorem 4]). The best that can be expected when $p = 2$ is an APN function. For primes p, the quest for PN and APN functions $\mathbf{Z}_p^a \to \mathbf{Z}_p^a$ focusses on polynomial power functions in the Galois Field $\mathrm{GF}(p^a)$, see for example [6,9,11]. For contributions to the more general theory of differential uniformity for block ciphers see for example [4,10,2,3].

K.G. Paterson (Ed.): Cryptography and Coding 2003, LNCS 2898, pp. 115–124, 2003.
© Springer-Verlag Berlin Heidelberg 2003

Recent interest in encryption algorithms involving arrays, sparked by the choice of *Rijndael* as the AES algorithm, raises the problem of differential cryptanalysis of ciphertext which is genuinely array-encrypted (rather than, for example, encrypted by a set of key-dependent S-boxes each encrypting an input block). The purpose of this paper is to extend the ideas of differential uniformity to S-boxes accepting array inputs, and to determine properties, constructions and links with the one-dimensional case of block inputs.

Highly nonlinear functions of arrays may themselves also provide a potential source of key-dependent S-boxes for block inputs, or of mixer functions for iterative block ciphers, or of hash-based MACs.

2   Total Differential Uniformity for Arrays

Suppose hereafter that G is a finite group (not necessarily abelian), written multiplicatively, and C is a finite abelian group, written additively.

The function $\phi : G \to C$ is *normalised* if $\phi(1) = 0$. The set of normalised mappings $\phi : G \to C$ forms an abelian group $C^1(G, C)$ under pointwise addition. Given $\phi \in C^1(G, C)$, for each $a \in G$ and $c \in C$ we set $n_\phi(a, c) = |\{g \in G : \phi(ag) - \phi(g) = c\}|$ and say the *distribution* of ϕ is the *multiset* of frequencies $\mathcal{D}(\phi) = \{n_\phi(a, c) : a \in G, c \in C\}$. Then ϕ is *differentially Δ_ϕ-uniform* if $\max\{n_\phi(a, c) : a \in G, c \in C, a \neq 1\} = \Delta_\phi$.

We extend these ideas to two dimensions. Let $C^2(G, C)$ be the set of two-dimensional functions $\Phi : G \times G \to C$ which satisfy $\Phi(1, g) = \Phi(g, 1) = 0$, $g \in G$. Under pointwise addition $C^2(G, C)$ is an abelian group.

A differential attack on an S-box function Φ with two-dimensional array inputs would involve fixing an input pair $(a, b) \neq (1, 1) \in G \times G$ and looking for bias in the frequencies of output differences $\Phi(ag, hb) - \Phi(g, h)$, as (g, h) runs through $G \times G$. Consequently, the susceptibility of such a function to differential attack is minimised if the maximum of these frequencies is as small as possible.

Definition 1 For $\Phi \in C^2(G, C)$ and for each $(a, b) \in G \times G$ and $c \in C$, we set
$$n_\Phi(a, b\,;c) = |\{(g, h) \in G \times G : \Phi(ag, hb) - \Phi(g, h) = c\}|.$$
Define Φ to be *totally differentially m-uniform* if
$$\max\{n_\Phi(a, b\,;c) : (a, b) \neq (1, 1) \in G \times G, c \in C\} = m.$$

Optimal total differential uniformity will occur only when every element of C arises as a total differential equally often; that is, when $|C|$ divides $|G|^2$ and $m = |G|^2/|C|$.

Obviously, if G is abelian, which is usual for implementations, $hb = bh$ in the above definition and the 2D *directional derivative* $(\mathbf{d}\Phi)_{(a, b)}$ *of Φ in direction* (a, b) may be defined unambiguously: $(\mathbf{d}\Phi)_{(a, b)}(g, h) = \Phi(ag, bh) - \Phi(g, h)$. The notions above then naturally extend to n-dimensional functions $\Phi : G^n \to C$, but here we are only going to deal with the 2D case.

We will focus most attention on those elements of $C^2(G,C)$ which satisfy a further condition and are known as cocycles. A normalised (2-dimensional) *cocycle* is a mapping $\psi : G \times G \to C$ satisfying $\psi(1,1) = 0$ and

$$\psi(g,h) + \psi(gh,k) = \psi(g,hk) + \psi(h,k), \quad \forall g,h,k \in G. \tag{1}$$

We can usefully represent a cocycle ψ as a matrix $[\psi(g,h)_{g,h\in G}]$ with rows and columns indexed by the elements of G in some fixed ordering (beginning with the identity 1).

For fixed G and C, the set of cocycles forms a subgroup $Z^2(G,C)$ of $C^2(G,C)$. There is a natural homomorphism $\partial : C^1(G,C) \to Z^2(G,C)$ which, as we will show, preserves distribution properties. For $\phi \in C^1(G,C)$, define $\partial\phi : G \times G \to C$ to be

$$\partial\phi(g,h) = \phi(gh) - \phi(g) - \phi(h), \quad g,h \in G. \tag{2}$$

Then $\partial\phi(1,g) = \partial\phi(g,1) = 0$, $g \in G$. A function $\partial\phi$ satisfying (2) is called a *coboundary*, and measures the amount by which ϕ differs from a homomorphism of groups. The coboundaries form a subgroup $B^2(G,C)$ of $Z^2(G,C)$, and $\ker \partial$ is the group of homomorphisms $\mathrm{Hom}(G,C)$ from G to C, so the induced mapping $\partial : C^1(G,C)/\mathrm{Hom}(G,C) \to B^2(G,C)$ is an isomorphism.

The advantages of using cocycles are manifold: firstly, coboundaries are cocycles so the extension from 1D to 2D functions is very natural; secondly, a lot is known about the group of cocycles so the search for highly nonlinear 2D functions can be undertaken in a structured way rather than by exhaustive search; and thirdly, the frequency counts of the total differential of a cocycle may be simplified in terms of the frequency counts of the partial differentials in the row and column directions of the cocyclic matrix.

For $\Phi \in C^2(G,C)$, and for $k \in G$, the *left first partial derivative* $(\Delta_1\Phi)_k :$ $G \times G \to C$ of Φ in direction k is $(\Delta_1\Phi)_k(g,h) = \Phi(kg,h) - \Phi(g,h)$ and the *right first partial derivative* $(\nabla_1\Phi)_k : G \times G \to C$ of Φ in direction k is $(\nabla_1\Phi)_k(g,h) = \Phi(gk,h) - \Phi(g,h)$. Corresponding definitions apply for the second partial derivatives $(\Delta_2\Phi)_k$ and $(\nabla_2\Phi)_k$.

Lemma 1 *Let* $\psi \in Z^2(G,C)$. *Then*
$$n_\psi(a,b\,;c) = |\{(g,h) \in G \times G : (\Delta_1\psi)_{ag}(h,b) + (\nabla_2\psi)_h(a,g) = c\}|.$$

Proof. Assume $(a,b) \neq (1,1) \in G \times G$. Then
$$n_\psi(a,b\,;c) = |\{(g,h) \in G \times G : \psi(ag,hb) - \psi(g,h) = c\}| =$$
$$|\{(g,h) \in G \times G : \psi(ag,hb) - \psi(ag,h) + \psi(ag,h) - \psi(g,h) = c\}|$$
and by (1) this equals
$$|\{(g,h) \in G \times G : (\psi(agh,b) - \psi(h,b)) + (\psi(a,gh) - \psi(a,g)) = c\}|. \quad \square$$

3   Coboundaries and Differential Row Uniformity

In [12] the distribution of a 1D function is related to the row frequencies of the corresponding 2D coboundary matrix. There is a differential action of G on $C^1(G, C)$ which preserves distributions.

Definition 2   1. For $\Phi \in C^2(G, C)$, define $N_\Phi(g, c) = |\{h \in G : \Phi(g, h) = c\}|$, $g \in G, c \in C$ and $M_\Phi(h, c) = |\{g \in G : \Phi(g, h) = c\}|$, $h \in G, c \in C$. The *distribution* of Φ is the *multiset* of row frequencies

$$\mathcal{D}(\Phi) = \{N_\Phi(g, c) : g \in G, c \in C\}.$$

Define Φ to be *differentially Δ_Φ-row uniform* if

$$\Delta_\Phi = \max\{N_\Phi(g, c) : g \in G, c \in C, g \neq 1\}$$

and *differentially ∇_Φ-column uniform* if

$$\nabla_\Phi = \max\{M_\Phi(g, c) : g \in G, c \in C, g \neq 1\}$$

2. The *shift action* of G on $C^1(G, C)$ is defined for $k \in G$ and $\phi \in C^1(G, C)$ by $\phi \cdot k : g \mapsto (\nabla\phi)_g(k) = \phi(kg) - \phi(k)$, $g \in G$. The *shift action* of G on $C^2(G, C)$ is defined for $k \in G$ and $\Phi \in C^2(G, C)$ to be

$$(\Phi \cdot k)(g, h) = (\nabla_1\Phi)_g(k, h) = \Phi(kg, h) - \Phi(k, h), \quad g, h \in G.$$

Note that in [12], the differential row-uniformity property was termed differential uniformity. This should not be confused with total differential uniformity as described here.

Lemma 2 [12] *Let* $\phi \in C^1(G, C)$. *Then for any* $\gamma \in \mathrm{Aut}(C)$, $k \in G$, $\theta \in \mathrm{Aut}(G)$, $\mathcal{D}(\phi) = \mathcal{D}(\gamma \circ (\phi \cdot k) \circ \theta)$ *and* $\Delta_\phi = \Delta_{\gamma \circ (\phi \cdot k) \circ \theta}$.   □

Lemma 3 [12] *For* $\psi \in Z^2(G, C)$ *define* $\mathcal{B}(\psi) \subset Z^2(G, C)$ *to be*

$$\mathcal{B}(\psi) = \{\gamma \circ (\psi \cdot k) \circ (\theta \times \theta), \ \gamma \in \mathrm{Aut}(C), \ k \in G, \ \theta \in \mathrm{Aut}(G)\}.$$

Then
   (i) if $\phi \in C^1(G, C)$, *then* $\mathcal{D}(\phi) = \mathcal{D}(\partial\phi)$ *and* $\Delta_\phi = \Delta_{\partial\phi}$;
   (ii) $\mathcal{B}(\partial\phi) = \{\partial(\gamma \circ (\phi \cdot k) \circ \theta), \ \gamma \in \mathrm{Aut}(C), \ k \in G, \ \theta \in \mathrm{Aut}(G)\}$ *for* $\phi \in C^1(G, C)$;
   (iii) $\mathcal{D}(\psi) = \mathcal{D}(\varphi)$ *for all* $\varphi \in \mathcal{B}(\psi)$.   □

In this section, we describe properties of differentially row-uniform cocycles. We will use the term PN (*perfect nonlinear*) for differentially m-row-uniform cocycles when $|C|$ divides $|G|$ and $m = |G|/|C|$. It is known that the existence of a PN cocycle is equivalent to the existence of a $(|G|, |C|, |G|, |G|/|C|)$ relative difference set in some central extension of C by G. In particular, when $G = C$

is an elementary abelian p-group, the existence of a differentially 1-row-uniform (PN) cocycle corresponds to the existence of a central $(p^a, p^a, p^a, 1)$ relative difference set.

We will also use the term APN (*almost perfect nonlinear*) for differentially 2-row-uniform cocycles when $G = C$ is an elementary abelian p-group.

Canteaut's characterisation [2] of binary 1D APN functions in terms of 'second derivatives' is in fact a special case of a characterisation of arbitrary differentially m-row uniform cocycles: no more than m entries in any non-initial row of the matrix corresponding to any shift of the cocycle can be 0.

Theorem 1 $\psi \in Z^2(G, C)$ *satisfies* $\Delta_\psi \leq m$ *if and only if, for all* $g \neq 1$, $h \neq 1 \in G$ *and all* $k \in G$, $|\{h \in G : (\psi \cdot k)(g, h) = 0\}| \leq m - 1$.

Proof. If $\Delta_\psi \leq m$ then by Lemma 3, $\Delta_{\psi \cdot k} \leq m$ for any $k \in G$ and the row condition holds. Conversely, if $\Delta_\psi \geq m + 1$ then for some $g' \neq 1 \in G$, there exist $m + 1$ distinct elements $x, y_1, y_2, \ldots, y_m \in G$ such that $\psi(g', x) = \psi(g', y_1) = \psi(g', y_2) = \ldots = \psi(g', y_m)$. Write $y_i = x u_i$, $i = 1, \ldots, m$, and $g = x^{-1} g' x$. By Equation (1), $\psi(g', x u_i) - \psi(g', x) = (\psi \cdot x)(g, u_i) = 0$, $i = 1, \ldots, m$, giving m distinct values $h \neq 1$ for which $(\psi \cdot x)(g, h) = 0$. $\square$

Canteaut's characterisation, namely: ϕ is APN if and only if, for all non-zero $g \neq h \in G$, $\Delta((\Delta\phi)_h)_g(k) \neq 0$, $\forall k \in G$, is the case $m = 2$, $G = C = (\mathrm{GF}(2^a), +)$, $\psi = \partial\phi$ of Theorem 1, noting that $\Delta((\Delta\phi)_h)_g(k) = \phi(k + g + h) - \phi(k + g) - \phi(k + h) + \phi(k) = (\partial(\phi \cdot k))(g, h) = ((\partial\phi) \cdot k)(g, h)$.

Corollary 1 (Canteaut [2]) *Let* $G = (\mathrm{GF}(2^a), +)$. *Then* $\phi \in C^1(G, G)$ *is APN if and only if, for all* $g \neq 0$, $h \neq 0 \in G$ *and all* $k \in G$, $((\partial\phi) \cdot k)(g, h) \neq 0$ *unless* $h = g$. $\square$

A general family of differentially row-uniform coboundaries may be constructed from DO polynomials. Recall that a DO (Dembowski-Ostrom) polynomial [5, p.23] is a mapping $\phi : \mathrm{GF}(p^a)[x] \rightarrow \mathrm{GF}(p^a)[x]$ which, when reduced modulo $x^{p^a} - x$, is of the form

$$\phi(x) = \sum_{i=0}^{a-1} \sum_{j=0}^{a-1} \lambda_{ij} \, x^{p^i + p^j}, \ \lambda_{ij} \in \mathrm{GF}(p^a). \tag{3}$$

If ϕ is a DO monomial then there exists m such that $N_{\partial\phi}(g, c) = m$ or 0 for every $g \neq 0$ and every $c \in G$. (We are informed Ostrom has an unpublished geometric proof of this and calls ϕ with this property *quasi-planar*. Apparently, for $m > 2$, all of the known quasi-planar functions may be represented by DO monomials.)

Moreover, m is a power of p. This means that not only is $\partial\phi$ differentially m-row-uniform, it is actually a PN coboundary from $G \times G$ to a subgroup of G.

Theorem 2 (Construction A) *Let $G = C = (\mathrm{GF}(p^a), +)$ and let $\phi : G \to G$ be a DO monomial, $\phi(g) = g^{p^i + p^j}$, $g \in G$. Then there exists $0 \le b \le a$ such that $N_{\partial\phi}(g, c) = p^b$ or 0 for every $g \neq 0 \in G$ and every $c \in G$. Consequently $\partial\phi$ is differentially p^b-row-uniform and $\partial\phi : G \times G \to \mathrm{Im}(\partial\phi)$ is PN.*

Proof. Since $\partial\phi(g, h) = g^{p^i} h^{p^j} + h^{p^i} g^{p^j}$ and $\partial\phi(g, 0) = 0$, if we fix $g \neq 0$, we may set $N_{\partial\phi}(g, 0) = m \ge 1$. Since $\partial\phi(g, h_1 + h_2) = \partial\phi(g, h_1) + \partial\phi(g, h_2)$, if $N_{\partial\phi}(g, c) \neq 0$ then $N_{\partial\phi}(g, c) = N_{\partial\phi}(g, 0) = m$. This additivity also ensures that $\mathrm{Im}(\partial\phi(g, -))$ is a subgroup of G, so $\mathrm{Im}(\partial\phi(g, -)) \cong \mathbf{Z}_p^{a-b}$ for some $0 \le b \le a$. A counting argument shows $m = p^b$. Since $\partial\phi(kg, kh) = k^{p^i + p^j} \partial\phi(g, h)$, $N_{\partial\phi}(kg, 0) = N_{\partial\phi}(g, 0) = p^b$ for every $k \neq 0 \in G$. $\square$

In fact the parameter p^b in Theorem 2 is readily determined.

Lemma 4 *Let $G = C = (\mathrm{GF}(p^a), +)$ and let $\phi : G \to G$ be a DO monomial, $\phi(g) = g^{p^i + p^j}$, where $i \le j$.*

1. *When $p = 2$, $\Delta_{\partial\phi} = 2^a$ when $i = j$, $\Delta_{\partial\phi} = 2$ when $i < j$ and $(a, j - i) = 1$ and $\Delta_{\partial\phi} = 2^{(j-i)}$ when $i < j$ and $(a, j - i) > 1$.*
2. *When p is odd, $\Delta_{\partial\phi} = 1$ when $i = j$ or when $i < j$ and $a/(a, j - i)$ is odd and $\Delta_{\partial\phi} = p^{(a, j-i)}$ otherwise.*

Proof. By taking θ in Lemma 3 to be the Frobenius automorphism $\theta(g) = g^{p^{-i}}$ of $\mathrm{GF}(p^a)$, we see $\Delta_{\partial\phi} = \Delta_{\partial(\phi \circ \theta)}$ so we may assume that $\phi(g) = g^{1 + p^{(j-i)}}$. The cases $i = j$ follow directly. Suppose $0 < j - i = k$ and let $g = 1$ and $n_p = |\{h \in G : h \neq 0, h \neq 1, h + h^{p^k} = 0\}| = |\{h \in G : h \neq 0, h \neq 1, h^{p^k - 1} = -1\}|$. If $p = 2$, $N_{\partial\phi}(1, 0) = 2 + n_2$ and $n_2 = 0$ iff $(2^a - 1, 2^k - 1) = 1$ iff $(a, k) = 1$; $n_2 = 2^k - 2$ iff $(2^k - 1) | (2^a - 1)$ iff $(a, k) > 1$. If p is odd, $N_{\partial\phi}(1, 0) = 1 + n_p$ and $n_p = 0$ iff $a/(a, k)$ is odd (eg. see [6, Theorem 3.3]). If $n_p > 0$, α is a primitive element of $\mathrm{GF}(p^a)$ and $(\alpha^m)^{(p^k - 1)} = -1$ then all other solutions $(\alpha^n)^{(p^k - 1)} = -1$ satisfy $n = m + t(p^a - 1)^2 / 2(p^d - 1)^2$ for some integer t where $d = (a, k)$, and there are only $p^d - 1$ values of t giving distinct solutions. $\square$

This construction clearly cannot explain the existence [11] of APN power functions for odd p. However, it does account for the simplest known classes of PN and APN power functions.

Example 1 *Let $G = C = (\mathrm{GF}(p^a), +)$ and let $\phi : G \to G$ be a DO monomial, $\phi(g) = g^{p^i + p^j}$, $g \in G$.*

1. *If $i = j = 0$ and p is odd, $\partial\phi$ (and ϕ) is differentially 1-row-uniform (see [11, Theorem 1.1]).*
2. *If $i = 0$, $j \ge 1$ and p is odd, $\partial\phi$ (and ϕ) is differentially 1-row-uniform if and only if $a/(a, j)$ is odd (see [11, Theorem 1.2]).*
3. *If $i = 0$, $j \ge 1$ and $p = 2$, $\partial\phi$ (and ϕ) is APN if and only if $(a, j) = 1$ (see [5, Theorem 3.ii]).*

We now give a construction of $m\ell$-row-uniform functions from m-row-uniform functions.

Theorem 3 (Construction B) *Let $\Phi \in C^2(G, C)$ and let $f \in \mathrm{Hom}(C, C)$. If $\Delta_\Phi = m$ and $f(C)$ has index ℓ in C then $\Delta_{f \circ \Phi} \leq m\ell$. In particular, if $|G| = |C|$ and $\Delta_\Phi = 1$ then $\Delta_{f \circ \Phi} = \ell$.*

Proof. If $c' \notin \mathrm{Im} f$ then $N_{f \circ \Phi}(g, c') = 0$. If $c' \in \mathrm{Im} f$ and $f(c) = c'$ then $N_{f \circ \Phi}(g, c') = \sum_{d \in \mathrm{Ker} f} N_\Phi(g, c + d) \leq m\ell$. $\square$

Because of the extensive literature on existence of $(p^a, p^a, p^a, 1)$ relative difference sets, we know there are many infinite families of 2D functions ψ on $G = C = (\mathrm{GF}(p^a), +)$ satisfying $\Delta_\psi = 1$. For example, Horadam and Udaya [13, Corollary 3.6, Table 1] show that the number of equivalence classes of such relative difference sets is at least a and in general appears to be much greater.

So we can construct square arrays such that the difference of every pair of distinct rows (and also, distinct columns) meets every element of $\mathrm{GF}(p^a)$ exactly once. One of the most interesting features from a practical point of view is that families with $\Delta_\psi = 1$ exist even when $p = 2$, in contrast with the 1D functions ϕ where $\Delta_\phi = \Delta_{\partial \phi} = 2$ is the best possible.

Construction B may therefore be employed to construct binary APN functions, but again, cannot account for the existence of APN power functions for odd p. There is some evidence to suggest the latter functions may arise from symmetrisations of differentially 1-row-uniform functions, and we conjecture their form below.

A mapping $\Phi \in C^2(G, C)$ is *symmetric* if $\Phi(g, h) = \Phi(h, g)$ always. For example, if G is abelian, any coboundary $\partial \phi$ is symmetric. The *symmetrisation* $\widehat{\Phi} \in C^2(G, C)$ of Φ is

$$\widehat{\Phi}(g, h) = \Phi(g, h) + \Phi(h, g), \ g, h \in G. \tag{4}$$

The symmetrisation of a cocycle $\psi \in Z^2(G, C)$ is not necessarily a cocycle, but if G is abelian or ψ is symmetric, $\widehat{\psi}$ is a cocycle.

Conjecture: Suppose p is odd. Let ϕ be a DO monomial and set $\Phi(g, h) = \phi(g)h$, so $\Delta_\Phi = 1$. If $\widehat{\Phi} = \partial \varphi$ for some coboundary $\partial \varphi$ then $\Delta_{\widehat{\Phi}} = 2$.

4   Multiplicative Cocycles

In this section we deal with a class of cocycles for which the total differential frequencies are easily calculated.

A cocycle $\psi \in Z^2(G, C)$ is *multiplicative* if it is a homomorphism in either coordinate (and hence, by (1), in both coordinates).

Lemma 5 *If $\psi \in Z^2(G, C)$ is multiplicative, and $(a, b) \neq (1, 1) \in G \times G$, then $n_\psi(a, b; c) = \sum_{e \in C} N_\psi(a, e) M_\psi(b, c - e)$.*
Consequently, if ψ is multiplicative and PN, then ψ has optimal total differential uniformity.

Proof. Assume $(a, b) \neq (1, 1) \in G \times G$. Then $n_\psi(a, b; c) =$
$$|\{(g, h) \in G \times G : \psi(ag, hb) - \psi(g, h) = c\}| =$$
$$|\{(g, h) \in G \times G : \psi(ag, b) + \psi(a, h) = c\}| =$$
$$\sum_{e \in C} |\{h \in G : \psi(a, h) = e\}| \cdot |\{g \in G : \psi(ag, b) = c - e\}|.$$
If ψ is PN, then $N_\psi(a, e) = M_\psi(b, c - e) = |G|/|C|$ for all e and $n_\psi(a, b; c) = |G|^2/|C|$. □

Chen (cf. [13, Lemma 2.11]) has shown that if a multiplicative cocycle is PN there is a prime p such that both G and C are elementary abelian p-groups.

The multiplicative PN cocycles with $G = C$ may be characterised in terms of presemifields. Recall that $(F, +, *)$ is a presemifield if $(F, +)$ is an abelian group (with additive identity 0), $(F \backslash \{0\}, *)$ is a quasigroup (that is, for any non-zero g, h in F, there are unique solutions in F to $g * x = h$ and $y * g = h$, and both distributive laws hold.

Theorem 4 *Let G be abelian and $\psi \in Z^2(G, G)$ be multiplicative. Then ψ is PN if and only if $(G, +, \psi)$ is a presemifield.*

Proof. If $(G, +, \psi)$ is a presemifield with multiplication ψ (that is, $\psi(g, h) = g * h$) then by [13, Lemma 2.8] ψ is PN. Conversely, if ψ is PN then $G = C = (\mathbf{Z}_p)^a$ and for any $g, h \neq 0 \in G$, there are unique solutions in G to $\psi(g, x) = h$ and $\psi(y, g) = h$. Finally, both distributive laws hold by multiplicativity. □

Multiplicative cocycles are wholly determined by their values on pairs of elements from a minimal generating set for G. As any other values are found by additions only, they are fast to compute by comparison with non-multiplicative cocycles, for which a generating set may be difficult to determine and from which other entries must be computed using (1).

Since the multiplicative cocycles are so highly structured, we always have to balance their potential utility as array-input S-box functions against the ease of recovering them. However, in the most likely case that $G = (\mathbf{Z}_p)^a$, there are $|C|^{a^2}$ multiplicative cocycles, so for example in the binary case we only need, say, $a = 32$ for a search space of size 2^{1024}, prohibitively large for exhaustion.

5   Application to Block Cipher Design

We close with some observations about the application of 2D functions to the design of (binary) iterated block ciphers. In *Rijndael*, as in its predecessor *Square*, arrays are formed in a round by dividing a 128-bit input block into 16 bytes and writing them, in order, into a 4×4 array of bytes. Round transformation then involves specified row and column permutations of the array of bytes as well as byte substitutions. A stated aim of this design feature was to minimise susceptibility to linear and differential cryptanalysis [7,8].

A 2D function, acting on array inputs, provides a model which captures essential features of this design. It is not necessary that every possible ordered pair be an input to the 2D function. (In the case of *Rijndael* and Square, the 16 bytes appear as 4 rows of entries from $\mathrm{GF}(2^8)$, for instance, so the round encryption can be regarded as a mapping $X \times \mathrm{GF}(2^8) \to \mathrm{GF}(2^8)$ where X is a subset of size 4 in $\mathrm{GF}(2^8)$.) If a 2D function $\Phi : \mathrm{GF}(2^a) \times \mathrm{GF}(2^a) \to \mathrm{GF}(2^a)$ with low total differential uniformity is used, this property will be inherited by any subset $X \subseteq \mathrm{GF}(2^a)$.

Another design feature which might easily be incorporated into a 2D model is the L-R splitting of input typical in Feistel ciphers. Input blocks of even length $2a$, written as usual as concatenated blocks $x = x_1 x_2$ each of length a, form input pairs (x_1, x_2) to a round function $\Phi : \mathrm{GF}(2^a) \times \mathrm{GF}(2^a) \to \mathrm{GF}(2^b)$. If $b < 2a$ there is clearly compression of output; this has been acceptable in the design of DES S-boxes, for example, because the input blocks had been suitably expanded. As pointed out above, the case $b = a$ has had the most attention in the literature.

However, to truly exploit the idea of array inputs involved in both the above design features, it seems reasonable in future to look for optimal totally differentially m-uniform functions where $m < 2^a$ and, indeed, where $m = 1$; that is, $b = 2a$.

We can also consider applying good 2D functions to the design of key-dependent or randomised S-boxes for iterated block ciphers. Here, for each input block the S-box function is selected from a large set $\{S_k, k \in K\}$ of S-box functions, to spread any biases inherent in individual S-box functions. In effect, this replaces a monoalphabetic substitution in a given round by a polyalphabetic substitution. This can be thought of as a 2D S-box, with (say) the first coordinate being the identifier of the 1D S-box to be used and the second coordinate being the input block for the round. The keyed S-box is then a 2D function $S(k, x) = S_k(x)$, $x \in M$ of the key and message spaces.

If instead, a 2D function Φ with low total differential uniformity is chosen, and the collection of 1D S-boxes derived as $S_k = \Phi(k, -)$, it should be possible to guarantee improved robustness to differential attacks on auto or cross-correlated output blocks, by a measurable amount. For example, if $K = M = (\mathbf{Z}_p)^a$ and $\psi : K \times M \to M$ is a cocycle with optimal total differential uniformity then the differences $S_{k+\ell}(x) - S_k(x)$, $\ell \neq 0 \in K$ are unbiased for each input block x, and the differences $S_k(x+b) - S_k(x)$, $b \neq 0 \in M$ are unbiased for each keyed S-box S_k.

Finally we emphasise that in the binary case, a 2D point of view can allow us to construct cryptographic functions which are better than their 1D counterparts. The following instance is probably the simplest demonstration of this statement.

Let $G = C = (\mathbf{Z}_p)^a$, considered as the additive group of $\mathrm{GF}(p^a)$. We know the quadratic function $\phi(g) = g^2, g \in G$ and the corresponding coboundary $\partial\phi(g, h) = 2gh$, $g, h \in G$ have the same differential row uniformity, by Lemma 3.(i). When p is odd, these functions are PN, that is, optimal, and hence so is the multiplicative cocycle $\psi(g, h) = gh = 2^{-1}\partial\phi(g, h)$, which is field multiplication on $\mathrm{GF}(p^a)$.

When $p = 2$, however, ϕ has worst-possible differential uniformity $\Delta_\phi = 2^a$ by Lemma 4.1, and $\partial\phi = 0$. Nonetheless, the field multiplication $\psi(g, h) = gh$ on $GF(2^a)$ is PN; in fact it has optimal total differential uniformity.

References

1. E. Biham and A. Shamir, Differential cryptanalysis of DES-like cryptosystems, *J. Cryptology* **4** (1991) 3–72.
2. A. Canteaut, Cryptographic functions and design criteria for block ciphers, IN-DOCRYPT 2001, eds. C. Pandu Rangan, C. Ding, LNCS 2247, Springer 2001, 1–16.
3. A. Canteaut and M. Videau, Degree of composition of highly nonlinear functions and applications to higher order differential cryptanalysis, EUROCRYPT-02, LNCS 2332, Springer, New York, 2002, p.518ff.
4. F. Chabaud and S. Vaudenay, Links between linear and differential cryptanalysis, EUROCRYPT-94, LNCS 950, Springer, New York, 1995, 356–365.
5. R. S. Coulter and M. Henderson, A class of functions and their application in constructing semi-biplanes and association schemes, *Discrete Math.* **202** (1999) 21–31.
6. R. S. Coulter and R. W. Matthews, Planar functions and planes of Lenz-Barlotti Class II, *Des., Codes and Cryptogr.* **10** (1997) 167–184.
7. J. Daemen, L. R. Knudsen and V. Rijmen, The block cipher Square, Fast Software Encryption '97, ed. E. Biham, LNCS 1267, Springer, 1997, 149–165.
8. J. Daemen and V. Rijmen, The Design of Rijndael: AES - The Advanced Encryption Standard, Springer, Berlin, 2002.
9. H. Dobbertin, Almost perfect nonlinear power functions on $GF(2^n)$: the Welch case, *IEEE Trans. Inform. Theory* **45** (1999) 1271–1275.
10. P. Hawkes and L. O'Connor, XOR and non-XOR differential probabilities, EUROCRYPT-99, LNCS 1592, Springer, New York, 1999, p.272ff.
11. T. Helleseth, C. Rong and D. Sandberg, New families of almost perfect nonlinear power mappings, *IEEE Trans. Inform. Theory* **45** (1999) 475–485.
12. K. J. Horadam, Differentially 2-uniform cocycles - the binary case, AAECC-15, eds. M. Fossorier, T. Hoeholdt, A. Poli, LNCS 2643, Springer, Berlin, 2003, 150–157.
13. K. J. Horadam and P. Udaya, A new construction of central relative $(p^a, p^a, p^a, 1)$ difference sets, *Des., Codes and Cryptogr.* **27** (2002) 281–295.
14. K. Nyberg, Perfect nonlinear S-boxes, EUROCRYPT-91, LNCS 547, Springer, New York, 1991, 378–385.
15. K. Nyberg, Differentially uniform mappings for cryptography, EUROCRYPT-93, LNCS 765, Springer, New York, 1994, 55–64.

Uses and Abuses of Cryptography
(Invited Paper)

Richard Walton

Visiting Professor,
Information Security Group,
Royal Holloway, University of London, Egham, Surrey TW20 0EX, UK.

1   Introduction

1. I have been associated with Information Security for the last 30 years. I started as a mathematician working on the design and assessment of cryptographic algorithms for the UK Government at a time when cryptography was a Government monopoly. Since then I have been involved in the broader managerial and policy work with my GCHQ career culminating as Director of CESG from 1999 to 2002. During that time I have had ample opportunity to observe the evolution of cryptography as an academic discipline and also as a practical science (the two are often very different). My main conclusion from my 30 years experience is:

Theorem 1. *(Empirical) If a user of a cryptographic system can find a way of abusing it he will do so.*

This is actually just a special case of the more general observation:

Theorem 2. *(Also empirical) Cryptography provides security subject to the satisfaction of various boundary conditions. In practical applications these conditions are ignored.*

2. In this talk I shall illustrate the above theorems and give some pointers towards avoiding their inherent dangers.

2   Cryptography for COMSEC

3. The traditional application for Cryptography is to provide Communications Security (COMSEC), the purpose being to protect the content of communications which have been intercepted by unauthorised third parties. The cryptography provides:

a. A practical algorithm for the originator of a message to encrypt the message.
b. A practical algorithm for the legitimate recipient to decrypt the message.
c. Some assurance that no unauthorised third party can find a practical algorithm to decrypt the message.

K.G. Paterson (Ed.): Cryptography and Coding 2003, LNCS 2898, pp. 125–132, 2003.

Cryptography is sufficient to provide the wanted COMSEC if in addition:

d. Unauthorised parties can't intercept the communication before the encryption takes place or after the decryption takes place.
e. Unauthorised parties can't masquerade as the legitimate recipient.
f. Practices are not introduced to compromise the security offered.
g. Equipment malfunction does not compromise the security.
h. Unauthorised parties cannot interfere with the equipment or transmission in ways that compromise the security.

4. These conditions form the basis of COMSEC assessment, yet they are 'so obvious' that they are often forgotten and actual implementation or deployment of cryptography can introduce subtle breaches. For example in World War II the German Enigma didn't quite satisfy c, but most importantly in order to implement b, the Germans breached f.

5. This leads us to one of the main properties of Cryptography.

Cryptography does not solve the security problem but changes its nature. The aim is to transform the problem into one that is solvable. (Just like Mathematics). Consider the original COMSEC problem: how to communicate securely in the absence of a physically secure channel. The cryptographic solution is to prepare in advance to enable the secure communications to take place. A physically secure channel is still needed to distribute machines and keys to support cryptographic protection but this has now been separated from the actual communication of information. The communications are sent over an open (insecure) channel when required protected by cryptography that has been pre-established.

6. An example is the one time pad. It can be proven that the one time principle offers absolute security. However the encryption consumes exactly the same quantity of key as there is plain text to be sent. Thus there is no saving at all in the need to communicate through a secure channel. The saving is that the key can be distributed at a time and manner of the communicators' choice. So a secure channel can be established (eg by using armed couriers) to distribute the one time key that can then be used when it is required (usually in circumstances when a physically secure channel is unavailable). A theoretically interesting refinement was proposed by Massey in the early 1980's which he dubbed the Rip Van Winkle cipher. In this system the one time key was transmitted in clear in advance. However the transmission was so long that the cost of searching for the key being used for any particular encryption was prohibitive (there being potentially infinite time elapsing between transmission of the key and its use to encrypt a message – hence Rip Van Winkle).

7. Later refinements have generally been an attempt to approximate the security advantages of the one-time key while minimizing the key distribution overhead. The solutions replace the genuine randomness of the one time key with a pseudo-random generator seeded with a much shorter key. For example modern stream ciphers use a short key (of the order of 50- 250 bits) to seed a pseudo-random generator that generates a long sequence of bits to encrypt a message directly. The Data Encryption Standard (DES) uses a slightly different encryption rule (64 bit block cipher), but essentially the principle is the same with 56 bits of key seeding an algorithm that provides a pseudo-random encryption method.

8. In general the cryptographic transformation is to convert the problem of secure communication to one of secure key distribution.

3   PKC and The Importance of Authentication

9. Physical key distribution is all very well for small networks or for well-disciplined communications, but can become prohibitively expensive (and inconvenient) for many modern applications. Attempts to solve the associated problems in the 1970's led to the discovery of Public Key Cryptography (PKC). The important property of PKC is that there is no need for communicators to share any secret information to enable them to set up a secure channel. (Hence the original name for PKC was Non-Secret Encryption (NSE)). This obviated the need for any pre-existing secure channel to distribute keys.

10. Although some methods of PKC can be used to encrypt message directly, they tend to be inconvenient to use in this way and in practise traditional ciphers are still used to encrypt actual messages and PKC is used to distribute (or establish) keys to seed symmetric encryption algorithms. This illustrates a fundamental point about cryptography. Mathematical cryptography is all very well but above all this is a practical science and cryptography must be compatible with technology in order to be of any use. Even today, implementations of PKC are not generally suitable direct encryption of messages. Other techniques are necessary to enable a practical cryptographic system to be implemented and the choice of cryptography is inextricably entwined with the underlying technology. The real reason why old cryptographic principles are superseded is more to do with technology than mathematical security. For example, despite ENIGMA being insecure, the wired-wheel technology on which it is based is fully able to support secure designs – but today's preferred implementation technologies are ill-suited to the wired-wheel rotor principles and now the whole of rotor-based cryptography has been consigned to the dustbin of history. Similarly the cryptographic principles underlying the hardware electronic designs of the last quarter of the twentieth century are dying out in the face of the need to implement cryptography in newer technologies. Back in the 1970's one of the problems with PKC was that it could not be implemented cost-effectively with the technology of the day. This was the main reason why it was the late 1980's before PKC became a practical option even for key management.

11. But PKC brought with it another new issue. In the days when access to specific cryptography involved secure distribution of key (and even equipments) a cryptographic system was considered self-authenticating. Only authorised personnel could make legitimate use of an encryption system so it became an automatic assumption that an encrypted message was authentic. There were some dangers associated with this assumption but it was more-or-less OK and no-one really thought very much about it except in very exceptional circumstances. The advent of PKC made this assumption very dangerous indeed and one of the earliest objections in CESG to the new technology was for this reason. However it was soon realised that the same technology that gives us PKC can also be used to provide secure authentication. So by the time that practical applications of PKC came on stream the necessary authentication was also available. However, convincing users of the need for such authentication has been a struggle.

4   More General Issues for Cryptographic Science

12. COMSEC remains the application to which cryptography is best suited but modern applications of cryptography extend far beyond traditional COMSEC. I have already mentioned Authentication but beyond that we employ cryptographic mechanisms throughout information systems to support a wide range of assurance services. It has been claimed that cryptography is the key enabler for the information age. This exaggerates its importance but nevertheless cryptography can be a valuable tool in providing the trust necessary for the information age to work.

13. Trust (rather than cryptography) is the key that unlocks the power of the Internet and other information technology. Susceptibility to interception is not really the killer weakness. After all, society has been comfortable with telephony and the postal services for years and, in general, these services are trusted by users despite having relatively weak protection against unauthorised interception (far weaker than internet transactions). Furthermore, although privacy and confidentiality are important concerns, there has been general acceptance of the weaknesses of open telecommunications systems in this regard to enable normal personal and business affairs to proceed. There is no need for extra paranoia over interception of internet communications – and this is recognised by most individuals and organisations who have looked at the issues dispassionately. Paranoia has been fed by special interest groups (often academics, commercial interests or conspiracy theorists as well as Government security experts seeking to bring commercial systems up to military security standards). The paranoid have been heeded because of genuine and visible security failures that have demonstrated the untrustworthiness of the technology.

14. The underlying cause of these failures is the inherent complexity of the technology and the difficulties of integrating a heterogeneous set of applications in a dynamic environment. Classical computer security is based on the twin premises that you can't trust people but you can trust a machine. So (the theory goes) you get the machine to mediate all transactions involving humans and subject them to strict regulation. GIGO applies! In reality the correct premises are that you can trust most people most of the time (subject to a few provisos – including not hacking them off through excessive regulation) and that you can trust a machine to do what it has been told to do. The difficulties arise in achieving the right balance of regulation of human beings and in knowing what the machines have been told to do. While technology (including cryptography) has an important part to play, this part is primarily supportive rather than controlling – 'on tap' not 'on top'. The most important contribution to engendering trust is in human factors. Principal among these is the ability to recover from errors. Unfortunately the history of human experience with computers is against us. Too often our culture is based on the machine being right. If the computer says your Gas Bill is £1 million this quarter then you must have used that much gas!

15. So, one of the prime requirements is to understand what the computer is doing. For security applications this leads to the importance of Evaluation. Unfortunately experience has shown that Evaluation is not straightforward – indeed in strict terms it is impossible! Building trust in this way is necessary but fraught with difficulties – an art rather than a science – requiring all sorts of embellishments and compromises.

16. Returning to technological mechanisms in general and cryptography in particular we find that there is no silver bullet and we again need to understand the actual role of the technology quite closely. The starting point has to be an analysis of the requirement and the way in which the technology can support it. Doing this brings us to the need to establish a clear model of the threat and then to build up the need for and design of countermeasures. (What are you trying to achieve and how are you going to achieve it?) In far too many cases these stages are skipped leading to abuse of the technology and failure to achieve the necessary security goals. This in turn leads to continued failure to engender trust.

17. One area of concern to me is the inappropriate reliance on cryptography. Cryptography (when properly applied) is good for providing confidentiality of transmissions. And there is a requirement for such confidentiality. But against the actual level of threat to an internet based application this requirement is relatively low in terms of its overall value as a defensive measure. The most serious threats are not aimed against confidentiality and, those that are, exploit different weaknesses from interception of a transmission within the communications supporting the net. The actual benefit of the cryptographic mechanism is small. Yet in many cases this is the only security measure considered. In answer to my question about the security of a popular commercial share service I was told that the system was secure because it employed 256 bit cryptography. In posing my question I was envisaging a wide variety of security failures, none of which would have been blocked by cryptography regardless of the number of bits used. Yet this is a common error in cyberspace. Small wonder that hackers thrive and the systems are not trusted. Another instance of inappropriate reliance on cryptography is in the nomenclature 'Secure Site' to denote those sites which will communicate using SSL. All this means is that the content of transmissions is secure from intercept in the physical transmission path. It still allows intercept at switches or of red tails – both of which are much more vulnerable than the transmission paths. Furthermore malicious transmissions are also protected.

There is no guarantee or even indication of the security of the actual site. Malware can be present on the site and delivered to you through the SSL-protected links. The site may maintain vulnerable, on-line records of your personal data. Your data may be sold to others or abused in a whole legion of ways – but the re-assuring padlock will appear on your screen. So that's all right isn't it?

18. Another issue is the acceptance of encrypted communications from untrusted sources. Acceptance of en clair communications from an untrusted source has its risks that need to be managed but these risks are magnified several-fold when the communications are encrypted. What is worse is that our culture is unprepared for the concept of untrusted sources of encrypted communications. Deep down we are still conditioned to a paradigm where access to your cryptography is limited to authorised entities. This paradigm has changed and it is dangerous to accept encrypted communications without secure authentication AND positive re-assurance that the authenticated originator is trusted. When GCHQ began to allow staff to access the internet through the Government Secure Intranet (GSI) – not connected to more highly classified internal GCHQ networks – the security policy included a blanket ban on connections involving encrypted communications because the security authorities in GCHQ had seen the dangers. In unmitigated form this policy is inimical to the conduct of some regular business and ways of approving specific sites have been found – but the important point is that when so-called 'secure sites' are involved sound security requires a white list approach.

19. On the more positive side cryptography in its traditional form is important for providing confidentiality against specific threats to transmitted and stored data. The case of stored data carries a number of special risks that need to be addressed. In no particular order these include: protection of loss of data through loss of cryptographic keys; management of decrypted files – no good relying on cryptographic protected if there are lots of copies of the decrypt lying about; access control – no good having cryptographic protection if anyone can access the data with the 'system' decrypting it for them; key management in an environment of long-term protection.

20. I now want to make a short digression into the fraught field of social responsibility. Like most other science and technology, cryptology (both cryptography and cryptanalysis) is value-neutral. It can be applied for both good and ill and the decision as to which is which is an ethical one and is independent of the actual technology. There has been an unfortunate tendency (born of American inherent distrust of Government) to adopt a simplistic white hat/black hat approach. Cryptographers are Good Guys protecting information from snooping by Bad Guys. Inconsistently, however, investigative journalists are also Good Guys uncovering information that the Bad Guys want to keep hidden. Whether it is 'Good' to maintain the secrecy of information and 'Bad' to uncover it or vice-versa is not absolute but dependent on the particular circumstances and viewpoint of the protagonists. Thus from the point of view of the British in 1939 (and the Americans in 1942) the work at Bletchley to uncover German secrets through the cryptanalysis of Enigma was morally right and clearly the work of Good Guys. German attempts to keep their military secrets, like all other German military activities, were similarly Bad. History agrees. Had the cryptanalysts failed history might have given a different verdict. The situation has not changed. Cryptography in the service of evil remains evil and in the service of good, good. Ditto cryptanalysis. And now that cryptography has spread beyond the service of Government to business and private individuals the ethical issues have also broadened. Cryptography can support trusted business and personal privacy. It can also conceal illegal business activities, cheating customers, Governments, and business competitors. Similarly it can enable the establishment of networks to support paedophilia and terrorism. The challenge for society is to ensure that cryptography can deliver the business and personal benefits through the maintenance of trust in the internet technology while limiting the ability of the same technology to support criminal activities. The traditional American approach based on distrust of their own Government remains one of the greatest obstacles to responsible use of the technology. (This parallels the American gun laws – the constitutional purpose of their 'right' to bear arms being to enable Americans to change their Government through armed revolution).

21. To summarise this section: cryptography provides some potentially strong mechanisms to enforce limited aspects of security. In itself it is insufficient and only delivers its promise if soundly implemented and supported by disciplined management processes. Otherwise the situation is like having strong doors and weak walls. Dependence on modern electronic technology is now so great that we can no longer ignore the inadequacies of piecemeal security solutions. The ability to exploit information and communications technology for both personal and business benefit is now limited by the trust we can place in the systems – human and technological. That trust depends on the assurance offered by our security systems. Cryptography of itself cannot provide that assurance and this will become apparent as serious breaches continue to hit the headlines despite claims that the latest cryptography is being used. The power of misapplication of cryptography to undermine confidence and trust should not be underestimated. Assurance will come from appropriate application of cryptography as part of a balanced set of security measures providing security in depth designed to address assessed threats while focussing on realisation of business benefits.

5   Future Directions

22. Today the main applications for cryptography remain COMSEC, authentication and protection of stored data. All these are important and all require improvements in their application. However in some sense these are chiefly concerned with perimeter defence. It is becoming increasingly apparent that perimeter defence is inadequate and that deeper defence is needed. The main part played by cryptography to date is through some form of limited compartmentation (essentially one aspect of protection of stored data). But the major challenges concern integrity of both data and process – with the latter being the more challenging. Cryptography can (and to a limited extent does) provide some integrity mechanisms but actual applications are patchy and not well integrated. In the pre-electronic world integrity of process is provided mainly by transparency. Transparency remains important for computer-controlled processes but cannot be achieved to the degree necessary to provide adequate assurance. We need assurance that our computers are doing the right thing, are doing the thing right and furthermore are continuing to do so under threat of subversion by hostile agents. How can cryptography help?

23. The biggest security challenges come from the most hidden attacks. Some attacks are apparent – cyber vandals leave their calling card, incompetent attacks get discovered, actual damage gets noticed (although a competent agent would ensure there was no/poor evidence that the damage was deliberately caused). But the most worrying attacks are those that continue without alerting the victim. Detection is the greatest challenge facing security authorities and system owners. What support, if any, can cryptography supply?

A Designer's Guide to KEMs

Alexander W. Dent

Information Security Group,
Royal Holloway, University of London,
Egham Hill, Egham, Surrey, U.K.
alex@fermat.ma.rhul.ac.uk
http://www.isg.rhul.ac.uk/~alex/

Abstract. A generic or KEM-DEM hybrid construction is a formal method for combining asymmetric and symmetric encryption techniques to give an efficient, provably secure public-key encryption scheme. This method combines an asymmetric key encapsulation mechanism (KEM) with a symmetric data encapsulation mechanism (DEM). A KEM is a probabilistic algorithm that produces a random symmetric key and an asymmetric encryption of that key. A DEM is a deterministic algorithm that takes a message and a symmetric key and encrypts the message under that key. Each of these components must satisfy its own security conditions if the overall scheme is to be secure. In this paper we describe generic constructions for provably secure KEMs based on weak encryption algorithms. We analyse the two most popular techniques for constructing a KEM and note that they are either overly complex or based on needlessly strong assumptions about the security of the underlying trapdoor function. Hence we propose two new, simple methods for constructing a KEM where the security of the KEM is based on weak assumptions about the underlying function. Lastly we propose a new KEM based on the Rabin function that is both efficient and secure, and is the first KEM to be proposed whose security depends upon the intractability of factoring.

1 Introduction

Whilst most dedicated public-key encryption algorithms are fine for sending short messages, many schemes have problems sending long or arbitrary length messages. Most of the normal "modes of operation" which might allow a sender to send a long message using a public-key encryption algorithm directly are cripplingly inefficient.

One particular way to solve these problems is to use symmetric encryption with a randomly generated key to encrypt a message, and then use asymmetric cryptography to encrypt that (short) random key. This method has been cryptographic folklore for years and, as such, was not formally studied. This led to papers such as [3] which can be used to attack schemes in which the set of symmetric keys is significantly smaller than the message space of the asymmetric scheme used to encrypt them. This folklore has recently been formalised in terms

K.G. Paterson (Ed.): Cryptography and Coding 2003, LNCS 2898, pp. 133–151, 2003.
© Springer-Verlag Berlin Heidelberg 2003

of a generic or KEM-DEM construction [4]. In this construction the encryption scheme is divided into two parts: an asymmetric KEM and a symmetric DEM. A KEM (or key encapsulation mechanism) is a probabilistic algorithm that produces a random symmetric key and an encryption of that key. A DEM (or data encapsulation mechanism) is a deterministic algorithm that encrypts a message of arbitrary length under the key given by the KEM.

This approach to the construction of hybrid ciphers has quickly become popular. Not only have several KEM schemes been proposed in the research literature [4, 8] but this approach has been adopted by the ISO standardisation body [13]. However KEMs are still proposed in an *ad hoc* fashion. Currently, if one wishes to propose a KEM based on one particular trapdoor problem then it is necessary to design such a KEM from scratch.

In this paper we examine generic methods for constructing KEMs from weak encryption schemes. We analyse the two methods for constructing a KEM based on existing schemes and show that either they require the underlying encryption scheme to have security properties which are stronger than they need to be or they are overly complex. We also provide two new generic construction methods which overcome these problems. Essentially this paper gives a toolbox to allow an algorithm designer to construct a KEM from almost any cryptographic problem. To demonstrate the power of the results we will also propose a new KEM, Rabin-KEM, that is as secure as factoring.

It should be noted that most of the results contained in this paper can be easily adapted to simple, "one-pass" key-agreement protocols like the Diffie-Hellman key agreement scheme [5].

2   The Security of a KEM

A KEM is a triple of algorithms:

- a key generation algorithm, *KEM.Gen*, which takes as input a security parameter 1^λ and outputs a public/secret key-pair (pk, sk),
- a encapsulation algorithm, *KEM.Encap*, that takes as input a public-key pk and outputs an encapsulated key-pair (K, C) (C is sometimes said to be an encapsulation of the key K),
- a decapsulation algorithm, *KEM.Decap*, that takes as input an encapsulation of a key C and a secret-key sk, and outputs a key K.

Obviously if the scheme is to be useful we require that, with overwhelming probability, the scheme is sound, i.e. for almost all $(pk, sk) = KEM.Gen(1^\lambda)$ and almost all $(K, C) = KEM.Encap(pk)$ we have that $K = KEM.Decap(C, sk)$. We also assume that the range of possible keys K is some set of fixed length binary strings, $\{0, 1\}^{KEM.KeyLen(\lambda)}$.

We choose to approach provable security from an asymptotic/complexity theoretic point of view and suggest that a scheme is secure if the probability of breaking that scheme is negligible as a function of the security parameter.

Definition 1. *A function f is said to be negligible if, for all polynomials p, there exists a constant N_p such that*

$$f(x) \leq \frac{1}{p(x)} \text{ for all } x \geq N_p. \tag{1}$$

A KEM is considered secure if there exists no attacker with a significant advantage in winning the following game played against a mythical challenger.

1. The challenger generates a public/secret key-pair $(pk, sk) = KEM.Gen(1^\lambda)$ and passes pk to the attacker.
2. The attacker runs until it is ready to receive a challenge encapsulation pair. During this time the attacker may repeatedly query a decapsulation oracle to find the key associated with any encapsulation.
3. The challenger prepares a challenge encapsulated key-pair as follows:
   a) The challenger generates a valid encapsulated key-pair

   $$(K_0, C) = KEM.Encap(pk). \tag{2}$$

   b) The challenger selects an alternate key K_1 chosen uniformly at random from the set $\{0, 1\}^{KEM.Gen(\lambda)}$.
   c) The challenger selects a bit σ uniformly at random from $\{0, 1\}$.
   The challenger then passes (K_σ, C) to the attacker.
4. The attacker is allowed to run until it outputs a guess σ' for σ. During this time the attacker may repeatedly query a decapsulation oracle to find the key associated with any encapsulation except the challenge encapsulation C.

The attacker is said to win this game if $\sigma' = \sigma$. We define an attacker's advantage *Adv* to be

$$Pr[\sigma' = \sigma] - 1/2. \tag{3}$$

If the maximum advantage of any attacker against a KEM is negligible (as a function of λ) then the KEM is said to be IND-CCA2 secure.

A KEM is only useful when coupled with a DEM (a *data encapsulation mechanism*) to form a hybrid public-key encryption scheme. A DEM is a symmetric algorithm that takes a message and a key, and encrypts the message under that key. In order for the overall hybrid encryption scheme to be secure, the KEM and the DEM must satisfy certain security properties. Happily these properties are independent (i.e. the security properties that a KEM must have are independent of the security properties of the DEM). For the overall encryption scheme to be IND-CCA2 secure (in the sense given below) the KEM, in particular, must be IND-CCA2 secure. For further details of hybrid constructions using KEMs and DEMs, and their security properties, the reader is referred to [4].

3 The Security of an Encryption Scheme

We will require formal definitions for an asymmetric encryption scheme. It will suit our purposes to draw a distinction between a deterministic and probabilistic encryption schemes as they present slightly different challenges to the KEM designer. We will start by considering deterministic encryption schemes.

Definition 2. *A deterministic encryption scheme is a triple* $(\mathcal{G}, \mathcal{E}, \mathcal{D})$ *where*

- $\mathcal{G}$ *is the key-generation algorithm which takes as input a security parameter* 1^λ *and outputs a public/secret key-pair* (pk, sk),
- $\mathcal{E}$ *is the encryption algorithm which takes as input a message* $m \in \mathcal{M}$ *and the public-key* pk *and outputs a ciphertext* $C \in \mathcal{C}$,
- $\mathcal{D}$ *is the decryption algorithm which takes as input a ciphertext* $C \in \mathcal{C}$ *and the secret-key* sk *and outputs either a message* $m \in \mathcal{M}$ *or the error symbol* $\bot$.

The weakest notion of security for a deterministic encryption scheme is one-way security.

Definition 3. *A deterministic encryption scheme* $(\mathcal{G}, \mathcal{E}, \mathcal{D})$ *is said to be one-way if the probability that a polynomial time attacker* $\mathcal{A}$ *can invert a randomly generated ciphertext* C *is negligible as a function of* λ. *Such a cryptosystem is often said to be secure in the OW-CPA model[1].*

A deterministic encryption scheme $(\mathcal{G}, \mathcal{E}, \mathcal{D})$ *is said to be secure in the OW-RA model if the scheme is one-way even when the attacker has access to an oracle that, when given a ciphertext* $C \in \mathcal{C}$, *determines whether* C *is a valid ciphertext or not, i.e. whether* C *is the correct encryption of some message or not. This is sometimes called a* reaction attack.

The idea of allowing an attacker access to an oracle that correctly determines if a ciphertext is valid was first used in a paper by Joye, Quisquater and Yung [7]. The paper used such an oracle to attack an early version of the EPOC-2 cipher.

For our purposes, a probabilistic encryption scheme will be viewed as a deterministic scheme whose encryption algorithm takes some random seed as an extra input.

Definition 4. *A probabilistic encryption scheme is a triple* $(\mathcal{G}, \mathcal{E}, \mathcal{D})$ *where*

- $\mathcal{G}$ *is the key-generation algorithm which takes as input a security parameter* 1^λ *and outputs a public/secret key-pair* (pk, sk),
- $\mathcal{E}$ *is the encryption algorithm which takes as input a message* $m \in \mathcal{M}$, *a random seed* $r \in \mathcal{R}$ *and the public-key* pk *and outputs a ciphertext* $C \in \mathcal{C}$,
- $\mathcal{D}$ *is the decryption algorithm which takes as input a ciphertext* $C \in \mathcal{C}$ *and the secret-key* sk *and outputs either a message* $m \in \mathcal{M}$ *or the error symbol* $\bot$.

To cement the idea that this is a probabilistic system we require that, for all public keys pk *that can be obtained from the key generation algorithm with an input* 1^λ *and for all* $m \in \mathcal{M}$ *we have that*

$$|\{r \in \mathcal{R} : \mathcal{E}(m, r, pk) = C\}| \leq \gamma(\lambda)/|\mathcal{R}| \tag{4}$$

where $C \in \mathcal{C}$ *and* $\gamma(\lambda)/|\mathcal{R}|$ *is negligible as a function of* λ.[2]

---

[1] OW for "one-way" and CPA for "chosen plaintext attack". The term "chosen plaintext attack" is used because the attacker is not allowed to make decryption queries.

[2] This condition basically states that for any public key and ciphertext, there cannot be two many choices for r that encrypt a message to that ciphertext.

Analogous notions of OW-CPA and OW-RA security can be defined for probabilistic encryption schemes. However, there is another issue that will affect our ability to design KEMs based on probabilistic encryption schemes - the need for a plaintext-ciphertext checking oracle.

Definition 5. *For a asymmetric encryption scheme $(\mathcal{G}, \mathcal{E}, \mathcal{D})$, a plaintext-ciphertext checking oracle is an oracle that, when given a pair $(m, C) \in \mathcal{M} \times \mathcal{C}$, correctly determines whether C is an encryption of m or not.*

Obviously, if $(\mathcal{G}, \mathcal{E}, \mathcal{D})$ is a deterministic algorithm then there exists an efficient plaintext-ciphertext checking oracle, however the situation is more complicated for a probabilistic encryption scheme. There are several ways in which a plaintext-ciphertext checking oracle for a probabilistic encryption scheme can be made be available to all parties in a security proof. In particular, it might be possible to construct such an oracle because of the nature of underlying intractability assumption (such as in the case of an encryption scheme based on the gap Diffie-Hellman problem, see [10]). Alternatively, it might be possible to simulate such an oracle using, say, knowledge of the hash queries an attacker has made in the random oracle model [2].

4   Analysing RSA-KEM

We present a method to construct a KEM from almost all one-way public-key encryption schemes; this generalises the ideas used in RSA-KEM [13].

The construction of a KEM from a deterministic encryption scheme $(\mathcal{G}, \mathcal{E}, \mathcal{D})$ is given in Table 1. This construction uses a key derivation function KDF. This function is intended to do more than simply format the random number correctly as a key: it is meant to remove algebraic relations between inputs. It is usually constructed from a hash function and will be modelled as a random oracle.

Table 1. A KEM derived from a deterministic encryption scheme

- Key generation is given by the key generation algorithm of the public-key encryption scheme (i.e. $KEM.Gen = \mathcal{G}$).

- Encapsulation is given by:
    1. Generate an element $x \in \mathcal{M}$ uniformly at random.
    2. Set $C := \mathcal{E}(x, pk)$.
    3. Set $K := KDF(x)$.
    4. Output (K, C).

- Decapsulation of an encapsulation C is given by:
    1. Set $x := \mathcal{D}(C, sk)$. If $x = \perp$ then output $\perp$ and halt.
    2. Set $K := KDF(x)$.
    3. Output K.

Theorem 1. *Suppose* $(\mathcal{G}, \mathcal{E}, \mathcal{D})$ *is a deterministic asymmetric encryption scheme that is secure in the OW-RA model. Then the KEM derived from* $(\mathcal{G}, \mathcal{E}, \mathcal{D})$ *in Table 1 is, in the random oracle model, IND-CCA2 secure.*

The proof of this theorem is similar to that of Theorem 2.

This style of KEM can be easily extended to the case where the underlying encryption scheme is probabilistic and not deterministic. The construction is given in Table 2. Note that the encapsulation is included in the input to the key derivation function to prevent the attacker from breaking the scheme by finding a second ciphertext C' that decrypts to the same value as the challenge ciphertext C.

Table 2. A KEM derived from a probabilistic encryption scheme

- Key generation is given by the key generation algorithm of the public-key encryption scheme (i.e. $KEM.Gen = \mathcal{G}$).
- Encapsulation is provided by the following algorithm.
    1. Generate elements $x \in \mathcal{M}$ and $r \in \mathcal{R}$ uniformly at random.
    2. Set $C := \mathcal{E}(x, r, pk)$.
    3. Set $K := KDF(\bar{x}||C)$, where $\bar{x}$ is a fixed length representation of the element $x \in \mathcal{M}$.
    4. Output (K, C).

- Decapsulation of an encapsulation C is given by the following algorithm.
    1. Set $x := \mathcal{D}(C, sk)$. If $x = \perp$ then output $\perp$ and halt.
    2. Set $K := KDF(\bar{x}||C)$ where $\bar{x}$ is a fixed length representation of the element $x \in \mathcal{M}$.
    3. Output K.

Theorem 2. *Suppose* $(\mathcal{G}, \mathcal{E}, \mathcal{D})$ *is a probabilistic asymmetric encryption scheme*

- *that is secure in the OW-RA model, and*
- *for which there exists a plaintext-ciphertext checking oracle.*

Then the KEM derived from $(\mathcal{G}, \mathcal{E}, \mathcal{D})$ *in Table 2 is, in the random oracle model, IND-CCA2 secure.*

Proof. See Appendix A

5   Analysing PSEC-KEM

Obviously it would be advantageous if we were able to remove the reliance on these non-optimal security criteria, i.e. produce a generic method for constructing a KEM from an OW-CPA encryption scheme rather than from an OW-RA encryption scheme and requiring a plaintext-ciphertext checking oracle. In this section, we present a method to construct a KEM from a OW-CPA encryption scheme; this generalise the ideas used in PSEC-KEM [13].

Table 3 gives a construction for a KEM from a (deterministic or probabilistic) asymmetric encryption scheme $(\mathcal{G}, \mathcal{E}, \mathcal{D})$. In the construction $Hash$ is a hash function and MGF is a mask generating function. A mask generating function is similar to a key derivation function, in fact the same constructions are used for both, but a mask generating function is used to create a bit string that is used to mask a data value. We will model these function as random oracles, hence care must be taken to ensure that the outputs of the hash function and the mask generating function are suitably independent. We also use a "smoothing function" $\phi : \{0,1\}^{n_2} \to \mathcal{M}$, where $\mathcal{M}$ is the message space of the encryption scheme $(\mathcal{G}, \mathcal{E}, \mathcal{D})$. This function must have the property that for Y' drawn uniformly at random from $\{0,1\}^{n_2}$ and $X \in \mathcal{M}$ we have

$$Pr[\phi(Y') = X] - \frac{1}{|\mathcal{M}|} \qquad (5)$$

is negligible.

For security, it is necessary that n is suitably large. Certainly $n \geq \lambda$ would be sufficient. Of the other lengths, n_1 should equal $KEM.Keylen$ and n_2 merely has to be large enough so that there exists a function ϕ which is suitably smooth.

Table 3. A KEM derived from a OW-CPA secure encryption scheme

- Key-generation is given by $\mathcal{G}$, i.e. $KEM.Gen = \mathcal{G}$.

- Encapsulation is given by:
  1. Generate a suitably large bit-string $y \in \{0,1\}^n$.
  2. Set $Y := Hash(y)$.
  3. Split Y into two strings $K \in \{0,1\}^{n_1}$ and $Y' \in \{0,1\}^{n_2}$ where $Y = K||Y'$.
  4. Set $X := \phi(Y')$.
  5. Set $C_1 := \mathcal{E}(X, pk)$. (If $\mathcal{E}$ is probabilistic then generate a random seed $r \in \mathcal{R}$ and set $C_1 := \mathcal{E}(X, r, pk)$.)
  6. Set $C_2 := y \oplus MGF(X)$.
  7. Set $C = (C_1, C_2)$.
  8. Output (K, C).

- Decapsulation of an encapsulation C is given by:
  1. Parse C as (C_1, C_2).
  2. Set $X := \mathcal{D}(C_1, sk)$. If $x = \perp$ then output $\perp$ and halt.
  3. Set $y = C_2 \oplus MGF(X)$.
  4. Set $Y = Hash(y)$.
  5. Split Y into two strings $K \in \{0,1\}^{n_1}$ and $Y' \in \{0,1\}^{n_2}$ where $Y = K||Y'$.
  6. Check that $\phi(Y') = X$. If not, output $\perp$ and halt.
  7. Output K.

Theorem 3. *Suppose $(\mathcal{G}, \mathcal{E}, \mathcal{D})$ is an asymmetric encryption scheme that is secure in the OW-CPA model. Then the KEM derived from $(\mathcal{G}, \mathcal{E}, \mathcal{D})$ in Table 3 is, in the random oracle model, IND-CCA2 secure.*

The proof of this theorem is similar to the proof given in [13].

6   A New Construction for a Deterministic Encryption Scheme

Although the construction given in Section 5 (the generalisation of PSEC-KEM) is based on weak assumptions, it is very complex and there are many places where a weakness may be introduced in an implementation. We now propose a simpler construction for designing a KEM based on a deterministic encryption scheme with similarly weak security assumptions. In other words, we build a secure KEM from a deterministic encryption scheme that is secure in the OW-CPA model, as opposed to the OW-RA model as in Section 4. The construction can be viewed as a simpler version of the REACT construction [11].

Table 4 gives a construction of a KEM based on a deterministic asymmetric encryption scheme $(\mathcal{G}, \mathcal{E}, \mathcal{D})$. The scheme makes use of a key derivation function *KDF* and a hash function *Hash*. These functions will be modelled as random oracles and so care must be taken that their outputs are suitably independent.

Table 4. A KEM derived from an OW-CPA secure, deterministic encryption scheme

- Key-generation is given by $\mathcal{G}$, i.e. *KEM.Gen* $= \mathcal{G}$.

- Encapsulation is given by:
   1. Generate a suitably large bit-string $x \in \mathcal{M}$.
   2. Set $C_1 := \mathcal{E}(x, pk)$.
   3. Set $C_2 := Hash(x)$.
   4. Set $C := (C_1, C_2)$.
   5. Set $K := KDF(x)$.
   6. Output (K, C).

- Decapsulation of an encapsulation C is given by:
   1. Parse C as (C_1, C_2).
   2. Set $x := \mathcal{D}(C_1, sk)$. If $x = \perp$ then output $\perp$ and halt.
   3. Check that $C_2 = Hash(x)$. If not, output $\perp$ and halt.
   4. Set $K := KDF(x)$.
   5. Output K.

Theorem 4. *Suppose that $(\mathcal{G}, \mathcal{E}, \mathcal{D})$ is a deterministic encryption algorithm that is secure in the OW-CPA model. Then the KEM derived from $(\mathcal{G}, \mathcal{E}, \mathcal{D})$ in Table 4 is, in the random oracle model, IND-CCA2 secure.*

Proof. See Appendix B

This construction also has the advantage that the decryption algorithm need not return a unique solution but need only return a small subset of the message space that includes the original message, as, with high probability, the original message will be the only message in the subset that hashes to give the correct value of C_2. We will make heavy use of this fact in the specification of Rabin-KEM (see Sect. 8).

7   A New Construction for a Probabilistic Encryption Scheme

Although the previous KEM construction can be generalised to be used with a probabilistic encryption scheme, the security proof still relies on the existence of a plaintext-ciphertext checking oracle (which is always easily constructed for a deterministic encryption algorithm). We now give a construction for a probabilistic encryption scheme, loosely based on the ideas of [6], that does not require a plaintext-ciphertext checking oracle. It is interesting to note, however, that this construction *cannot* be used for a deterministic scheme.

Table 5 gives the construction of a KEM based on a OW-CPA secure, probabilistic encryption scheme. Furthermore the proof of security for this construction does not require there to exist a plaintext-ciphertext checking oracle. The scheme makes use of a key derivation function *KDF* and a hash function *Hash*. These functions will be modelled as random oracles and so care must be taken that their outputs are suitably independent.

Table 5. A KEM derived from an OW-CPA secure, probabilistic encryption scheme

– Key-generation is given by $\mathcal{G}$, i.e. *KEM.Gen* = $\mathcal{G}$.

– Encapsulation is given by:
1. Generate a suitably large bit-string $x \in \mathcal{M}$.
2. Set $r := Hash(x)$.
3. Set $C := \mathcal{E}(x, r, pk)$.
4. Set $K := KDF(x)$.
5. Output (K, C).

– Decapsulation is given by:
1. Set $x := \mathcal{D}(C, sk)$. If $x = \perp$ then output $\perp$ and halt.
2. Set $r := Hash(x)$.
3. Check that $\mathcal{E}(x, r, pk) = C$. If not, output $\perp$ and halt.
4. Set $K := KDF(x)$.
5. Output K.

Theorem 5. *Suppose that* $(\mathcal{G}, \mathcal{E}, \mathcal{D})$ *is a probabilistic encryption algorithm that is secure in the OW-CPA model. Then the KEM derived from* $(\mathcal{G}, \mathcal{E}, \mathcal{D})$ *in Table 5 is, in the random oracle model, IND-CCA2 secure.*

Proof. See Appendix C

8   Case Study: Rabin-KEM

We demonstrate the power of these results by proposing a new KEM whose security is equivalent to factoring: Rabin-KEM. The Rabin-KEM construction will be based on the generic construction given in Sect. 6 and the Rabin trapdoor permutation [12,9]. The algorithm is described in Table 6.

Table 6. Rabin-KEM

Key Generation. On input of 1^λ for some integer $\lambda > 0$,

1. Randomly generate two distinct primes p and q of bit length λ.
2. Set $n := pq$.
3. Set $pk := (n)$ and $sk := (p, q)$.
4. Output (pk, sk).

Encapsulation. On input of a public key PK,

1. Randomly generate an integer $x \in [0, n)$.
2. Set $C_1 := x^2 \bmod n$.
3. Set $C_2 := Hash(x)$.
4. Set $C := (C_1, C_2)$.
5. Set $K := KDF(x)$.
6. Output (K, C).

Decapsulation. On input of an encapsulated key C and a secret key sk.

1. Parse C as (C_1, C_2).
2. Check that C_1 is a square modulo n. If not, output $\perp$ and halt.
3. Compute the four square roots x_1, x_2, x_3, x_4 of C_1 modulo n using the secret key sk.
4. If there exists no value $1 \leq i \leq 4$ such that $Hash(x_i) = C_2$ then output $\perp$ and halt.
5. If there exists more than one value $1 \leq i \leq 4$ such that $Hash(x_i) = C_2$ then output $\perp$ and halt.
6. Let x be the unique square root of C_1 modulo n for which $Hash(x) = C_2$.
7. Set $K := KDF(x)$.
8. Output K

Theorem 6. *Providing the factoring problem is hard, Rabin-KEM is, in the random oracle model, IND-CCA2 secure.*

Proof. It is well known that the Rabin trapdoor function is one-way providing that the factoring assumption is hard [12]. Therefore, given that the factoring problem is intractable, the given KEM is IND-CCA2 secure in the random oracle model by Theorem 4.   □

This KEM is both and efficient and secure, being the first KEM ever proposed whose security depends on the assumption that factoring is intractable. Of course there is a chance that the decryption algorithm will fail, i.e. that $KEM.Decap(C, sk) = \perp$ even though C is actually a valid encapsulation of a key K. However this will only happen if there is a collision in the hash function, which, as we model the hash function as a random oracle, only happens with probability $2^{-Hash.Len}$ (where $Hash.Len$ is the length of the output of the hash function).

9 Conclusion

This paper has provided four generic constructions for key encapsulation mechanisms (KEMs): two generalisations of existing KEMs and two new KEMs. These results show that KEMs can be constructed from almost any trapdoor function. We also proposed a new KEM: Rabin-KEM. This is a new fast, secure KEM based on the intractability of factoring large numbers.

Acknowledgements. I would like to thank Victor Shoup, Louis Granboulan and Kenny Paterson for some very useful discussions in this area. As always the help of Christine Swart has been invaluable.

References

1. M. Bellare, A. Desai, D. Pointcheval, and P. Rogaway. Relations among notions of security for public-key encryption schemes. In H. Krawczyk, editor, *Advances in Cryptology – Crypto '98*, LNCS 1462, pages 26–45. Springer-Verlag, 1998.
2. M. Bellare and P. Rogaway. Random oracles are practical: A paradigm for designing efficient protocols. In *Proc. of the First ACM Conference on Computer and Communications Security*, pages 62–73, 1993.
3. D. Boneh, A. Joux, and A. Nguyen. Why textbook ElGamal and RSA encryption are insecure. In T. Okamoto, editor, *Advances in Cryptology – Asiacrypt 2000*, LNCS 1976, pages 30–43. Springer-Verlag, 2000.
4. R. Cramer and V. Shoup. Design and analysis of practical public-key encryption schemes secure against adaptive chosen ciphertext attack. Available from http://shoup.net/, 2002.
5. W. Diffie and M. Hellman. New directions in cryptography. *IEEE Transactions on Information Theory*, 22:644–654, 1976.
6. E. Fujisaki and T. Okamoto. How to enhance the security of public-key encryption at minimal cost. In H. Imai and Y. Zheng, editors, *Public Key Cryptography*, LNCS 1560, pages 53–68. Springer-Verlag, 1999.
7. M. Joye, J. Quisquater, and M. Yung. On the power of misbehaving adversaries and security analysis of the original EPOC. In D. Naccache, editor, *Topics in Cryptography – CT-RSA 2001*, LNCS 2020, pages 208–222. Springer-Verlag, 2001.
8. S. Lucks. A variant of the Cramer-Shoup cryptosystem for groups of unknown order. In Y. Zheng, editor, *Advances in Cryptology – Asiacrypt 2002*, LNCS 2501, pages 27–45. Springer-Verlag, 2002.

9. A. J. Menezes, P. van Oorschot, and S. Vanstone. *Handbook of Applied Cryptography*. CRC Press, 1997.
10. T. Okamoto and D. Pointcheval. The gap problems: A new class of problems for the security of cryptographic schemes. In K. Kim, editor, *Public Key Cryptography*, LNCS 1992, pages 104–118. Springer-Verlag, 2001.
11. T. Okamoto and D. Pointcheval. REACT: Rapid enhanced-security asymmetric cryptosystem transform. In D. Naccache, editor, *Proceedings of CT-RSA 2001*, LNCS 2020, pages 159–175. Springer-Verlag, 2001.
12. M. O. Rabin. Digitalized signatures and public-key functions as intractable as factorization. Technical Report MIT/LCS/TR-212, MIT Laboratory for Computer Science, 1979.
13. V. Shoup. A proposal for the ISO standard for public-key encryption (version 2.0). Available from `http://shoup.net/`, 2001.

A   Proof of Theorem 2

This is a simple result. We will use standard techniques to prove a more detailed result.

Theorem 7. *Let $(\mathcal{G}, \mathcal{E}, \mathcal{D})$ be an encryption scheme and let KEM be the KEM derived from $(\mathcal{G}, \mathcal{E}, \mathcal{D})$ using the construction described in Table 2. If there exist an attacker $\mathcal{A}$ that, in the random oracle model, breaks KEM in the IND-CCA2 model*

- *with advantage ϵ,*
- *in time t,*
- *makes at most q_D queries to the decapsulation oracle,*
- *and at most q_K queries to the random oracle that represents the key derivation function,*

then there exists an algorithm that inverts the underlying encryption function in the OW-RA model (and makes use of a plaintext-ciphertext checking oracle) with probability ϵ' and in time t' where

$$\epsilon' \geq \epsilon, \tag{6}$$

$$t' = t. \tag{7}$$

Proof. We assume that there exists an attacker for the KEM and use this to construct an algorithm that can break the underlying encryption scheme. Suppose $\mathcal{A}$ is an attacker that breaks the KEM with the properties stated above.

Consider the following algorithm that takes as input a public-key pk for the underlying encryption scheme and a challenge ciphertext C^*. This algorithm makes use of two lists: $KDFList$ which stores the answers to queries made to the KDF oracle and $DecList$ which stores the answers to queries made to the decapsulation oracle.

1. Prepare two empty lists $KDFList$ and $DecList$.
2. Randomly generate a bit-string K^* of length $KEM.KeyLen$.

3. Pass the public key to pk to $\mathcal{A}$.
4. Allow $\mathcal{A}$ to run until it requests a challenge encapsulation. If the attacker requests the evaluation of the key derivation function KDF on an input z then the following steps are performed:
   a) Check to see if $(z, K) \in KDFList$ for some value of K. If so, return K.
   b) Check to see if z can be parsed as $\bar{x}||C$ for some fixed length representation of a message $x \in \mathcal{M}$ and a ciphertext $C \in \mathcal{C}$. If not, randomly generate an appropriately sized K, add (x, K) to $KDFList$ and return K.
   c) Check to see if C is an encryption of x using the plaintext-ciphertext checking oracle. If not, randomly generate an appropriately sized K, add (x, K) to $KDFList$ and return K.
   d) Randomly generate an appropriately sized K, add (x, K) to $KDFList$, add (C, K) to $DecList$ and return K.

   If the attacker requests the decapsulation of an encapsulation C then the following steps are performed:
   a) Check to see if $C = C^*$. If so, return $\perp$.
   b) Check to see if $(C, K) \in DecList$ for some value K. If so, return K.
   c) Check to see if C is a valid ciphertext or not (using the oracle provided by the RA model). If not, add $(C, \perp)$ to $DecList$ and return $\perp$.
   d) Otherwise randomly generate an appropriately sized K, add (C, K) to $DecList$ and return K.
5. When the attacker requests a challenge encapsulation, return (K^*, C^*)
6. Allow the attacker to run until it outputs a bit σ'. Answer all oracle queries as before.
7. Check to see if there exists a pair $(z, K) \in KDFList$ such that z can be decomposed as $\bar{x}||C^*$ where $\bar{x}$ is the fixed length representation of a message $x \in \mathcal{M}$ and C^* is an encryption of x (using the plaintext-ciphertext checking oracle). If so, output x and halt.
8. Otherwise randomly generate $x \in \mathcal{M}$. Output x and halt.

This algorithm perfectly simulates the attack environment for $\mathcal{A}$ up until the point where $\mathcal{A}$ queries the key derivation function on $\bar{x}||C^*$ where $\bar{x}$ is the fixed length representation of $\mathcal{D}(C^*, sk)$. Since $\mathcal{A}$ can have no advantage in breaking the KEM unless this event occurs, we can show that this event must occur with probability at least ϵ. However if this event occurs then we will successfully recover $\mathcal{D}(C^*, sk)$. Hence the theorem holds.   □

B   Proof of Theorem 4

We use standard techniques to prove the following, slightly more detailed result.

Theorem 8. *Let $(\mathcal{G}, \mathcal{E}, \mathcal{D})$ be a deterministic encryption scheme and let KEM be the KEM derived from $(\mathcal{G}, \mathcal{E}, \mathcal{D})$ using the construction described in Table 4. If there exists an attacker $\mathcal{A}$ that, in the random oracle model, breaks KEM in the IND-CCA2 model*

- *with advantage ϵ,*
- *in time t,*
- *and makes at most q_D decapsulation queries,*
- *at most q_H queries to the random oracle that represents the hash function,*
- *and at most q_K queries to the random oracle that represents the key deriva-tion function,*

then there exists an algorithm that inverts the underlying encryption function with probability ϵ' and in time t' where

$$\epsilon' \geq \epsilon - \frac{q_D}{2^{Hash.Len}}\left(1 + \frac{1}{|\mathcal{C}|}\right), \tag{8}$$

$$t' \leq t + (q_H + q_K + q_D)T, \tag{9}$$

where Hash.Len is the length of the output of the hash function Hash and T is the time taken to evaluate the encryption function $\mathcal{E}$.

Proof. We assume that there exists an attacker for the KEM and use this to construct an algorithm that can break the underlying encryption scheme. Suppose $\mathcal{A}$ is an attacker that breaks the KEM with the properties stated above.

First we slightly change the environment that $\mathcal{A}$ operates in. Let game 1 be the game in which $\mathcal{A}$ attacks the KEM as described in the IND-CCA2 environment described in Sect. 3. Let game 2 be similar to game 1 except that

- the challenge encapsulation (K^*, C^*) is chosen at the beginning of the algorithm and if the attacker ever requests the decapsulation of $C^* = (C_1^*, C_2^*)$ then the decapsulation algorithm returns $\perp$,
- instead of allowing the attacker $\mathcal{A}$ access to the "real" decapsulation oracle, hash function oracle and KDF oracle we only allow $\mathcal{A}$ to have access to the "partially simulated" versions of these oracles described below.

The simulated oracles make use of two lists $HashList$ and $KDFList$, both of which are initially empty. If the attacker requests the evaluation of the hash function $Hash$ on an input x then the following steps are performed:

1. If $(x, hash) \in HashList$ for some value of $hash$ then return $hash$.
2. If $x = \mathcal{D}(C_1^*, sk)$ then return $Hash(x)$.
3. Otherwise randomly generate an appropriately sized $hash$, add $(x, hash)$ to $HashList$ and return $hash$.

Hence the hash function is changed to a random function with the proviso that it must agree with the original hash function on the input $\mathcal{D}(C_1^*, sk)$. This simulation is equivalent to some random oracle and every random oracle can be represented by this simulation. Similarly, if the attacker requests the evaluation of the key derivation function KDF on an input x then the following steps are performed:

1. If $(x, K) \in KDFList$ for some value of K then return K.
2. If $x = \mathcal{D}(C_1^*, sk)$ then return $KDF(x)$.
3. Otherwise randomly generate an appropriately sized K, add (x, K) to $KDFList$ and return K.

If the attacker requests the evaluation of decapsulation function on the encapsulated key (C_1, C_2) then the following steps are performed:

1. If $C_1 = C_1^*$ then return $\perp$.
2. Check that there exists a unique $x \in \mathcal{M}$ such that $(x, C_2) \in HashList$ and $\mathcal{E}(x, pk) = C_1$. If not, return $\perp$.
3. Compute $K := KDF(x)$ using the KDF algorithm described above.
4. Return K.

To analyse the effects of only allowing $\mathcal{A}$ to have access to the simulated oracles we require the following simple lemma [1, 4].

Lemma 1. *If A, B and E are events is some probability space and that $Pr[A|\neg E] = Pr[B|\neg E]$ then $|Pr[A] - Pr[B]| \leq Pr[E]$.*

Let A be the event that $\mathcal{A}$ succeeds in breaking the KEM with access to the real oracles and let B be the event that $\mathcal{A}$ succeeds in breaking the KEM with access to the simulated oracles. Let E be the event that either

1. $\mathcal{A}$ queries the decapsulation oracle on the challenge encapsulation before the challenge encapsulation is given to $\mathcal{A}$, or
2. $\mathcal{A}$ queries the decapsulation oracle on some encapsulation (C_1, C_2) where $Hash(\mathcal{D}(C_1, sk)) = C_2$ but $\mathcal{A}$ has not queried the hash function simulator on the input $\mathcal{D}(C_1, sk)$.

If E does not occur then $\mathcal{A}$ will receive the same responses to his queries regardless of whether it is querying the real oracles or the simulated oracles. Hence $Pr[A|\neg E] = Pr[B|\neg E]$.

Since the challenge ciphertext has to be chosen completely at random, the probability that E occurs because $\mathcal{A}$ queries the decapsulation oracle on the challenge encapsulation before it has been issued is bounded above by $q_D/2^{\text{H ash.}Len}|\mathcal{C}|$. Since the hash function is modelled as a random oracle, the probability that $\mathcal{A}$ queries the decapsulation oracle on some encapsulation (C_1, C_2) where $Hash(\mathcal{D}(C_1, sk)) = C_2$ but $\mathcal{A}$ has not queried the hash function $Hash$ on the input $\mathcal{D}(C_1, sk)$ is at most $q_D/2^{\text{H ash.}Len}$. Hence the advantage of $\mathcal{A}$ in game 2 is least

$$\epsilon - \frac{q_D}{2^{\text{H ash.}Len}}(1 + \frac{1}{|\mathcal{C}|}). \tag{10}$$

Let E' be the event that, in game 2, the attacker queries either the hash function simulator or the key derivation function oracle with the input $x^* = \mathcal{D}(C_1^*, sk)$. Since the attacker can have no knowledge of the whether $KDF(x^*) = K^*$ or not unless E' occurs we have that

$$Pr[E'] \geq \epsilon - \frac{q_D}{2^{\text{H ash.}Len}}(1 + \frac{1}{|\mathcal{C}|}). \tag{11}$$

Consider the following algorithm that takes as input a public key pk for the underlying encryption scheme and a challenge ciphertext C_1^*.

1. Prepare two empty lists $HashList$ and $KDFList$.
2. Generate random bit strings C_2^* of length $Hash.Len$ and K^* of length $KEM.KeyLen$. Set $C^* := (C_1^*, C_2^*)$.
3. Pass the public key pk to $\mathcal{A}$.
4. Allow the attacker $\mathcal{A}$ to run until it requests a challenge encapsulation. If the attacker requests the evaluation of the hash function $Hash$ on an input x then the following steps are performed:
   a) If $(x, hash) \in HashList$ for some value of $hash$ then return $hash$.
   b) Otherwise randomly generate an appropriately sized $hash$, add $(x, hash)$ to $HashList$ and return $hash$.

   If the attacker requests the evaluation of the KDF KDF on an input x then the following steps are performed:
   a) If $(x, K) \in KDFList$ for some value of K then return K.
   b) Otherwise randomly generate an appropriately sized K, add (x, K) to $KDFList$ and return K.

   If the attacker requests the evaluation of decapsulation function on the encapsulated key (C_1, C_2) then the following steps are performed:
   a) If $C_1 = C_1^*$ then return $\perp$.
   b) Check that there exists a unique $x \in \mathcal{M}$ such that $(x, C_2) \in HashList$ and $\mathcal{E}(x, pk) = C_1$. If not, return $\perp$.
   c) Compute $K := KDF(x)$ using the simulator described above.
   d) Return K.
5. When the attacker requests a challenge encapsulation pass the pair (K^*, C^*) to the attacker.
6. Allow the attacker to run until it outputs a bit σ'. Answer all oracle queries with the simulators described above.
7. Check to see if there exists some $(x, hash) \in HashLish$ or $(x, K) \in KDFList$ such that $\mathcal{E}(x, pk) = C_1^*$. If so, output x and halt.
8. Randomly generate $x \in \mathcal{M}$. Output x and halt.

This algorithm perfectly simulates the attack environment for the attacker $\mathcal{A}$ in game 2, up until the point where event E' occurs. However, if E' occurs then the above algorithm will correctly output $x^* = \mathcal{D}(C_1^*, sk)$. Hence the above algorithm will correctly invert a randomly generated ciphertext with probability at least

$$\epsilon - \frac{q_D}{2^{Hash.Len}}\left(1 + \frac{1}{|\mathcal{C}|}\right). \tag{12}$$

This value is negligible providing ϵ is negligible, hence the KEM is secure in the IND-CCA2 model providing the underlying encryption scheme is secure in the OW-CPA model. $\square$

C   Proof of Theorem 5

Again, we prove a slightly more detailed result.

Theorem 9. *Let $(\mathcal{G}, \mathcal{E}, \mathcal{D})$ be a probabilistic encryption scheme and let KEM be the KEM derived from $(\mathcal{G}, \mathcal{E}, \mathcal{D})$ using the construction described in Table 5. If there exists an attacker $\mathcal{A}$ that, in the random oracle model, breaks KEM in the IND-CCA2 model*

- *with advantage ϵ,*
- *in time t,*
- *and makes at most q_D decapsulation queries,*
- *at most q_H queries to the random oracle that represents the hash function,*
- *and at most q_K queries to the random oracle that represents the key derivation function.*

then there exists an algorithm that inverts the underlying encryption function with probability ϵ' and in time t' where

$$\epsilon' \geq \frac{1}{q_D + q_H + q_K}\left(\epsilon - \frac{q_D}{|\mathcal{M}|} - \frac{\gamma q_D}{|\mathcal{R}|}\right), \tag{13}$$

$$t' \approx t, \tag{14}$$

where γ is defined in Definition 4.

Proof. The proof is similar to that given in Appendix B.

Let game 1 be the game in which $\mathcal{A}$ attacks the KEM as described in the IND-CCA2 environment described in Sect. 3. Let game 2 be similar to game 1 except that

- the challenge encapsulation (K^*, C^*) is chosen at the beginning of the algorithm and if the attacker ever requests the decapsulation of C^* then the decapsulation algorithm returns $\perp$,
- instead of allowing the attacker $\mathcal{A}$ access to the "real" decapsulation oracle, hash function oracle and KDF oracle we only allow $\mathcal{A}$ to have access to the "partially simulated" versions of these oracles described below.

We simulate the hash function oracle and the KDF oracle exactly as before, making use of two lists, *HashList* and *KDFList*, both of which are initially empty. If the attacker requests the evaluation of the hash function *Hash* on an input x then the following steps are performed:

1. If $(x, hash) \in HashList$ for some value of $hash$ then return $hash$.
2. If $x = \mathcal{D}(C^*, sk)$ then return $Hash(x)$.
3. Otherwise randomly generate an appropriately sized $hash$, add $(x, hash)$ to *HashList* and return $hash$.

If the attacker requests the evaluation of the key derivation function *KDF* on an input x then the following steps are performed:

1. If $(x, K) \in KDFList$ for some value of K then return K.
2. If $x = \mathcal{D}(C^*, sk)$ then return $KDF(x)$.
3. Otherwise randomly generate an appropriately sized K, add (x, K) to $KDFList$ and return K.

If the attacker requests the evaluation of the decapsulation function on the encapsulated key C then the following steps are performed:

1. If $C = C^*$ then return $\perp$.
2. For each pair $(x, hash) \in HashList$, check whether $\mathcal{E}(x, hash, pk) = C$. If no such pair exists then return $\perp$.
3. If there exists such a pair $(x, hash)$ then run the simulator for the key derivation function on the input x to get a key K.
4. Return K.

As before, we note that game 1 and game 2 are identical except if either of the following events occur:

1. $\mathcal{A}$ queries the decapsulation oracle on the challenge encapsulation before the challenge encapsulation is given to $\mathcal{A}$, or
2. $\mathcal{A}$ queries the decapsulation oracle on some encapsulation C where $x = \mathcal{D}(C, sk)$ and $C = \mathcal{E}(x, Hash(x), pk)$ but $\mathcal{A}$ has not queried the hash function simulator on the input x.

The probability that the first event occurs is bounded above by $q_D/|\mathcal{M}|$ (as there exists $|\mathcal{M}|$ valid encapsulations). The probability that the second event occurs is bounded above by $q_D\gamma/|\mathcal{R}|$ (where γ is defined in Definition 4). Hence the advantage of $\mathcal{A}$ in game 2 is at least

$$\epsilon - q_D\left(\frac{1}{|\mathcal{M}|} + \frac{\gamma}{|\mathcal{R}|}\right). \tag{15}$$

Let E' be the event that, in game 2, the attacker queries either the hash function simulator or the key derivation function oracle with the input $x^* = \mathcal{D}(C^*, sk)$. Again, we have

$$Pr[E'] \geq \epsilon - q_D\left(\frac{1}{|\mathcal{M}|} + \frac{\gamma}{|\mathcal{R}|}\right). \tag{16}$$

Now, consider the following algorithm that takes as input a public key pk for the underlying encryption scheme and a challenge ciphertext C^* (which is the encryption of some randomly chosen message $x^* \in \mathcal{M}$).

1. Prepare two empty lists $HashList$ and $KDFList$.
2. Generate a random bit strings K^* of length $KEM.KeyLen$.
3. Pass the public key pk to $\mathcal{A}$.
4. Allow the attacker $\mathcal{A}$ to run until it requests a challenge encapsulation. If the attacker requests the evaluation of the hash function $Hash$ on an input x then the following steps are performed:

a) If $(x, hash) \in HashList$ for some value of $hash$ then return $hash$.

b) Otherwise randomly generate an appropriately sized $hash$, add $(x, hash)$ to $HashList$ and return $hash$.

If the attacker requests the evaluation of the KDF KDF on an input x then the following steps are performed:

a) If $(x, K) \in KDFList$ for some value of K then return K.

b) Otherwise randomly generate an appropriately sized K, add (x, K) to $KDFList$ and return K.

If the attacker requests the evaluation of decapsulation function on the encapsulated key C then the following steps are performed:

a) If $C = C^*$ then return $\perp$.

b) Check that there exists a unique $x \in \mathcal{M}$ such that $(x, hash) \in HashList$ and $\mathcal{E}(x, hash, pk) = C$ for some value of $hash$. If not, return $\perp$.

c) Run the simulator for the key derivation function on the input x to get a key K.

d) Return K.

5. When the attacker requests a challenge encapsulation pass the pair (K^*, C^*) to the attacker.

6. Allow the attacker to run until it outputs a bit σ'. Answer all oracle queries with the simulators described above.

7. Pick, uniformly at random, some value x from the set of x such that either $(x, hash) \in HashList$ or $(x, K) \in KDFList$. Output x as the inverse of C^*.

This algorithm perfectly simulates the environment for the attacker in game 2 up until the point in which E' occurs. However if E' occurs then the above correctly output x^* with probability $1/(q_D + q_H + q_K)$. Hence the above algorithm will correctly invert the encryption of a randomly generated message with probability at least

$$\frac{1}{q_D + q_H + q_K}(\epsilon - \frac{q_D}{|\mathcal{M}|} - \frac{\gamma q_D}{|\mathcal{R}|}). \tag{17}$$

This value is negligible providing ϵ is negligible, hence the KEM is secure in the IND-CCA2 model providing the underlying encryption scheme is secure in the OW-CPA model.   □

A General Construction of IND-CCA2 Secure Public Key Encryption[*]

Eike Kiltz[1] and John Malone-Lee[2]

[1] Lehrstuhl Mathematik & Informatik, Fakultät für Mathematik,
Ruhr-Universität Bochum, Germany.
kiltz@lmi.rub.de
http://www.rub.de/lmi/kiltz
[2] University of Bristol, Department of Computer Science,
Merchant Venturers Building, Woodland Road,
Bristol, BS8 1UB, UK.
malone@cs.bris.ac.uk

Abstract. We propose a general construction for public key encryption schemes that are IND-CCA2 secure in the random oracle model. We show that the scheme proposed in [1,2] fits our general framework and moreover that our method of analysis leads to a more efficient security reduction.

1 Introduction

Since Diffie and Hellman proposed the idea of public key cryptography [13], one of the most active areas of research in the field has been the design and analysis of public key encryption schemes [4,5,11,14,15,16,17,20,21]. Initially this research followed two separate paths: practical and theoretical. In [14,21] efficient primitives were suggested from which to build encryption schemes. Formal models of security were developed in [16,17,20]. Schemes were designed in these models using tools from complexity theory to provide proofs of security. While the ideas were ground breaking, the schemes that were proposed where of theoretical interest only since they were not at all efficient.

In recent years much research has been done into methods for designing encryption schemes that are both practical and that may be analyzed formally [5, 11]. One approach that has enjoyed a great deal of success is the *random oracle model* proposed by Bellare and Rogaway [5]. In this model cryptographic hash functions are assumed to be perfectly random. Although a proof of security in this model is a heuristic argument, it is generally accepted as a demonstration of sound design, so much so that several schemes analyzed in this way have enjoyed widespread standardization [4,6].

[*] This work was done during a stay of the two authors at BRICS in Aarhus, Denmark. The stay was supported by the Marie-Curie fellowship scheme of the European Union. Both authors wish to thank BRICS and the EU for making this visit possible.

K.G. Paterson (Ed.): Cryptography and Coding 2003, LNCS 2898, pp. 152–166, 2003.
© Springer-Verlag Berlin Heidelberg 2003

In this paper we propose a very general construction for public key encryption. We prove, in the random oracle model, that our construction yields schemes with indistinguishable encryptions under adaptive chosen ciphertext attack (IND-CCA2) [20]. Our security result is tight. It makes use of the work of Fujisaki and Okamoto [15]. We show that our scheme is a generalization of that proposed in [1,2], moreover our method of analysis results in a more efficient security reduction.

The paper is organized as follows. In Section 2 we begin by discussing some security notions before defining an abstract computational problem that we call the *Y-computational problem* (YC). Many number-theoretic primitives in the literature fit our abstract definition.

In Section 3 we propose an encryption scheme which is secure in the sense of indistinguishable encryptions under chosen plaintext attack (IND-CPA). Our result is in the random oracle model. We recall the technique of Fujisaki and Okamoto [15] to transform an IND-CPA secure cryptosystem into one that is IND-CCA2 secure in Section 4. In Section 5 we apply this technique to our construction. Some concrete examples of our cryptosystem are given in Section 6. One example, based on the computational Diffie-Hellman problem, turns out to be the cryptosystem of [1,2]. Our method of security analysis provides an improved reduction than that of [1,2] however. We give a second example based on Rabin [19]. In this case encryption consists of one squaring and the security is equivalent to factoring.

2   Definitions

2.1   Security Notions

A public key encryption scheme Π consists of three algorithms $(\mathcal{K}, \mathcal{E}, \mathcal{D})$ with the following properties.

- The *key generation algorithm*, $\mathcal{K}$, is a probabilistic algorithm that takes a security parameter $1^k \in \mathbb{N}$, represented in unary, and returns a pair (pk, sk) of matching public and secret keys.
- The *encryption algorithm*, $\mathcal{E}$, is a probabilistic algorithm that takes a public key pk and a message $m \in \{0,1\}^*$ to produce a ciphertext $c \in \{0,1\}^*$.
- The *decryption algorithm*, $\mathcal{D}$, is a deterministic algorithm that takes a secret key sk and a ciphertext $c \in \{0,1\}^*$ to produce either a message $m \in \{0,1\}^*$ or a special symbol $\bot$. The symbol $\bot$ is used to indicate that the ciphertext was invalid in some way.

The first formal security definitions for public key encryption appeared in [16]. In this work Goldwasser and Micali proved that, if an adversary running in polynomial time can not distinguish which of two chosen messages has been encrypted, then it can learn no information about a message from its ciphertext. This idea underpins all accepted definitions of security used today. The adversary of a public key encryption scheme is modeled as a probabilistic polynomial time algorithm $\mathcal{A}$ that runs in two stages: $\mathcal{A}_1$ and $\mathcal{A}_2$. In the first stage of its attack $\mathcal{A}_1$ is given a public key pk to attack. At the end $\mathcal{A}_1$ outputs two messages m_0 and m_1 of equal length. A bit b is chosen at random and m_b is encrypted under pk to produce a ciphertext c^*. In the second stage of the attack $\mathcal{A}_2$ is given c^* and asked to determine the bit b.

During the first stage $\mathcal{A}_1$ may be given a decryption oracle for the secret key corresponding to the public key it is attacking. This attack is a *non-adaptive chosen ciphertext attack* [17], or CCA1 for short. An attack in which the adversary $\mathcal{A}_2$ is also given the decryption oracle in the second stage is an *adaptive chosen ciphertext attack* (CCA2) [20]. If the adversary has no access to such oracles we call the attack a *chosen plaintext attack* (CPA).

We formalize these notions in Definition 1 below. Here we denote the fact that the encryption of one of two messages must be indistinguishable to the adversary by IND.

Definition 1. *Let* $\Pi = (\mathcal{K}, \mathcal{E}, \mathcal{D})$ *be an encryption scheme and let* $\mathcal{A} = (\mathcal{A}_1, \mathcal{A}_2)$ *be an adversary. For* $atk \in \{cpa, cca1, cca2\}$ *and* $1^k \in \mathbb{N}$ *let*

$$\mathrm{Adv}_{\mathcal{A},\Pi}^{ind-atk}(1^k) = 2 \cdot \Pr \begin{bmatrix} (\mathrm{pk}, \mathrm{sk}) \leftarrow \mathcal{K}(1^k); \\ (m_0, m_1, state) \leftarrow \mathcal{A}_1^{\mathcal{O}_1}(\mathrm{pk}); \\ b \leftarrow \{0,1\}; \\ c^* \leftarrow \mathcal{E}_{\mathrm{pk}}(m_b); \\ \mathcal{A}_2^{\mathcal{O}_2}(m_0, m_1, c^*, state) = b \end{bmatrix} - 1$$

where

$$atk = cpa \Rightarrow \mathcal{O}_1(\cdot) = \epsilon \quad \text{and } \mathcal{O}_2(\cdot) = \epsilon,$$
$$atk = cca1 \Rightarrow \mathcal{O}_1(\cdot) = \mathcal{D}_{\mathrm{sk}}(\cdot) \text{ and } \mathcal{O}_2(\cdot) = \epsilon,$$
$$atk = cca2 \Rightarrow \mathcal{O}_1(\cdot) = \mathcal{D}_{\mathrm{sk}}(\cdot) \text{ and } \mathcal{O}_2(\cdot) = \mathcal{D}_{\mathrm{sk}}(\cdot).$$

We insist that $\mathcal{A}_1$ *outputs* m_0 *and* m_1 *with* $|m_0| = |m_1|$. *Also,* $\mathcal{A}_2$ *is not permitted make the query* $\mathcal{O}_2(c^*)$.

The encryption scheme Π *is IND-ATK secure if* $\mathcal{A}$ *being polynomial-time implies that* $\mathrm{Adv}_{\mathcal{A},\Pi}^{ind-atk}(1^k)$ *is negligible.*

We define the advantage function for the scheme

$$\mathrm{Adv}_{\Pi}^{ind-atk}(1^k, t, q_d) = \max\{\mathrm{Adv}_{\mathcal{A},\Pi}^{ind-atk}(1^k)\},$$

where the maximum is taken over all adversaries that run for time t *and make at most* q_d *queries to the decryption oracle.*

NOTE: In the ROM we also consider the number of RO queries made by an adversary in the advantage function.

Our construction will make use of a symmetric encryption scheme SE. This consists of two algorithms (E, D) with the following properties.

- The *encryption algorithm*, E, is a deterministic algorithm that takes a key $\kappa \in \{0,1\}^l$ and a message $m \in \{0,1\}^*$ to produce a ciphertext $c \in \{0,1\}^*$.
- The *decryption algorithm*, D, is a deterministic algorithm that takes a key $\kappa \in \{0,1\}^l$ and a ciphertext $c \in \{0,1\}^*$ to produce a message $m \in \{0,1\}^*$.

As in the public key case, the security definition for SE that we use is based on indistinguishability of encryptions. We give a formal definition in Definition 2 below. Here OTE means "one time encryption". Our definition is similar to the notion of find-then-guess security from [3]; however, in [3] an adversary may be able to access an encryption oracle for the key that it is attacking. We require security in a weaker sense where no such oracle is considered.

Definition 2. *Let $SE = (E, D)$ be a symmetric encryption scheme. Let $\mathcal{A} = (\mathcal{A}_1, \mathcal{A}_2)$ be an adversary that runs in two stages. Define*

$$\text{Adv}^{ote}_{\mathcal{A},SE} = 2 \cdot \Pr \begin{bmatrix} \kappa \leftarrow \{0,1\}^l; \\ (m_0, m_1, state) \leftarrow \mathcal{A}_1(); \\ b \leftarrow \{0,1\}; \\ c^* \leftarrow \mathcal{E}_\kappa(m_b); \\ b \leftarrow \mathcal{A}_2(m_0, m_1, c^*, state) \end{bmatrix} - 1.$$

We insist that $\mathcal{A}_1$ outputs m_0 and m_1 with $|m_0| = |m_1|$.

The encryption scheme SE is OTE secure if $\mathcal{A}$ being polynomial-time implies that $\text{Adv}^{ote}_{\mathcal{A},SE}$ is negligible.

We define the advantage function for the scheme

$$\text{Adv}^{ote}_{SE}(t) = \max\{\text{Adv}^{ote}_{\mathcal{A},SE}\},$$

where the maximum is taken over all adversaries that run for time t.

2.2   Computational Problems

In Definition 3 below we define the general computational problem that our construction will use. Our formalization captures many of the most widely used cryptographic primitives such as RSA [21] and Diffie-Hellman [13]. Some illustrative examples are given following the definition. We call our general problem the Y-computational problem (YC). The reason for this can be seen in the shape of Figure 1.

Definition 3. *An instance generator $\mathcal{I}_{YC}(1^k)$ for YC outputs a description of (S_1, S_2, f_1, f_2, t). Here S_1 and S_2 are sets with $|S_1| = k$,*

$$f_1, f_2 : S_1 \to S_2$$

are functions and $t : S_2 \to S_2$ is a (trapdoor) function such that for all $x \in S_1$, $t(f_1(x)) = f_2(x)$. The functions f_1, f_2 and t should be easy to evaluate and it should be possible to sample efficiently from S_1.

Let $\mathcal{A}$ be an adversary and define

$$\text{Adv}_{\mathcal{A},\mathcal{I}_{\text{YC}}}(1^k) = \Pr\left[\begin{array}{c}(S_1, S_2, f_1, f_2, t) \leftarrow \mathcal{I}_{\text{YC}}(1^k); \\ x \leftarrow S_1; \\ f_2(x) \leftarrow A(S_1, S_2, f_1, f_2, f_1(x))\end{array}\right].$$

We define the advantage function

$$\text{Adv}_{\mathcal{I}_{\text{YC}}}(1^k, t) = \max\{\text{Adv}_{\mathcal{A},\mathcal{I}_{\text{YC}}}(1^k)\}$$

where the maximum is taken over all adversaries that run for time t. We say that YC is hard for $\mathcal{I}_{\text{YC}}(1^k)$ if t being polynomial in k implies that the advantage function $\text{Adv}_{\mathcal{I}_{\text{YC}}}(1^k, t)$ is negligible in k. Figure 1 illustrates the hard Y-computational problem.

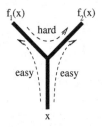

Fig. 1. The Y-computational problem: Given $f_1(x)$, compute $f_2(x)$.

2.3   Examples of Hard Y-Computational Problems

In this subsection we show that many known cryptographic primitives fit the general definition of YC problems.

EL GAMAL. For the El Gamal cryptosystem, the instance generator $\mathcal{I}_{\text{YC}}(1^k)$ computes a random k-bit prime p and a random generator g_1 of the multiplicative group $\mathbb{Z}_p^*$. The sets S_1 and S_2 are $\mathbb{Z}_p^*$ together with the generator g_1. A random value $s \in \{1, \ldots, p-1\}$ is chosen and $g_2 \in \mathbb{Z}_p^*$ is computed as $g_2 = g_1^s$. The functions f_1 and f_2 are defined as $f_1(x) = g_1^x$ and $f_2(x) = g_2^x$. The trapdoor function is $t(x) = x^s$.
Obviously, $t(f_1(x)) = (g_1^x)^s = g_1^{xs} = g_2^x = f_2(x)$ holds and YC is hard if the computational Diffie-Hellman assumption [13] holds.

POINTCHEVAL [18]. For the Pointcheval cryptosystem, the instance generator $\mathcal{I}_{\text{YC}}(1^k)$ computes a random k-bit composite $n = pq$. The sets S_1 and S_2 are $\mathbb{Z}_n^*$. A random exponent e is chosen with $\gcd(e, \varphi(n)) = 1$ and its inverse $d = e^{-1}$

modulo $\varphi(n)$ is computed. The functions f_1 and f_2 are defined as $f_1(x) = x^e$ and $f_2(x) = (x+1)^e$. The trapdoor function is $t(x) = (x^d + 1)^e$.

Obviously, $t(f_1(x)) = f_2(x)$ holds and YC is hard if the *computational dependent RSA problem* (see also [18]) is hard.

ARBITRARY TRAPDOOR ONEWAY FUNCTIONS. Let $\mathcal{I}_{towf}$ be an instance generator for trapdoor oneway functions. Informally speaking, on input 1^k, $\mathcal{I}_{towf}$ outputs the description of two sets S_1 and S_2, together with a oneway function $f_1 : S_1 \rightarrow S_2$, and its trapdoor t such that $t(f_1(x)) = x$, for all $x \in S_1$. The functions f_1 and t should be easy to evaluate.

The instance generator $\mathcal{I}_{YC}(1^k)$ runs $\mathcal{I}_{towf}$ on input 1^k, sets $f_2(x) = x$ (f_2 is the identity), and outputs (S_1, S_2, f_1, f_2, t) as an instance of YC. Then obviously YC is hard if inverting the oneway function f_1 is hard.

The two most important examples of trapdoor oneway functions are the RSA [21] and the Rabin [19] function. The latter is especially interesting for cryptographic purposes because its onewayness is provably equivalent to factoring.

RSA-PAILLIER [9]. For the RSA-Paillier cryptosystem, the instance generator $\mathcal{I}_{YC}(1^k)$ computes a random k-bit composite $n = pq$. It outputs the sets $S_1 = \mathbb{Z}_n$ and $S_2 = \mathbb{Z}_n^*$. Then it computes a random exponent e with $\gcd(e, \varphi(n)) = 1$ and its inverse $d = e^{-1}$ modulo $\varphi(n)$. For $x = aN + b \in \mathbb{Z}_{N^2}^*$, we define $[x]_1$ as $b \in \mathbb{Z}_n$ and $[x]_2$ as $a \in \mathbb{Z}_n^*$. The functions f_1 and f_2 are defined as $f_1(x) = [x^e \bmod n^2]_1 = x^e \bmod n$ and $f_2(x) = [x^e \bmod n^2]_2$. The trapdoor function $t(x) = [(x^d \bmod N)^e \bmod n^2]_2$.

Obviously, $t(f_1(x)) = f_2(x)$ holds. In [10] it was shown that YC is hard if the RSA problem is hard.

3   IND-CPA under YC in the RO Model

In this section we present a general construction of an IND-CPA secure cryptosystem based on the hardness of YC. The method uses a hash function which is modelled as a random oracle [5] in the security analysis.

Definition 4 (The Cryptosystem Π_0).

- *The key generator $\mathcal{K}(1^k)$ runs the instance generator $\mathcal{I}_{YC}(1^k)$ for YC as in Definition 3 and outputs the description of (S_1, S_2, f_1, f_2) as the public key pk. The corresponding secret key sk is the trapdoor $t : S_2 \rightarrow S_2$.*
- *The cryptosystem uses a symmetric encryption scheme $SE = (E, D)$ with keys of length l. It also uses a hash function*

$$G : S_2 \rightarrow \{0,1\}^l.$$

- *The encryption function works as follows. Choose $x \stackrel{r}{\leftarrow} S_1$. Compute $\kappa = G(f_2(x))$.*

$$\mathcal{E}_{\mathrm{pk}}(m, x) = \big(f_1(x), E_\kappa(m)\big) = (\alpha, \beta).$$

- *To decrypt (α, β) one computes $\kappa = G(t(\alpha))$ and outputs*

$$D_\kappa(\beta).$$

The following Theorem proves the IND-CPA security of the encryption scheme Π_0 in the random oracle model.

Theorem 1 (IND-CPA security of Π_0). *For the encryption scheme Π_0 we have*

$$\mathrm{Adv}_{\Pi_0}^{ind-cpa}(t, q_g) \leq 2q_g \cdot \mathrm{Adv}_{\mathcal{I}_{\mathrm{YC}}}(1^k, t') + \mathrm{Adv}_{SE}^{ote}(t'),$$

where $t' \approx t$.

Proof. We prove the theorem by constructing algorithms using an adversary $\mathcal{A}$ as a subroutine to show that if $\mathcal{A}$ is to have any advantage then, with overwhelming probability, it must either solve an instance of YC or it must break the symmetric encryption scheme $SE = (E, D)$.

We begin by constructing an algorithm $\mathcal{B}$ to solve YC. Let us assume that $\mathcal{I}_{\mathrm{YC}}(1^k)$ has been run to produce (S_1, S_2, f_1, f_2, t) and that we are given the description of (S_1, S_2, f_1, f_2) and $X = f_1(x)$ for some $x \in S_1$. We make (S_1, S_2, f_1, f_2) the public key pk which $\mathcal{A}$ attacks.

The task of $\mathcal{B}$ is to compute $f_2(x)$. We run $\mathcal{B}$ responding to its hash queries with a simulator G_{sim}. Simulator G_{sim} keeps a list G_L of query/response pairs (y, κ) to maintain consistency between calls.

We may now describe $\mathcal{B}$.

Algorithm $\mathcal{B}(X)$
$(m_0, m_1, state) \leftarrow \mathcal{A}_1^{G_{sim}}(\mathrm{pk})$
$b \stackrel{r}{\leftarrow} \{0, 1\}$
$\kappa^* \stackrel{r}{\leftarrow} \{0, 1\}^l$
$\alpha^* \leftarrow X$
$\beta^* \leftarrow E_{\kappa^*}(m_b)$
$c^* = (\alpha^*, \beta^*)$
$b' \leftarrow \mathcal{A}_2^{G_{sim}}(m_0, m_1, c^*, state, \mathrm{pk})$
$(y, \kappa) \stackrel{r}{\leftarrow} G_L$
Return y

Let us now analyse our simulation. We will consider how $\mathcal{A}$ runs in a real run $(real)$ and in our simulation (sim). We define an event ERR to be one that would cause $\mathcal{A}$'s view to differ in $real$ and sim. We have

$$[t] \Pr[\mathcal{A} \text{ wins} \wedge \neg\text{ERR}]_{sim} = \Pr[\mathcal{A} \text{ wins} \wedge \neg\text{ERR}]_{real}$$
$$\geq \Pr[\mathcal{A} \text{ wins}]_{real} - \Pr[\text{ERR}]_{real}$$
$$= \frac{1}{2} + \frac{1}{2}\text{Adv}_{\mathcal{A},\Pi}^{ind-cpa} - \Pr[\text{ERR}]_{real}$$
$$= \frac{1}{2} + \frac{1}{2}\text{Adv}_{\mathcal{A},\Pi}^{ind-cpa} - \Pr[\text{ERR}]_{sim}. \quad (1)$$

The final equality follows from that fact that, by definition of ERR, $\mathcal{A}$'s view in *real* and in *sim* are identical up until ERR occurs.

We now consider $\Pr[\text{ERR}]_{sim}$. The event can only be caused by an error in G_{sim}. The only possible error here is caused by $\mathcal{A}$ making the query $t(X) = f_2(x)$ to which G_{sim} should respond κ^*. Moreover, if such a query is made algorithm $\mathcal{B}$ succeeds with probability $1/q_g$. We infer that

$$\Pr[\text{ERR}]_{sim} \leq q_g \cdot \text{Adv}_{\mathcal{B},\mathcal{I}_{\text{YC}}}(1^k) \quad (2)$$

Let us now reconsider $\Pr[\mathcal{A} \text{ wins} \wedge \neg\text{ERR}]_{sim}$. We show that $\mathcal{A}$ can have no advantage in this situation unless it can break the one-time security of the symmetric encryption function SE. To do this we construct an adversary $\mathcal{C} = (\mathcal{C}_1, \mathcal{C}_2)$ of SE. This adversary will again run $\mathcal{A}$ as a subroutine. The simulator to respond to $\mathcal{A}$'s queries to G will be as above.

$$\text{Algorithm } \mathcal{C}_1()$$
$$(m_0, m_1, state) \leftarrow \mathcal{A}_1^{G_{sim}}(\text{pk})$$
$$\text{Return } (m_0, m_1, state)$$

Now outside of $\mathcal{C}$'s view a random bit b is chosen and m_b is encrypted under a random key κ^* to produce c^*.

$$\text{Algorithm } \mathcal{C}_2(m_0, m_1, c^*, state)$$
$$\alpha^* \xleftarrow{r} S_2$$
$$\beta^* \leftarrow c^*$$
$$c^* \leftarrow (\alpha^*, \beta^*)$$
$$b' \leftarrow \mathcal{A}_2^{G_{sim}}(m_0, m_1, c^*, state)$$
$$\text{Return } b'.$$

The important things to note are first of all, in the event $\neg\text{ERR}$, adversary $\mathcal{C}$ runs $\mathcal{A}$ in exactly the same way that the latter would be run in *sim*. Secondly, if $\mathcal{A}$ wins in *sim* then $\mathcal{C}$ wins. We infer that

$$\Pr[\mathcal{A} \text{ wins} \wedge \neg\text{ERR}]_{sim} \leq \frac{1}{2} + \frac{1}{2}\text{Adv}_{\mathcal{C},SE}^{ote}. \quad (3)$$

The result now follows from (1), (2), (3) and the construction of $\mathcal{B}$ and $\mathcal{C}$. $\square$

Now consider $\Pr[\mathsf{ERR}]_{sim}$ from equation (2). If we had access to an efficient verify algorithm $\mathcal{V}$ that on input $f_1(x_1)$ and $f_2(x_2)$, checks if $x_1 = x_2$, then we could drop this error probability to

$$\Pr[\mathsf{ERR}]_{sim} \leq \mathrm{Adv}_{\mathcal{B}, \mathcal{I}_{\mathrm{YC}}}(1^k).$$

This is done by simply running $\mathcal{V}$ on input (y, X) for all queries y from the list G_L (that contains all queries made to the oracle G_{sim}). Indeed, if such an algorithm $\mathcal{V}$ exists (we say that YC has the "easy to verify" property), we get the improved result in Remark 1 below.

Remark 1. If YC has the "easy to verify" property, then for the encryption scheme Π_0 we have

$$\mathrm{Adv}_{\Pi_0}^{ind-cpa}(t, q_g) \leq 2 \cdot \mathrm{Adv}_{\mathcal{I}_{\mathrm{YC}}}(1^k, t') + \mathrm{Adv}_{SE}^{ote}(t'),$$

where $t' = t + q_G(T_{\mathcal{V}}(1^k) + O(k))$. Here $T_{\mathcal{V}}(1^k)$ denotes the running time of the verify algorithm $\mathcal{V}$.

Note that, with the exception of El Gamal, all the Y-computational problems presented in subsection 2.3 have the easy to verify property.

Remark 2. If one removes the symmetric encryption algorithm in the IND-CPA scheme of Definition 4, and merely output the symmetric key, then the scheme becomes a key encapsulation mechanism as introduced by Cramer and Shoup [12]. This is another approach to the problem of desiging IND-CCA2 secure encryption schemes.

4   The Fujisaki-Okamoto Transform

In [15] Fujisaki-Okamoto (FO) described a method to transform a cryptosystem with IND-CPA security into one with IND-CCA2 security. The method uses a hash function which is modelled as a random oracle [5] in the security analysis. The reduction is very tight. In this section we define the necessary notions and state the FO result.

Definition 5. *Let $\Pi = (\mathcal{K}, \mathcal{E}, \mathcal{D})$ be an IND-CPA secure cryptosystem. We define the transformed scheme $\Pi' = (\bar{\mathcal{K}}, \bar{\mathcal{E}}^H, \bar{\mathcal{D}}^H)$ as follows.*

- *The key generator $\bar{\mathcal{K}}(1^k)$ runs the key generator $\mathcal{K}(1^k)$.*
- *The cryptosystem uses a hash function*

$$H : \{0,1\}^* \to \{0,1\}^{k_0}$$

- *The encryption function works as follows. Choose $x \stackrel{R}{\leftarrow} \{0,1\}^{k_0}$ and compute*

$$\bar{\mathcal{E}}_{\mathrm{pk}}^H(m, x) = \mathcal{E}_{\mathrm{pk}}\left((m\|x), H(m\|x)\right).$$

- *To decrypt the ciphertext c, one computes $m'||x' = \mathcal{D}_{sk}(c)$ and outputs*

$$\bar{\mathcal{D}}_{sk}^H = \begin{cases} m' & \text{if } \bar{\mathcal{E}}_{pk}^H(m', x') = c \text{ and} \\ \perp & \text{otherwise.} \end{cases}$$

Definition 6 (λ-uniformity). *Let $\Pi = (\mathcal{K}, \mathcal{E}, \mathcal{D})$ be a public-key cryptosystem taking random input from $\{0,1\}^{k_0}$ and messages from $\{0,1\}^{\text{mlen}}$. For given $x \in \{0,1\}^{\text{mlen}}$ and $y \in \{0,1\}^*$, define*

$$\lambda(x,y) := \Pr_{x \xleftarrow{R} \{0,1\}^{k_0}} [y = \mathcal{E}_{pk}(m, x)].$$

We say that Π is λ-uniform if, for any $x \in \{0,1\}^{\text{mlen}}$ and $y \in \{0,1\}^*$, $\lambda(x,y) \leq \lambda$.

Fujisaki and Okamoto proved the following result about Π'.

Theorem 2 (IND-CCA2 security [15]). *Suppose that the encryption scheme Π' is λ-uniform. Then we have*

$$\text{Adv}_{\Pi'}^{ind-cca2}(1^k, t, q_d, q_h) \leq \text{Adv}_{\Pi}^{ind-cpa}(1^k, t') \cdot (1 - \lambda)^{-q_d} + q_h \cdot 2^{-k_0 - 1}.$$

where $t' = t + q_h(T_{\mathcal{E}}(1^k) + O(k))$. Here $T_{\mathcal{E}}(1^k)$ denotes the running time of $\mathcal{E}_{pk}(\cdot)$.

5   IND-CCA2 under YC in the RO Model

As proved in Section 3, the cryptosystem Π_0 is IND-CPA secure in the random oracle model if the Y-computational problem YC is hard and the symmetric encryption function is OTE secure. It is now natural to apply the FO construction from the last section to this cryptosystem to get a cryptosystem Π_1 that is IND-CCA2 secure in the random oracle model. The construction uses two hash functions which are modelled as random oracles [5] in the security analysis. The reduction is very tight.

Definition 7 (The Cryptosystem Π_1).

- *The key generator $\mathcal{K}(1^k)$ runs the instance generator $\mathcal{I}_{YC}$ for YC as in Definition 3 and outputs the description of (S_1, S_2, f_1, f_2) as the public key pk. The corresponding secret key sk is the trapdoor $t : S_2 \rightarrow S_2$.*
- *The cryptosystem uses a symmetric encryption scheme $SE = (E, D)$ with keys of length l. It also uses two hash functions*

$$G : S_2 \rightarrow \{0,1\}^l \text{ and } H : \{0,1\}^* \rightarrow S_1.$$

- *The encryption function works as follows. Choose $x \xleftarrow{r} \{0,1\}^{k_1}$. Compute $h = H(m||x)$ and $\kappa = G(f_2(h))$.*

$$\mathcal{E}_{pk}(m, x) = \big(f_1(h), E_\kappa(m||x)\big) = (\alpha, \beta).$$

– *To decrypt (α, β) one computes $\kappa = G(t(\alpha))$, $m'||x' = D_\kappa(\beta)$, $h' = H(m'||x')$.*

$$D_{\text{sk}}(\alpha, \beta) = \begin{cases} m' & \text{if } \alpha = f_1(h') \\ \bot & \text{otherwise} \end{cases}$$

The symbol $\bot$ denotes the fact that the ciphertext was rejected.

The following Theorem proves the IND-CCA2 security of the encryption scheme Π_1 in the random oracle model.

Theorem 3 (IND-CCA2 security of Π_1). *For the encryption scheme Π_1 we have*

$$\text{Adv}_{\Pi_1}^{ind-cca2}(1^k, t, q_d, q_h) \leq$$

$$(2q_g \cdot \text{Adv}_{\mathcal{I}_{\text{YC}}}(1^k, t') + \text{Adv}_{SE}^{ote}(t')) \cdot (1 - 2^{-k})^{-q_d} + \frac{q_h}{2^{k_0+1}}.$$

where $t' = t + q_h(T_{\mathcal{E}}(1^k) + O(k))$. Here $T_{\mathcal{E}}(1^k)$ denotes the running time of $\mathcal{E}_{\text{pk}}(\cdot)$.

The proof of this Theorem directly follows by applying the Theorem 2 and Theorem 1, and the following Lemma about the λ-uniformity of the cryptosystem Π_0.

Lemma 1. *The cryptosystem Π_0 from Definition 4 is 2^{-k}-uniform.*

Proof. By definition of λ-uniformity (see Definition 6), we have that

$$\begin{aligned} \lambda(m, \alpha, \beta) \quad &= \quad \Pr_{x \xleftarrow{R} \{0,1\}^{k_0}} \begin{bmatrix} \alpha = f_1(x) \\ \beta = E_\kappa(m) \end{bmatrix} \\ &\leq \quad \Pr_{x \xleftarrow{R} \{0,1\}^{k_0}} [\alpha = f_1(x)] \\ &= \quad \frac{1}{|S_1|} = \frac{1}{2^k}. \end{aligned}$$

$\square$

Remark 3. If YC has the "easy to verify" property, then for the encryption scheme Π_1 we have

$$\text{Adv}_{\Pi_1}^{ind-cca2}(1^k, t, q_d, q_h) \leq$$

$$(2 \cdot \text{Adv}_{\mathcal{I}_{\text{YC}}}(1^k, t') + \text{Adv}_{SE}^{ote}(t')) \cdot (1 - 2^{-k})^{-q_d} + \frac{q_h}{2^{k_0+1}}.$$

where $t' = t + q_h(T_{\mathcal{E}}(1^k) + O(k)) + q_G(T_{\mathcal{V}}(1^k) + O(k))$. Here $T_{\mathcal{E}}(1^k)$ denotes the running time of $\mathcal{E}_{\text{pk}}(\cdot)$. and $T_{\mathcal{V}}(1^k)$ denotes the running time of $\mathcal{V}(\cdot)$.

6   Examples

In this section we apply our construction of the cryptosystem Π_1 from section 5 to two important examples of instances of YC mentioned in Section 2.3, the El Gamal function and the Rabin function.

ENHANCED EL GAMAL ENCRYPTION SCHEME.

- The key generator $\mathcal{K}(1^k)$ runs the instance generator $\mathcal{I}_{YC}$ for the El Gamal case and gets S_1 and S_2 as $\mathbb{Z}_p^*$ together with a generator g_1. Furthermore it gets $f_1(x) = g_1^x$ and $f_2(x) = g_2^x$. (S_1, S_2, f_1, f_2) form the public key pk. The corresponding secret key sk is the trapdoor $t(x) = x^s$ (where $s = \log_{g_1} g_2$).
- The cryptosystem uses a symmetric encryption scheme $SE = (E, D)$ with keys of length l. It also uses two hash functions

$$G : S_2 \to \{0,1\}^l \text{ and } H : \{0,1\}^* \to \mathbb{Z}_p^*.$$

- The encryption function works as follows. Choose $x \xleftarrow{r} \{0,1\}^{k_1}$. Compute $h = H(m||x)$ and $\kappa = G(g_2^h)$.

$$\mathcal{E}_{pk}(m, x) = \left(g_1^h, E_\kappa(m||x)\right) = (\alpha, \beta).$$

- To decrypt (α, β) one computes $\kappa = G(\alpha^s)$, $m'||x' = D_\kappa(\beta)$, $h' = H(m'||x')$.

$$D_{sk}(\alpha, \beta) = \begin{cases} m' & \text{if } \alpha = g_1^{h'} \\ \perp & \text{otherwise} \end{cases}$$

The symbol $\perp$ denotes the fact that the ciphertext was rejected.

Corollary 1. *In the random oracle model, the enhanced El Gamal encryption scheme is IND-CCA2 secure if the computational Diffie-Hellman problem is intractable and the symmetric encryption scheme SE is OTE secure.*

We note that our Enhanced El Gamal scheme is exactly that proposed in [1, 2] (we refer to it as the BKL-scheme henceforth), when we use the onetime pad as the symmetric encryption function SE (i.e., $E_\kappa(m) = \kappa \oplus m$). Moreover, our method of security reduction is *tight* (linear in terms of both, time and probability), opposed to that of [1,2] which gives a reduction that is cubic in the time parameter.

In [1,2] a comparison of the BKL-scheme is made with El Gamal encryption once the FO-transform has been applied. It is claimed that the BLK-scheme is preferable since its security is guaranteed by the computational Diffie-Hellman problem rather than the, possibly easier, decisional Diffie-Hellman problem [7]. This argument is misleading since, if G is a random oracle, the distributions $\left(g, g^a, g^b, G(g^{ab})\right)$ and $\left(g, g^a, g^b, G(g^r)\right)$ are indistinguishable if and only if the computational Diffie-Hellman problem is hard. It is easy to see that, with a random oracle a decisional problem comes for free from a computational problem.

ENHANCED RABIN ENCRYPTION SCHEME. As an example of how to use our scheme with a trapdoor-oneway function we use the Y-computational problem induced by the Rabin function.

Applying our result to the Rabin function requires care because square roots modulo $n = pq$ are not unique. To this end we slightly modify our decryption algorithm and make the trapdoor function t act from $S_2 \to S_2 \times S_2$.

– The key generator $\mathcal{K}(1^k)$ runs the instance generator $\mathcal{I}_{YC}$ for Rabin and gets a modulus $n = pq$ where n is a $(k-1)$ bit number and p and q are two primes of roughly the same size with $p = q = 1 \bmod 4$. The set S_1 is

$$S_1 = \{1, \dots, (n-1)/2\} \cap \mathbb{Z}_n^*,$$

and S_2 is the set of quadratic residues modulo n. The function f_1 is $f_1(x) = x^2$ and $f_2(x) = x$. The quadruple (S_1, S_2, f_1, f_2) forms the public key pk. The corresponding secret key sk is the trapdoor $t(x)$ which maps $y \in S_2$ to a pair $(z_1, z_2) \in S_1 \times S_1$ such that $z_1^2 = z_2^2 = y$ (z_1 and z_2 differ in their Jacobi symbol).
– The cryptosystem uses a symmetric encryption scheme $SE = (E, D)$ with keys of length l. It also uses two hash functions

$$G : S_2 \to \{0,1\}^l \text{ and } H : \{0,1\}^* \to S_1.$$

– The encryption function works as follows. Choose $x \xleftarrow{r} \{0,1\}^{k_1}$. Compute $h = H(m||x)$ and $\kappa = G(h)$.

$$\mathcal{E}_{\mathrm{pk}}(m, x) = \big(h^2, E_\kappa(m||x)\big) = (\alpha, \beta).$$

– To decrypt (α, β) one computes $(z_1, z_2) = t(\alpha)$ and for $i \in \{1, 2\}$, $\kappa_i = G(z_i)$, $m_i'||x_i' = D_{\kappa_i}(\beta)$, $h_i' = H(m_i'||x_i')$.

$$D_{\mathrm{sk}}(\alpha, \beta) = \begin{cases} m_1' & \text{if } \alpha = (h_1')^2 \\ m_2' & \text{if } \alpha = (h_2')^2 \\ \bot & \text{otherwise} \end{cases}$$

The symbol $\bot$ denotes the fact that the ciphertext was rejected.

Corollary 2. *In the random oracle model, the enhanced Rabin encryption scheme is IND-CCA2 secure if the factoring problem is intractable and the symmetric encryption scheme SE is OTE secure.*

The encryption procedure of the Enhanced Rabin encryption scheme seems to be very efficient. When we neglect the cost of using the hash functions G and H, and the symmetric encryption scheme SE, the scheme uses only one squaring modulo n. Decryption requires two exponentiation, one modulo p and the other modulo q.

As already noted before, the "easy to verify" property (see Remark 1) is true in the case of the Rabin function, since $f_2(x) = x$ is the identity function. Therefore, the running time of the reduction algorithm is tight (Remark 3).

7   Conclusions

We have introduced a general construction for public key encryption schemes that are IND-CCA2 secure in the random oracle model. Our construction may be used with many of the number theoretic primitives in the literature. The scheme generalises that of [1,2] and we have provided an improved security reduction.

There is some doubt concerning the meaning of a proof of security in the random oracle model. In [8] it is demonstrated that there exist cryptosystems that are provably secure in the random model, but insecure when the random oracle is instantiated with any hash function. Following from our remark in Section 6, it may be interesting to investigate the possibility of Y-computational problems where there is a separation between the computational problem and the decisional analogue no matter what hash function is used to instantiate the random oracle.

Acknowledgments. We would like to thank Ronald Cramer for useful advice and the anonymous referees for their suggestions of how to improve the paper.

References

1. J. Baek and B. Lee and K. Kim. Provably secure length-saving public-key encryption scheme under the computational Diffie-Hellman assumption. Electronics and Telecommunications Research Institute (ETRI) Journal, Vol 22, No. 4, Dec., pages 25–31, 2000.
2. J. Baek, B. Lee, and K. Kim. Secure Length-Saving El Gamal Encryption Under the Computational Diffie-Hellman Assumption. In Proceedings of the Fifth Australian Conference on Information Security and Privacy (ACISP 2000), volume 1841 of Lecture Note in Computer Science, pages 49–58. Springer-Verlag, 2000.
3. M. Bellare and A. Desai and E. Jokipii and P. Rogaway. A Concrete Security Treatment of Symmetric Encryption. In 38th Annual Symposium on Foundations of Computer Science, pages 394–403. IEEE Computer Science Press, 1997.
4. M. Bellare and P. Rogaway. Optimal Asymmetric Encryption - How to Encrypt with RSA. In Advances in Cryptology - EUROCRYPT '94, volume 950 of Lecture Notes in Computer Science, pages 92–111. Springer-Verlag, 1994.
5. M. Bellare and P. Rogaway. Random Oracles are Practical: A Paradigm for Designing Efficient Protocols. In Proceedings of the First ACM Conference on Computer and Communications Security, pages 62–73. 1993.
6. M. Bellare and P. Rogaway. The Exact Security of Digital Signatures - How to sign with RSA and Rabin. In Advances in Cryptology - EUROCRYPT '96, volume 1070 of Lecture Notes in Computer Science, pages 399–416. Springer-Verlag, 1996.
7. D. Boneh. The Decisional Diffie-Hellman Problem. In proceedings of the 3rd Algoritmic Number Theory Symposium, volume 1423 of Lecture Notes in Computer Science, pages 48–63. Springer-Verlag, 1998.
8. R. Canetti, O. Goldreich, and S. Halevi. The Random Oracle Methodology, Revisited. In Proceedings of the Thirtieth Annual ACM Symposium on the Theory of Computing - STOC '98, pages 209–218. ACM, 1998.

9. D. Catalano, R. Gennaro, N. Howgrave-Graham, and P. Q. Nguyen. Paillier's Cryptosystem Revisited. Proceedings of the 8th ACM Conference on Computer and Communications Security, 2001.

10. D. Catalano, P. Q. Nguyen, and J. Stern. The Hardness of Hensel Lifting: The Case of RSA and Discrete Logarithm. In Advances in Cryptology - ASIACRYPT 2002, volume 2501 of Lecture Notes in Computer Science, pages 299–310. Springer Verlag, 2002.

11. R. Cramer and V. Shoup. A Practical Public Key Cryptosystem Provably Secure Against Adaptive Chosen Ciphertext Attack. In Advances in Cryptology - CRYPTO '98, volume 1462 of Lecture Notes in Computer Science, pages 13–25. Springer-Verlag, 1998.

12. R. Cramer and V. Shoup. Design and Analysis of Practical Public-Key Encryption Schemes Secure against Adaptive Chosen Ciphertext attack. To appear, SIAM Journal of Computing.

13. W. Diffie and M. E. Hellman. New Directions in Cryptography. In IEEE Transactions on Information Theory, volume IT-22(6), pages 644–654. 1976.

14. T. ElGamal, A Public Key Cryptosystem and a Signature Scheme Based on Discrete Logarithms. In IEEE Transactions on Information Theory, volume IT-31, pages 469–472. 1985.

15. E. Fujisaki and T. Okamoto. How to Enhance the Security of Public-Key Encryption at Minimum Cost. In Public Key Cryptography - PKC '99, volume 1560 of Lecture Notes in Computer Science, pages 53–68. Springer-Verlag, 1999.

16. S. Goldwasser and S. Micali. Probabilistic Encryption. In Journal of Computer and System Sciences, volume 28, pages 270–299. 1984.

17. M. Naor and M. Yung. Public-key Cryptosystems Provably Secure Against Chosen Ciphertext Attack. In Proceedings of 22nd ACM Symposium on Theory of Computing, pages 427–437. ACM Press, 1990.

18. D. Pointcheval. New Public Key Cryptosystems based on the Dependent-RSA Problems. In Advances in Cryptology - Proceedings of EUROCRYPT '99, volume 1592 of Lecture Notes in Computer Science, Pages 239–254, Springer-Verlag, 1999.

19. M. O. Rabin. Digitalized signatures and public key cryptosystems as intractable as factorization. MIT/LCS/TR-212, Technical Report MIT. 1979.

20. C. Rackoff and D. Simon. Non-Interactive Zero-Knowledge Proof of Knowledge and Chosen Ciphertext Attack. In Advances in Cryptology - CRYPTO '91, volume 576 of Lecture Notes in Computer Science, pages 433–444. Springer-Verlag, 1992.

21. R. L. Rivest, A. Shamir, and L. Adleman. A Method for Obtaining Digital Signatures and Public-Key Cryptosystems. In Communications of the ACM, volume 21(1), pages 120–126. 1978.

22. Y. Tsiounis and M. Yung. On the Security of El Gamal Based Encryption. In Public Key Cryptography '98, volume 1431 of Lecture Notes in Computer Science, pages 117–134, Springer-Verlag, 1998.

Efficient Key Updating Signature Schemes Based on IBS*

Dae Hyun Yum and Pil Joong Lee

Department of Electronic and Electrical Engineering, POSTECH
Hyoja-dong, Nam-Gu, Pohang, Kyoungbuk, 790-784, Rep. of Korea
{dhyum,pjl}@postech.ac.kr
http://islab.postech.ac.kr

Abstract. To mitigate the damage of secret key exposure, key updating signature schemes can be used such as a key-insulated signature scheme and an intrusion-resilient signature scheme. We propose efficient key updating signature schemes based on a secure identity-based signature (IBS) scheme. KUS-SKI is a strong $(N-1, N)$ key-insulated signature scheme with random-access key updates, and KUS-IR is a Type (I) intrusion-resilient signature scheme. We also provide an equivalence theorem between a secure identity-based signature scheme and a strong $(N-1, N)$ key-insulated signature scheme with random-access key up-dates.

Keywords: Key insulated signature, intrusion resilient siganture, identity based signature.

1 Introduction

Exposure of Secret Key. To achieve secure communication, the cryptographic community has conducted much research based on Kerckhoffs' assumption that all details of an encryption function except the secret key are known to adversaries. Even though this assumption seems theoretically reasonable, the real world shows that this often fails. Careless users store their secret keys in an insecure place and frequent losses even occur. It commonly occurs that an adversary gains access to a user's secret key without breaking the underlying cryptographic schemes. As portable devices are spreading widely, this threat has increased considerably.

To reduce the damage of secret key exposure, secret sharing [17] and threshold cryptography [6] can be used. In these models, the secret key is shared in a distributed manner and the attacker should compromise more than a predefined number of share holders. However, distributed computation is required to generate a valid signature or to decrypt a ciphertext, and this is undesirable in many circumstances. While secret sharing and threshold cryptography can be

---

* This research was supported by University IT Research Center Project, the Brain Korea 21 Project, POSTECH PIRL and Com²MaC-KOSEF.

K.G. Paterson (Ed.): Cryptography and Coding 2003, LNCS 2898, pp. 167–182, 2003.

considered as a separation of secret information in location, there is another approach, i.e., a separation in time. A forward-secure public key cryptosystem [2][5] evolves the secret key at the beginning of every time period while standard cryptographic computation is performed by a single device. Note that the public key remains for the lifetime of the scheme. A forward-secure public key cryptosystem prevents an adversary with a secret key for one time period from breaking the scheme for previous time periods. This model is conceptually the basis of other key updating (or evolving) public key cryptosystems, such as key-insulated cryptosystems [8][9] and intrusion-resilient cryptosystems [11][7].

Key-Insulated Cryptosystems. In a key-insulated public key cryptosystem [8][9], a user's secret key is stored in an insecure device and updated regularly through collaboration with a physically-secure device which stores a master key. When a user generates a public key PK which remains for the lifetime of the scheme, a master key SK^* is stored on a physically-secure device. If the lifetime of the scheme is divided into distinct periods 1, 2, $\cdots$, N, the user interacts with the secure device to derive a temporary secret key SK_i which will be used by the insecure device to perform cryptographic operations like signing and deciphering for the time period i. The user executes this key updating procedure at the beginning of every time period. When an adversary who compromises the insecure device up to $t < N$ periods cannot break the remaining $N - t$ periods, we call the scheme (t, N) key-insulated. Additionally, an adversary who compromises the physically-secure device cannot break the scheme for any time period in a "strong" (t, N) key-insulated scheme. Hence, the strong key-insulated cryptosystem is immune to break-ins either on the secure device or on the insecure device (not both). The first key-insulated public key cryptosystem [8] had a drawback that the length of the public key depends on the parameter t and removal of this dependency was attained in [3][9].

Intrusion-Resilient Cryptosystems. Key-insulated cryptosystems preserve the security of past and future time periods when either a user (an insecure device) or a base (a secure device) is compromised. This feature can be extended by adopting a proactivation paradigm [14]. Intrusion-resilient cryptosystems [11][12][7] are designed such that an adversary who obtains keys stored at both the user and the base on multiple occasions (but not at the same time) cannot break the scheme for any time period other than that for which the user's key was exposed. Therefore, the intrusion-resilient model appears to provide maximum possible security in the face of corruptions that often occur. However, there are some trade-offs between the two models, since intrusion-resilient cryptosystems are inefficient and more complex than key-insulated cryptosystems.

Our Contribution. The generic constructions of key-insulated encryption and key-insulated signature schemes do not show good performance in practical implementation. More practical key-insulated encryption scheme in [8] and signature scheme in [9] are constructed based on the DDH assumption and the CDH assumption, respectively. These schemes employ Pedersen commitment [16] and the length of the public key is linearly dependent on a system parameter. This dependency in a key-insulated signature scheme can be removed by using a trapdoor signature scheme [9] and the same dependency in a key-insulated encryption scheme is eliminated with an identity-based encryption scheme [3]. Bellare and Palacio also showed that the existence of a secure identity-based encryption scheme is equivalent to that of an $(N-1, N)$ key-insulated encryption scheme with random-access key updates [3]. In this paper, we show that a similar existence equivalence result holds between a secure identity-based signature scheme and an $(N-1, N)$ key-insulated signature scheme with random-access key updates. Additionally, we show that this result can be extended to the case for a "strong" $(N-1, N)$ key-insulated signature scheme with random-access key updates. Afterwards, we construct an efficient example of a "strong" $(N-1, N)$ key-insulated signature scheme with random-access key updates based on a secure identity-based signature scheme due to J. C. Cha and J. H. Cheon [4]. Another contribution of our paper is that we revisit the definition of the intrusion-resilient cryptosystems and classify three types of intrusion-resilient model according to the security requirements. We construct a Type (I) intrusion-resilient signature scheme based on a secure identity-based signature scheme, which is more efficient than previous intrusion-resilient signature schemes.

2   Key-Insulated Signature and Identity-Based Signature

In this section, we review key-insulated signature schemes [9] and identity-based signature schemes [1][4][10][15]. After reviewing their definition and security, we provide an equivalence result between the two signature schemes.

2.1   Key-Insulated Signature

Definition 1. *A key-insulated signature scheme is a 5-tuple of poly-time algorithms* (Gen, Upd*, Upd, Sign, Vrfy) *such that:*

- Gen, *the key generation algorithm, is a probabilistic algorithm that takes as input a security parameter 1^k and the total number of time periods N. It returns a public key PK, a master key SK^*, and an initial key SK_0.*
- Upd*, *the device key-update algorithm, is a probabilistic algorithm that takes as input indices i, j for time periods (throughout, we assume $1 \leq i, j \leq N$) and the master key SK^*. It returns a partial secret key $SK'_{i,j}$.*
- Upd, *the user key-update algorithm, is a deterministic algorithm that takes as input indices i, j, a secret key SK_i, and a partial secret key $SK'_{i,j}$. It returns the secret key SK_j for time period j.*
- Sign, *the signing algorithm, is a probabilistic algorithm that takes as input an index i of a time period, a message M, and a secret key SK_i. $\mathsf{Sign}_{SK_i}(i, M)$ returns a signature $\langle i, s \rangle$ consisting of the time period i and a signature value s.*
- Vrfy, *the verification algorithm, is a deterministic algorithm that takes as input the public key PK, a message M, and a pair $\langle i, s \rangle$. $\mathsf{Vrfy}_{PK}(M, \langle i, s \rangle)$ returns a bit b, where $b = 1$ means the signature is accepted.*

If $\mathsf{Vrfy}_{PK}(M, \langle i, s \rangle) = 1$, we say that $\langle i, s \rangle$ is a valid signature of M for period i. We require that all signatures output by $\mathsf{Sign}_{SK_i}(i, M)$ are accepted as valid by Vrfy.

In a key-insulated signature scheme, a user begins by generating $(PK, SK^*, SK_0) \leftarrow \mathsf{Gen}(1^k, N)$, registering PK in a central location (just as he would for a standard public key scheme), storing SK^* on a physically-secure device, and storing SK_0 himself. When the user who currently holds SK_i wants to obtain SK_j, the user requests $SK'_{i,j} \leftarrow \mathsf{Upd}^*(i, j, SK^*)$ from the secure device. Using SK_i and $SK'_{i,j}$, the user computes $SK_j \leftarrow \mathsf{Upd}(i, j, SK_i, SK'_{i,j})$. SK_j may then be used to sign messages during the time period j without further access to the secure device. After computation of SK_j, the user erases SK_i and $SK'_{i,j}$. Note that verification is always performed with respect to a fixed public key PK which is never changed. If it is possible to update the secret key from SK_i to SK_j in one step, we say that the scheme supports random-access key updates. The above definition includes this property implicitly.

For security considerations, we define a key exposure oracle $\mathsf{Exp}_{SK^*,SK_0}(\cdot)$ that performs the following on input i: (1) The oracle first checks whether period i has been activated; if so, the oracle returns the value already stored for SK_i. Otherwise, (2) the oracle runs $SK'_{0,i} \leftarrow \mathsf{Upd}^*(0, i, SK^*)$ followed by $SK_i \leftarrow \mathsf{Upd}(0, i, SK_0, SK'_{0,i})$, returns and stores the value SK_i, and labels period i as activated. We also give the adversary access to a signing oracle $\mathsf{Sign}_{SK^*,SK_0}(\cdot, \cdot)$ that does the following on inputs (i, M): (1) The oracle first checks whether period i has been activated; if so, the oracle returns $\mathsf{Sign}_{SK_i}(i, M)$ where a value for SK_i is already stored. Otherwise, (2) the oracle runs $SK'_{0,i} \leftarrow \mathsf{Upd}^*(0, i, SK^*)$ followed by $SK_i \leftarrow \mathsf{Upd}(0, i, SK_0, SK'_{0,i})$, stores SK_i, returns $\mathsf{Sign}_{SK_i}(i, M)$, and labels period i as activated. Here we give the formal definition of an $(N - 1, N)$ key-insulated signature scheme.

Definition 2. *Let Π be a key-insulated signature scheme and fix t. For any adversary A, we may perform the following experiment:*

$$(PK, SK^*, SK_0) \leftarrow \mathsf{Gen}(1^k, N);$$
$$(M, \langle i, s \rangle) \leftarrow A^{\mathsf{Sign}_{SK^*,SK_0}(\cdot,\cdot), \mathsf{Exp}_{SK^*,SK_0}(\cdot)}(PK).$$

We say that A succeeds if $\mathsf{Vrfy}_{PK}(M, \langle i, s \rangle) = 1$, (i, M) was never submitted to the signing oracle, i was never submitted to the key exposure oracle, and A made at most t calls to the key-exposure oracle. Denote the probability of A's success by $\mathsf{Succ}_{A,\Pi}(k)$. We say that Π is (t, N) key-insulated if for any PPT A, $\mathsf{Succ}_{A,\Pi}(k)$ is negligible. We say that Π is perfectly key-insulated if Π is $(N-1, N)$ key-insulated.

In a key-insulated signature scheme, an adversary can compromise the user's storage while a key is being updated from SK_i to SK_j; we call this a key-update exposure at (i, j). When this occurs, the adversary receives SK_i, $SK'_{i,j}$, and SK_j. We say a scheme has secure key updates if a key-update exposure at (i, j) is of no more help to the adversary than key exposures at both periods i and j.

Definition 3. *A key-insulated signature scheme Π has secure key updates if the view of any adversary A making a key-update exposure at $(i,\ j)$ can be perfectly simulated by an adversary A' making key exposure requests at periods i and j.*

Finally, we address attacks that compromise the physically-secure device (this includes attacks by the device itself, in case it cannot be trusted). The definition is similar to Definition 2 except that instead of having access to the key exposure oracle, the adversary is simply given the master key SK^*. A $(t,\ N)$ key-insulated scheme which is secure in this sense is termed strong $(t,\ N)$ key-insulated.

Definition 4. *Let $\Pi\ =\ (\mathsf{Gen}, \mathsf{Upd}^*, \mathsf{Upd}, \mathsf{Sign}, \mathsf{Vrfy})$ be a signature scheme which is (t, N) key-insulated. For any adversary B, we perform the following experiment:*

$$(PK,\ SK^*,\ SK_0) \leftarrow \mathsf{Gen}(1^k, N);$$
$$(M, \langle i, s \rangle) \leftarrow B^{\mathsf{Sign}_{SK^*, SK_0}(\cdot, \cdot)}(PK, SK^*).$$

We say that B succeeds if $\mathsf{Vrfy}_{PK}(M, \langle i, s \rangle) = 1$ and $(i,\ M)$ was never submitted to the signing oracle. Denote the probability of B's success by $\mathsf{Succ}_{B,\Pi}(k)$. We say that Π is strong (t, N) key-insulated if for any PPT B, $\mathsf{Succ}_{B,\Pi}(k)$ is negligible.

2.2 Identity-Based Signature

Shamir's original motivation for the identity-based cryptosystem [18] was to simplify certificate management. To securely communicate with Bob, Alice does not need to obtain Bob's public key certificate in identity-based cryptosystems; instead, she need only know Bob's e-mail address. Since the notion was introduced in 1984, there were several proposals for identity-based cryptosystems [19][20][13]. However, practical implementation with provable security was achieved only recently [1][4][10][15]. Here, we review the definition of identity-based signature schemes and their security.

Definition 5. *An identity-based signature scheme is a 4-tuple of poly-time algorithms (IBGen, Extract, IBSign, IBVrfy) such that:*

- IBGen, *the key generation algorithm, is a probabilistic algorithm that takes as input a security parameter 1^k. It returns a master key $IBSK^*$ and a parameter list $IBPK$.*
- Extract, *the signing key issuance algorithm, is a probabilistic algorithm that takes as input a user identity i and a master key $IBSK^*$. It returns the user i's secret signing key $IBSK_i$.*
- IBSign, *the signing algorithm, is a probabilistic algorithm that takes as input a message M and a secret key $IBSK_i$. $\mathsf{IBSign}_{IBSK_i}(M)$ returns a signature s.*
- IBVrfy, *the verification algorithm, is a deterministic algorithm that takes as input a parameter list $IBPK$, a message M, a user identity i, and a signature s. $\mathsf{IBVrfy}_{IBPK}(M, i, s)$ returns a bit b, where $b = 1$ means the signature is accepted.*

If $\mathsf{IBVrfy}_{IBPK}(M, i, s) = 1$, we say that s is a valid signature of M by a user i. We require that all signatures output by $\mathsf{IBSign}_{IBSK_i}(\cdot)$ are accepted as valid by $\mathsf{IBVrfy}_{IBPK}(\cdot, i, \cdot)$.

In an identity-based signature scheme, IBGen and $\mathsf{Extract}$ are performed by a trusted center. A secret key $IBSK_i$ is given to a user i by the center (through a secure channel). Note that key escrow of user's secret key is inherent in an identity-based signature scheme.

For security considerations, we define a key exposure oracle $\mathsf{IBExp}_{IBSK^*}(\cdot)$ that returns a secret key $IBSK_i$ on input i. We also give the adversary access to a signing oracle $\mathsf{IBSign}_{IBSK^*}(\cdot, \cdot)$ that returns $\mathsf{IBSign}_{IBSK_i}(M)$ on input (i, M). The security goal of an identity-based signature scheme is existential unforgeability. This means that any PPT adversary A should have a negligible probability of generating a valid signature of a new message given access to the key exposure oracle $\mathsf{IBExp}_{IBSK^*}(\cdot)$ and the signing oracle $\mathsf{IBSign}_{IBSK^*}(\cdot, \cdot)$. Naturally, A is considered successful if it forges a valid signature s of M by a user i where i was not queried to the key exposure oracle and (i, M) was not queried to the signing oracle.

2.3   Equivalence Result

An $(N-1, N)$ key-insulated encryption scheme with random-access key updates is effectively the same as a secure identity-based encryption scheme [3]. We show that a similar result holds between a secure identity-based signature scheme and an $(N-1, N)$ key-insulated signature scheme with random-access key updates. Note that these equivalence results do not guarantee the existence of a "strong" $(N-1, N)$ key-insulated cryptosystem with random-access key updates. Hence, we go one step forward and show that these equivalent results can be extended to "strong" $(N-1, N)$ key-insulated cryptosystems with random-access key updates.

Theorem 1. *A secure identity-based signature scheme exists if and only if there is an $(N-1, N)$ key-insulated signature scheme with random-access key updates.*

Proof. Let $\Pi_{\mathsf{IBS}} = (\mathsf{IBGen}, \mathsf{Extract}, \mathsf{IBSign}, \mathsf{IBVrfy})$ be a secure identity-based signature scheme. From Π_{IBS}, we build an $(N-1, N)$ key-insulated signature scheme with random-access key updates $\Pi_{\mathsf{KIS}} = (\mathsf{Gen}, \mathsf{Upd}^*, \mathsf{Upd}, \mathsf{Sign}, \mathsf{Vrfy})$. The underlying idea is that two implementations of Π_{IBS} are used for the construction. The notation $i - 1$ means the previous period of i and $\&$ denotes the bit-wise AND operation. We assume that an update is immediately followed by a key generation.

Conversely, let $\Pi_{\mathsf{KIS}} = (\mathsf{Gen}, \mathsf{Upd}^*, \mathsf{Upd}, \mathsf{Sign}, \mathsf{Vrfy})$ be an $(N-1, N)$ key-insulated signature scheme with random-access key updates. From Π_{KIS}, we build a secure identity-based signature scheme $\Pi_{\mathsf{IBS}} = (\mathsf{IBGen}, \mathsf{Extract}, \mathsf{IBSign}, \mathsf{IBVrfy})$. The input N of Gen can be determined by the security parameter k.

Algorithm Gen($1^k, N$)
  $(IBSK_1^*, IBPK_1) \leftarrow$ IBGen(1^k);
  $(IBSK_2^*, IBPK_2) \leftarrow$ IBGen(1^k);
  $SK^* \leftarrow IBSK_1^*$;
  $SK'_{-1,0} \leftarrow$ Extract($0, SK^*$);
  $SK_{0,0} \leftarrow$ Extract($0, IBSK_2^*$);
  $SK_0 \leftarrow (IBSK_2^*, SK_{0,0}, SK'_{-1,0})$;
  $PK \leftarrow (IBPK_1, IBPK_2)$;
  Return (PK, SK^*, SK_0)

Algorithm Upd*(i, j, SK^*)
  $SK'_{i,j} \leftarrow$ Extract(j, SK^*)
  Return $SK'_{i,j}$

Algorithm Upd($i, j, SK_i, SK'_{i,j}$)
  $(IBSK_2^*, SK_{i,i}, SK'_{i-1,i}) \leftarrow SK_i$;
  $SK_{j,j} \leftarrow$ Extract($j, IBSK_2^*$);
  $SK_j \leftarrow (IBSK_2^*, SK_{j,j}, SK'_{i,j})$;
  Return SK_j

Algorithm Sign(i, M, SK_i)
  $(IBSK_2^*, SK_{i,i}, SK'_{i-1,i}) \leftarrow SK_i$;
  $s_1 \leftarrow$ IBSign$_{SK'_{i-1,i}}(M)$;
  $s_2 \leftarrow$ IBSign$_{SK_{i,i}}(M)$;
  $s \leftarrow (s_1, s_2)$;
  Return $\langle i, s \rangle$

Algorithm Vrfy($PK, M, \langle i, s \rangle$)
  $(IBPK_1, IBPK_2) \leftarrow PK$;
  $(s_1, s_2) \leftarrow s$;
  $b_1 \leftarrow$ IBVrfy$_{IBPK_1}(M, i, s_1)$;
  $b_2 \leftarrow$ IBVrfy$_{IBPK_2}(M, i, s_2)$;
  $b \leftarrow b_1 \& b_2$;
  Return b

Algorithm IBGen(1^k)
  $(PK, SK^*, SK_0) \leftarrow$ Gen($1^k, N$);
  $IBSK^* \leftarrow (SK^*, SK_0)$;
  $IBPK \leftarrow PK$;
  Return $(IBSK^*, IBPK)$

Algorithm Extract($i, IBSK^*$)
  $(SK^*, SK_0) \leftarrow IBSK^*$;
  $SK'_{0,i} \leftarrow$ Upd*($0, i, SK^*$);
  $SK_i \leftarrow$ Upd($0, i, SK_0, SK'_{0,i}$);
  $IBSK_i \leftarrow SK_i$;
  Return $IBSK_i$

Algorithm IBSign($M, IBSK_i$)
  $\langle i, s \rangle \leftarrow$ Sign$_{IBSK_i}(i, M)$;
  Return s

Algorithm IBVrfy($IBPK, M, i, s$)
  $b \leftarrow$ Vrfy$_{IBPK}(M, \langle i, s \rangle)$;
  Return b

The security of derived signature schemes can be proved by the security of the underlying signature schemes.

Q.E.D. □

Theorem 2. *The existence of one of the following signature schemes entails the other two signature schemes: a secure identity-based signature scheme, an $(N-1, N)$ key-insulated signature scheme with random-access key updates, and a "strong" $(N-1, N)$ key-insulated signature scheme with random-access key updates.*

Proof. In the proof of Theorem 1, Π_{KIS} derived from Π_{IBS} is actually a strong $(N-1, N)$ key-insulated signature scheme with random-access key updates, since $IBSK_1^*$ and $IBSK_2^*$ are from two different implementations of Π_{IBS}. In addition, the existence of a strong $(N-1, N)$ key-insulated signature scheme with random-access key updates implies by definition that of an $(N-1, N)$

key-insulated signature scheme with random-access key updates. Along with Theorem 1, we can obtain Theorem 2.

Q.E.D. □

Since this paper concentrates on key updating signature schemes, we provided Theorem 1 and Theorem 2 for signature schemes. Theorem 1 and Theorem 2 can be easily converted to the case for encryption schemes. Hence, a secure identity-based cryptosystem can be used to construct a strong $(N-1, N)$ key-insulated cryptosystem with random-access key updates. In the next section, we give an efficient example for the case of a signature scheme.

3   KUS-SKI: Key Updating Signature with Strong Key Insulation

A strong $(N-1, N)$ key-insulated signature scheme based on a secure identity-based signature scheme can be constructed by employing the method presented in the proof of Theorem 1. However, more efficient constructions can be achieved if we utilize the algebraic properties of the underlying identity-based signature scheme. We will devise an efficient strong $(N-1, N)$ key-insulated signature scheme with random-access key updates based on the identity-based signature scheme of J. C. Cha and J. H. Cheon [4]. We will call this scheme KUS-SKI (Key Updating Signature with Strong Key Insulation). Note that other identity-based signature schemes [10][15] can be used to construct strong $(N-1, N)$ key-insulated signature schemes with different characteristics.

3.1   The Scheme

We assume that readers are familiar with the identity-based signature scheme from gap Diffie-Hellman groups. Let G be a group of prime order q in which DDH problems can be solved – w.l.o.g., we write G additively.

1. Gen: On inputting 1^k, Gen chooses a generator P of G, picks random numbers SK^1 and SK^2 in $\mathbf{Z}_q$, sets $PK = (SK^1 + SK^2)P$, $SK^* = SK^1$, $SK_0 = (SK_{0,0}, SK_{0,1}) = (SK^2, \phi)$, and chooses cryptographic hash functions H_1: $\{0,1\}^* \times G \to \mathbf{Z}_q$ and H_2: $\{0,1\}^* \to G$.
2. Upd*: On inputting indices i, j and the master key SK^*, Upd* computes and returns a partial secret key $SK'_{i,j} = SK^* H_2(j)$.
3. Upd: On inputting indices i, j, a secret key SK_i, and a partial secret key $SK'_{i,j}$, Upd parses SK_i as $(SK_{i,0}, SK_{i,1})$, sets $SK_{j,0} = SK_{i,0}$, $SK_{j,1} = SK_{i,0}H_2(j) + SK'_{i,j}$, erases SK_i, $SK'_{i,j}$, and returns $SK_j = (SK_{j,0}, SK_{j,1})$.
4. Sign: On inputting an index i of a time period, a message M, and a secret key SK_i, Sign picks a random number $r \in \mathbf{Z}_q$, computes $U = rH_2(i)$, $h = H_1(M, U)$, $V = (r + h)SK_{i,1}$, and returns $\langle i, s \rangle$ where $s = (U, V)$.

5. Vrfy: On inputting the public key PK, a message M, and a pair $\langle i, s \rangle$, Vrfy parses s as (U, V) and checks whether $(P, PK, U + hH_2(i), V)$, where $h = H_1(M, U)$, is a valid Diffie-Hellman tuple. If so, Vrfy returns $b = 1$ and returns $b = 0$, otherwise.

Note that $SK_{i,0} = SK^2$ and $SK_{i,1} = (SK^1 + SK^2)H_2(i)$ hold for all i. The consistency of KUS-SKI can be proved by the following equations:

$$(P, PK, U + hH_2(i), V)$$
$$= (P, PK, (r + h)H_2(i), (r + h)SK_{i,1})$$
$$= (P, (SK^1 + SK^2)P, (r + h)H_2(i), \{(r + h)(SK^1 + SK^2)\}H_2(i))$$

The key length of KUS-SKI is not dependent on a parameter and the computational cost of signing and verifying is comparable to ordinary signature schemes.

3.2   The Security Analysis

We follow the proof technique in [3]. Nonetheless, our simulation is different from that in [3] because we adopt the original security model of key-insulated signature schemes in [9]. For example, we do not allow an adversary compromising a period i secret key SK_i to access the channel between the user and the secure device for the period i with no cost.

Theorem 3. *KUS-SKI has secure key updates and supports random key updates.*

Proof. Let A be an adversary who makes a key-update exposure at (i, j). This adversary can be perfectly simulated by an adversary A' who makes key exposure requests at periods i and j. Since A' can get SK_i and SK_j, he can compute $SK'_{i,j}$ by $SK'_{i,j} = SK_{j,1} - SK_{i,0}H_2(j)$. The property of random-access key updates follows from the definition of Upd* and Upd in KUS-SKI.

$$\text{Q.E.D. } \square$$

Theorem 4. *KUS-SKI is an $(N - 1, N)$ key-insulated signature scheme.*

Proof. Let CC-IBS $=$ (IBGen, Extract, IBSign, IBVrfy) be the identity-based signature scheme in [4]. Note that CC-IBS is a secure signature scheme, assuming the hardness of CDH problems. We will show that an adversary A, who breaks KUS-SKI $=$ (Gen, Upd*, Upd, Sign, Vrfy), can be translated into an adversary B who can break CC-IBS. After IBGen generates a master key $IBSK^* \in Z_q$ and a parameter list $IBPK = (P, P_{pub}, H_1, H_2)$ where $P_{pub} = (IBSK^*)P$, B randomly selects $SK^2 \in Z_q$ and runs A. The adversary B, who is given access to $\text{IBSign}_{IBSK^*}(\cdot, \cdot)$ and $\text{IBExp}_{IBSK^*}(\cdot)$, has to answer A's queries to $\text{Sign}_{SK^*, SK_0}(\cdot, \cdot)$ and $\text{Exp}_{SK^*, SK_0}(\cdot)$. In response to a query (i, M) to the signing oracle $\text{Sign}_{SK^*, SK_0}(\cdot, \cdot)$, B forwards the query to its signing oracle

$\mathsf{IBSign}_{IBSK^*}(\cdot,\cdot)$ and returns the oracle's answer s with i as $\langle i, s \rangle$ to A. In response to a query i to the key exposure oracle $\mathsf{Exp}_{SK^*, SK_0}(\cdot)$, B forwards the query to its key exposure oracle $\mathsf{IBExp}_{IBSK^*}(\cdot)$ and obtains the oracle's answer $IBSK_i$. B returns $SK_i = (SK_{i,0}, SK_{i,1}) = (SK^2, IBSK_i)$. Since B simulates A's environment in its attack against KUS-SKI perfectly, B can forge a valid signature of a new message based on A's answer.

Q.E.D. □

Theorem 5. *KUS-SKI is a strong $(N-1, N)$ key-insulated signature scheme.*

Proof. Assume that there is an adversary A, who breaks KUS-SKI = (Gen, Upd*, Upd, Sign, Vrfy) with access to the secure device. We will show that A can be translated into an adversary B who can break CC-IBS = (IBGen, Extract, IBSign, IBVrfy). After IBGen generates a master key $IBSK^* \in \mathbf{Z}_q$ and a parameter list $IBPK = (P, P_{pub}, H_1, H_2)$ where $P_{pub} = (IBSK^*)P$, B randomly selects $SK^* \in \mathbf{Z}_q$ and gives SK^* to A. The adversary B, who is given access to $\mathsf{IBSign}_{IBSK^*}(\cdot,\cdot)$ and $\mathsf{IBExp}_{IBSK^*}(\cdot)$, has to answer A's queries to $\mathsf{Sign}_{SK^*, SK_0}(\cdot,\cdot)$. In response to a query (i, M) to the signing oracle $\mathsf{Sign}_{SK^*, SK_0}(\cdot,\cdot)$, B forwards the query to its signing oracle $\mathsf{IBSign}_{IBSK^*}(\cdot,\cdot)$ and returns the oracle's answer s with i as $\langle i, s \rangle$ to A. Since B simulates A's environment in its attack against KUS-SKI perfectly, B can forge a valid signature of a new message based on A's answer.

Q.E.D. □

4   Intrusion-Resilience Signature

The notion of signer-base intrusion-resilient (SiBIR) signatures was proposed in [11] and extended in [12]. The case for the encryption scheme was studied in [7]. As in key-insulated schemes, the user in SiBIR signature schemes has two modules, the signer (an insecure device) and the base (a physically-secure device). The main strength of intrusion-resilient schemes is that they remain secure even after many arbitrary compromises of both modules, as long as the compromises are not simultaneous. In this section, we review the definition of SiBIR signature schemes and their security. Afterwards, we classify SiBIR signature schemes into three types.

4.1   SiBIR Signature Schemes

The system's secret key may be modified in two different ways, called *update* and *refresh*. Updates change the secrets from one time period to the next while refreshes affect only the internal secrets of the system. The notation $SK_{t.r}$ (respectively, $SK^*_{t.r}$) denotes the signer's (respectively, the base's) secret key for time period t following r refreshes. We assume for convenience that a key update occurs immediately after key generation to obtain keys for $t = 1$, and that a key refresh occurs immediately after every key update to obtain keys for $r = 1$.

Definition 6. *A SiBIR signature scheme Π is a 7-tuple of poly-time algorithms* (Gen, Sign, Vrfy, UB, US, RB, RS) *such that:*

- Gen, *the key generation algorithm, takes as input security parameter 1^k and the total number of time periods N. It returns the initial signer key $SK_{0.0}$, the initial base key $SK^*_{0.0}$, and the public key PK.*
- Sign, *the signing algorithm, takes as input the current signer key $SK_{t.r}$ and a message M. It returns a signature $\langle t, s \rangle$ consisting of the time period t and a signature value s.*
- Vrfy, *the verifying algorithm, takes as input a message M, a signature $\langle t, s \rangle$ and the public key PK. It returns a bit b where $b = 1$ means the signature is accepted.*
- UB, *the base key update algorithm, takes as input the current base key $SK^*_{t.r}$. It returns a new base key $SK^*_{(t+1).0}$ for the next time period as well as a key update message SKU_t.*
- US, *the signer key update algorithm, takes as input the current signer key $SK_{t.r}$ and a key update message SKU_t. It returns the new signer key $SK_{(t+1).0}$ for the next time period.*
- RB, *the base key refresh algorithm, takes as input the current base key $SK^*_{t.r}$. It returns a new base key $SK^*_{t.(r+1)}$ as well as a key refresh message $SKR_{t.r}$.*
- RS, *the signer key refresh algorithm, takes as input the current signer key $SK_{t.r}$ and a key refresh message $SKR_{t.r}$. It returns a new signer key $SK_{t.(r+1)}$.*

Let $RN(t)$ denote the number of refreshes that occur in time period t. RN is used for notational convenience. Consider the following "thought experiment," which generates all keys for the entire run of the signature scheme.

Experiment Generate-Keys(k, N, RN)
$\quad t \leftarrow 0, r \leftarrow 0;$
$\quad (SK_{0.0}, SK^*_{0.0}, PK) \leftarrow$ Gen($1^k, N$);
$\quad$ for $t = 1$ to N
$\quad\quad (SK^*_{t.0}, SKU_{t-1}) \leftarrow$ UB($SK^*_{(t-1).r}$);
$\quad\quad SK_{t.0} \leftarrow$ US($SK_{(t-1).r}, SKU_{t-1}$);
$\quad\quad$ for $r = 1$ to $RN(t)$
$\quad\quad\quad (SK^*_{t.r}, SKR_{t.(r-1)}) \leftarrow$ RB($SK^*_{t.(r-1)}$);
$\quad\quad\quad SK_{t.r} \leftarrow$ RS($SK_{t.(r-1)}, SKR_{t.(r-1)}$);

Let $\widehat{SK}$, $\widehat{SK^*}$, $\widehat{SKU}$ and $\widehat{SKR}$ denote the sets of singer keys, base keys, update messages, and refresh messages generated in the course of the above experiment. For security considerations, we define the following oracles available to the adversary:

- Osig, the signing oracle, which on input (M, t, r) outputs Sign($SK_{t.r}, M$).
- Osec, the key exposure oracle (based on the sets $\widehat{SK}$, $\widehat{SK^*}$, $\widehat{SKU}$ and $\widehat{SKR}$), which

1. On input ("s", $t.r$) outputs $SK_{t.r}$;
2. On input ("b", $t.r$) outputs $SK^*_{t.r}$;
3. On input ("u", t) outputs SKU_t and $SKR_{(t+1).0}$;
4. On input ("r", $t.r$) outpus $SKR_{t.r}$.

Queries to Osec correspond to compromise of the signer or base, or to intercepting update or refresh messages. Note that a key exposure at (i, j) in key-insulated signature schemes can be realized by two queries ("s" and "u") to Osec in SiBIR signature schemes. We assume that queries to the oracles always have t, r within the appropriate bounds.

For any set Q of key exposure queries, we say that $SK_{t.r}$ is Q-exposed if at least one of the following is true:

- ("s", $t.r$) $\in Q$
- $r > 1$, ("r", $t.(r-1)$) $\in Q$, and $SK_{t.(r-1)}$ is Q-exposed
- $r = 1$, ("u", $t-1$) $\in Q$, and $SK_{(t-1).RN(t-1)}$ is Q-exposed

Replacing SK with SK^* throughout the above definition yields the definition of base key exposure.

We say that the scheme is (t, Q)-compromised, if either $SK_{t.r}$ is Q-exposed (for some r) or if both $SK_{t'.r}$ and $SK^*_{t'.r}$ are Q-exposed (for some r and $t' < t$). The following experiment captures adversary's functionality.

Experiment Run-Adversary(A, k, N, RN)
    Generate-Keys(k, N, RN);
    $(M, \langle t, s \rangle) \leftarrow A^{\text{Osig,Osec}}(1^k, N, PK, RN)$;
    $b \leftarrow \text{Vrfy}(M, \langle t, s \rangle, PK)$;
    Let Q be the set of queries made by A to Osec;
    if $b = 0$ or (M, t, r) was queried by A to Osig
            or the scheme is (t, Q)-compromised
      then return 0
      else return 1

We define A's probability of success as the probability that 1 is output in the above experiment. We denote this probability by $\text{Succ}_{A,\Pi}(k)$.

Definition 7. *A signature scheme Π is said to be an existentially unforgeable SiBIR signature scheme if, for any PPT adversary A and all N, $RN(\cdot)$ polynomial in k, we have $\text{Succ}_{A,\Pi}(k) < \epsilon(k)$ for some negligible function $\epsilon(\cdot)$.*

SiBIR signature schemes treat all compromises in one definition, as opposed to defining security against different kinds of compromises separately in key-insulated signature schemes.

4.2   Classification of SiBIR Signature Schemes

SiBIR signature schemes remain secure in the face of multiple compromises of the signer and the base, as long as they are not both compromised simultaneously. Furthermore, in case both are compromised simultaneously, prior time periods remain secure, as in forward-secure signature schemes. While the first requirement is desirable without doubt, we find that the latter requirement can be adjusted according to the security requirements. We introduce three types of SiBIR signature schemes.

- Type (I) SiBIR signature schemes do not guarantee the security of other time periods in case of a simultaneous compromise.
- Type (II) SiBIR signature schemes guarantee the security of prior time periods in case of a simultaneous compromise.
- Type (III) SiBIR signature schemes guarantee the security of prior and posterior time periods in case of a simultaneous compromise.

Type (I) SiBIR signature schemes are appropriate for the case that the singer and the base can be at least one time period apart at a given time. In this case, it is difficult for an adversary to compromise both the signer and the base simultaneously. Type (II) SiBIR signature schemes correspond to those in Section 4.1. Type (III) SiBIR signature schemes provide the highest security level. However, it seems that we need additional secret information such as passwords or biometric information to achieve this level of security.

5 KUS-IR: Key Updating Signature with Type (I) Intrusion Resilience

Previous Type (II) SiBIR signature schemes [11] are less efficient than ordinary signature schemes. In addition, cumbersome techniques are needed to avoid the limit on the total number of periods in Type (II) SiBIR signature schemes [12]. In this section, we introduce an efficient Type (I) SiBIR signature scheme, called KUS-IR (Key Updating Signature with type (I) Intrusion Resilience). Even though the security level of Type (I) is lower than that of Type (II), KUS-IR is very efficient and has no limit on the total number of periods. The computational cost of signing and verifying is comparable to ordinary signature schemes. Furthermore, KUS-IR enables the honest user to request "old" keys thereby allowing, e.g., the signing of documents for prior time periods when needed. Note that this property can be offered in key-insulated cryptosystems [9] and is especially invaluable for encryption schemes.

5.1 The Scheme

KUS-IR is constructed on KUS-SKI and employs proactive security techniques. Let G be a group of prime order q in which DDH problems can be solved.

1. Gen: On inputting 1^k, Gen chooses a generator P of G, picks random numbers SK^1 and SK^2 in $\mathbf{Z}_q$, sets $PK = (SK^1 + SK^2)P$, $SK_{0.0}^* = SK^1$, $SK_{0.0} = (SKA_{0.0}, SKB_{0.0}) = (SK^2, \phi)$, and chooses cryptographic hash functions $H_1: \{0,1\}^* \times G \to \mathbf{Z}_q$ and $H_2: \{0,1\}^* \to G$.

2. Sign: On inputting the current signer key $SK_{t.r}$ and a message M, Sign picks a random number $r \in \mathbf{Z}_q$, computes $U = rH_2(t)$, $h = H_1(M, U)$, $V = (r + h)SKB_{t.r}$, and returns $\langle t, s \rangle$ where $s = (U, V)$.

3. Vrfy: On inputting the public key PK, a message M, and a pair $\langle t, s \rangle$, Vrfy parses s as (U, V) and checks whether $(P, PK, U + hH_2(t), V)$, where $h = H_1(M, U)$, is a valid Diffie-Hellman tuple. If so, Vrfy returns $b = 1$ and returns $b = 0$, otherwise.

4. UB: On inputting the current base key $SK_{t.r}^*$, UB returns a new base key $SK_{(t+1).0}^* = SK_{t.r}^*$ as well as a key update message $SKU_t = SK_{t.r}^* H_2(t+1)$.

5. US: On inputting the current signer key $SK_{t.r}$ and a key update message SKU_t, US sets $SKA_{(t+1).0} = SKA_{t.r}$, $SKB_{(t+1).0} = SKA_{t.r}H_2(t+1) + SKU_t$ and returns the new signer key $SK_{(t+1).0} = (SKA_{(t+1).0}, SKB_{(t+1).0})$.

6. RB: On inputting the current base key $SK_{t.r}^*$, RB chooses a random $w_{t.(r+1)} \in \mathbf{Z}_q$ and returns $SK_{t.(r+1)}^* = SK_{t.r}^* + w_{t.(r+1)}$ as well as $SKR_{t.r} = w_{t.(r+1)}$.

7. RS: On inputting the current signer key $SK_{t.r}$ and a key refresh message $SKR_{t.r}$, RS sets $SKA_{t.(r+1)} = SKA_{t.r} - SKR_{t.r}$, $SKB_{t.(r+1)} = SKB_{t.r}$ and returns a new signer key $SK_{t.(r+1)} = (SKA_{t.(r+1)}, SKB_{t.(r+1)})$.

Recall that each update is immediately followed by a refresh and keys with refresh index 0 are never actually used.

5.2   The Security Analysis

Since we did not present a formal definition of three types of SiBIR signature schemes, we will merely give an intuitive clue to the security analysis. Firstly, KUS-IR maintains the security condition of KUS-SKI, i.e., a strong $(N, N-1)$ key-insulated signature scheme with random-access key updates. Next, assume that an adversary obtains z different keys. If we let $W_{t.r} = \sum_{i=1}^{t-1} \sum_{j=1}^{RN(i)} w_{i.j} + \sum_{j=1}^{r} w_{t.j}$, then $SK_{t.r}^* = SK^1 + W_{t.r}$ and $SKA_{t.r} = SK^2 - W_{t.r}$. When C denotes a column matrix whose z elements are composed of the exposed keys of SK^* and SKA, we get $C = A \times B$ where

$$A = \begin{vmatrix} \alpha_1 & \beta_1 & E_1 & 0 & 0 & 0 & 0 \\ \alpha_2 & \beta_2 & 0 & E_2 & 0 & 0 & 0 \\ \alpha_3 & \beta_3 & 0 & 0 & E_3 & 0 & 0 \\ & & & \cdots & & & \\ \alpha_z & \beta_z & 0 & 0 & 0 & 0 & E_z \end{vmatrix}, \quad B = \begin{vmatrix} SK^1 & SK^2 & W_{t_1,r_1} & W_{t_2,r_2} & W_{t_3,r_3} & \cdots & W_{t_z,r_z} \end{vmatrix}^{\mathrm{T}}$$

s.t. $\alpha_i, \beta_i \in \{0, 1\}$, $\alpha_i + \beta_i = 1$, and $E_i \in \{-1, +1\}$ for $\forall i \in [1, z]$.

Note that $W_{t_i.r_i}$'s have all different values (with very high probability) in a Type (I) SiBIR signature scheme and A is a z by $z+2$ matrix whose rows are linearly independent. In addition, each row of A contains two unknown values. Hence, the adversary cannot obtain any information on an element of B. This argument can be extended to include the exposed messages of SKU and SKR.

Theorem 6. *KUS-IR is an existentially unforgeable Type (I) SiBIR signature scheme.*

6   Concluding Remarks

The danger of secret key exposure can be significantly reduced by key updating cryptosystems. In this paper, we proposed efficient key updating signature schemes based on a secure identity-based signature scheme. We showed that the existence of a secure identity-based signature scheme implies that of a "strong" $(N-1, N)$ key-insulated signature scheme with random-access key updates. Hence, identity-based cryptosystems can play an important role in research on key updating cryptosystems. Starting from a secure identity-based signature scheme, we could reach a strong $(N-1, N)$ key-insulated signature scheme with random-access key updates and a Type (I) intrusion-resilient signature scheme. We leave the efficient construction of a Type (II) intrusion-resilient signature scheme based on a secure identity-based signature scheme as a challenging topic for future research.

Acknowledgements. We thank the anonymous referees for their valuable comments.

References

1. D. Boneh and M. Franklin, "Identity based encryption from the Weil pairing," Crypto 2001, LNCS Vol. 2139, pp. 213–229, 2001.
2. M. Bellare and S. Miner, "A forward-secure digital signature scheme," Crypto 1999, LNCS Vol. 1666, pp. 431–448, 1999.
3. M. Bellare and A. Palacio, "Protecting against key exposure: strong key-insulated encryption with optimal threshold," Cryptology ePrint archive 2002/064, http://eprint.iacr.org/, 2002.
4. J. C. Cha and J. H. Cheon, "An identity-based signature from gap Diffie-Hellman groups," PKC 2003, LNCS Vol. 2567, pp. 18–30, 2003.
5. R. Canetti, S. Halevi and J. Katz, "A forward-secure public-key encryption scheme," Eurocrypt 2003, LNCS Vol. 2656, pp. 256–271, 2003.
6. Y. Desmedt and Y. Frankel, "Threshold cryptosystems," Crypto 1989, LNCS Vol. 435, pp. 307–315, 1989.
7. Y. Dolis, M. Franklin, J. Katz, A. Miyaji, and M. Yung, "Intrusion-resilient public-key encryption," CT-RSA 2003, LNCS Vol. 2612, pp. 19–32, 2003.
8. Y. Dolis, J. Katz, S. Xu and M. Yung, "Key-insulated public key cryptosystems," Eurocrypt 2002, LNCS Vol. 2332, pp. 65–82, 2002.
9. Y. Dolis, J. Katz, S. Xu and M. Yung, "Strong key-insulated signature schemes," PKC 2003, LNCS Vol.2567, pp.130–144, 2003.
10. F. Hess, "Efficient Identity Based Signature Schemes Based on Pairings," SAC 2002, LNCS Vol. 2595, pp. 310–324, 2003.
11. G. Itkis and L. Reyzin, "SiBIR: Siner-base intrusion-resilient signatures," Crypto 2002, LNCS Vol. 2442, pp. 499–514, 2002.
12. G. Itkis, "Intrusion-resilient signatures: generic constructions, or defeating strong adversary with minimal assumptions," SCN 2002, LNCS Vol. 2576, pp.102–118, 2003.
13. U. Maurer and Y. Yacobi, "Non-interactive public-key cryptography," Crypto 1991, LNCS Vol. 547, pp. 498–507, 1991.

14. R. Ostrovsky and M. Yung, "How to withstand mobile virus attacks," PODC 1991, pp 51–59, 1991.
15. K. G. Paterson, "ID-based signatures from pairings on elliptic curves," Electronics Letters Vol. 38 (18), pp. 1025–1026, 2002.
16. T. Pedersen, "Non-interactive and information-theoretic secure verifiable secret sharing," Crypto 1991, LNCS Vol. 576, pp. 129–140, 1991.
17. A. Shamir, "How to share a secret," Comm. ACM, 22(11), pp. 612–613, 1979.
18. A. Shamir, "Identity-based cryptosystems and signature schemes," Crypto 1994, LNCS Vol. 196, pp. 47–53, 1984.
19. H. Tanaka, "A realization scheme for the identity-based cryptosystem," Crypto 1987, LNCS Vol. 293, pp. 341–349, 1987.
20. S. Tsuji and T. Itoh, "An ID-based cryptosystem based on the discrete logarithm problem," IEEE Journal on Selected Areas in Communication, Vol. 7, pp. 467–473, 1989.

Periodic Sequences with Maximal Linear Complexity and Almost Maximal k-Error Linear Complexity

Harald Niederreiter[1] and Igor E. Shparlinski[2]

[1] Department of Mathematics, National University of Singapore,
2 Science Drive 2, Singapore 117543, Republic of Singapore
nied@math.nus.edu.sg
[2] Department of Computing, Macquarie University,
NSW 2109, Australia
igor@comp.mq.edu.au

Abstract. C. Ding, W. Shan and G. Xiao conjectured a certain kind of trade-off between the linear complexity and the k-error linear complexity of periodic sequences over a finite field. This conjecture has recently been disproved by the first author, by showing that for infinitely many period lengths N and some values of k both complexities may take very large values (contradicting the above conjecture). Here we use some recent achievements of analytic number theory to extend the class of period lengths N and the number of admissible errors k for which this conjecture fails for rather large values of k. We also discuss the relevance of this result for stream ciphers.

1 Introduction

Let $S = (s_i)_{i=0}^{\infty}$ be a sequence of elements of a finite field $\mathbb{F}_q$ of q elements. For an integer $N \geq 1$, we say that S is N-*periodic* if $s_{i+N} = s_i$ for all $i \geq 0$. Since an N-periodic sequence is determined by the terms of one period, we can completely describe S by the notation $S \leftrightarrow (s_0, s_1, \ldots, s_{N-1})$.

The *linear complexity* $L(S)$ of an N-periodic sequence $S \leftrightarrow (s_0, s_1, \ldots, s_{N-1})$ of elements of $\mathbb{F}_q$ is the smallest nonnegative integer L for which there exist coefficients $c_1, \ldots, c_L \in \mathbb{F}_q$ such that

$$s_j + c_1 s_{j-1} + \cdots + c_L s_{j-L} = 0, \qquad j = L, L+1, \ldots.$$

It is obvious that for any N-periodic sequence S over $\mathbb{F}_q$ we have $L(S) \leq N$.

A natural generalisation of the notion of linear complexity has been introduced in [17] by defining, for an integer $k \geq 0$, the k-*error linear complexity* $L_{N,k}(S)$ to be

$$L_{N,k}(S) = \min_{T:\, d(S,T) \leq k} L(T),$$

K.G. Paterson (Ed.): Cryptography and Coding 2003, LNCS 2898, pp. 183–189, 2003.
© Springer-Verlag Berlin Heidelberg 2003

where the minimum is taken over all N-periodic sequences $T \leftrightarrow (t_0, t_1, \dots, t_{N-1})$ over $\mathbb{F}_q$ for which the Hamming distance $d(S, T)$ of the vectors $(s_0, s_1, \dots, s_{N-1})$ and $(t_0, t_1, \dots, t_{N-1})$ is at most k. It is useful to remark that we do not insist that N be the smallest period of S, however the value of $L(S)$ does not depend on N but rather only on the sequence S itself, that is, $L_{N,0}(S) = L(S)$ for any multiple N of the smallest period of S. We also remark that the concept of k-error linear complexity is closely related to k-*error sphere complexity*, see [2, 3].

The linear complexity, k-error linear complexity and related complexity measures for periodic sequences over a finite field play an important role for stream ciphers in cryptology (see [12,13,15,16]). Periodic sequences that are suitable as keystreams in stream ciphers should possess a large linear complexity to thwart an attack by the Berlekamp-Massey algorithm. Moreover, a cryptographically strong sequence should not only have a large linear complexity, but also changing a few terms should not cause a significant decrease of the linear complexity. In other words, the k-error linear complexity of the sequence should also be large for a reasonable range of small values of k.

In [3, Section 7.1] a conjecture has been made that that there may be a trade-off between the linear complexity and the k-error linear complexity (or rather the closely related k-error sphere complexity). In particular, it has been conjectured that for any binary N-periodic sequence S and for any positive integer $k \leq N$ we have

$$L_{N,k}(S) + L(S) \leq \left(1 + \frac{1}{k}\right) N - 1, \tag{1}$$

thus that $L_{N,k}(S)$ and $L(S)$ cannot be simultaneously large. In fact, as we have mentioned, the conjecture has been made in terms of the k-error sphere complexity which however would immediately imply (1).

This conjecture has recently been disproved in [14], see also [11]. In particular, it has been shown in [14] that for almost all primes p (that is, for all primes $p \leq x$ except maybe at most $o(x/\ln x)$ of them, for $x \to \infty$) there is a sequence S over $\mathbb{F}_2$ of period $N = p$ and with

$$L(S) = N \quad \text{and} \quad L_{N,k}(S) \geq N - 1 \tag{2}$$

simultaneously for all $k \leq N^{1/2-\varepsilon}$ for any fixed $\varepsilon > 0$.

Here we use some recent advances in analytic number theory [1] to show that in fact for infinitely many primes p and $N = p$, one can achieve (2) for a substantially larger range of values of k. In fact, we obtain our result for an arbitrary (but fixed) value of the prime power q.

We also show that for almost all integers N coprime to q there is a sequence S with $L(S) = N$ and $L_{N,k}(S) \sim N$ for a large range of values of k.

Acknowledgement. The research of the first author was partially supported by the grant R-394-000-011-422 with Temasek Laboratories in Singapore. The research of the second author was partially supported by the ARC grant DP0211459.

2   Preparations

Here we collect some, mainly well known, facts which we use in the sequel.

Let $\mathbb{Z}_N = \{0, 1, \ldots, N-1\}$ be the residue ring modulo N. We assume from now on that $\gcd(N, q) = 1$. We say that two integers $a, b \in \mathbb{Z}_N$ are equivalent (relative to powers of q) if $a \equiv bq^u \pmod{N}$ for some integer $u \geq 0$. Let $\mathcal{C}_1 = \{0\}, \mathcal{C}_2, \ldots, \mathcal{C}_h$ be the different equivalence classes with respect to this equivalence relation. We note that the sets $\mathcal{C}_1, \ldots, \mathcal{C}_h$ are called the *cyclotomic cosets* modulo N (relative to powers of q). We assume that they are ordered in such a way that $\ell_1 \leq \ldots \leq \ell_h$ where $\ell_i = \#\mathcal{C}_i$, $1 \leq i \leq h$. We also remark that $\gcd(a, N)$ is the same for all elements from the same cyclotomic coset $\mathcal{C}_i$ which we denote by d_i, $1 \leq i \leq h$. In particular $\ell_1 = 1$ and $\ell_h = t_N$, where for a positive integer n with $\gcd(n, q) = 1$ we denote by t_n the multiplicative order of q modulo n (that is, the smallest positive integer k with $q^k \equiv 1 \pmod{n}$). More generally, $\ell_i = t_{N/d_i}$ for $1 \leq i \leq h$.

The following result is a special partial case of [11, Theorem 1].

Lemma 1. *Let* $\gcd(N, q) = 1$ *and* $N \geq 2$. *Then for any positive integers* $\ell \leq \ell_h$ *and* $k \leq N$ *with*

$$\sum_{j=0}^{k} \binom{N}{j} (q-1)^j < q^\ell$$

there is a sequence S *of elements of* $\mathbb{F}_q$ *of period* N *with*

$$L(S) = N \qquad and \qquad L_{N,k}(S) \geq N - \sum_{\ell_i < \ell} \ell_i.$$

We need a lower bound on $t_{N/d}$ which is a special case of [4, Lemma 2], combined with the inequality $\varphi(mn) \leq m\varphi(n)$ which holds for any positive integers m, n, where as usual φ is the Euler totient function.

Lemma 2. *For any divisor* $d | N$ *we have* $t_{N/d} \geq t_N/d$.

We are mainly interested in integers N for which t_N is large enough.

Lemma 3. *Let* q *be fixed. Then for infinitely many primes* $N = p$ *we have* $t_N \geq N^{0.677}$.

Proof. Let $x > 0$ be sufficiently large and let $\mathcal{Q}$ be the set of primes $p \leq x$ for which $t_p < p^{2/5}$. Obviously

$$\prod_{p \in \mathcal{Q}} p \mid \prod_{j \leq x^{2/5}} (q^j - 1).$$

Therefore

$$2^{\#\mathcal{Q}} \leq \prod_{p \in \mathcal{Q}} p \leq \prod_{j \leq x^{2/5}} (q^j - 1) \leq q^{x^{4/5}}.$$

Thus $\#\mathcal{Q} = O(x^{4/5})$.

On the other hand, it is shown in [1] that there exist a constant $A > 0$ and a set $\mathcal{R}$ of cardinality $\#\mathcal{R} \geq x/\ln^A x$, consisting of primes $p \leq x$ for which $p-1$ has the largest prime factor $P(p-1) > p^{0.677}$. It is clear that $t_p \geq P(p-1) > p^{0.677}$ for each prime p from the set $\mathcal{R} \backslash \mathcal{Q}$ of cardinality

$$\# (\mathcal{R} \backslash \mathcal{Q}) \geq x/\ln^A x + O(x^{4/5}) \geq x/2\ln^A x$$

for sufficiently large x.   □

The famous *Artin's conjecture* asserts that for any q which is not a perfect square, q is a primitive root modulo p, that is, $t_p = p - 1$, for infinitely many primes p. This conjecture remains unproved, however Heath-Brown [6] has shown that it holds for most of the prime powers q which are not perfect squares. In particular, by [6] we have the following statement.

Lemma 4. *For all but at most three primes r and any fixed power $q = r^{2m+1}$ of r with odd exponent, there are infinitely many primes p such that $t_p = p - 1$.*

We also need the following result from [9].

Lemma 5. *For any fixed $\varepsilon > 0$ and sufficiently large $x > 0$, $t_N \geq N^{1/2-\varepsilon}$ for all but $o(x)$ positive integers $N \leq x$ coprime to q.*

Finally, we recall that the number of integer divisors $\tau(N)$ of an integer N satisfies the estimate

$$\tau(N) = N^{o(1)}, \tag{3}$$

see [5, Theorem 317].

3   Main Results

Theorem 1. *For any fixed prime power q, there are infinitely many primes p such that for $N = p$ there is a sequence S over $\mathbb{F}_q$ of period $N = p$ and such that (2) holds for all positive integers $k \leq N^{0.677}/2\ln N$.*

Proof. Let $N = p$ be a prime described in Lemma 3, that is, $t_N \geq N^{0.677}$.

We note that there is only one cyclotomic coset with $\ell_i < t_N$, namely $C_1 = \{0\}$. Trivially, for $k \leq N^{0.677}/2 \ln N$ and sufficiently large N, we have

$$\sum_{j=0}^{k} \binom{N}{j} (q-1)^j < kq^k N^k < q^{t_N}.$$

We now apply Lemma 1 (or [14, Theorem 1]) with $\ell = \ell_h = t_N$ which finishes the proof.   □

Theorem 2. *For all but at most three primes r and any fixed power $q = r^{2m+1}$ of r with odd exponent, there are infinitely many primes p such that for $N = p$ there is a sequence S over $\mathbb{F}_q$ of period $N = p$ and such that (2) holds for all positive integers $k \leq (N-3)/2$.*

Proof. Let $N = p$ be a prime described in Lemma 4, that is, $t_p = p - 1$. The rest of the proof repeats the arguments of [14, Example 1].   □

Theorem 3. *For any fixed prime power q, any $\varepsilon > 0$ and sufficiently large $x > 0$, for all but $o(x)$ positive integers $N \leq x$ coprime to q, there is a sequence S over $\mathbb{F}_q$ of period N and such that*

$$L(S) = N \qquad and \qquad L_{N,k}(S) = N + o(N)$$

for all positive integers $k \leq N^{1/2-\varepsilon}$.

Proof. Let N be an integer described in Lemma 5 but applied with $\varepsilon/4$ instead of ε, that is, $t_N \geq N^{1/2-\varepsilon/4}$.

We now apply Lemma 1 with $\ell = \lceil t_N N^{-\varepsilon/4} \rceil$. By Lemma 2 we see that if $\ell_i < \ell$ for some cyclotomic coset C_i, $1 \leq i \leq h$, then

$$t_N N^{-\varepsilon/4} \geq \ell - 1 \geq \ell_i = t_{N/d_i} \geq t_N/d_i,$$

thus $d_i \geq N^{\varepsilon/4}$. Therefore,

$$\sum_{\ell_i < \ell} \ell_i \leq \sum_{\substack{a \in \mathbb{Z}_N \\ \gcd(a,N) \geq N^{\varepsilon/4}}} 1 \leq \sum_{\substack{d \mid N \\ d \geq N^{\varepsilon/4}}} \sum_{\substack{a \in \mathbb{Z}_N \\ \gcd(a,N)=d}} 1$$

$$\leq \sum_{\substack{d \mid N \\ d \geq N^{\varepsilon/4}}} \frac{N}{d} \leq \tau(N) N^{1-\varepsilon/4} = o(N)$$

by the inequality (3).

Trivially, for $k \leq N^{1/2-\varepsilon}$ and sufficiently large N, we have

$$\sum_{j=0}^{k} \binom{N}{j} (q-1)^j < kq^k N^k < q^{\ell}$$

which finishes the proof.   □

4   Remarks

We remark that using the full power of [11, Theorem 1] one can show that there are exponentially many sequences satisfying the bounds of Theorem 1, Theorem 2 and Theorem 3. It is an interesting open question to find explicit constructions of sequences satisfying the bounds of these theorems (note that the proof of [11, Theorem 1] does not lead to explicit constructions).

We have seen that our results depend drastically on the size of the multiplicative order of q modulo N. In particular, as it has been noticed in [14], Artin's conjecture becomes relevant, see also Theorem 2. Hooley [7] has shown that the Extended Riemann Hypothesis (ERH) is enough to prove an asymptotic formula for the number of such primes (which shows that they form a set of positive density in the set of all primes). It has also been shown in [9] that, for any fixed q and $\varepsilon > 0$, for almost all N coprime to q the bound $t_N \geq N^{1/2-\varepsilon}$ holds. Furthermore, under the ERH, for any fixed $q > 1$ and $\varepsilon > 0$ we have $t_N \geq N^{1-\varepsilon}$ for almost all N coprime to q, see [8]. In fact, this bound has recently been improved in [10, Corollary 2] as

$$t_N \geq N/(\ln N)^{2\ln\ln\ln N}$$

for almost all N coprime to q, again under the ERH.

References

1. R. C. Baker and G. Harman, 'Shifted primes without large prime factors', *Acta Arithmetica*, **83** (1998), 331–361.
2. T. W. Cusick, C. Ding and A. Renvall, *Stream Ciphers and Number Theory*, Elsevier, Amsterdam, 1998.
3. C. Ding, G. Xiao and W. Shan, *The Stability Theory of Stream Ciphers*, Lecture Notes in Comp. Sci., vol. 561, Springer-Verlag, Berlin, 1991.
4. J. B. Friedlander, C. Pomerance and I. E. Shparlinski, 'Period of the power generator and small values of Carmichael's function', *Math. Comp.*, **70** (2001), 1591–1605 (see also **71** (2002), 1803–1806).
5. G. H. Hardy and E. M. Wright, *An Introduction to the Theory of Numbers*, Oxford Univ. Press, Oxford, 1979.
6. D. R. Heath-Brown, 'Artin's conjecture for primitive roots', *Quart. J. Math.*, **37** (1986), 27–38.
7. C. Hooley, 'On Artin's conjecture', *J. Reine Angew. Math.*, **225** (1967), 209–220.
8. P. Kurlberg, 'On the order of unimodular matrices modulo integers', *Acta Arithmetica*, **110** (2003), 141–151.
9. P. Kurlberg and Z. Rudnick, 'On quantum ergodicity for linear maps of the torus', *Comm. Math. Phys.*, **222** (2001), 201–227.
10. S. Li and C. Pomerance, 'On generalizing Artin's conjecture on primitive roots to composite moduli', *J. Reine Angew. Math.*, **556** (2003), 205–224.
11. W. Meidl and H. Niederreiter, 'Periodic sequences with maximal linear complexity and large k-error linear complexity', *Appl. Algebra in Engin., Commun. and Computing*, (to appear).

12. A. J. Menezes, P. C. van Oorschot and S. A. Vanstone, *Handbook of Applied Cryptography*, CRC Press, Boca Raton, FL, 1996.

13. H. Niederreiter, 'Some computable complexity measures for binary sequences', *Sequences and Their Applications* (C. Ding, T. Helleseth and H. Niederreiter, eds.), Springer-Verlag, London, 1999, pp. 67–78.

14. H. Niederreiter, 'Periodic sequences with large k-error linear complexity', *IEEE Trans. Inform. Theory*, **49** (2003), 501–505.

15. R. A. Rueppel, *Analysis and Design of Stream Ciphers*, Springer-Verlag, Berlin, 1986.

16. R. A. Rueppel, 'Stream ciphers', *Contemporary Cryptology: The Science of Information Integrity* (G. J. Simmons, ed.), IEEE Press, New York, 1992, pp. 65–134.

17. M. Stamp and C. F. Martin, 'An algorithm for the k-error linear complexity of binary sequences with period 2^n', *IEEE Trans. Inform. Theory*, **39** (1993), 1398–1401.

Estimates for Discrete Logarithm Computations in Finite Fields of Small Characteristic

Robert Granger

University of Bristol, Department of Computer Science,
Merchant Venturers Building,
Woodland Road,
Bristol, BS8 1UB, UK.
granger@cs.bris.ac.uk

Abstract. We give estimates for the running-time of the function field sieve (FFS) to compute discrete logarithms in $\mathbb{F}_{p^n}^{\times}$ for small p. Specifically, we obtain sharp probability estimates that allow us to select optimal parameters in cases of cryptographic interest, without appealing to the heuristics commonly relied upon in an asymptotic analysis. We also give evidence that for any fixed field size some may be weaker than others of a different characteristic or field representation, and compare the relative difficulty of computing discrete logarithms via the FFS in such cases.

1 Introduction

The importance of the discrete logarithm problem (DLP) to modern cryptography is difficult to overestimate. It has been for more than two decades now, the subject of intense research activity, and failing an imminent development of superior computer technology (see [19]), it appears unlikely that an efficient process for computing discrete logarithms will be forthcoming in the near future. As such, it is a basic imperative for cryptographers to obtain a practical knowledge of the computational security that this primitive bestows.

Currently the best publically known algorithm for the DLP over a finite field of size p^n, with p small, has asymptotic complexity $L_{p^n}[1/3, (32/9)^{\frac{1}{3}}] = \exp(((32/9)^{\frac{1}{3}} + o(1)) \log(p^n)^{\frac{1}{3}} \log(\log(p^n))^{\frac{2}{3}})$. This result was first obtained by Adleman and Huang [2] using the function field sieve (FFS) with the restriction that $p^6 \leq n$ as $p^n \to \infty$, and later by Joux and Lercier [10], with a slightly more practical algorithm. As we show however, these asymptotics do not necessarily yield reliable estimates of running-times for the DLP in finite fields of cryptographic interest, since the assumptions they rely upon are either heuristic, or are valid only in the asymptotic case. Moreover this result appears to be blind to the characteristic of the finite field under consideration; we show this is not the case for practical field sizes.

The problem we refer to is as follows. Let $\mathbb{F}_q$ denote the finite field of q elements, where $q = p^n$, and let g be a generator for the multiplicative group $\mathbb{F}_q^{\times}$.

K.G. Paterson (Ed.): Cryptography and Coding 2003, LNCS 2898, pp. 190–206, 2003.

The DLP over $\mathbb{F}_q^\times$ is to compute, for a given element $h \in \mathbb{F}_q^\times$ the unique integer $x \bmod q - 1$ such that $g^x = h$. Note that in contrast to the analogous problem of integer factorisation, which has approximately the same asymptotic complexity [11] as the DLP in a finite field, an exact expression exists for the solution in this case [14],

$$\log_g h = \sum_{m=0}^{n-1} \left\{ \sum_{i=1}^{q-2} h^i (1 - g^i)^{-p^m} \right\} p^m,$$

though this is clearly less effective than a naive enumeration of powers of g.

In any large-scale implementation of the FFS it is essential to solve many small example DLP's to gain an idea of what parameter choices are most effective. This initial understanding of the performance of an implementation can then be used to extrapolate optimal parameter choices to the larger, more costly example. To date there are only very few published accounts of discrete logarithm computations undertaken in reasonably large fields in characteristic two [9,10,20], and none in fields of other small characteristics. Considering that the current record for the field $\mathbb{F}_{2^{607}}$ [20] was the result of over a year's worth of highly parallelised computation, rather than implementing another example, we develop a detailed model of the behaviour of the FFS that allows one to make practical estimates of optimal parameters and expected running-times for any $\mathbb{F}_q$ where the FFS applies.

In recent years fields of characteristic three have gained cryptographic interest. For instance, to optimise security parameters in the use of supersingular elliptic curves in identity based encryption schemes, it is recommended that finite fields with $p = 3$ should be used [4,7]. Furthermore fast implementations in hardware of these fields have been studied [16,3] where arithmetic need not significantly under-perform comparable characteristic two alternatives. For this reason alone, it is prudent to consider how hard is it to solve discrete logarithms in characteristic three, and how this compares with the characteristic two case for fields of approximately the same size. We also consider the possible effect of different field representations, when the extension degree is composite, and indicate that in some cases these may offer an advantage to the cryptanalyst.

Currently the linear algebra step is considered to be a bottleneck in the index calculus method, due the difficulty of parallelising this stage efficiently. Although the situation is improving, the heuristically optimum factor base sizes are still beyond what can adequately be accommodated in the matrix step. One question we answer is how varying the size of the factor base, which is the main constraint in the linear algebra step, affects the efficiency of the FFS. It is the notion of 'smoothness' that allows us to assess the probability that a random element will factor in the factor base, and this is based solely on the degree of that element. Consequently if the factor base is chosen to consist of all monic irreducible polynomials up to a certain degree bound, then although we can make good estimates of smoothness probabilities, we are restricted in the choice of the size of the factor base. In higher characteristics this problem renders typical estimates of smoothness probabilities useless.

Ideally for an optimisation analysis we would like to allow the size of the factor base to vary over $\mathbb{N}$, but it is essential that we have highly accurate estimates for the probabilities involved. Such an interpolation is often used in practice, yet no formal analysis has been given. We give such an analysis, and allow implementors to decide whether such an interpolation is useful in practice.

The model we assume in our running-time estimates is the version of the FFS as detailed in [2]. In order to assess these times accurately, it is necessary to provide in detail an analysis of how we expect a basic implementation of the FFS to behave in practice. In the next section we briefly explain the general principle of the FFS. In Section 3 we give the revised probability estimates we need and describe our model. In Section 4 we present our results, and in Section 5 we draw some conclusions.

2   The Index Calculus Method

Currently the most effective principle for calculating discrete logarithms in finite fields of small characteristic is the 'index-calculus' method. It can be applied if we choose as our representation of $\mathbb{F}_q$ the set of equivalence classes in the quotient of the ring $\mathbb{F}_p[t]$ by one of its maximal ideals $f(t)\mathbb{F}_p[t]$, where $f(t)$ is an irreducible polynomial of degree n. Each equivalence class can then be uniquely identified with a polynomial in $\mathbb{F}_p[t]$ of degree strictly less than n. Arithmetic is performed as in $\mathbb{F}_p[t]$, except that when we multiply two elements we reduce the result modulo $f(t)$. The utility of this representation is that its elements are endowed with a notion of size (the degree), and hence 'smoothness', where we call an element of $\mathbb{F}_p[t]$ m-smooth if all its irreducible factors have degrees $\leq m$.

The index calculus method essentially consists of two stages: one first fixes a subset $F = \{b_1, \ldots, b_{|F|}\}$ of $\mathbb{F}_q^{\times}$ called the *factor base* and one then generates multiplicative relations between elements of $\mathbb{F}_q^{\times}$ in the hope that they factor completely within F. For this reason the factor base is usually chosen to exhaust all the monic irreducible polynomials up to a certain degree bound, since the size of the subset of $\mathbb{F}_q^*$ generated by this set is larger than that generated by any other set of the same cardinality. With this choice, heuristically the probability that a random polynomial factors wholly within the factor base is maximised. The notion of degree therefore is essential to this method.

Upon taking logarithms with respect to a generator g one obtains a set of equations of the form

$$\sum_{(j_i, b_i) \in \mathbb{Z} \times F} j_i \log_g b_i \equiv 0 \mod q - 1. \tag{1}$$

Once sufficiently many such relations have been collected, one inverts the corresponding linear system mod $q - 1$ to obtain the quantities $\log_g b_i$ for each $b_i \in F$. If the prime factors of $q - 1$ are known this computation can be simplified by solving for the b_i modulo each prime divisor of $q - 1$, and then reconstructing their values mod $q - 1$ using the Chinese Remainder Theorem.

The second stage then allows one to find the discrete logarithm of an arbitrary element $h \in \mathbb{F}_q^\times$ not in F. One looks for an integer l such that $g^l h$ is a product of elements of F, yielding the relation

$$\log_g h = \left\{ -l + \sum_{(k_i, b_i) \in \mathbb{Z} \times F} k_i \log_g b_i \right\} \bmod q - 1.$$

The main issues to address in such a calculation are: by what method should we collect relations? What is the most time-efficient size for the factor base? And can we do better than randomly guessing in the second stage? The first question is dealt with effectively by the FFS, see Section 2.1, [2,10]; the second has been answered in the asymptotic case [2,10]; and the third, while also answered in the asymptotic case using the FFS [1,2,18], or Coppersmith's method [5,6,15], is relatively easy in practice. For this reason, we do not consider the third problem here.

2.1   The Function Field Sieve

The idea of using a function field to obtain relations of the form (1) is originally due to Adleman [1] and generalises Coppersmith's early algorithm [5,6] for the characteristic two case. Here we give a brief overview of the algorithm.

One selects a polynomial μ in $\mathbb{F}_p[t]$, and an absolutely irreducible bivariate polynomial $H(t, X)$ such that $H(t, \mu(t)) = 0 \bmod f(t)$, where $f(t)$ is the irreducible polynomial in $\mathbb{F}_p[t]$ which defines $\mathbb{F}_q$. This condition provides a map ϕ from the ring $\mathbb{F}_p[t, X]/(H)$ to $\mathbb{F}_q^\times$ induced by sending $X \mapsto \mu$. Given $H(t, X)$, its function field L is defined as Quotient$(F_p[t, X]/(H))$.

In the FFS we have two factor bases. The small degree monic irreducible polynomials in $\mathbb{F}_q$ form the rational factor base F_R, while those places in L that lie above the primes of F_R and are of degree $\leq$ the maximum degree of elements of F_R, form the algebraic factor base F_A. For a given coprime pair $(r, s) \in \mathbb{F}_p[t]^2$, we check if $r\mu + s$ decomposes on the rational side as a product of primes from F_R, and also whether the divisor associated to the function $rX + s$ decomposes as a sum of places in F_A on the algebraic side. To verify the second condition we need only compute the norm of $rX + s$ over $\mathbb{F}_p[t]$, $r^d H(t, -s/r)$, where d is the degree in X of H, and check this for smoothness in F_R.

Provided that $H(t, X)$ satisfies eight technical conditions as given by Adleman, we obtain a relation in $\mathbb{F}_q^\times$ by raising the function $rX + s$ to the power h_L, the class number of L, and applying the morphism ϕ to the components of its associated divisor, and also to $rX + s$, giving $r\mu + s$. One point to note is that for each element of the algebraic factor base F_A occuring in the decomposition of the divisor associated to $(rX + s)^{h_L}$, the previous computation gives an additional logarithm for an element of $\mathbb{F}_q^\times$ (which we need not compute explicitly), and so with regard to the linear algebra step of the algorithm, we expect the matrix to have about $F_R + F_A$ rows.

The sieving stage of the FFS refers to various methods which may be employed to enhance the probability that the pairs (r, s) will be coprime and 'doubly smooth', see [15,6,8]. For simplicity we do not incorporate these methods into our probabilistic model.

In Section 3.2, we describe the properties of the curve $H(t,X)$ that are essential to our model.

3   Methodology of Our Analysis

In this section we describe the methodology of our calculations. Our goal here is twofold: primarily we want to ensure that our model portrays as accurately as possible the behaviour of a practical implementation of the FFS; and secondly to ensure that the full spectrum of choice for the relevant variables is thoroughly investigated. With regard to the former, we first give a refinement of the well-known smoothness probability function [15].

3.1   Some Factorisation Probabilities

Let $\eta = |F_R|$ and let

$$I_p(j) = \frac{1}{j} \sum_{d|j} \mu(d) p^{j/d} \tag{2}$$

be the number of monic irreducible polynomials in $\mathbb{F}_p[t]$ of degree j, where μ is the mobius function. This then determines a unique $m \in \mathbb{N}$ such that

$$\sum_{j=1}^{m} I_p(j) \leq \eta < \sum_{j=1}^{m+1} I_p(j)$$

so that while we have all irreducible polynomials in $\mathbb{F}_p[t]$ of degrees $\leq m$ in our factor base, we are also free to select a fraction α of primes of degree $m+1$, with

$$\alpha = (\eta - \sum_{j=1}^{m} I_p(j))/I_p(m+1).$$

Such a proportion α is used in some implementations of the FFS, and in the following we investigate the practical implications of such a parameter. We assume that the additional degree $m + 1$ members of the factor base have been chosen at random, before sieving begins. In an implementation these would probably be selected dynamically, as the sieve progressed, though we do not believe this would affect these estimates significantly.

Definition 1. *Let $\rho_{p,\alpha}(k,m)$ be the probability that a given monic polynomial in $\mathbb{F}_p[t]$ of degree k has all of its irreducible factors of degrees $\leq m+1$, and that those of degree $m + 1$ are contained in a proportion α of preselected irreducible polynomials of degree $m + 1$.*

We implicitly assume throughout that elements of $\mathbb{F}_p[t]$ behave like independent random variables with respect to the property of being smooth. Provided we have a process to generate elements uniformly, this is reasonable, and as do all authors on this subject, we take this assumption for granted. Note that the case $p = 2, \alpha = 0$ is the definition of $\rho(k, m)$ in [15], which can be computed using the counting function we introduce in Definition 2, and with which we derive an exact expression for $\rho_{p,\alpha}(k, m)$. When a given polynomial factors within this extended factor base we say it is (m, α)-smooth.

Definition 2. *Let $N_p(k, m)$ be the number of monic polynomials $e(t) \in \mathbb{F}_p[t]$ of degree k such that $e(t)$ has all of its irreducible factors of degrees $\leq m$, ie.*

$$e(t) = \prod_i e_i(t)^{\beta_i}, \ \ deg(e_i(t)) \leq m.$$

For exactness we further define $N_p(k, 0) = 0$ for $k > 0$, $N_p(k, m) = p^k$ if $k \leq m$, and $N_p(k, m) = 0$ if $k < 0$ and $m \geq 0$.

We are now ready to state

Theorem 1.

i. $$N_p(k, m) = \sum_{n=1}^{m} \sum_{r \geq 1} N_p(k - nr, n - 1) \binom{r + I_p(n) - 1}{r},$$

ii. $$\rho_{p,\alpha}(k, m) = \sum_{r \geq 0} \frac{N_p(k - r(m + 1), m)}{p^k} \binom{r + I_p(m + 1) - 1}{r} \alpha^r.$$

See the Appendix for a proof.

3.2   Model of the Function Field Sieve

In this section we describe the details of the model we use in our estimates, which is based on the FFS as set forth in [2].

Let d be the degree in X of the curve $H(t, X)$, and let $d' = \lceil n/d \rceil$, where n is the extension degree of $\mathbb{F}_q$ over $\mathbb{F}_p$. Let $\delta(\cdot)$ be the degree function. The irreducible polynomial f of degree n which defines $\mathbb{F}_q$ is chosen to be of the form $f(t) = t^n + \hat{f}(t)$, where $\delta(\hat{f}) \approx \log_p n$. We make this last assertion since by (2) the proportion of degree n monic polynomials in $\mathbb{F}_p[t]$ that are irreducible is about $1/n$, and so we expect to be able to find one satisfying $\delta(\hat{f}) \leq \log_p n$. A further condition on $\hat{f}$ is that it has at least one root of multiplicity one, so that the curve $H(t, X) = X^d + t^{dd'-n} \hat{f}(t)$ is absolutely irreducible. We let $\mu(t) = t^{d'}$ and it is easily checked then that $H(t, \mu(t)) \equiv 0 \bmod f$. Lastly we assume that we can easily find such curves for each set of parameter values we analyse, that furthermore satisfy the remaining technical requirements.

Within the model itself, we are slightly conservative in some approximations. This means our final estimates for the running-times will tend to err on the upper side of the average case, so that we know with a fair degree of certainty that given a final estimate, we can compute discrete logarithms within this time. We take as the unit run-time of our calculations the time to perform a basic field operation in $\mathbb{F}_p$, but do not discriminate between which ones as this will only introduce a logarithmic factor into the estimates. Furthermore, for a real implementation the computational difference for different p will often be irrelevant, since all small p will fit in the word size, and for characteristics two and three, there will be only a small constant difference in performance.

Given a factor base size η as in Section 3.1, we determine the corresponding m, and α. We assume that F_A has the same cardinality as the set of primes we use in the factor base F_R on the rational side. So for a given size of matrix that we can theoretically handle in the linear algebra step, the set of useful logarithms that we obtain is only half of this.

We now consider the number of relations we expect to obtain. Since we wish to minimise the degree of the polynomials we generate, we consider those pairs $(r, s) \in \mathbb{F}_p[t]^2$ where $\delta(r) = R, \delta(s) = S$ are as small as possible, and $R \leq S$. We refer to this collection of elements as the sieve base, and it is typically chosen to consist of all relatively prime pairs of polynomials with degrees bounded by l, a parameter to be selected. We observe though that since we are looking to generate pairs (r, s) such that $r\mu + s$ and $rX + s$ give useful relations, we may assume that either r or s is a monic polynomial, as otherwise we will obtain $p-1$ copies of each useful pair $r\mu + s$ and $rX + s$.

With this in mind let $a_{R,S}$ be the number of coprime pairs of polynomials (r, s) of degrees $0 \leq R \leq S$ with r monic. Then (see the appendix) we have that

$$a_{R,S} = \begin{cases} (p-1)^2 p^{R+S-1} & R, S > 0 \\ (p-1)p^S & \text{otherwise.} \end{cases} \quad (3)$$

For a given $\mu \in \mathbb{F}_q^\times$ of degree $\lceil n/d \rceil$ and a given pair (r, s) with degrees (R, S), $0 \leq R, S \leq l$, the degrees of the rational and algebraic sides are respectively bounded by

$$\delta_{RAT_d}(R, S) \leq \max\{R + \lceil n/d \rceil, S\},$$
$$\delta_{ALG_d}(R, S) \leq \max\{dR + d + \log_p n, dS\}.$$

As discussed above we assume both these bounds are equalities.

The probability that each of these is (m, α)-smooth, assuming they are independent, is

$$\rho_{p,\alpha}(\delta_{RAT_d}(R, S), m).\rho_{p,\alpha}(\delta_{ALG_d}(R, S), m),$$

and the number of suitable pairs (r, s) such that this can occur is $a_{R,S}$. Since $0 \leq R \leq S \leq l$, the number of (m, α)-smooth relations we expect to obtain is

$$M(l) = \sum_{S=0}^{l} \sum_{R=0}^{S} \rho_{p,\alpha}(\delta_{RAT_d}(R, S), m).\rho_{p,\alpha}(\delta_{ALG_d}(R, S), m)a_{R,S}.$$

For each such set of relations we assume that we obtain the same number of linearly independent equations amongst the corresponding logarithms. To obtain a matrix of full rank we require that this number exceeds $|F_R| + |F_A|$. A basic constraint therefore is

$$M(l) \geq 2\eta. \tag{4}$$

When this is the case, we calculate the expected running-time as follows. We estimate the running-time of a gcd calculation for a pair r, s simply as RS, as we do not presume any fast multiplication algorithm. The time for this stage is therefore

$$\sum_{S=1}^{l} \sum_{R=1}^{S} RSp^{R+S}, \tag{5}$$

as for each (R, S) there are p^{R+S} such pairs of monic polynomials (r, s). Furthermore, with regard to factoring both the rational and algebraic sides, it only makes sense to factor both if the first one factored possesses factors lying entirely within the factor base, which cuts down the number factorisations needed to be performed considerably. With this in mind and noting the form of δ_{ALG_d} and δ_{RAT_d}, we choose the rational side to be the first factored. Again we suppose a naive polynomial factoring algorithm and simply estimate the running time as δ^3. For this stage the running-time is therefore

$$\sum_{S=0}^{l} \sum_{R=0}^{S} a_{R,S} \left\{ \delta_{RAT_d}(R, S)^3 + \rho_{p,\alpha}(\delta_{RAT_d}(R, S), m)\delta_{ALG_d}(R, S)^3 \right\}. \tag{6}$$

This is essentially all that we need. The process of minimising the expected running-time for each factor base size is to compute, for various pair values (d, l), the value of (5) and (6) provided the constraint (4) is satisfied, and simply choose the pair which gives the smallest result. Preliminary runs of our model indicated however that the estimated running-times altered considerably when the optimal value for l was changed by just one. A little reflection will reveal that the process of varying l, the bound of the degrees of the sieve base elements, is analogous to the way in which we initially regarded m, the bound of the degrees of the factor base elements. To remedy this problem we interpolate between successive values of l. Let

$$L_l(i) = \lfloor p^{l+1}i/100 \rfloor, \ i = 0, ..., 100,$$

and suppose that we have $L_l(i)$ additional monic sieve base polynomials of degree $l+1$. The choice of 100 here is arbitrary, but we did not want to extend running-times excessively, and we can view this number simply as the percentage of monic polynomials of degree $l + 1$ in the sieve base, since there are p^{l+1} monic polynomials of degree $l + 1$. This gives us an additional number of expected doubly-smooth relations $M_+(L_l(i))$ equal to

$$\sum_{R=0}^{l+1} \rho_{p,\alpha}(\delta_{RAT_d}(R, l+1), m).\rho_{p,\alpha}(\delta_{ALG_d}(R, l+1), m)a_{R,l+1}L_l(i)/p^{l+1},$$

where here we have implicitly assumed that the proportion of pairs that are coprime is propogated uniformly for each interpolation value. Our constraint (4) then becomes

$$M(l) + M_+(L_i(l)) \geq 2\eta,$$

and the total expected running-time is then

$$T(m, \alpha, l, d, i) = \sum_{S=0}^{l} \sum_{R=0}^{S} \left\{ RSp^{R+S} + \delta_{RAT_d}(R, S)^3 a_{R,S} + \right.$$

$$+ \rho_{p,\alpha}(\delta_{RAT_d}(R, S), m)\delta_{ALG_d}(R, S)^3 a_{R,S} \bigg\}$$

$$+ \sum_{R=0}^{l+1} \left\{ R(l+1)p^R L_l(i) + \delta_{RAT_d}(R, l+1)^3 a_{R,l+1}L_l(i)/p^{l+1} + \right.$$

$$+ \rho_{p,\alpha}(\delta_{RAT_d}(R, l+1), m)\delta_{ALG_d}(R, l+1)^3 a_{R,l+1}L_l(i)/p^{l+1} \bigg\}.$$

$$(7)$$

This equation accounts for the effect of the differing degrees in sieve base pairs, the necessary gcd and factorisation computations, and for the l value interpolations. Note that this differs considerably from the

$$2\eta/\rho_{p,0}(dl + d + l + \lceil n/d \rceil + \log_p n, m)$$

typically assumed in an asymptotic analysis.

4   Empirical Results

In the following computations we looked at fields of characteristics 2,3 and 107, with bitsizes of 336, 485, 634, and 802. We let the size of the factor base vary from 100,000 to 2,000,000, with increments of 50,000. The correspondence was as follows:

$$\mathbb{F}_{2^{336}} \sim \mathbb{F}_{3^{212}} \sim \mathbb{F}_{107^{50}}$$
$$\mathbb{F}_{2^{485}} \sim \mathbb{F}_{3^{306}} \sim \mathbb{F}_{107^{72}}$$
$$\mathbb{F}_{2^{634}} \sim \mathbb{F}_{3^{400}} \sim \mathbb{F}_{107^{94}}$$
$$\mathbb{F}_{2^{802}} \sim \mathbb{F}_{3^{506}} \sim \mathbb{F}_{107^{119}}$$

These bit-lengths were chosen so that for each pair of characteristics p, r and for each corresponding pair of extension degrees n, b, $|\log(p^n)/\log(r^b) - 1| \leq 10^{-2}$ holds. This restriction on the field sizes is superficial and was chosen only to permit a fair comparison of the characteristics, and as such, these fields may not be directly of cryptographic interest. This is because in practice the group order $p^n - 1$ should possess a large prime factor to prevent the Pohlig-Hellman attack [17]. For all three characteristics, we let d, the degree of the function field, vary from 3 to 7. The maximal degree of the sieve base elements was 70 for characteristic two, 60 for characteristic three, and 20 for characteristic 107. In none of the calculations we performed were these bounds reached, and so we can be confident that the time-optimal parameters we found in each case were optimal over all feasible parameters. The figures below compare the minimum expected running-times for the three characteristics, for each factor base size tested.

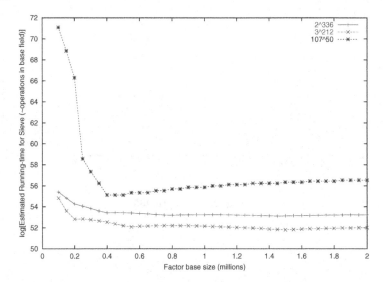

Fig. 1. Minimum sieving times for characteristics 2,3 and 107: bit-length 336

4.1   Discussion

We observe first that for each field size investigated, there appears to be a negligible difference between the difficulty of computing discrete logarithms in characteristics 2 and 3. For characteristic 107 however, the difference is significant. Furthermore, for the two smaller characteristics, the sieving time continues to fall quite smoothly as the factor base size is increased, whereas for $p = 107$, the optimum factor base size is reached for $\eta \approx 400,000$, with the sieving time then remaining almost constant beyond this number. The asymptotic analysis of [2] applied to the field $\mathbb{F}_{107^{50}}$ implies that the optimum factor base size is about 4 million, and for the larger extension degrees is far larger than this still. Note however that for this p and n, there is not yet a proven $L[1/3]$ algorithm for the DLP, and so the results of [2] do not properly apply. Clearly though, if we are faced with this problem, we can not rely solely on asymptotic arguments if we are trying to optimise time-efficiency in using the FFS.

The question of whether or not increasing η improves the efficiency of a discrete logarithm computation can be seen from the figures to depend on the the field size, its characteristic, and the factor base size. The case $p = 107$ just mentioned shows that it is not necessarily beneficial to increase the factor base

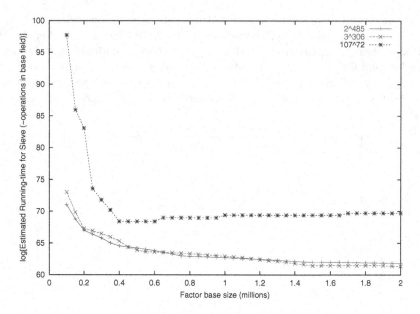

Fig. 2. bit-length 485

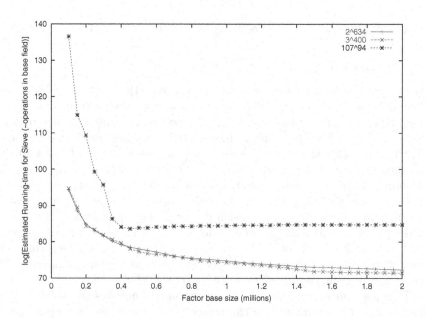

Fig. 3. bit-length 634

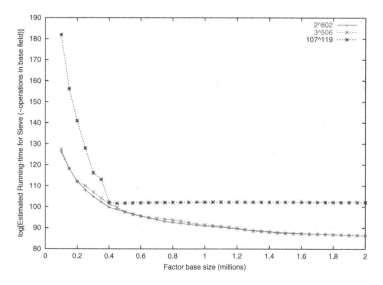

Fig. 4. bit-length 802

size as far as possible, though with the two smaller characteristics, we probably should increase η. However to make a fair assessment we need to analyse not only the sieving time but also the running-time for the linear algebra step. There are no references however which for sparse matrices give a precise complexity estimate, although it is commonly estimated to be $\mathcal{O}(\eta^2)$. We note though that the linear algebra step for larger characteristics will in general be faster than for smaller ones; the degree of the relations generated among field elements is bounded by the extension degree of $\mathbb{F}_q$ over $\mathbb{F}_p$, and hence will on average be smaller in the former case, and so the corresponding matrix will be sparser and more easy to compute with. Moreover the probability of an element having a square factor is also diminished (it is $\approx 1/p$), and so most non-zero entries will be ± 1, further reducing the time needed for this step. The cut-off point at which the reward for increasing η becomes negligible will naturally depend on the implementation being used.

These results raise the obvious question of why in the fields $\mathbb{F}_{107^n}$ is the optimum factor base size seemingly much lower than that predicted by the asymptotic analysis, presuming it is adequate? To this end, we define a simple measure of the smoothness of a field in the usual representation. For a given factor base size η, let $S(\eta, \mathbb{F}_q^\times)$ be the proportion of η-smooth monic polynomials in $\mathbb{F}_q^\times$, where here η-smooth means the (m, α)-smoothness corresponding to a factor base of size η. The point of this measure is to obtain an idea of how increasing η affects the 'total' smoothness of a given field. In Figure 5 we show the computations of this measure for some η in our range for fields of bit-length 802 and characteristics 2, 3, 107, and also 199.

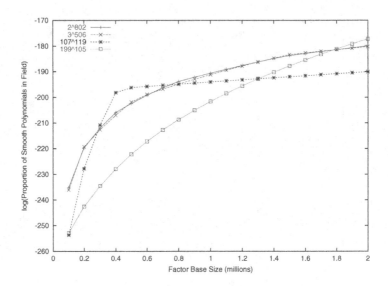

Fig. 5. $S(\eta, \mathbb{F}_{p^n}^{\times})$

We notice immediately that ignoring the characteristic 199 example, this graph is almost Figure 4 inverted. This is intuitive since the time needed for sieving is inversely related to the smoothness probabilities involved. It also suggests a reason why within the range of factor base sizes we ran estimates for, increasing this size did not yield a significant reduction in the running-time for $\mathbb{F}_{107^{119}}$: the total smoothness of $\mathbb{F}_{107^{119}}$ does not increase much for η from 4×10^5 to 2×10^6. Since this measure is independent of any particular implementation of the FFS, we can be fairly confident that for any index calculus computation there would be very little benefit in increasing the size of the factor base beyond 4×10^5 for this field, as a huge saving of time is gained in the linear algebra step.

In contrast, for $p = 199$ we found a significant reduction in our estimates for the minimum running-times over the same range of η. The improvement of running-time estimates for $\eta = 2 \times 10^6$ over $\eta = 4 \times 10^5$ for $p = 199$ is in fact about 1.5×10^{15}. For $p = 2, 3,$ and 107, the corresponding improvements are 1.0×10^4, 3.2×10^4, and 1.1. The running-time estimates themselves for $\mathbb{F}_{199^{105}}$ are however still slower than those for $\mathbb{F}_{2^{802}}$ by a factor of 300 for $\eta = 2 \times 10^6$. Again, this may just reflect the improper application of the FFS to the DLP with these parameters.

Figure 5 indicates though that we may obtain useful information regarding the use of the FFS simply from this smoothness measure, in that while it does not take into account the details of any particular implementation the FFS, it may be used as a good heuristic for its behaviour.

Of course, given a system whose security is based on the DLP in a finite field, one can not choose the characteristic, and so these results may not seem applicable. However, for fields with a composite extension degree, we are free

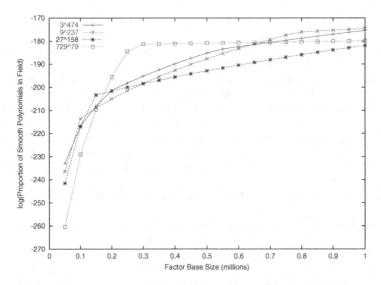

Fig. 6. Comparison of $S(\eta, \mathbb{F}_{3^{474}}^{\times})$ for different field representations

to choose which representation to work with, since every representation can be efficiently converted to any other [13,12]. Figure 6 gives a comparison of $S(\eta)$ for four representations of the field $\mathbb{F}_{3^{474}}$. The graphs suggest that for fields with a composite extension degree, it may be the case that some representations are more susceptable to attack than others. Whether this translates into a noticible speed-up in practice remains to be seen. It would clearly be useful to have an efficient implementation of the FFS in small characteristics to verify these results and their implications, and this is currently a work-in-progress.

5   Concluding Remarks

The purpose of this paper was to gain an idea of how different parameter choices affect the performance of a basic probabilistic FFS model. The model we have developed permits the easy computation of optimal parameter estimates that should be used in implementations, for any finite field where the FFS applies. This model may be developed or specialised to a particular implementation, if needed, but already there is some indication that there are perhaps some subleties of the FFS that are enshrouded by the broad form of its asymptotic complexity.

Furthermore the question of whether finite fields with a composite extension degree are potentially weaker than others due to the multiple representations available was shown, with some heuristic evidence, to be an area worthy of future research.

References

1. Adleman L. M. The function field sieve. In *Algorithmic number theory*, volume 877 of *LNCS*, pages 108–121. Springer, 1994.
2. Adleman L. M. and Huang M. A. Function field sieve method for discrete logarithms over finite fields. *Inform. and Comput.*, 151(1-2):5–16, 1999.
3. Bertoni G., Guajardo J., Kumar S., Orlando G., Paar C. and Wollinger T. Efficient $GF(p^m)$ arithmetic architectures for cryptographic applications. In *Topics in Cryptology, CT-RSA 2003*, volume 2612 of *LNCS*, pages 158–175. Springer, 2003.
4. Boneh D. and Franklin M. Identity-based encryption from the Weil pairing. In *Advances in cryptology, CRYPTO 2001*, volume 2139 of *LNCS*, pages 213–229. Springer, 2001.
5. Coppersmith D. Evaluating logarithms in $GF(2^n)$. In *16th ACM Symp. Theory of Computing*, pages 201–207, 1984.
6. Coppersmith D. Fast evaluation of logarithms in fields of characteristic two. *IEEE Transactions in Information Theory*, 30(4):587–594, July 1984.
7. Galbraith S. D. Supersingular curves in cryptography. In *Advances in cryptology, ASIACRYPT 2001*, volume 2248 of *LNCS*, pages 495–513. Springer, 2001.
8. Gao S. and Howell J. A general polynomial sieve. *Designs, Codes and Cryptography*, 18:149–157, 1999.
9. Gordon D. M. and McCurley K. S. Massively parallel computation of discrete logarithms. In *Proc. CRYPTO'92*, volume 740 of *LNCS*, pages 312–323. Springer, 1993.
10. Joux A. and Lercier R. The function field sieve is quite special. In *Proceedings of the Fifth Symposium on Algorithmic Number Theory*, volume 2369 of *LNCS*, pages 431–445. Springer, 2002.
11. Lenstra A. K. and Lenstra H. W. *The development of the number field sieve*, volume 1554 of *Lecture Notes in Mathematics*. Springer, 1993.
12. Lenstra H. W. Jr. Finding isomorphisms between finite fields. *Mathematics of Computation*, 56(193):329–347, 1991.
13. Lidl R. and Niederreiter H. *Finite Fields*, volume 20 of *Enclyclopedia of Mathematics and its Applications*. Addison-Wesley, 1983.
14. Meletiou G. C. Explicit form for the discrete logarithm over the field $GP(p, k)$. *Archivum Mathematicum (BRNO)*, 29:25–28, 1993.
15. Odlyzko A. M. Discrete logarithms in finite fields and their cryptographic significance. In *Advances in cryptology*, volume 209 of *LNCS*, pages 224–314. Springer, 1985.
16. Page D. and Smart N. Hardware implementation of finite fields of characteristic three. In *Proceedings CHES 2002*, volume 2523 of *LNCS*, pages 529–539. Springer, 2003.
17. Pohlig S. and Hellman M. An improved algorithm for computing logarithms over GF(p) and its cryptographic significance. *IEEE Transactions on Information Theory*, 24:106–110, 1978.
18. Schirokauer O. The special function field sieve. *SIAM Journal on Discrete Mathematics*, 16(1):81–98, 2002.
19. Shor P. W. Polynomial-time algorithms for prime factorization and discrete logarithms on a quantum computer. *SIAM Rev.*, 41(2):303–332, 1999.
20. Thomé E. Computation of discrete logarithms in $\mathbb{F}_{2^{607}}$. In *Advances in cryptology, ASIACRYPT 2001*, volume 2248 of *LNCS*, pages 107–124. Springer, 2001.

A   Proofs of Assertions

Theorem 1.

i. $N_p(k, m) = \sum_{n=1}^{m} \sum_{r \geq 1} N_p(k - nr, n - 1) \binom{r + I_p(n) - 1}{r},$

ii. $\rho_{p,\alpha}(k, m) = \sum_{r \geq 0} \frac{N_p(k - r(m+1), m)}{p^k} \binom{r + I_p(m+1) - 1}{r} \alpha^r.$

Proof. The proof of (i) is virtually identical to the derivation of $\rho_{2,0}(k, m)$ in [15], since in $\mathbb{F}_2[t]$ all non-zero polynomials are monic. For (ii) let $e(t)$ be any monic polynomial in $\mathbb{F}_p[t]$ of degree k, all of whose irreducible factors are of degrees $\leq m + 1$. Such a polynomial can be written uniquely as

$$e(t) = g(t) \prod_{u(t)} u(t)^{\beta_{u(t)}},$$

where the $u(t)$ are monic and of degree $m + 1$, $\sum \beta_{u(t)} = r$ for some $r \in \mathbb{N}$, and $g(t)$ is a monic polynomial of degree $k - r(m + 1)$, all of whose prime factors are of degrees $\leq m$. Given $m + 1$ and r, there are $N_p(k - r(m + 1), m)$ such $g(t)$, and the number of such $\prod u(t)^{\beta_{u(t)}}$ is the number of $I_p(m + 1)$-tuples of non-negative integers which sum to r (since we have $I_p(m + 1)$ possibilities for the $u(t)$), which is just

$$\binom{r + I_p(m+1) - 1}{r}.$$

So for a given r, the probability that a monic polynomial $e(t) \in \mathbb{F}_p[t]$ of degree exactly k has all its irreducible factors of degrees $\leq m + 1$, exactly r of its irreducible factors having degree $m + 1$, and that these are in the preselected factor base is then

$$\frac{N_p(k - r(m+1), m)}{p^k} \binom{r + I_p(m+1) - 1}{r} \alpha^r,$$

since there are p^k monic polynomials of degree k. Hence the total probability that $e(t)$ has all its irreducible factors in the chosen factor base is

$$\sum_{r \geq 0} \frac{N_p(k - r(m+1), m)}{p^k} \binom{r + I_p(m+1) - 1}{r} \alpha^r,$$

which is our theorem.

$\square$

Remark. We have noted already that for $\alpha = 0$, we obtain $\rho_p(k, m)$; while for $\alpha = 1$, we obtain

$$\rho_p(k, m) + \sum_{r \geq 1} \frac{N_p(k - r(m+1), m)}{p^k} \binom{r + I_p(m+1) - 1}{r},$$

which, by the recurrence (i) is equal to

$$\rho_p(k,m) + \frac{1}{p^k}\{N_p(k,m+1) - N_p(k,m)\} = \rho_p(k,m+1),$$

verifying our calculation.

With regard to equation (3) we reason as follows. Considering first just monic polynomials, let $0 \leq R \leq S$. Since there are p^R monic polynomials of degree R, and p^S monic polynomials of degree S, there are p^{R+S} pairs of monic polynomials in this range. Let $\hat{a}_{R,S}$ be the number of monic polynomial pairs (r,s) with degrees R and S such that $\gcd(r,s) = 1$, and for each pair (r,s), let $h = \gcd(r,s)$ where $0 \leq k = \delta(h) \leq R$. There are p^k possible such monic h. Furthermore since $\gcd(r/h, s/h) = 1$, there are $\hat{a}_{R-k,S-k}$ possibilities for the pair $(r/h, s/h)$. Summing these possibilities over k we obtain the recurrence relation

$$p^{R+S} = \sum_{k=0}^{R} \hat{a}_{R-k,S-k}p^k.$$

Noting that $\hat{a}_{0,S-R} = p^{S-R}$ we see this has the solution

$$\hat{a}_{R,S} = \begin{cases} (p-1)p^{R+S-1} & 0 < R \leq S \\ p^S & R = 0 \end{cases}$$

If we allow s to be non-monic then we simply multiply $\hat{a}_{R,S}$ by $|\mathbb{F}_p^{\times}|$, giving the stated result.

Resolving Large Prime(s) Variants for Discrete Logarithm Computation

Andrew J. Holt and James H. Davenport

Dept. of Computer Science, University of Bath, BA2 7AY, England
{ajh,jhd}@cs.bath.ac.uk

Abstract. We conduct an investigation of large prime variant modifications to the well-known index calculus method for discrete logarithm computation modulo a prime p. We highlight complications in this technique that do not occur in its analogue application in factoring, and show how a simple adaption of the methods of [16] can be used to resolve relations such that the yield of the technique is within some ϵ of that obtained in factoring. We then consider how these techniques scale to the use of more large primes, and demonstrate how the techniques of [10], [16] allow us to resolve relations involving more large primes, such that yield is generally the same as that obtained in factoring. Finally, we consider the impact that 'large prime' relations have on both the linear algebra stage of the index calculus method, and on actual discrete logarithm computation – a problem with no analogue in factoring.

1 Introduction

Computation of discrete logarithms over finite fields is of great importance due to their widespread use as the foundations of a variety of cryptographic protocols [9], [20]. We consider computation of discrete logarithms modulo a prime p via the well-known index calculus method. In particular, we are interested in gaining a better understanding of the application of 'large prime variants' of this method, as have been successfully applied to factoring [10], [16], [18]. Such methods have also been used for discrete logarithm computation [25], [26], but few details are given concerning the resolution of large prime relations and the problems one can encounter in this latter application.

Index calculus-type methods can take a variety of forms [6], [21], [23], but in their most basic form[1] can be described as follows. Define a factor base P of all primes up to some smoothness bound B_1. Choose some random value $a \in \mathbb{Z}/(p-1)\mathbb{Z}$, and compute $g^a \bmod p$ where g is a generator of $(\mathbb{Z}/p\mathbb{Z})^*$. If the result of this calculation is B_1-smooth, i.e. if

$$g^a = \prod q_i^{e_i}, \quad q_i \in P \tag{1}$$

---

[1] The method as outlined here is often denoted by 'Hafner-McCurley variant' following the method described in [12].

K.G. Paterson (Ed.): Cryptography and Coding 2003, LNCS 2898, pp. 207–222, 2003.

then

$$a = \sum e_i DL(q_i) \qquad (2)$$

where the $DL(q_i)$ are the discrete logarithms of the prime factors q_i to the base g. Once we have enough of these expressions – some number greater than the number of factor base elements – we can solve them modulo $p - 1$ (or rather modulo the prime factors of $p-1$) to obtain values for the discrete logarithms of the elements of our factor base P. We thus build a database of known discrete logarithms.

In order to compute the discrete logarithm of a given β, we now pick a random a and compute βg^a. If this number is B-smooth, then

$$\beta g^a \bmod p = \prod q_i^{e_i}, \quad q_i \in P \qquad (3)$$

and so

$$DL(\beta) + a = \sum e_i DL(q_i) \qquad (4)$$

and we can use our values for $DL(q_i)$ to compute $DL(\beta)$. Runtime for the basic method is dominated by the time taken to build the database of known values, which takes roughly $L^{1.581}$ [21] as $p \to \infty$, where

$$L = \exp((1 + o(1))(\log p \log \log p)^{1/2}) \qquad (5)$$

The organisation of the paper is as follows: in section 2 we recall how one may introduce 'partial' relations to the precomputation phase of the index calculus method, and consider how such techniques from factoring adapt to the discrete logarithm case. In section 3 we discuss how best to resolve relations involving more than two large primes. Section 4 briefly discusses the impact of such techniques on the linear algebra step of the index calculus method, and in section 5 we look at making use of 'large prime' data to gain substantial speedup in the final step of actual logarithm computation.

2   Large Prime Variants

A simple modification to the basic method is to relax the smoothness bound by allowing one large prime factor in the factorisation of $g^a \bmod p$ which is larger than the factor base bound B_1 but less than some second bound B_2. We shall refer to such a relation as a '1-partial', and a relation involving factor base elements only as a 'full'. Choosing $B_2 \le B_1{}^2$ simplifies processing in that one can then identify a 1-partial if the remainder of $g^a \bmod p$ after division by factor base elements is less than B_2. In practice, one generally takes B_2 to be around $100B_1$ in order to restrict the amount of data generated [2].

It is reasonably straightforward to extend this idea to the use of two large primes [16]. We consider the remainder after division by elements of P. If this remainder is both less than some bound $B_3 \le B_1 B_2$, and is composite, it is the product of exactly two large primes Q_1 and Q_2, with $B_1 < Q_1, Q_2 \le B_2$.

Application of a factoring method such as Pollard's 'rho' method [13] can be used to obtain the Q_i and produce what we shall refer to as a '2-partial' relation.

Going further, one can of course consider relations involving 3 or more large primes. Such relations can be found either by using one or two large primes per relation over multiple factor bases (as used in the Number Field Sieve for both factoring and discrete logarithm computation [10], [17], [26]), or extending the technique as described above to use three (or more) large primes [4], [18]. In order to generate such data for testing purposes, we used the so-called *Waterloo variant* of the basic index calculus method [1], [5]. Here, instead of testing a single value $A \equiv g^a \bmod p$ for smoothness, we use the extended Euclidean algorithm [13] on A and p to obtain values d, x and y such that

$$d = xA + yp \tag{6}$$

where $d = \gcd(A, p)$. Of course, on termination of the extended Euclidean algorithm, d will be 1 since p is prime; however, the absolute sizes of *both* d and x are smallest when $x \approx \sqrt{p}$, and at this point we exit extended Euclid[2]. If both d and x are smooth, we obtain a full relation by computing

$$A = g^a \equiv dx^{-1} \bmod p \tag{7}$$

Use of the Waterloo variant does not improve the asymptotic complexity of the method, but does give practical speedup. Also, since we are now testing two values for smoothness, we may accept partial relations from each one. Allowing up to two large primes per test gives us 3 and 4-partial relations in addition to the 1 and 2-partials from the original method; however, we now need to take account of the exponent of the large primes when resolving. We assume that any cancellation among the large primes in a relation is completed when the relation is created. The authors of [10] note that the additional costs of using a '2 + 2' large prime approach is prohibitively high when using a basic index calculus approach, but can be made more practical with improved sieving methods. The generation of our data is then similar to that described in [10], [26] in that we use two smoothness tests and allow up to two large primes in each. However, the subsequent cycle finding is more like that in [18] (which allows up to three large primes in a single smoothness test), as we have a *single* set of large primes. Using the Number Field Sieve approach of [10], [26], one obtains two sets of large prime ideals, and values cannot generally be matched between one set and the other.

Various papers give estimates for the number of partial relations one can expect to obtain after a given number of smoothness tests – see, for example, [4], [15]. Due to the large amounts of data generated by the use of partial relations, it is advisable to 'prune' datasets prior to attempting to eliminate the large primes – keeping only those relations whose large primes all occur in at least

---

[2] We note that we could consider several values from successive iterations of extended Euclid, while both $|d|$ and $|x|$ are less than some bound. Here we tested for smoothness at up to 3 successive iterations.

one other relation ensures that all remaining relations form part of some cycle. We now look at the practicalities of resolving such relations by eliminating the large primes.

2.1   Resolving 1 and 2-Partial Relations

To resolve partial relations we must eliminate the large primes in order to ensure that all relations consist only of elements from the factor base. Resolving 1-partials is simply a matter of storing them in some ordered way, such as by the hash of their large prime, and dividing relations if the exponents of the large primes match, or multiplying them otherwise[3]. For k 1-partial relations having the same large prime, we can obtain $k - 1$ independent full relations. The total yield for this technique can be estimated reasonably accurately – see [2], [16] for details.

In order to resolve 2-partial relations, the authors of [16] use a graph theoretic approach. The large primes are represented as the vertices of a graph, and a relation involving two large primes is considered as an edge in this graph. For the purposes of factoring, we are looking for relations of the form $x^2 \equiv y^2 \bmod n$ with $x \neq \pm y$. A cycle in the graph is thus sufficient to 'eliminate' the large primes from a relation, since each large prime will appear an even number of times in a cycle. As we are solving modulo 2 in the exponent, the product of the partial relations representing the edges in the cycle may be considered as a full relation for factoring purposes – likewise for resolving 1-partials for factoring purposes, we can always simply multiply relations with matching large primes, rather than having to ensure we divide out the large prime.

Even and Odd Cycles. The crucial difference for discrete logarithm calculation is that we are not solving modulo 2, but over a large finite ring. As a result, we must divide out the large primes. [26] accomplishes this by firstly building cycles as described in [16], and subsequently solving a small linear system in order to eliminate the large primes [27]. However, if we take account of both the nature of the cycles and the order of edges in the cycles as they are built, we can avoid this latter step, as we now discuss.

Consider the graphic at the left of Fig. 1, which shows an even and an odd cycle. The Q_i correspond to large primes and the E_i correspond to relations such that, for example, E_1 represents the relation $g^{a_1} = Q_1 Q_2 f_1$, where f_1 is the factorisation using elements from the factor base only. It can be seen that evaluating

$$\frac{E_1 E_3}{E_2 E_4} = g^{a_1 - a_2 + a_3 - a_4} = \frac{Q_1 Q_2 Q_3 Q_4}{Q_2 Q_3 Q_4 Q_1} \frac{f_1 f_3}{f_2 f_4} = \frac{f_1 f_3}{f_2 f_4} \tag{8}$$

(in a similar manner to that described in [17, sect. 2.12]) eliminates the large primes Q_i. However, such an operation cannot be carried out on the odd-length

---

[3] Of course, relations as not multiplied or divided per se; we simply add or subtract the exponents of the factor base elements found during the smoothness tests.

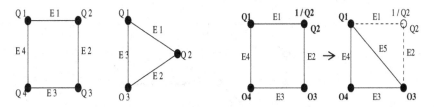

Fig. 1. Even and odd cycles (left) and taking account of exponents (right)

cycle without one of the large primes remaining – for example

$$\frac{E_1 E_3}{E_2} = g^{a_1 - a_2 + a_3} = \frac{Q_1 Q_2 Q_3 Q_1}{Q_2 Q_3} \frac{f_1 f_3}{f_2} = Q_1{}^2 \frac{f_1 f_3}{f_2} \tag{9}$$

is still a 2-partial (with repeated large prime Q_1). We thus cannot eliminate all the large primes in an odd cycle. However, by considering 1-partials as pseudo 2-partials (with 1 as the second 'large prime'), we permit one of the vertices in the graph to be 1. This is standard practice anyway[4], but for discrete logarithm computation it has further importance. If this vertex occurs in an odd cycle, we can order edges to ensure that all other vertices cancel, leaving 1 as the repeated – but irrelevant – 'large prime'.

Since our data come via a Waterloo variant approach, we must also take account of the exponent of the large primes. This is accomplished quite simply by 'reducing' cycles as shown in the right hand diagram in Fig. 1. As we build the cycles, we can 'flip' edges such that the exponents of the vertices match. We then consider the final edge in the cycle, corresponding to edge $E_1 = Q_1 Q_2{}^{-1}$ in this example. Either the exponents of the final vertex in the cycle will match – in which case we do nothing – or they will differ. In our example, suppose we are linking $Q_1{}^{-1}$ in E_1 to $Q_1{}^{+1}$ in E_2. We can simply multiply edges E_1 and E_2 to eliminate Q_1 (and effectively create the new edge E_5), and subsequently resolve the cycle as before.

We thus have the additional criteria that for discrete logarithm purposes, we must either restrict ourselves to looking for even cycles in our graph, or else we *must* add the special vertex 1 and process 1 and 2-partials together. The nature of cycles in the graph and the order in which the edges are processed are consequently more important for discrete logarithm computation than for factoring. However, the manner in which cycles are constructed in the method of [16] makes identifying and resolving even and odd cycles quite straightforward.

As a brief summary of this approach, the union-find algorithm is used to determine the number of edges E, vertices V and components C in the graph, where the root of each component is taken to be the smallest vertex. Determining the graph structure allows one to compute the number of fundamental cycles as

---

[4] Adding 1 as a vertex brings a variety of benefits: increased yield as a result of more relations in the graph, shorter cycle length, etc.

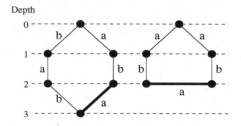

Fig. 2. Identifying cycles and ordering edges

$E + C - V$ and thus estimate how much longer one needs to generate relations. A breadth first search is then used to build the cycles.

We start by adding the component root vertices at depth zero. Edges are added to the graph until one comes across an edge whose vertices are both already held in the graph – we then have a cycle. If both vertices of the new edge occur at the same depth, we have an odd cycle. If the depths differ, we have an even cycle, as shown in Fig. 2. In each case, multiplying the edges labelled a and dividing by the product of the edges labelled b will eliminate all large primes for the even cycle, and all but the root vertex for the odd cycle. So long as the root vertex is the special vertex 1, we can make use of all these cycles. Large primes can thus be eliminated as the cycle is built, avoiding the need to build and solve a small linear system.

Maximising Yield. In practice, we find that the largest component in the graph – generally containing all cycles – is that having the special vertex 1 as its root. Since the method of [16] tends to create cycles which include the root vertex, we find that almost all odd and even cycles can be resolved for the discrete log situation. The exceptions correspond to any odd cycles which do not pass through the root vertex. Occasionally one comes across 2-partial relations with a repeated large prime. These form loops in the graph, which are odd cycles but do not include the special vertex 1. One can however look for 1-partial relations having this large prime, and eliminate by dividing (or multiplying) twice. For factoring purposes, a loop is not a problem since the repeated prime forms a square, and thus cancels directly modulo 2. It is also possible to find other odd cycles which do not include the special vertex 1; perhaps in a secondary component of the graph. It may still be possible to resolve these, but only if we can find a second odd cycle having at least one vertex in common. We can then 'join' these cycles to create an even cycle. The downside of this technique is that it creates longer cycles, which usually leads to more factor base elements being represented in the final full relation. This in turn reduces the sparsity of the ensuing matrix and has a detrimental effect on algorithms to solve for the discrete logarithms of the factor base. However, by allowing for all these possibilities we can try to maximise the yield of the technique, and be within some ϵ of the yield obtained in factoring. As an alternative strategy, one could partially resolve these

Fig. 3. Simple yet unresolvable hypercycles

'isolated' odd cycles without 1, and simply add the unresolvable vertex Q to the factor base. This will of course add to the size of the linear system, but would generally give a sparser row than if we had created a larger even cycle. Extra factor base values would also improve the efficiency of final discrete logarithm computation; although we show later how one can easily gain further factor base values by back substituting over the original partial relations. The author of [26] reports that, in certain cases, some 2.8% of cycles could not be resolved. Our experiments are much smaller, but using the above techniques allowed us to recover a full relation for all cycles in the graph.

3   Towards n Large Primes

Partial relations involving up to 4 large primes are discussed from a factoring perspective in [10], [18], and are mentioned in the context of discrete logarithm computation in [26]. All take essentially the same 'reduction' approach – one firstly resolves 1-partials, and subsequently removes all primes in the set of 1-partial relations from the set of relations having more than one large prime. We then separate these remaining relations into 1-partials and '>1'-partials, and repeat the procedure until no further reduction is possible. In order for this procedure to be effective, we firstly 'prune' the datasets as before, such that any given large prime occurs in at least two relations. This gives us the set of all relations occurring in cycles. It also allows us to determine the number of vertices V and edges E in the graph (or rather hypergraph), from which we can estimate the number of 'hypercycles' present by computing $E - V$.

The method is then almost the opposite of the graph resolve method, in that one effectively 'deconstructs' the cycles in order to build them. This procedure implies that all cycles built will contain at least one 1-partial relation. For factoring purposes, again, any subset of relations where all large primes occur an even number of times will give a full relation on multiplying all relations in the subset. In order to divide out all large primes for discrete logarithm purposes, we can now have complications. It is simple to find examples of hypercycles which cannot be resolved – consider Fig. 3. On the left we have a 3-partial with repeated large prime Q_1, together with a 2-partial. The two relations form a simple hypercycle, yet we cannot eliminate all large primes – for factoring or for discrete log – since Q_1 occurs an odd number of times (assuming that these are 'genuine' partial relations and that the vertex 1 is not present).

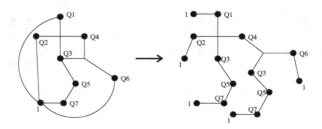

Fig. 4. Hypercycle as a 'tree'

Of course, such a hypercycle would not be built by the procedure outlined above, as it does not contain a 1-partial. However, even if a hypercycle does contain a 1-partial, as on the right, it too may not be built, as noted by Leyland [19] – in this example, we need another 1-partial with large prime Q_3 or Q_4. This hypercycle differs from the other example in that we *can* resolve it by dividing the 3-partial by the product of the 1-partial and the 2-partial. This, however, assumes that all exponents are 1 – in reality, for Waterloo-derived data, either one or two of Q_2, Q_3, Q_4 will have exponent -1. If we consider the possibilities for the exponents of the primes, we find that, even had we built it, we only have a 50% chance of resolving a full relation. For factoring purposes, solving modulo 2 would always allow us to resolve such a hypercycle.

Using 1-partials to reduce higher-order partial relations means that we *always* find an even number of edges incident at any vertex in the hypercycle, with the exception of the special vertex 1. Thus, for factoring purposes one can eliminate the large primes modulo 2 simply by multiplying all edges in the hypercycle. If we allow vertices to be repeated for diagrammatic purposes, we can consider the hypercycles found by such a procedure as trees, as shown in Fig. 4. It can now be seen that we can also divide out all large primes in the hypercycle.

The actual hypercycle is shown on the left. What is built by recursively removing 1-partials is the 'tree' on the right. Now every vertex (except maybe the vertex 1) occurs in an even number of edges, and tracking round the tree one can eliminate all the large primes. The question of order is again all-important – if we do not maintain the order of edges as built, we are in danger of hitting a problem at a vertex such as Q_3 in the example, in that we may have a choice of direction to take. If we follow the order in which the hypercycle was built, we will always take a 'depth first' route through the tree and end up at a 1-partial.

Since this method of building hypercycles means that the vertex 1 is always present at the end of every 'branch', we do not have the unresolvable cases of odd cycles without the vertex 1 (including loops) which we had previously. We can thus resolve *all* hypercycles that are actually built by this technique, both for factoring and for discrete logarithm computation. We did find that, as noted in [10], after the build procedure, occasionally one is left with a small subset of partials (having more than one large prime) which together form a cycle. In practice, for our implementations, these remaining relations were always 2-

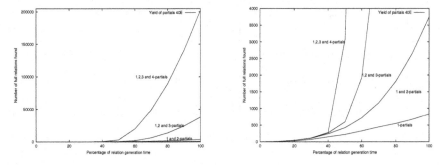

Fig. 5. Yield of 1, 2, 3 and 4-partials

partials. Some of these were loops, which could not be resolved (since no 1-partials match the repeated prime, and unsurprisingly we did not find two loops at the same prime). The others were pairs of 2-partials, having the same primes Q_1 and Q_2. As these are even cycles, these could be resolved. Any odd cycles among the remaining edges would probably cause problems, since 1 is not present in the graph component under consideration. If relations remain, then, after recursively removing 1-partials, we may not quite reach the same yield as in factoring.

Another point to note about Fig. 4 is that, in the case of processing 1 and 2-partials only, the methods of [10], [18] will often create slightly longer cycles than the method of [16] – the repeated path from Q_3 to the special vertex 1 is unnecessary as we could have stopped at vertex Q_3 where the paths join. For higher order partial relations, however, we do need to follow paths all the way to vertex 1 in order to guarantee removal of all large primes. This suggests that, if using 1 and 2-partials only, the method of [16] would be preferable. The length of cycles can be reduced further by a strategy such as that described in [8].

We implemented these techniques to compute discrete logarithms modulo primes of up to 40 digits on a 1.8GHz Pentium IV desktop. Yield for different combinations of partials for computation modulo a prime of 40 digits is shown in Fig. 5. The left hand graphic shows overall yield as more partial relations are added to the hypergraph. The right hand graphic highlights the explosive increase in yield as noted by [10], [18]. The yield of the 1-partials here is roughly quadratic in the number of relations taken, while the yield when processing 1 and 2-partials together is roughly cubic.

It is interesting to note the weight – i.e. number of factor base elements represented – in the full relations we obtain from partials. This is directly related to the length of the cycles which we find. Figure 6 shows (left) the average length of (hyper)cycles found, using the same dataset as in Fig. 5. As noted by [10], the explosive growth in cycles is accompanied by a massive increase in cycle length. This in turn causes the actual time taken to build and resolve these cycles to increase (right). However, if we continue adding to the hypergraph, the

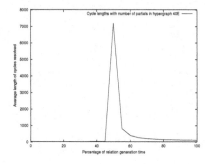 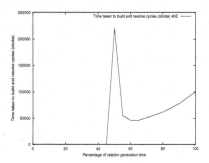

Fig. 6. Length of hypercycles (left) and time taken to resolve (right)

average length of cycles reduces quite quickly. The resulting full relations have considerably fewer entries, as noted by Fig. 7.

Here we show the number of nonzeros per relation for full relations resolved from 1, 2 and 3-partials (left) and 1, 2, 3 and 4-partials (right). Thus, had we processed 50% of the 1, 2, 3 and 4-partial relations for this dataset, the 3000 or so resulting fulls would have had an average of some 2600 nonzeros. Had we processed 60%, this average would have dropped to around 1000 nonzeros, while processing all relations would have lead to fulls with an average of around 400 nonzeros per row. Processing 1, 2 and 3-partials only, we see a similar effect, although the 'explosion' comes with slightly more relations than before. It therefore pays to go *beyond* the 'big bang' in resolving partial relations with more than two large primes, as the resulting full relations are considerably less dense and consequently take less time to resolve[5].

4   Impact on Linear Algebra

The linear algebra step of the index calculus method is not important asymptotically, but forms a practical bottleneck. To obtain the discrete logarithms of the factor base elements, we build a matrix with rows corresponding to full relations and columns corresponding to factor base elements (or vice versa). Nonzero entries occur only when a given factor base element occurs in a full relation, so the matrix is extremely sparse. Further, most of the nonzeros occur in the columns corresponding to smaller factor base elements. Techniques for solving such a system try to exploit this sparseness and structure. It is well known that full relations derived from partial relations lead to a denser matrix, and in general, one tries to avoid using the densest relations. Further, the absolute sizes of the

---

[5] Compared with other full relations, of course, these are still far more dense – for this dataset, the average full has some 15 nonzero entries, fulls via 1-partials have around 24, and fulls via 1 and 2-partials (cycles of length 3 or more) have around 45, compared with 160 and 405 nonzeros for fulls via 1, 2 and 3-partials and 1, 2, 3 and 4-partials respectively.

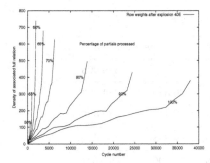 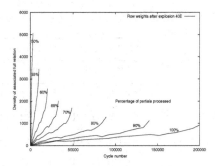

Fig. 7. Density of fulls resolved from 1, 2 and 3-partials (left) and using 1, 2, 3 and 4-partials (right)

coefficients in such relations are generally larger, which encourages the numerical blowup in pivoting strategies noted in [11].

Strategies have been developed to reduce the density of relations obtained via partials, and to reduce the size of the linear system prior to solving [3], [8], [14], [22]. The former builds on [7] and foregoes a graph-theoretic resolve to eliminate large primes; instead using linear algebra on both full and partial relations, over a factor base which now includes those large primes occurring with high frequency. Steps are taken to control the weight of the system and maximise the efficiency of the block Lanczos algorithm[6].

The index calculus method for discrete logarithms and its analogue for factoring now diverge, however. While it remains the case that we wish to minimise the time taken to solve the linear system, for the discrete log case we also want to solve for as many of the factor base elements as possible, to maximise the size of our database of known values. In general, not all factor base elements are represented in the matrix, even when one takes many excess rows as recommended in [14]. Using a certain amount of denser relations will allow us to increase this coverage of values. We can then recover more values and improve the performance of the second phase of the method; but this compromises the efficiency of methods for solving sparse linear systems, as shown in Table 1.

Here we used a basic structured Gaussian elimination (SGE)/Lanczos algorithm method [14] to solve a system of 19,783 full relations in 17,985 unknowns, where the 10% excess rows were either further direct fulls, or else fulls derived from 1, 2, 3 or 4-partials. Timings are in hundredths of a second. Adding these denser relations increases the number of values in our solution vector, but takes its toll on the time taken to resolve the system. We had hoped that the SGE routine would reject the small number of dense rows without taking too much

---

[6] In a discrete logarithm application, this approach would still suffer from failing to eliminate certain large primes modulo q for $q > 2$ (corresponding to the 'isolated' odd cycles). In this case, all but one large prime could be eliminated, and the number of unknowns in the system would increase by 1.

Table 1. Increased factor base coverage using fulls via partials – 40 digit p

17,985 fulls	Coverage	Loss	Nonzeros	Time to solve
+1,798 fulls	17,066	5.11%	270,996	330,908
+1,798 via 1P	17,194	4.40%	286,389	384,922
+1,798 via 2P	17,282	3.91%	305,490	442,635
+1,798 via 3P	17,435	3.06%	335,388	488,242
+1,798 via 4P	17,737	1.38%	450,496	502,723

more time, but that the increased coverage would allow us to recover more values. This, however, was not the case. We are better off concentrating on making the matrix as sparse as possible, as in factoring, and increasing coverage in a secondary step prior to computation, as we now discuss.

5   Discrete Logarithm Computation

Using large prime variants for discrete logarithm computation has an important secondary benefit which does not occur in factoring applications. For the final phase of the index calculus method – actual computation of an arbitrary discrete logarithm – one can make use of the 'partial' data to extend the database of known discrete logarithms. Assuming we have recovered the discrete logarithms of all factor base elements in the linear algebra step, we can back substitute over the 1-partial relations to obtain further values. We can subsequently back substitute over the other partial relations and recover still more values. As one would expect, this allows us to take considerably fewer attempts at finding a smooth value, as shown in Fig. 8. This also shows the effect of an incomplete solution vector for the factor base elements – we see that losing some 30% of the factor base values causes a doubling in the number of attempts needed to find a smooth element. Using an extended factor base allows us to improve dramatically on the number of attempts needed. It also gives us more resistance to any loss in the linear algebra step; but without having to resort to using denser rows in the matrix.

The maximum number of extra values we can find by using partials is simple to find in the case of 1-partials – it is the number of unique large primes in the dataset. This is easily found by computing

$$(\# \text{ 1-partials}) - (\# \text{ fulls resolved from 1-partials}) \qquad (10)$$

If, for some reason, an estimate is required before we have resolved the 1-partials, one can adapt the yield estimate from [16, sect. 3][7]. For the case of using 2-

---

[7] For further information on computing this approximation, see [2].

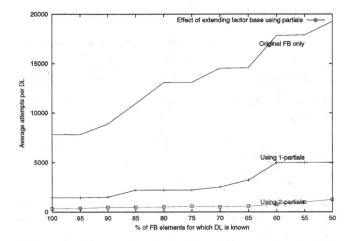

Fig. 8. Average attempts needed to compute a given discrete logarithm using extended factor base – 30 digit p

partials, if we assume that we have solved for all factor base elements and obtained the maximum possible further values from the 1-partials, then we should be able to obtain the logarithms of the large primes contained in all the 2-partial relations which form a connected component in the graph of both 1 and 2-partials (as used in resolving the 2-partial relations). Here, however, the datasets are *not* pruned beforehand, as in this situation 'singletons' are also of use. We note that, once we have back substituted over some set of partials to gain further known discrete logarithms, we can then back substitute over any partials for which we know the discrete logarithm of the large prime(s) but which contain as yet unknown factor base discrete logarithms. Further back substitution is of course also possible over the sets of 3 and 4-partials. As with using the 2-partials, processing is much slower since we need to take several passes through the data (in Table 2, we required 9 passes through the 2-partial data). This may be made more efficient by re-using the (hyper)graph structure used to resolve full relations, and then iterating only over the singletons dropped in pruning – timings in Table 2 are from extremely basic implementations.

In actual discrete logarithm computation, testing for smoothness using a trial division approach with such an extended factor base is obviously impractical, but we can of course use the same approach used in relation generation, by considering the remainder after removing factors from the original factor base. Table 2 shows the average number of attempts needed to compute an arbitrary logarithm modulo a 30 digit prime using the basic index calculus technique with trial division (TD), the single large prime variant (1PV) and the double large prime variant (2PV). Here we assume that we have solved for all factor base values. It was found that allowing at most one large prime per smoothness test, with a table lookup to determine the discrete logarithm of the large prime (if

Table 2. Effect of back substitution over partials – 30 digit p

	Fulls only	+1-partials	+2-partials
FB size	25,998	+719,261	+1,110,254
Time to solve	161,867	+3,105	+74,185
Attempts via TD	7816	1448	348
Time per attempt	0.29	12.62	31.13
Time per log	2,266	18,273	10,833
Attempts via 1PV	n/a	2578	1371
Time per attempt	n/a	0.29	0.29
Time per log	n/a	747	397
Attempts via 2PV	n/a	n/a	406
Time per attempt	n/a	n/a	1.30
Time per log	n/a	n/a	430

known), was faster than using the double large prime variant with a lookup on both primes due to the additional effort the latter entails in factoring the remainder. These examples are again too small for anything but illustration purposes – for larger datasets, the fewer attempts required by allowing two large primes should make such an approach more competitive.

6   Conclusions

Using 'large prime variant' index calculus techniques for discrete logarithm computation is rather more complicated than their use in factoring, due to the fact that we are working in a large finite ring rather than modulo 2. We have shown that, by considering the ordering of edges within cycles, it is possible to resolve single and double large prime relations without the need to solve a small linear system. By considering all possibilities in the graph representation of the partial relations, one can try to maximise the yield of the technique. In practice, we found that all cycles in such a graph could be resolved, but it remains the case that, unlike in factoring, certain situations – namely 'isolated' odd cycles – cannot be resolved for discrete logarithm computation without increasing the size of the linear system which must subsequently be solved.

Moving on to consider relations involving more than two large primes, we found empirically that the techniques of [10], [18] can be adapted such that all cycles which are suitable for factoring purposes can also be resolved for discrete logarithm computation. The downside of this technique is that it can create slightly longer cycles; so if one is using a maximum of two large primes per relation, it would probably be advisable to use the method of [16].

Subsequent consideration of the second phase of the index calculus method showed that using partial relations in a simple back substitution step allowed us to substantially increase our database of known values. Using these values, coupled with the single (rather than double or more, for the sizes implemented here) large prime variant in discrete logarithm computation, leads to substantial speedup – a factor of 5 or more for our examples – in computation. While large prime variant derived relations do allow more values to be recovered in the linear algebra step, the increased time needed to solve the system means that one is better off using as sparse a linear system as possible, as in factoring, and subsequently back substituting over partial relations to increase performance when computing an arbitrary discrete logarithm.

Acknowledgements. The authors would like to thank Dr. P. Leyland and Prof. D. Weber for their advice during the course of this study. All implementations were carried out in C++ using Victor Shoup's NTL library [24] and the GMP multiprecision package – we note that code was not optimised by hand, and could without doubt be made more efficient.

References

1. I. F. Blake, R. Fuji-Hara, R. C. Mullin, S. A. Vanstone, Computing Logarithms in Finite Fields of Characteristic Two, *SIAM Journal of Discrete Methods* **5** (1984) pp. 276–285
2. H. Boender, H. J. J. te Riele, Factoring Integers with Large Prime Variations of the Quadratic Sieve, *Technical Report NM-R9513* CWI Amsterdam (1995)
3. S. Cavallar, Strategies in Filtering in the Number Field Sieve, *Technical Report MAS-R0012* CWI Amsterdam (2000)
4. S. Cavallar, The Three Large Prime Variant of the Number Field Sieve, *Technical Report MAS-R0219* CWI Amsterdam (2002)
5. D. Coppersmith, Fast Evaluation of Logarithms in Fields of Characteristic Two, *IEEE Transactions on Information Theory* **IT-30** (1984) pp. 587–593
6. D. Coppersmith, A. M. Odlyzko, R. Schroeppel, Discrete Logarithms in GF(p), *Algorithmica* **1** (1986) pp. 1–15
7. J. Cowie, B. Dodson, R. M. Elkenbracht-Huizing, A. K. Lenstra, P. L. Montgomery, J. Zayer, A World Wide Number Field Sieve Factoring Record: On to 512 Bits, *Proc. AsiaCrypt '96* **LNCS 1163** (1996) pp. 382–394
8. T. Denny, V. Müller, On the Reduction of Composed Relations from the Number Field Sieve, *Proc. ANTS-II* **LNCS 1122** (1996) pp. 75–90
9. W. Diffie, M. E. Hellman, New Directions in Cryptography, *IEEE Transactions on Information Theory* **22** (1976) pp. 644–654
10. B. Dodson, A. K. Lenstra, NFS with Four Large Primes: An Explosive Experiment, *Proc. Crypto '95* **LNCS 963** (1995) pp. 372–385
11. D. M. Gordon, K. S. McCurley, Massively Parallel Computation of Discrete Logarithms, *Proc. Crypto '92* **LNCS 740** (1993) pp. 312–323
12. J. Hafner, K. S. McCurley, A Rigorous Subexponential Algorithm for Computation of Class Groups, *Journal of the American Mathematical Society* **2** (1989) pp. 837–850
13. D. E. Knuth, The Art of Computer Programming (vol II), *2nd edition* , Addison-Wesley (1981)

14. B. A. LaMacchia, A. M. Odlyzko, Solving Large Sparse Linear Systems Over Finite Fields, *Proc. Crypto '90* **LNCS 537** (1991) pp. 109–133
15. R. Lambert, Computational Aspects of Discrete Logarithms, *Ph.D. Thesis* University of Waterloo (1996)
16. A. K. Lenstra, M. S. Manasse, Factoring With Two Large Primes, *Proc. EuroCrypt '90* **LNCS 473** (1990) pp. 72–82
17. A. K. Lenstra, H. W. Lenstra (eds.), The Development of the Number Field Sieve, *1st edition*, Springer Verlag (1993)
18. P. Leyland, A. K. Lenstra, B. Dodson, A. Muffet, S. Wagstaff, MPQS with Three Large Primes, *Proc. ANTS-V* **LNCS 2369** (2002) pp. 446–460
19. P. Leyland, Personal communication, April 2003
20. K. S. McCurley, The Discrete Logarithm Problem, *Cryptology and Computational Number Theory, Proc. Symposium in Applied Mathematics* **42** (1990) pp. 49–74
21. A. M. Odlyzko, Discrete Logarithms in Finite Fields and their Cryptographic Significance, *Proc. EuroCrypt '84* **LNCS 209** (1985) pp. 224–314
22. C. Pomerance, J. W. Smith, Reduction of Huge, Sparse Linear Systems over Finite Fields via Created Catastrophes, *Experimental Mathematics* **1** (1992) pp. 89–94
23. O. Schirokauer, D. Weber, T. Denny, Discrete Logarithms: The Effectiveness of the Index Calculus Method, *Proc. ANTS-II* **LNCS 1122** (1996) pp. 337–361
24. V. Shoup, NTL – A Library for Number Theory, http://www.shoup.net/ntl/
25. E. Thomé, Computation of Discrete Logarithms in GF(2^{607}), *Proc. AsiaCrypt '01* **LNCS 2248** (2001) pp. 107–124
26. D. Weber, Computing Discrete Logarithms with the General Number Field Sieve, *Proc. ANTS-II* **LNCS 1122** (1996) pp. 391–403
27. D. Weber, Personal communication, February 2003

Computing the $M = UU^t$ Integer Matrix Decomposition

Katharina Geißler and Nigel P. Smart

Dept. Computer Science,
University of Bristol,
Merchant Venturers Building,
Woodland Road,
Bristol, BS8 1UB
{kathrin,nigel}@cs.bris.ac.uk

Abstract. The cryptanalysis of Gentry and Szydlo of the revised NTRU signature scheme requires the computation of the integer matrix decomposition $M = UU^t$. We propose a heuristic algorithm to compute this decomposition and investigate its properties. Our test implementation of this algorithm in Magma is able to deal with matrices up to 158 rows and columns.

Keywords: Lattice based cryptography, NTRU.

1 Introduction

The hardness of the shortest and closest vector problem in integer lattices can be used for cryptographic purposes and several encryption techniques have been proposed on that basis [11]. Among these the NTRU encryption system [8] appears to be the most practical and efficient. It is based on a special lattice, related to and defined by certain rings, which facilitates the easy description of the public and private keys, as well as efficient encryption and decryption.

The task of defining a signature scheme based on the NTRU lattice appears more difficult. The scheme NSS as proposed in [9] and its successor "R-NSS" defined in the CEES standard [4], have been proven insecure by Gentry and Szydlo [5]. Conceptually different from these schemes is the latest scheme NTRUSign [7] which is more clearly related to the closest vector problem in lattices. NTRUSign is itself closely related to the Goldreich, Goldwasser and Halevi (GGH) signature scheme [6]. However, NTRUSign allows much smaller key sizes and signing/verification speeds. The GGH signature scheme requires $N^2 \log N$ bits to specify both the keys and the signature, where as for NTRUSign these are only $N \log N$, where N is the dimension of the underlying lattice.

The basic idea is that a signer has to prove knowledge of the secret key, which is a "short" basis, by solving an approximate closest vector problem related to a message using this basis. The verifier on the other hand has to check that indeed the right approximate closest vector problem has been solved, by using his public

K.G. Paterson (Ed.): Cryptography and Coding 2003, LNCS 2898, pp. 223–233, 2003.
© Springer-Verlag Berlin Heidelberg 2003

key, which is a "long" basis. A first cryptanalysis of this scheme was carried out by Gentry and Szydlo in [5]. Besides various other results it was observed that if one could solve a certain matrix decomposition problem, then one could recover the private key given a sufficiently long transcript of signatures.

The precise matrix decomposition problem is as follows. Suppose M is a unimodular, symmetric positive definite matrix of dimension n. The matrix M is the Gram matrix of a basis, given by a unimodular matrix U, of the standard lattice $\mathbb{Z}^n$. As such we can write

$$M = UU^t.$$

Our goal is to write down such a matrix U. This problem of matrix factorisation is interesting in its own right, as it is closely related to the problem of lattice basis reduction and lattice distinction [12]. Since M is a Gram matrix of the standard lattice $\mathbb{Z}^N$ the problem is essentially asking for a change of basis matrix from the basis represented by M to the standard basis given by the identity matrix.

In this paper we investigate this matrix decomposition problem further. We use a heuristic algorithm to reduce coefficient sizes in addition to LLL lattice reduction. In [7] a different approach using lattice reduction is discussed and it is concluded that it too does not lead to an easier way to recover the secret key than previous lattice attacks on the public basis of the NTRU lattice. Our conclusion is that with current recommended parameters for NTRUSign the problem of matrix decomposition is hard, since extrapolation suggests that our method is not a threat for the recommended dimension of $n = 2N = 2 \cdot 251$.

2   NTRUSign and the $M = UU^t$ Decomposition

For details on NTRUSign and its cryptanalysis we refer to [5,7]. The public and private keys can be represented by two matrices with integer coefficients

$$B_{priv}, B_{pub} \in \mathbb{Z}^{2N \times 2N}$$

of a special "bicirculant" form. In addition, B_{pub} is in row upper Hermite normal form. For the precise definitions of these matrices see later

The matrices B_{priv} and B_{pub} are related by a unimodular matrix U via the formula

$$B_{pub} = UB_{priv}.$$

Our goal is to recover the matrix U, and hence determine the private matrix B_{priv}.

An "averaging" attack (cf. [5], section 4.3) on a transcript of signatures can provide us with $R = B_{priv}^t B_{priv}$, see [5] for details on how this is accomplished. This leads to the question, whether it is possible to determine B_{priv} given

$$R = B_{priv}^t B_{priv} \text{ and } B_{pub} = UB_{priv}.$$

This is equivalent to decomposing

$$
\begin{aligned}
UU^t &= (B_{pub}B_{priv}^{-1})(B_{pub}B_{priv}^{-1})^t \\
&= B_{pub}(B_{priv}^t B_{priv})^{-1} B_{pub}^t \\
&= B_{pub}R^{-1}B_{pub}^t \\
&=: M.
\end{aligned}
$$

Observe that such a U is uniquely determined up to multiplication by a signed permutation matrix from the right. For if $M = UU^t$ and $M = UT(UT)^t$ then $TT^t = 1$ and $T^{-1} = T^t$, i.e. T is orthogonal. But since T is integral and unimodular, it must be a signed permutation matrix.

We will need some properties of lattices and lattice reduction, so we recap on these here. A lattice is given by a set of $\mathbb{R}$-linear independent vectors $\{b_1, \ldots, b_m\}$, $b_i \in \mathbb{R}^n$, $(1 \le i \le m)$, and is defined to be the $\mathbb{Z}$-module

$$
L = \left\{ \sum_{i=1}^{m} a_i b_i : a_i \in \mathbb{Z} \right\} = \{Ba : a \in \mathbb{Z}^m\}.
$$

The set $\{b_1, \ldots, b_m\}$ is called the set of basis vectors and the matrix B is called the basis matrix. A basis $\{b_1, \ldots, b_m\}$ is called LLL reduced [10] if the associated Gram–Schmidt basis $\{b_1^*, \ldots, b_m^*\}$, $b_i^* := b_i - \sum_{j=1}^{i-1} \mu_{i,j} b_j^*$, $(1 \le i \le m)$ with $\mu_{i,j} := b_i b_j^* / b_j^* b_j^*$, $(1 \le j < i \le m)$ satisfies

$$
|\mu_{i,j}| \le \frac{1}{2} \text{ for } 1 \le j < i \le m,
$$
$$
\|b_i^*\|^2 \ge \left(\delta - \mu_{i,i-1}^2\right) \|b_{i-1}^*\|^2 \text{ for } 1 < i \le m, \tag{1}
$$

where δ is a constant chosen to satisfy $\frac{1}{4} \le \delta < 1$. The larger the value of δ the better the resulting lattice reduction, however this improvement comes at the expense of cost of the reduction.

If B is a basis matrix of a lattice, then the matrix BB^t is called the Gram matrix of the basis. The LLL algorithm can either be applied to a lattice basis or to the Gram matrix associated to a basis, in both cases it produces a new matrix which is "smaller" than the previous one, although the quality of the reduction decreases as the dimension increases.

One can phrase our problem above as given the Gram matrix M, find a change of basis matrix U such that

$$
UMU^t = I_n,
$$

in which case we have

$$
M = U^{-1}(U^{-1})^t.
$$

Hence, our problem is essentially one of lattice reduction. Repetition of lattice basis reduction with increasing values of δ seems to eventually always recover U, but this can take a large number of iterations. The goal of this paper is to heuristically reduce the number of such LLL iterations and therefore minimise the total amount of time.

3   The Algorithm

In this section we describe the strategy and techniques which lead to our algorithm. The basic intuition of our approach is to apply LLL reduction, modify and reorder the basis of the Gram matrix M and to repeat this procedure. The fact that the corresponding lattice is completely reducible, e.g. M is a Gram matrix of the standard lattice $\mathbb{Z}^N$ will render this approach successful.

In order to describe our algorithm we need a few tools from matrix theory. We recall that matrix multiplication from the right (left) yields an operation on the columns (rows) of the given matrix. Especially we are interested in multiplications by elementary matrices of $\mathbb{Z}^{n \times n}$. For an easy description we introduce the following matrix types $(1 \leq i, j \leq n)$.

The matrix E_{ij} contains exactly one entry 1 in column j and row i, otherwise zeros; hence

$$I_n := \sum_{i=1}^{n} E_{ii}$$

$$S_{ij} := \sum_{k=1, k \neq i, j}^{n} E_{kk} + E_{ij} + E_{ji} \ (i \neq j)$$

$$T_{ij}(a) := I_n + a E_{ij} \ (a \in \mathbb{Z}, \ i \neq j)$$

Multiplication of a matrix $M \in \mathbb{Z}^{n \times n}$ from the right (left) by S_{ij} swaps the columns (rows) i and j. Multiplication of a matrix M from the right (left) by $T_{ij}(a)$ adds a-times column i (row j) to column j (row i). Moreover, the inverse and transpose matrices of $S_{ij}, T_{ij}(a)$ are easily seen to be

$$S_{ij}^{-1} = S_{ij}^{t} = S_{ij}, \ T_{ij}(a)^{-1} = T_{ij}(-a) \text{ and } T_{ij}(a)^{t} = T_{ji}(a).$$

We use these matrices and other unimodular transformations to transform the given matrix M into the identity matrix, in a sequence of steps. Denote by $U_i, (i \in \mathbb{N}_{>0})$ the unimodular transformations applied in each step, this yields

$$M_0 = M,$$
$$M_1 = U_1 \cdot M_0 \cdot U_1^t,$$

$$\vdots \quad \vdots$$

$$M_{n-1} = U_{n-1} \cdot M_{n-2} \cdot U_{n-1}^t,$$
$$M_n = U_n \cdot M_{n-1} \cdot U_n^t = I_n,$$

Recursively replacing the M_is results in

$$I_n = U_n \cdot M_{n-1} \cdot U_n^t$$
$$= (U_n \cdots U_1) \cdot M \cdot (U_1^t \cdots U_n^t)$$

and it follows that

$$M = (U_n \cdots U_1)^{-1} \cdot (U_1^t \cdots U_n^t)^{-1}$$
$$= (U_n \cdots U_1)^{-1} \cdot ((U_n \cdots U_1)^{-1})^t$$
$$= UU^t,$$

where
$$U = (U_n \cdots U_1)^{-1}.$$

So the question is, how do we apply these transformations? The goal appears to be to produce a matrix whose entries are as small as possible, i.e. the identity matrix. Hence, the following heuristic strategy seems to be a good one. We apply the following ideas in turn, returning to the start once a single operation is performed.

1. Apply the S_{ij} for various values of i and j so as to make the matrix have diagonal entries in increasing order.
2. If the absolute value of the entries m_{ji}, $(1 \leq i \leq n - 1, i + 1 \leq j \leq n)$ (e.g. the entries below the diagonal) are equal or bigger then the m_{ii} reduce them by an appropriate multiple $a \in \mathbb{Z}$ of m_{ii}. In other words apply $T_{ij}(a), T_{ij}(a)^t$ to the matrix M. The effect is to keep the entries of the matrix as small as possible. If $m_{ii} = 1$ we can reduce m_{ji} to zero.
3. Compute the LLL reduced Gram matrix of M in order to try to minimise the entries on the diagonal.

Part 2 of the above strategy leads to the following definition.

Definition 1. *We call a symmetric matrix*
$$M = (m_{ij})_{1 \leq i,j \leq n} \in \mathbb{Z}^{n \times n}$$

diagonally-reduced, if and only if
$$|m_{ji}| < |m_{ii}|, \ (1 \leq i \leq n - 1, i + 1 \leq j \leq n).$$

Before we state the actual algorithm consider the following example which just involves parts 1 and 2 of the above strategy. Suppose we are given
$$M = \begin{pmatrix} 6 & 7 & 1 \\ 7 & 13 & -7 \\ 1 & -7 & 14 \end{pmatrix}.$$

Then we apply $T_{21}(-1)$ since $\lceil -6/7 \rfloor = -1$, and S_{12} to obtain
$$M_2 = S_{12} T_{21}(-1) M T_{21}^t S_{12}^t = \begin{pmatrix} 5 & 1 & -8 \\ 1 & 6 & 1 \\ -8 & 1 & 14 \end{pmatrix}.$$

Then we apply, since $\lceil 8/5 \rfloor = 2$, $T_{31}(2)$ and then $S_{13}, S_{23}, T_{21}(-1), T_{31}(-1)$ and S_{23} to obtain
$$M_8 = \begin{pmatrix} 2 & 1 & 0 \\ 1 & 2 & -2 \\ 0 & -2 & 3 \end{pmatrix}.$$

We now apply $T_{32}(1), S_{13}, T_{31}(-1), S_{23}$ and finally $T_{32}(-1)$ to obtain

$$M_{13} = \begin{pmatrix} 1 & 0 & 0 \\ 0 & 1 & 0 \\ 0 & 0 & 1 \end{pmatrix}.$$

Hence, we have

$$U = (\, T_{32}(-1) \cdot S_{23} \cdot T_{31}(-1) \cdot S_{13} \cdot T_{32}(1) \cdot S_{23} \cdot T_{31}(-1) \cdot$$
$$T_{21}(-1) \cdot S_{23} \cdot S_{13} \cdot T_{31}(2) \cdot S_{12} \cdot T_{21}(-1)\,)^{-1}$$
$$= \begin{pmatrix} 1 & 2 & 1 \\ 3 & 2 & 0 \\ -3 & 1 & 2 \end{pmatrix}.$$

It is readily checked that

$$UU^t = M.$$

We now explain the full algorithm in detail. Essentially our heuristic philosophy is to use the above naive technique to aid the LLL algorithm in trying to reduce the Gram matrix to the identity.

Algorithm 2 *($M = UU^t$)*

Input: An unimodular, symmetric positive definite matrix

$$M = (m_{ij})_{1 \le i,j \le n} \in \mathbb{Z}^{n \times n}.$$

Output: Unimodular matrix $U \in \mathbb{Z}^{n \times n}$ with $M = UU^t$.

1. *(Initialise) $n \leftarrow rank(M)$, $U \leftarrow I_n$.*

2. *(Hermite normal form) Compute the Hermite normal form $H \in \mathbb{Z}^{n \times n}$ of M and an invertible transformation matrix $T = (t_{ij})_{1 \le i,j \le n} \in \mathbb{Z}^{n \times n}$ with $T \cdot M = H$*

3. *(Choose T?) If $\prod_{i=1}^{n} t_{ii} < \prod_{i=1}^{n} m_{ii}$ then*

$$M_{new} = (m_{ij})_{1 \le i,j \le n} \leftarrow T \text{ and } inv \leftarrow 1$$

   else

$$M_{new} = (m_{ij})_{1 \le i,j \le n} \leftarrow M \text{ and } inv \leftarrow 0$$

4. *(LLLGram) Compute the LLL-reduced Gram matrix G of M_{new} with $\delta := 0.75$ and a unimodular matrix $T \in \mathbb{Z}^{n \times n}$ with $G = T \cdot M_{new} \cdot T^t$. Set*

$$M_{new} \leftarrow G \text{ and } U \leftarrow T \cdot U.$$

5. (Main Loop) While $M_{new} \neq I_n$ do

   5.1 (Arrange diagonal entries) Apply the unimodular transformations S_{ij} so as to arrange the diagonal entries of M_{new} in increasing order. Set

   $$M_{new} \leftarrow S_{ij} \cdot M_{new} \cdot S_{ij} \text{ and } U \leftarrow S_{ij} \cdot U.$$

   5.2 (Reduce matrix) While M_{new} is not diagonally-reduced, apply suitable unimodular transformations $T_{ij}(a)$, set

   $$M_{new} \leftarrow T_{ij}(a) \cdot M_{new} \cdot T_{ij}(a)^t \text{ and } U \leftarrow T_{ij}(a) \cdot U, \, (a \in \mathbb{Z}).$$

   Arrange the diagonal entries of M_{new} in increasing order, by applying, for suitable choices of S_{ij},

   $$M_{new} \leftarrow S_{ij} \cdot M_{new} \cdot S_{ij} \text{ and } U \leftarrow S_{ij} \cdot U.$$

   5.3 (Further reduction of diagonal entries) If it is possible to reduce the diagonal entries of M_{new} by applying $T_{ij}(a)$ for $a = \pm 1$ by setting

   $$M_{new} \leftarrow T_{ij}(a) \cdot M_{new} \cdot T_{ij}(a)^t \text{ and } U \leftarrow T_{ij}(a) \cdot U$$

   and return to the start of the while loop.

   5.4 Set $d \leftarrow \prod_{i=1}^{n} m_{ii}$, $d_{new} \leftarrow d$.

   5.5 While $M_{new} \neq I_n$ and $d_{new} \geq d$ do

   a) Compute LLL-reduced Gram matrix G of M_{new} with $\delta := 0.99$ and a unimodular matrix $T \in \mathbb{Z}^{n \times n}$ with $G = T \cdot M_{new} \cdot T^t$. Set

   $$M_{new} \leftarrow G \text{ and } U \leftarrow T \cdot U.$$

   b) Produce a diagonally reduced matrix as in step 5.2 and try to reduce diagonal entries of $M_{new} = (m_{ij})_{1 \leq i,j \leq n}$ as in step 5.3.

   c) Set $d_{new} \leftarrow \prod_{i=1}^{n} m_{ii}$.

6. (Reduced T?) If $inv = 1$ then set $U = U^t$, else if $inv = 0$ then set $U = U^{-1}$.

7. (End) Return U.

Some remarks are in order here:

1. The given matrix $M = (m_{ij})_{1 \leq i,j \leq n} \in \mathbb{Z}^{n \times n}$ will always be positive definite, since it is the Gram matrix of the standard scalar product. We also note that $m_{ii} > 0$, $(1 \leq i \leq n)$.

2. Step 2 is important to overcome problems relating to the size of the entries of the given matrix, which would slow down computations considerably. The Hermite normal form computation essentially computes the inverse of the matrix M. Due to Proposition 1 below, we can use the inverse of M in place of M within the calculation if it is advantageous to do so. In turns out that often the inverse has much smaller entries, which enables us to deal with larger dimensions than would otherwise be the case.

3. The δ constant used in the LLL-reduced Gram matrix computation is usually set to be 0.75, see [2], Algorithm 2.6.3, Step 3 or equation (1) of this paper, but can be replaced by any real number δ in the range $0.25 < \delta < 1$. A large value of δ tends to yield a higher quality result while taking extra time; whereas a small value of δ will give lower quality while taking less time. In practice we have obtained the best running times if the constant δ is varied in the course of the algorithm in the following way: starting the reduction with a lower constant, so that the reduction is relatively fast and increasing it to $\delta = 0.99$ during the following steps, so that the quality of the reduced basis is as good as possible.

4. During the whole computation (see step 5.2) we keep the matrix always in diagonally-reduced form with the values of the diagonal entries increasing. The value $a \in \mathbb{Z}$ is chosen in a way that $|m_{ji}| - a \cdot m_{ii}$ is positive and has smallest possible value. This has the following positive side effect, that the number of required LLL-reduced Gram matrix computations will decrease considerably. For example in dimension $n = 142$ ($N = n/2 = 71$) the number of LLL-reduced Gram matrix computations was decreased by 4606 computations.

For a given integer matrix A with $\det(A) \neq 0$ the Hermite Normal form H is a matrix $H = (h_{ij})_{1 \leq i,j \leq n} \in \mathbb{Z}^{n \times n}$ with $h_{ii} > 0$ and $h_{ij} = 0, (1 \leq i \leq n, \ j < i \leq n)$. It is unique (provided we also insist on the condition $0 \leq h_{ij} < h_{ii}, (1 \leq i < j \leq n)$) and can be computed along with a unimodular transformation matrix $T \in \mathbb{Z}^{n \times n}$ with $T \cdot A = H$. See [2] for details.

Proposition 1. *Let $M = (m_{ij})_{1 \leq i,j \leq n} \in \mathbb{Z}^{n \times n}$ be a unimodular, symmetric positive definite matrix. Then the Hermite Normal Form satisfies $H = I_n$ and $T = M^{-1}$ and T is a unimodular, symmetric positive definite matrix.*

Proof. Since H is an upper triangular matrix with $h_{ii} > 0$, we have $\det(T) \cdot \det(M) = \det(H) = 1$. According to the uniqueness condition of H it follows $H = I_n$. From Cramer's rule for matrix inversion it is easily deduced that $T = M^{-1}$ is a symmetric matrix. Since M is positive definite, the matrix M^{-1} is also positive definite according to the spectral theorem for Euclidean vector spaces.

4   Timings

We first introduce some notation for the NTRUSign algorithm.

– The main security parameter is an integer N which is public and chosen to be prime.

– Two other public integer parameters are p and q where q is coprime to N, and is usually chosen to be a power of two, and p is coprime to q and is usually chosen to be three.
– Operations take place on polynomials in the rings

$$R = \mathbb{Z}[X]/(X^N - 1) \text{ and } R_q = \mathbb{Z}_q[X]/(X^N - 1).$$

Multiplication in these rings we denote by $\otimes$, the precise context in which we use $\otimes$ we shall make clear.
– We also require other integers d_u, d_g and d_f to define various sets of polynomials.

If d_1 and d_2 are integers then we define $\mathcal{L}(d_1, d_2)$ to be the set of polynomials in R_q, with d_1 coefficients equal to 1, d_2 coefficients equal to -1 and all other coefficients are set to zero.

To define the public and private keys we set

$$f = u + p \cdot f_1,$$
$$g = u + p \cdot g_1,$$

where

$$u \in \mathcal{L}(d_u, d_{u+1}),$$
$$f_1 \in \mathcal{L}(d_f, d_f),$$
$$g_1 \in \mathcal{L}(d_g, d_g)$$

for our choices of d_u, d_f and d_g. From f and g one also derives polynomials F and G such that

$$f \otimes G - g \otimes F = q$$

as an equation in R. Note, that F and G are not unique.

The tuple (f, g, F, G) represents the private basis and the polynomial h is the public key, where

$$f \otimes h = g$$

as an equation in R_q.

The matrices $B_{priv}, B_{pub} \in \mathbb{Z}^{2N \times 2N}$, mentioned earlier, are of the form

$$B_{priv} = \begin{pmatrix} M_f & M_g \\ M_F & M_G \end{pmatrix} \text{ and } B_{pub} = \begin{pmatrix} I_N & M_h \\ 0 & qI_N \end{pmatrix},$$

The $N \times N$ matrix M_f is derived from the polynomial

$$f = \sum_{i=0}^{N-1} f_i x^i$$

by placing the coefficients $f_0, \ldots f_{N-1}$ of f in the first row and then rotating each row to the right by one place to obtain the next row. The same method is used to produce M_g, M_F, M_G and M_h. This cyclic structure of the matrices arises from the convolution nature of the $\otimes$ operation.

Table 1 shows the results from Algorithm 2. Column 6 in Table 1 denotes the number of Gram LLL reductions to obtain the matrix U from

$$UU^t$$

using Algorithm 2; the last column gives the number of Gram LLL reductions to solve the UU^t problems using only the Gram LLL algorithm without further improvements. All tests of our algorithm were run on a 1.5GHz machine, Linux Redhat 7.1, using the computer algebra package Magma [1,3].

Table 1. Gram LLL reductions needed to decompose the matrix $M = UU^t$

n	N=n/2	p	q	d_u	d_{f_1}	d_{g_1}	Number of LLL red. using Algorithm 2	Number of LLL red. using only LLL
106	53	3	64	26	26	25	13	16
118	59	3	64	29	26	25	58	93
122	61	3	64	30	26	25	134	142
134	67	3	64	32	32	31	968	1980
142	71	3	64	32	32	31	3013	7619
146	73	3	64	32	32	31	10880	32378
158	79	3	64	32	32	31	75154	261803

Assuming the trend shown in the previous table continues for higher key sizes and using an optimistic assumption as to how fast one can perform LLL at higher dimensions, we predict how long it will take to break more higher security levels of NTRUSign using the decomposition of the matrix $M = UU^t$. We estimate the time, in seconds, to apply our method to a key of size N as

$$1.3 \cdot \exp\left[0.22 \cdot (2N - 106.9)\right]$$

Hence, our estimates for the effort required to break larger keys can be found in Table 2. We conclude that attacking NTRUSign via decomposing $M = UU^t$ is harder than trying to attack the public key directly.

Table 2. Estimated time needed to decompose a matrix at higher security levels

Security	Time to compute $M = UU^t$
Low (N=107, q=64)	714y
Moderate (N=167, q=128)	$2.08 \cdot 10^{14}$y
Standard (N=251, q=128)	$8.45 \cdot 10^{32}$y
High (N=347, q=128)	$1.86 \cdot 10^{51}$y
Highest (N=503, q=256)	$1.20 \cdot 10^{81}$y

5   Conclusion

We have presented a heuristic method which enables the LLL algorithm to solve the matrix decomposition problem $M = UU^t$ for higher dimensions than would otherwise be the case. The main improvement is to keep the diagonal elements in increasing order and to use the input matrix or its inverse as input to the whole procedure. This led to a decrease in the number of times we had to apply the Gram LLL reduction. Our experiment shows that NTRUSign keys of moderate security can be considered secure.

References

1. W. Bosma and J. Cannon and C. Playoust. *The Magma algebra system I: The user language*, J. Symbolic Comput., **24**, 235–265, 1997.
2. H. Cohen. *A Course in Computational Algebraic Number Theory*, Springer-Verlag, 1996.
3. Computational algebra group, Uni. Sydney. Magma, 2002.
   See http://www.maths.usyd.edu.au:8000/u/magma/.
4. Consortium for Efficient Embedded Security. Efficient Embedded Security Standard (EESS).
   See http://www.ceesstandards.org/.
5. C. Gentry and M. Szydlo. Cryptanalysis of the Revised NTRU Signature Scheme In *Advances in Cryptology - EUROCRYPT 2002*, Springer-Verlag LNCS 2332, 299–320, 2002.
6. O. Goldreich, S. Goldwasser and S. Halevi. Public-Key Cryptosystems from Lattice Reduction Problems. In *Advances in Cryptology - CRYPTO '97*, Springer-Verlag LNCS 1294, 112–131, 1997.
7. J. Hoffstein, N. Howgrave-Graham, J. Pipher, J. H. Silverman, W. Whyte. NTRUSign: Digital Signatures Using the NTRU Lattice, 2002.
   See http://www.ntru.com/.
8. J. Hoffstein, J. Pipher, J. H. Silverman. NTRU: A ring-based public key cryptosystem. In *Algorithmic Number Theory, ANTS-III*, Springer-Verlag LNCS 1423, 267–288, 1998.
9. J. Hoffstein, J. Pipher, J. H. Silverman. NSS: The NTRU Signature Scheme. In *Advances in Cryptology - EUROCRYPT 2001*, Springer-Verlag LNCS 2045, 211–228, 2001.
10. A.K. Lenstra, H.W. Lenstra, and L. Lovász. Factoring polynomials with rational coefficients. *Math. Ann.*, **261**, 515–534, 1982.
11. J. Stern and P. Nguyen. Lattice reduction in cryptology: an update. In *Algorithmic Number Theory - ANTS IV*, Springer-Verlag LNCS 1838, 85–112, 2000.
12. M. Szydlo. Hypercubic Lattice Reduction and Analysis of GGH and NTRU Signatures. In *Advances in Cryptology - EUROCRYPT 2003*, Springer-Verlag LNCS 2656, 433–448, 2003.

Cryptanalysis of the Public Key Cryptosystem Based on the Word Problem on the Grigorchuk Groups

George Petrides*

Department of Mathematics, UMIST
P.O. Box 88, Sackville Street, Manchester M60 1QD, UK.
george.petrides@student.umist.ac.uk

Abstract. We demonstrate that the public key cryptosystem based on the word problem on the Grigorchuk groups, as proposed by M. Garzon and Y. Zalcstein, is insecure. We do this by exploiting information contained in the public key in order to construct a key which behaves like the private key and allows successful decryption of ciphertexts. Before presenting our attack, we briefly describe the Grigorchuk groups and the proposed cryptosystem.

1 Introduction

In 1991, M. Garzon and Y. Zalcstein proposed a public key cryptosystem based on the word problem on the Grigorchuk groups [1]. This cryptosystem was similar to a cryptosystem proposed in 1984 by Wagner and Magyarik [5] and was the first application of the Grigorchuk groups in cryptography.

1.1 A Quick Introduction to the Grigorchuk Groups

Leaving the details of the construction of the Grigorchuk groups until Sect. 2, we mention only that the Grigorchuk groups are infinite, and there are infinitely many of them: each group, denoted G_ω, is specified by a ternary sequence $\omega \in \{0, 1, 2, \}^{\mathbf{N}}$. The elements of G_ω are given as words in terms of four generators a, b_ω, c_ω and d_ω. There is an efficient and easy to implement algorithm (described in Sect. 2.2), which decides whether two words F_1 and F_2 are equal in G_ω, or, equivalently, whether $F = F_1 F_2^{-1} = 1$. In such case the word F is called a *relator* in G_ω. Notice that only $\lceil \log_2 n \rceil$ digits of the defining sequence ω are needed for checking whether a word of length n is a relator (Theorem 2).

The sets of relators are different for different sequences ω. Furthermore, if at least two of the symbols $0, 1, 2$ are repeated in ω infinitely often, the set of relators in G_ω cannot be deduced from any of its finite subsets [2, Theorem 6.2].

---

* The author was partially sponsored by the ORS Awards Scheme.

K.G. Paterson (Ed.): Cryptography and Coding 2003, LNCS 2898, pp. 234–244, 2003.

1.2   The Garzon–Zalcstein Cryptosystem

Description. The public key cryptosystem proposed in [1] is the following:

Alice chooses an efficient procedure which generates a ternary sequence ω such that at least two of its digits repeat infinitely often. Recall that in this case G_ω is not finitely presentable. This sequence ω is her **private key**.

She then publishes a finite subset of relators (that is words equal to 1 in G_ω) and two words w_0, w_1 representing distinct group elements of G_ω. These comprise the **public key**.

Bob encrypts a bit i, $i \in \{0, 1\}$, of his message as a word w_i^* obtained from w_i by a sequence of random additions and/or deletions of relators as found in the public key. Concatenation of the encrypted bits yields the encrypted message (with a separator between the encryption of successive bits). He then sends the encrypted message to Alice.

Alice decrypts the message by checking whether w_i^* is equal to w_0 or w_1. This can be done by checking whether, for example, $w_0^{-1} w_i^* = 1_{G_\omega}$ using the algorithm from Sect. 2.2.

Security Claims. When discussing the security of their cryptosystem, Garzon and Zalcstein [1, Sect. 4] claim that the public key will not contain enough information to uniquely determine the private key ω. They also claim that in order to establish in polynomial time that a guess of a key is correct, it is necessary to follow a chosen plaintext attack.

Cryptanalysis. In this paper we refute this claim. In particular we show that the words used as public key, in combination with an algorithm presented in Sect. 3, give enough information to allow construction of a key which behaves like the private key and is capable of successfully decrypting ciphertexts.

Cryptanalysis of this cryptosystem was first attempted by Hofheinz and Steinwandt in [4]. In their paper, they tried to exploit the public key to derive a secret key equivalent to the original one by brute-force search. However, not much detail was given.

In order to make this paper as self-contained as possible, first we will briefly describe the Grigorchuk groups as defined in [2] together with the word problem solving algorithm. In Sect. 3, we state and prove some results, which we use to describe our cryptanalysis. We finish by presenting an example of how the cryptanalysis works.

2   Grigorchuk Groups

This section gives a brief description of the Grigorchuk groups and their properties that we are going to need.

2.1   Definition of Grigorchuk Groups.

The Grigorchuk groups G_ω are certain groups of transformations of the infinite binary tree.

Denote by T the set of one way infinite paths from the root $\emptyset$ of the complete infinite binary tree. Each $j \in T$ can be regarded as an infinite binary sequence $(j_i)_{i \leq 1}$ of vertices. We refer to these as left turns (denoted by 0) and right turns (denoted by 1), according to orientation from the parent. The empty sequence denotes the root vertex $\emptyset$.

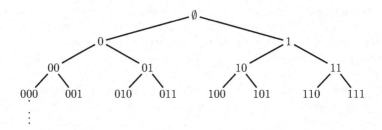

Fig. 1. The first three levels of the infinite binary tree

Given an infinite ternary sequence ω, the group G_ω is a group of permutations of T generated by four bijections a, b_ω, c_ω and d_ω. The action of a on a path $j \in T$ consists of complementing the first turn (that is making 0 into 1 and vice versa). The actions of b_ω, c_ω and d_ω depend on the sequence ω as follows:

We form three sequences U, V and W by substituting each digit of ω with a vector as shown below, depending on whether it is 0,1 or 2:

$$0 : \begin{cases} S \\ S \\ I \end{cases} \qquad 1 : \begin{cases} S \\ I \\ S \end{cases} \qquad 2 : \begin{cases} I \\ S \\ S \end{cases}$$

Here I = "Identity" and S = "Swap"; the latter means making 0 into 1 and vice versa.

We thus obtain three sequences:

$$U = u_1 u_2 \ldots u_n \ldots$$
$$V = v_1 v_2 \ldots v_n \ldots$$
$$W = w_1 w_2 \ldots w_n \ldots$$

The bijection b_ω (resp. c_ω, d_ω) leaves invariant all turns $j_1, \ldots, j_i$ including the first left turn j_i of j and complements j_{i+1} if u_i (resp. v_i, w_i) is S. Otherwise it leaves j invariant.

For example, if $\omega = 012012012\ldots$ then

$$U = SSISSISSI\ldots$$
$$V = SISSISSIS\ldots$$
$$W = ISSISSISS\ldots$$

and $b_\omega(1,1,1,0,1,\ldots) = 1,1,1,0,0,\ldots$ since $u_4 = S$.

It is clear that all four generators are involutions, that is

$$a^2 = b_\omega^2 = c_\omega^2 = d_\omega^2 = 1 \ . \tag{1}$$

We also have that

$$b_\omega c_\omega = c_\omega b_\omega = d_\omega, \ b_\omega d_\omega = d_\omega b_\omega = c_\omega \text{ and } c_\omega d_\omega = d_\omega c_\omega = b_\omega \ . \tag{2}$$

Therefore, any one of the generators b_ω, c_ω and d_ω can be dropped from the generating set $\langle a, b_\omega, c_\omega, d_\omega \rangle$.

Notice that every Grigorchuk group G_ω is a canonical homomorphic image of the *basis* group G, generated by four elements $a, b_\omega, c_\omega, d_\omega$ satisfying only the relations (1) and (2). When a sequence ω is specified, this introduces extra relations between the generators $a, b_\omega, c_\omega, d_\omega$ and maps G onto the Grigorchuk group G_ω.

The Grigorchuk groups are infinite and if at least two of the symbols 0,1 and 2 repeat infinitely often in ω then G_ω is not finitely presentable; that is, one needs infinitely many independent relators to define G_ω [2, Theorem 6.2].

In the cases when ω is periodic G_ω can be defined by a finite state automaton, which allows for an especially streamlined implementation. See [3, Sect. 4] for an example of the case $\omega = 012012012\ldots$.

2.2   Word Problem

A word in G is any product of the four generators $a, b_\omega, c_\omega, d_\omega$ and represents an element of G. Let F be such a word and let $\partial(F)$ denote the length of the word F and $\partial_k(F)$ (resp. $\partial_{p,q}(F)$ etc) denote the number of occurrences of symbol k (resp. both symbols p and q etc) in the word F (k, p and $q \in \{a, b_\omega, c_\omega, d_\omega\}$). Due to (1) and (2), each word can be uniquely reduced to 1_{G_ω}, $*$, or the form

$$* a * a \ldots a * a * \ . \tag{3}$$

where $* \in \{b_\omega, c_\omega, d_\omega\}$ and $*$ at the beginning or end of the word can be absent. We call words of this form *reduced*.

By word problem we mean finding whether a given word F in G_ω is equal to 1 or not. The word problem is solvable when the sequence ω is recursive [2, Sect. 5] and we shall shortly give a description of the algorithm used to solve it.

Words equal to 1 are called *relators*. By definition, relators act trivially on any path $j \in T$. Recalling the action of generator a on such a path j, we can deduce that all relators must have an even number of occurrences of the generator a.

Let σ denote the left shift operator defined on a sequence $\omega = \omega_1 \omega_2 \omega_3 \omega_4 \ldots$ by $\sigma(\omega) = \omega_2 \omega_3 \omega_4 \ldots$. We define the sequence of groups G_n, $n = 1, 2, \ldots$ by $G_n = G_{\sigma^{n-1}(\omega)}$ and denote by a, b_n, c_n and d_n the respective generators. In particular $G_1 = G_\omega$.

For each n, consider the subgroup H_n of G_n consisting of all elements representable by a word with an even number of occurrences of the generator a. Then $|G_n : H_n| = 2$ and H_n is generated by $b_n, c_n, d_n, ab_n a, ac_n a$ and $ad_n a$. Every element of H_n leaves setwise invariant the left and right halves of the tree.

Restricting the action of H_n from the whole of the tree to the two halves induces two homomorphisms, $\phi_0^{(n)}$ and $\phi_1^{(n)}$ of H_n. These act on the elements of H_n according to Table 1 and play a key role in the word problem solving algorithm.

Table 1. The actions of $\phi_0^{(n)}$ and $\phi_1^{(n)}$. Here $\tilde{u}_n$ (resp. $\tilde{v}_n, \tilde{w}_n$) denotes a if the nth digit of the sequence U (resp. V, W) is S, and 1 otherwise

	b_n	c_n	d_n	$ab_n a$	$ac_n a$	$ad_n a$
$\phi_0^{(n)}$	$\tilde{u}_n$	$\tilde{v}_n$	$\tilde{w}_n$	b_{n+1}	c_{n+1}	d_{n+1}
$\phi_1^{(n)}$	b_{n+1}	c_{n+1}	d_{n+1}	$\tilde{u}_n$	$\tilde{v}_n$	$\tilde{w}_n$

Word Problem Solving Algorithm (W.P.S.A.). Given a word F and a sequence ω, in order to decide whether F is equal to 1 in G_ω we do the following:

First Round

Step 1. Find $\partial_a(F)$. If it is odd *then* $F \neq 1_{G_\omega}$ and the algorithm **terminates**. *Otherwise* proceed to step **2**.

Step 2. Reduce F to obtain F_r^1 (the index denotes the current round). If it is equal to 1_{G_ω} then so is word F and the algorithm **terminates**. *Otherwise* proceed to step **3**. Note that $\partial_a(F_r^1)$ will also be even.

Step 3. Apply $\phi_0^{(1)}$ and $\phi_1^{(1)}$ to F_r^1 to obtain two words F_0^2 and F_1^2 of length at most $\left\lceil \frac{\partial(F_r)}{2} \right\rceil$ (see Theorem 1, Sect. 3).

Step 4. Proceed to next round.

i^{th} Round

Each word from the previous round which is not equal to 1 yields, in its third step, two new words. Therefore, at most 2^i words are obtained to be used in this round. However, the total length of these words is bounded by the length of the word they originated from, which is less than the length of the input word F.

Step 1. Find $\partial_a(F_0^i), \ldots, \partial_a(F_k^i)$, $k \leq 2^i$. *If* any one of them is odd *then* $F \neq 1_{G_\omega}$ and the algorithm **terminates**. *Otherwise* proceed to step **2**.

Step 2. Reduce $F_0^i, \ldots, F_k^i$ $(k \leq 2^i)$ to obtain $F_{r_0}^i, \ldots, F_{r_k}^i$. For each one of them which is equal to 1 or $*$ $(* \in \{b_\omega, c_\omega, d_\omega\})$, an *end node* has been reached. *If* an end node has been reached for all k words, the algorithm **terminates**.

Step 3. Apply $\phi_0^{(i)}$ and $\phi_1^{(i)}$ to any of $F_0^i, \ldots, F_k^i$ for which an end node has not been reached yet, to obtain at most 2^{i+1} words $F_{r_0}^{i+1}, \ldots, F_{r_{2k}}^{i+1}$. These will have lengths at most $\left\lceil \frac{\partial(F_0^i)}{2} \right\rceil, \ldots, \left\lceil \frac{\partial(F_k^i)}{2} \right\rceil$ respectively.

Step 4. Proceed to the next round.

Since the lengths of the words obtained are decreasing, the algorithm will eventually terminate.

While running the algorithm, a finite tree will be created having the word F as root and depth at most $\lceil \log_2 \partial(F) \rceil$ (see Theorem 2, Sect. 3). Each round of the algorithm corresponds to a different level of the tree. Left branching occurs after application of $\phi_0^{(i)}$ and right branching after application of $\phi_1^{(i)}$ on words on the ith level, with the resulting words as vertices. The word F will be equal to 1_{G_ω} if and only if all the end nodes of the tree are equal to 1.

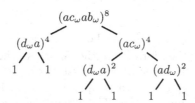

Fig. 2. The finite tree obtained after applying the W.P.S.A. on the word $(ac_\omega ab_\omega)^8 = 1_{G_\omega}$, $\omega = 012012\ldots$

A feasible implementation of this algorithm constructs the finite tree depthwise instead of level-wise, a branch at a time. In this way only $k \leq \lceil \log_2 \partial(F) \rceil$ words need to be stored instead of 2^k. The time complexity of the algorithm is $O(n \log n)$ in terms of the length $n = \partial(F)$ of the input word F.

Further information on the material in this subsection can be found in [2].

3   Cryptanalysis of the Cryptosystem

In this section we first prove some results which we later use to describe our cryptanalysis. The notation used has been carried over from Sect. 2.

3.1   Auxiliary Results

We begin with a small theorem, whose proof is trivial:

Theorem 1. *Given a sequence ω, let F be a reduced word in G_ω with even $\partial_a(F)$. If $\partial_a(F) < \partial_{b,c,d}(F)$ (resp. $>, =$), then after a successful round of the Word Problem Solving Algorithm (W.P.S.A.) we get two words of length at most $\lceil \partial(F)/2 \rceil$ (resp. $\lfloor \partial(F)/2 \rfloor$, $\partial(F)/2$).*

Proof. When $\partial_a(F) < \partial_{b,c,d}(F)$ then the word is of the same form as (3) and of odd length. It can then be broken into a product of quadruples $*a*a$ of total length $\partial(F) - 1$, and a single $*$ at the end. Under the action of $\phi_0^{(n)}$ and $\phi_1^{(n)}$ each quadruple goes into $*$, $*a$ or $a*$ and the asterisk at the end goes to $*, a$ or 1. There are $\frac{\partial(F)-1}{4}$ quadruples, and so the maximum possible length of the word obtained is

$$2 \cdot \frac{\partial(F) - 1}{4} + 1 = \frac{\partial(F) - 1}{2} + 1 = \frac{\partial(F) + 1}{2} = \left\lceil \frac{\partial(F)}{2} \right\rceil .$$

The proof for $>$ and $=$ is similar.   □

Theorem 2. *Given a sequence ω, for successful application of the W.P.S.A. to a word F in G_ω, at most only the first $\lceil \log_2 \partial(F) \rceil$ of its digits are needed. This number is also the maximum depth of the tree obtained when running the W.P.S.A.*

Proof. For the substitution of the symbols $\tilde{u}_i$, $\tilde{v}_i$ and $\tilde{w}_i$ determining the action of $\phi_0^{(i)}$ and $\phi_1^{(i)}$ during the ith round of the W.P.S.A., the ith digit of the ternary sequence ω is needed. Therefore, for successful application of the W.P.S.A., the maximum number of digits needed is equal to the maximum number of rounds possible. Denote this number by n. This maximum is attained if the algorithm terminates because an end node has been reached for all obtained words on the nth level.

Suppose that n rounds of the W.P.S.A. take place. By Theorem 1, after the first round of the W.P.S.A., the maximum length of the words obtained is $\lceil \partial(F)/2 \rceil$. Note that these words will have an even number of occurrences of the symbol a, otherwise the algorithm will terminate before all n rounds take place. After step 2 of the second round of the algorithm, the words will also be in reduced form. So, again by Theorem 1, after the second round of the algorithm we will obtain words of length at most $\lceil \lceil \partial(F)/2 \rceil /2 \rceil = \lceil \partial(F)/2^2 \rceil$. Continuing this way we see that after the nth round, the length of any words we obtain will not exceed $\lceil \partial(F)/2^n \rceil$.

By assumption, after the nth round of the algorithm, end nodes have been reached for all words on the nth level. Recall that each end node is a word of length 1. Thus, n is an integer such that $\left\lceil \frac{\partial(F)}{2^n} \right\rceil = 1$ or, equivalently, $\partial(F) < 2^n$. Taking logarithms we get $\log_2 \partial(F) < n$ which implies $n = \lceil \log_2 \partial(F) \rceil$.

Since each round corresponds to a different level of the tree obtained when running the algorithm, n is also the maximum depth of the tree.   □

For example to check whether a word of length 1000 is equal to 1_{G_ω} we will only need at most the $\lceil \log_2 1000 \rceil = 10$ first digits of the sequence ω.

Denote by M_F^ω the exact number of digits of the sequence ω needed to run the W.P.S.A. on a given word F in G_ω. Clearly, $M_F^\omega \leq \lceil \log_2 \partial(F) \rceil$.

Corollary 1. Suppose that a word F is a relator in the group G_ω defined by a sequence ω. Then it is also a relator in all groups defined by sequences that share the first M_F^ω digits with the sequence ω and differ in any of the rest.

Proof. By definition, only the first M_F^ω digits of the sequence ω are needed to verify that the word F is a relator in G_ω. $\square$

Denote by Ω_F the set of all sequences ω such that a given word F is a relator in G_ω and by M_F the number $\min\{M_F^\omega \mid \omega \in \Omega_F\}$.

Clearly, by Corollary 3, Ω_F contains sequences that share the first M_F^ω digits with a sequence $\omega \in \Omega_F$ and differ in any of the rest.

Remark 1. Ω_F might also contain sequences that are different in at least one of the first M_F digits.

One example is the following: Let $F = (ac_\omega ab_\omega)^8$. It can be checked using the W.P.S.A. that F is a relator for all ternary sequences having 012 and 021 as their first three digits.

Theorem 3. There exists an algorithm which, given a word F, finds the first M_F^ω digits of all sequences $\omega \in \Omega_F$, if any.

Proof. Let ω_* be an arbitrary infinite ternary sequence and let F be the given word. The algorithm is as follows:

Step 1. Set $\omega_1 = 0$.

Step 2. In case $\omega_1 = 3$, **terminate** the algorithm. Run the the first round of the W.P.S.A. on F. If it fails its first step *or* if the word F is equal to 1_{G_ω} ($\omega = \omega_1 \omega_*$), increase ω_1 by one and repeat step. In the latter case print out ω_1. *Otherwise*, set $\omega_2 = 0$ and proceed to step **3**.

Step i. ($i \geq 3$). If $\omega_{i-1} = 3$, delete it, increase ω_{i-2} by one and return to step *i*-1. Run the ith round of the W.P.S.A. on F. If it fails its first step *or* if the word F is equal to 1_{G_ω} ($\omega = \omega_1 \ldots \omega_{i-1} \omega_*$, $i-1 \leq n$), increase ω_{i-1} by one and repeat step. In the latter case, print out $\omega_1 \ldots \omega_{i-1}$. *Otherwise*, set $\omega_i = 0$ and proceed to step *i*+1.

The algorithm terminates when it eventually returns to step 2 with $\omega_1 = 3$.

Each output is obtained immediately after a successful application of the W.P.S.A. Thus only the first M_F^ω digits of all sequences $\omega \in \Omega_F$ are printed. Clearly, if no sequences are printed, Ω_F is empty. $\square$

An upper bound for the number of outputs of the algorithm is $3^{\lceil \log_2 \partial(F) \rceil}$, which corresponds to all ternary sequences of length $\lceil \log_2 \partial(F) \rceil$. Note that $3^{\lceil \log_2 \partial(F) \rceil} \leq \partial(F)^{\lceil \log_2 3 \rceil} = \partial(F)^2$. It is therefore computationally feasible to run the algorithm, even for large $\partial(F)$.

3.2   Cryptanalysis

We begin with an important observation, omitted by Garzon and Zalcstein:

Remark. It was not mentioned in [1] that the public key words w_0 and w_1 must share the following property, otherwise comparing them with the ciphertext becomes trivial: *If $\partial_a(w_0)$ is even (or odd) then so should $\partial_a(w_1)$ be.*

The reason for this is that relators have an even number of letters a and so, if for example $\partial_a(w_0)$ is even, addition of relators will result to a ciphertext word with an even number of a's. If $\partial_a(w_1)$ is odd then we can immediately tell that $w_i^* = w_0$.

In the sequel we will exploit the information provided by each relational word in the public key of the Cryptosystem. Suppose there are n relators in the public key, namely $r_1, \ldots, r_n$. Remark 4 suggests that each one of these relators could be a relator in groups defined by other sequences as well, not just by the private key. What we want to find are sufficiently many digits of a sequence ω such that all of $r_1, \ldots, r_n$ are indeed relators in G_ω, but $w_0 \neq w_1$ (or equivalently $w_0^{-1} w_1 \neq 1_{G_\omega}$). Then, by Corollary 3, we can just complete the sequence with random digits to obtain a key enabling us to decrypt messages.

Using the algorithm in Theorem 5 on $w_k = w_0^{-1} w_1$, we firstly obtain the first $M_{w_k}^\omega$ digits of every sequence $\omega \in \Omega_{w_k}$. This will be a list of unacceptable initial segments of sequences. Note that Ω_{w_k} could be empty in which case the list would be empty as well. The sequences in Ω_{w_k} are unacceptable because for them, $w_* w_i^{-1} = 1_{G_\omega}, i \in \{0, 1\}$ and hence we cannot distinguish between w_0 and w_1, as required for decryption.

We then apply the same algorithm on every r_i to obtain the first $M_{r_i}^\omega$ digits of all sequences $\omega \in \Omega_{r_i}$. In this way we get n lists of initial segments of sequences.

After obtaining the n lists, we compare them in order to find all initial segments which are common. This will be our list of candidates for a key. Notice that this list cannot be empty since it must contain the initial segment of Alice's private key. The number of digits of these common initial segments will be $\max\{M_{r_1}^\omega, \ldots, M_{r_n}^\omega\}$.

In case we have more than one candidates, we decide the suitability of the first candidate to be used for decryption by looking at the list of unacceptable initial segments of sequences. Let p denote the number of digits of the candidate. There are three possible cases:

Case 1. The unacceptable initial segments list contains at least one member with more digits than the candidate, with its first p digits being the candidate itself. If more than one such members exist, we consider the one with the fewest digits. Suppose this sequence has k digits, $k > p$. We can then make the candidate into a suitable sequence by adding arbitrary digits. However, we must ensure that the first $k - p$ arbitrary digits we add will not coincide with the last $k - p$ digits of this particular unacceptable initial segment or any others sharing the first p digits with it. If this is impossible to perform, the candidate is unsuitable and we should test the next candidate for suitability.

Case 2. An unacceptable initial segment has k digits, $k \leq p$, and these digits coincide with the first k digits of the candidate. Then, this candidate is unsuitable and we should test the next candidate for suitability.

Case 3. The unacceptable initial segments list is empty or none of the previous two cases occur. In this case the candidate is suitable.

As mentioned above, there is at least one suitable candidate, namely the initial segment of Alice's original private key. We can make this into an infinite sequence by adding random digits.

This is all we need for decryption since, on applying the W.P.S.A. to the word $w_j^{-1} w_i^*$, j, $i \in \{0,1\}$, the relators would vanish and we would get 1_{G_ω} only if $j = i$.

We conclude this section with a worked example.

3.3   Example

In this example we will omit the subscript ω from the generators b, c and d.
Suppose **Alice** publishes the following **public key**:
$r_1 = (ab)^4$, $r_2 = (acab)^8$, $r_3 = (bada)^8$,
$w_0 = (bacabacacaca)^2 bacab$,
$w_1 = acacacabacabacacaca$.
(Note that $\partial_a(w_0)$ and $\partial_a(w_1)$ are both even.)

Applying the algorithm from Theorem 5 on $w_k = w_0^{-1} w_1$ we get the list of **unacceptable** initial segments of sequences:

1. 01
2. 1
3. 21

Applying the algorithm on r_1, r_2 and r_3 we obtain 3 lists of initial segments of all sequences for which r_1, r_2 and r_3 are relators:

r_1	2							
r_2	012	021	101	11	121	202	212	22
r_3	00	010	020	102	120	202	212	22

We compare these in order to obtain the **candidates** list:

1. 202
2. 212
3. 22

Finally, by comparing the candidates and unacceptable lists, we deduce that by completing 202 (or 22) arbitrarily (e.g with zeroes to get 202000...), we will be able to successfully decrypt all messages sent by **Bob**.

4 Conclusion

The Public Key Cryptosystem proposed in [1] is unsuitable for any practical implementations due to lack of security. This is because, as demonstrated in Sect. 3, the public key provides enough information to easily obtain keys behaving like the private key and which are suitable for successful decryption of ciphertexts.

Acknowledgements. The author would like to thank Professor Alexandre V. Borovik whose support, valuable comments and suggestions made the preparation of this paper possible. He would also like to thank Dr. Dennis Hofheinz for making a preprint of the unpublished work [4] available.

References

1. M. Garzon and Y. Zalcstein: The Complexity of Grigorchuk Groups with Application to Cryptography. Theoretical Computer Science **88** (1991) 83–98, Elsevier.
2. R.I. Grigorchuk: Degrees of Growth of Finitely Generated Groups, and the Theory of Invariant Means. Math. USSR Izvestiya Vol. 25 (1985) No. 2 259–300
3. R.I. Grigorchuk, V.V. Nekrashevich and V.I. Sushchanskii: Automata, Dynamical Systems and Groups. Proc. of the Steklov Institute of Mathematics Vol. 231 (2000).
4. D. Hofheinz and R. Steinwandt: Cryptanalysis of a Public Key Cryptosystem Based on Grigorchuk Groups. Unpublished.
5. N. Wagner and M. Magyarik: A Public Key Cryprosystem Based on the Word Problem. Proc. CRYPTO'84, Lecture Notes in Computer Science Vol. 196 (1985) 19–36, Springer.

More Detail for a Combined Timing and Power Attack against Implementations of RSA

Werner Schindler[1] and Colin D. Walter[2]

[1] Bundesamt für Sicherheit in der Informationstechnik (BSI)
Godesberger Allee 185-189, 53175 Bonn, Germany
Werner.Schindler@bsi.bund.de
[2] Comodo Research Laboratory
10 Hey Street, Bradford, BD7 1DQ, UK
Colin.Walter@comodogroup.com

Abstract. Implementations of Montgomery's modular multiplication algorithm (MMM) typically make conditional subtractions in order to keep the output within register or modulus bounds. For some standard exponentiation algorithms such as m-ary, it has been shown that this yields enough information to deduce the value of the exponent. This has serious implications for revealing the secret key in cryptographic applications without adequate counter-measures. Much more detail is provided here about the distribution of output values from MMM when the output is only reduced to keep it within register bounds, about how implementations of sliding windows can be attacked, and about handling errors.

Keywords. RSA cryptosystem, side channel leakage, power analysis, timing attack, Montgomery modular multiplication, exponentiation, statistical decision problem.

1 Introduction

Side-channel leakage occurs through data dependent variation in the use of resources such as time and hardware. The former results from branching in the code or compiler optimisation [4], and the latter from data causing gate switching in the circuitry. Both manifest themselves measurably through overall current variation as well as local or global electro-magnetic radiation (EMR) [5,6,7]. Information leaking in these ways might be used by an attacker to deduce secret cryptographic keys which are contained securely inside a smart card.

Timing variation can be measured using overall delay during decryption or signing [1,4,11,13]. However, for successful attacks on keys with limited lifespans for which only a bounded number of decryptions with the same key is allowed, some finer detail may become necessary, such as the time variations for executing component procedures. This detail can be seen by observing delays between the characteristic power consumption wave form for loading instructions and data at the start of the procedure. Here this is applied to the version of Montgomery

K.G. Paterson (Ed.): Cryptography and Coding 2003, LNCS 2898, pp. 245–263, 2003.
© Springer-Verlag Berlin Heidelberg 2003

modular multiplication [9] which contains a final conditional subtraction for reducing the output by only enough to fit within register bounds. The aim is to recover a secret RSA [10] exponent from deductions of whether or not this conditional subtraction occurs.

An essential assumption is that there is no blinding of the exponent or randomisation in the exponentiation algorithm so that the same type of multiplicative operations are performed (but with different data) every time the key is used. This enables an attacker to capture the collection of instances of the extra final subtraction for each individual operation. We will show how to determine exponent digits from this data, recover from any errors, and hence obtain the secret key even if the input text (the base of the exponentiation) has been blinded.

The results emphasise the need for care in the implementation of RSA. Indeed, similar results apply to elliptic curve cryptography (ECC) [8] and other crypto-systems based on modular arithmetic. In all cases where a key is re-used, some counter-measures should be employed to ensure that the conditional subtraction is removed, or at least hidden, and that the same sequence of key-dependent operations is not re-used for each exponentiation. In elliptic curve cryptography, standard key blinding adds about 20% to the cost of a point multiplication. The temptation to avoid this overhead should be resisted.

Montgomery modular multiplication [9] is arguably the preferred method for hardware implementation. We treat this algorithm here, but equivalent results should hold for any method for which there are correlations between the data and the multiplication time. The generic timing attack methodology which we adopt was first described in [16]. This approach established the feasibility of recovering the key within its lifetime and also gave the theoretical justification behind the different frequencies of final subtractions observed between squarings and multiplications.

A closer study of such distributions reduces the number of exponentiations required to recover the key. This was done for various settings in [11,12,13] when the final reduction yields output less than the modulus. However, reducing output only as far as the same length as the modulus is more efficient, and is the case studied here. The mathematics is more involved and so requires numerical approximation methods, but this is easy for an attacker. In addition, more detail is given concerning the recovery of the key when sliding windows exponentiation is employed.

2   The Computational Model

In order to clarify the precise conditions of the attack, each assumption is numbered. First, the computational model requires several assumptions, namely:

i) secret keys are unblinded in each exponentiation;
ii) Montgomery Modular Multiplication (MMM) is used with a conditional subtraction which reduces the output to the word length of the modulus;
iii) the most significant bit of the modulus lies on a word boundary; and
iv) exponentiation is performed using the m-ary sliding windows method [2,3].

Exponent blinding might be deemed an unnecessary expense if other counter-measures are in place. So we still need better knowledge of how much data leaks out under the first assumption. For a sample of one exponentiation this particular assumption is irrelevant, but then, with data leaked from other sources, the techniques here might be just sufficient to make a search for the key computationally feasible.

As shown below, efficiency is good justification for the second hypothesis. The word length referred to there is the size of inputs to the hardware multiplier, typically 8-, 16- or 32-bit. However, standards tend to specify key lengths which are multiples of a large power of 2, such as 768, 1024, 1536 and 2048 bits. So the third hypothesis is almost certainly the case.

Because of its small memory requirements and greater safety if an attacker can distinguish squares from multiplications, a sliding window version of 4-ary exponentiation is the usual algorithm employed in smartcards. The exponent is re-coded from least to most significant bit. When a bit 0 occurs, it is re-coded as digit 0 with base 2 and when a bit 1 occurs, this bit and the next one are recoded as digit 1 or 3 with base 4 in the obvious way. This representation is then processed left to right, with a single modular squaring of the accumulating result for digit 0, and two modular squarings plus a modular multiplication for digits 1 and 3. According to the digit, the first or pre-computed third power of the initial input is used as the other operand in the multiplications.

An attacker can recover the secret exponent if he can i) distinguish squarings from multiplications, and ii) determine whether the first or third power was used as an operand in the multiplications. The classical square-and-multiply or binary algorithm ($m = 2$) is less secure against this type of attack as the second task is omitted. The attack here treats any m, and applies equally well when the Chinese Remainder Theorem (CRT) is used.

3   Initial Notation

Let R be the so-called "Montgomery factor" associated with an n-bit RSA modulus M. Then the implementation of Montgomery's algorithm satisfies the following specification [15]:

v) For inputs $0 \leq A$, $B < R$, MMM generates output $P \equiv A*B*R^{-1} \bmod M$ satisfying $ABR^{-1} \leq P < M+ABR^{-1}$ before any final, conditional subtraction.

Clearly this specifies P uniquely. The division by R is the result of shifting the accumulating product down by the multiplier word length for a number of iterations equal to the number of words in A. Since A, B and M will all have the same word length for the applications here, hypothesis (iii) yields

vi) $M < R < 2M$

i.e. R is the smallest power of 2 which exceeds M.

The output bound $P < M+R$ implied by (v) means that any overflow into the next word above the most significant of M has value at most 1. When this overflow bit is 1, the conditional subtraction can be called to reduce the output to less than R without fear of a negative result. Hence that bit can be used efficiently to produce an output which satisfies the pre-conditions of (v) for further applications of MMM. This is the case of MMM that is considered here.

Alternatively, the condition $P < M$ is often used to control the final subtraction. This is the standard version, but the test requires evaluating the sign of $P-M$, which is more expensive than just using the overflow bit. This case of MMM was discussed in [12]. A constant time version of MMM is possible by computing and storing both P and $P-M$, and then selecting one or other according to the sign of the latter. However, the subtraction can be avoided entirely if the inputs satisfy $A, B < 2M$ and R satisfies $4M < R$. This, too, is more expensive [15].

As usual, the private and public exponents are denoted d and e. For deciphering (or signing), ciphertext C is converted to plaintext $C^d \bmod M$. The m-ary sliding windows exponentiation algorithm using MMM requires pre-computation of a table containing $C^{(i)} \equiv C^i R \bmod M$ for each odd i with $1 \le i < m$. This is done using MMM to form $C^{(1)}$ from C and $R^2 \bmod M$, $C^{(2)}$ from $C^{(1)}$, and then iteratively $C^{(i+2)}$ from $C^{(i)}$ and $C^{(2)}$. By our assumptions, the word length of each $C^{(i)}$ is the same as that of M, but it may not be the least non-negative residue. We also define $b = \log_2 m$ and assume it to be integral.

4   The Threat Model

The security threat model is simply that:

vii) an attacker can observe and record every occurrence of the MMM conditional subtraction over a number of exponentiations with the same key.

Section 1 provided the justification for this. The attack still works, although less efficiently, if manufacturers' counter-measures reduce certainty about the occurrence or not of the subtraction.

Unlike many attacks in the past, we make no assumptions about the attacker having control or knowledge of input to, or output from, the exponentiation. Although he may have access to the ciphertext input or plaintext output of a decryption, random nonces and masking should be employed to obscure such knowledge so that he is unable to use occurrences of the conditional subtraction to determine whether a square or multiply occurred. It is therefore assumed that

viii) the attacker can neither choose the input nor read either input or output.

Indeed, even without masking this is bound to be the case for exponentiations when the Chinese Remainder Theorem has been used. Lastly, as the attacker may have legitimate access to the public key $\{M, e\}$, it is assumed that

ix) the correctness of a proposed value for the private exponent d can be checked.

The assumptions so far mean that the same sequence of multiplicative operations is carried out for every decryption. So, from a sample of N decryptions with the same key, the attacker can construct an array $Q = (q_{i,j})$ whose elements are 1 or 0 depending on whether or not the ith MMM of the jth decryption includes the conditional subtraction. The elements $q_{i,j}$ are called *extra reduction* or *er*-values. Similarly, initialisation gives a matrix $Q' = (q'_{i,j})$: if C_j is the input to the jth decryption then the er-value $q'_{i,j}$ is associated with the calculation of $C_j^{(i)} \equiv C_j{}^i R \bmod M$ for the digit i.

5   Some Limiting Distributions

The timing attack here was first reported in [16], but precise MMM output bounds now allow a much more accurate treatment of the probabilities and hence more reliable results. More exact figures should make it easier to determine the precise strength of an implementation against the attack. Important first aims are to determine the probability of extra reductions and to establish the probability distribution function for the MMM outputs P. To this end two reasonable assumptions are made about such distributions:

x) The ciphertext inputs C behave as realizations of independent random variables which are uniformly distributed over $[0, M)$.
xi) For inputs A and B to MMM during an exponentiation, the output prior to the conditional subtraction is uniformly distributed over the interval $[ABR^{-1}, M+ABR^{-1})$.

Assumption (x) is fulfilled if, for example, the ciphertext is randomly chosen (typical for RSA key exchange) or message blinding has been performed as proposed in [4]. Here it is also a convenient simplification. In fact, there may be some initial non-uniformity (with respect to the Euclidean metric) which might arise as a result of content or formatting, such as with constant prefix padding. We return to this topic in Remark 1.

In (xi), the multiples of M subtracted from the partial product P during the execution of MMM are certainly influenced by *all* bits of A, B and M. However, the probability for the final conditional subtraction is essentially determined by the *topmost* bits of the arguments.

Before the formal treatment we illustrate the limit behaviour of the distributions which will be considered and provide the informal reasoning behind their construction. In order to exhibit the difference between squares and multiplications software was written to graph the probability of the extra subtraction in the limit of three cases, namely the distribution of outputs after a long sequence of i) squarings of a random input, ii) multiplications of independent random inputs, and iii) multiplications of a random input by the same fixed constant A. The result is illustrated in Figure 1. In the second case, two independent values are chosen from the kth distribution in the sequence. They are used as the inputs to MMM and the output generates the $k+1$st distribution. In all three cases the convergence is very rapid, with little perceptible change after 10 or so iterations.

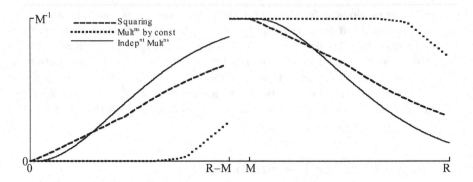

Fig. 1. Probability (vertical axis) of output in the range $0..R-1$ (horizontal axis) after a number of squarings, independent multiplications or multiplications by a constant A for the case $M = 0.525R$ and $A = 1.5M$.

The density functions have three distinct sections which correspond to the output belonging to one of the intervals $[0, R-M)$, $[R-M, M)$ or $[M, R)$. This is because outputs with a residue in $[R-M, M)$ have only one representative in $[0, R)$ and so, under hypothesis (xi), they occur in $[R-M, M)$ with probability M^{-1} whereas outputs with a residue in $[0, R-M)$ have two representatives in $[0, R)$ with combined probability M^{-1}. Note the discontinuities in the functions at $R-M$ and that probabilities tend to 0 as the output approaches 0.

Because M is large, the discrete probabilities are approximated accurately by the density functions of continuous random variables. If f is the limiting density function of a sequence of independent multiplications, then

1) $\int_0^R f(x)dx$ $= 1$
2) $f(x)$ $= 0$ for $x < 0$ and $R < x$
3) $f(x)$ $= M^{-1}$ for $R-M \leq x \leq M$
4) $f(x)+f(x+M) = M^{-1}$ for $0 \leq x \leq R-M$
5) $f(x) = \frac{1}{M}\int_0^x f(y)dy + \frac{1}{M}\int_x^R \int_0^{Rx/y} f(y)f(z)dz\, dy$ for $0 \leq x \leq R-M$

Properties 1 to 4 are already clear, and apparent in the figure. Property 5 encapsulates the restrictions imposed by MMM. To establish it, consider outputs x in the interval $[0, R-M)$. These do not involve a final conditional subtraction, and so, by (xi), they are derived from MMM inputs y and z for which $[yzR^{-1}, M+yzR^{-1})$ contains x. As the distribution is assumed to be uniform on this interval, there is a 1 in M chance of obtaining x if $yzR^{-1} \leq x < M+yzR^{-1}$, i.e. if $R(x-M)/y < z \leq Rx/y$. Since $R(x-M)/y \leq R(R-2M)/y < 0$, the lower bound on feasible z is 0 and the upper bound is $\min\{R, Rx/y\}$. Thus,

$$f(x) = \frac{1}{M}\int_0^R \int_0^{\min\{R, Rx/y\}} f(y)f(z)dz\, dy.$$

The integral over y can be split into the sub-intervals $[0, x]$ and $[x, R]$ in order to separate the cases of the upper limit on z being R or Rx/y. This yields

$$f(x) = \frac{1}{M} \int_0^x \int_0^R f(y)f(z)dz\,dy + \frac{1}{M} \int_x^R \int_0^{Rx/y} f(y)f(z)dz\,dy$$

in which the first integral simplifies to give the expression in property 5.

These properties determine f completely. Although an algebraic solution does not exist, numerical approximations are easy to obtain. The other two density functions satisfy the same properties 1 to 4. The analogues to property 5 are:

  5') $f(x) = \frac{1}{M} \int_0^{\sqrt{Rx}} f(y)dy$ for $0 \le x \le R{-}M$

for the limit of consecutive squarings, and

  5'') a) $f(x) = \frac{1}{M} \int_0^{Rx/A} f(y)dy$ for $0 \le x \le A \le R{-}M$
      b) $f(x) = \frac{1}{M}$ for $A \le x \le R{-}M$

for the limit of consecutive multiplications by a constant A.

Although there are differences between the distributions for squaring and multiplication, they are substantially the same. Figure 1 illustrates the largest possible differences, which occur for M close to $\frac{1}{2}R$. For M close to R they are essentially all equal to the same uniform distribution. The distributions which lead to these limiting cases are considered in more detail in the next section.

6   Conditional Probabilities

As before, a prime $'$ is used on quantities relating to the initialisation phase, and unprimed identifiers relate to the computation stage. The attacker wants to estimate the types of the Montgomery multiplications in the computation phase on basis of the observed extra reduction (er-) values $q'_{j,k}$ and $q_{j,k}$ within the initialization and computation phases respectively. These types will be denoted using the text characters of the set $\mathcal{T} := \{`S', `M_1', \ldots, `M_{m-1}'\}$ where 'S' corresponds to a square, and 'M_i' to multiplication by the precomputed table entry $C^{(i)}$ associated with digit i of the re-coded exponent. In this section we derive explicit formulas for the probabilities of these events given the observed er-values. We begin with some definitions. It is convenient to normalise the residues mod M to the unit interval through scaling by a factor M.

Definition 1. *A realization of a random variable X is a value assumed by X. For sample number k, the er-values $q'_{j,k}$ and $q_{j,k}$ are defined by $q'_{j,k} := 1$ if computation of table entry $C_k^{(j)}$ in the initialisation phase requires an extra reduction (this includes both $q'_{1,k}$ and $q'_{2,k}$), and similarly $q_{j,k} := 1$ if the jth Montgomery multiplication in the computation phase requires an extra reduction. Otherwise $q'_{j,k} := 0$ and $q_{j,k} := 0$. As abbreviations, let $q'_k := (q'_{1,k}, \ldots, q'_{m-1,k})$ and $q_{i,\ldots,i+f-1;k} := (q_{i,k}, \ldots, q_{i+f-1,k})$. For $A \subseteq B$ the indicator or characteristic function $1_A: B \to \mathbb{R}$ is defined by $1_A(x) := 1$ if $x \in A$ and 0 otherwise. Further, for $\gamma := M/R$, let $\chi: [0, 1 + \gamma^{-1}) \to [0, \gamma^{-1})$ be given by $\chi(x) := x$ if $x < \gamma^{-1}$ and $\chi(x) := x - 1$ otherwise; that is, $\chi(x) := x - 1_{x \ge \gamma^{-1}}$. Lebesgue measure is denoted by λ.*

Lemma 1. (i) $\frac{\mathrm{MMM}(A,B)}{M} = \chi \left(\frac{A}{M} \frac{B}{M} \frac{M}{R} + \frac{ABM^*(\bmod\ R)}{R} \right)$
where $M^* = (-M)^{-1}(\bmod\ R)$.

(ii) The extra reduction in MMM is necessary exactly when -1 is subtracted by the application of χ.

This lemma follows immediately from the definition of Montgomery's multiplication algorithm. Assertion (xi) is equivalent to saying that the second summand in the right-hand side of Lemma 1(i) is uniformly distributed on the unit interval $[0, 1)$. For a formal justification, the ideas from the proof of Lemma A.3(iii) in [13] can be adjusted in a straightforward way. To formulate a mathematical model we need further definitions which mirror the operations which comprise the exponentiation:

Definition 2. *Assume $T(i) \in \mathcal{T}$, $i = 1, 2, \ldots$, describes the sequence of multiplicative operations in the exponentiation. Let $F := \{i \mid 1 \leq i \leq m-1,\ i\ odd\}$. Suppose the random variables V_i' ($i \in F \cup \{2\}$) and $V_1, V_2, \ldots$ are independent and equidistributed on the unit interval $[0, 1)$. Define the random variables S_i' ($i \in F \cup \{0, 2\}$) so that S_0' assumes values in $[0, 1)$ and*

$$S_1' := \chi\left(S_0'(R^2 \,(\mathrm{mod}\ M)/M)\gamma + V_1'\right) \tag{1}$$

$$S_2' := \chi\left(S_1'^2\gamma + V_2'\right) \tag{2}$$

$$S_{2i-1}' := \chi\left(S_{2i-3}'S_2'\gamma + V_{2i-1}'\right) \qquad for\ 1 < i \leq \frac{m}{2} \tag{3}$$

Similarly, define $S_0 := S_r'$ where $r \in F$ is the left-most digit of the secret exponent d after recoding (cf. Sect. 2) and, for $i \geq 1$, let

$$S_i := \begin{cases} \chi\left(S_{i-1}^2\gamma + V_i\right) & if\ T(i) = `S' \\ \chi\left(S_{i-1}S_j'\gamma + V_i\right) & if\ T(i) = `M_j' \end{cases} \tag{4}$$

Lastly, define $\{0, 1\}$-valued random variables $W_1', \ldots, W_{m-1}'$ and $W_1, W_2, \ldots$ by

$$W_1' := 1_{S_1' < S_0'(R^2\,(\mathrm{mod}\ M)/M)\gamma} \tag{5}$$

$$W_2' := 1_{S_2' < S_1'^2\gamma} \tag{6}$$

$$W_{2i-1}' := 1_{S_{2i-1}' < S_{2i-3}'S_2'\gamma} \qquad for\ 1 < i \leq \frac{m}{2} \tag{7}$$

$$W_i := \begin{cases} 1_{S_i < S_{i-1}^2\gamma} & if\ T(i) = `S' \\ 1_{S_i < S_{i-1}S_j'\gamma} & if\ T(i) = `M_j' \end{cases} \tag{8}$$

Thus the distribution of S_i describes the random behaviour of the output from the ith multiplication, V_i' and V_i correspond to the variation described in assumption (xi), and W_i', W_i are the associated distributions of final subtractions recorded in Q' and Q:

Mathematical Model. We interpret the components of the er-vector $\boldsymbol{q}_k' = (q_{1,k}', \ldots, q_{m-1,k}')$, row subscripts in $F \cup \{2\}$, as realizations of the random variables $W_1', \ldots, W_{m-1}'$ with S_0' having the uniform distribution of the normed (random) input C_k/M to the kth exponentiation. Similarly, we interpret $q_{1,k}, q_{2,k}, \ldots$ as realizations of the random variables $W_1, W_2, \ldots$.

Consequently, we have to study the stochastic processes $W_1', W_2', W_3', W_5', \ldots$, W_{m-1}' and $W_1, W_2, \ldots$. However, the situation is much more complicated than in the case of the standard Montgomery algorithm considered in [12] because the random variables $S_1, S_2, \ldots$ are neither independent nor identically distributed. Their distribution depends on the sequence of operations $T(1), T(2), \ldots$.

Definition 3. *(i) For $i \in F \cup \{2\}$, $w \in \{0,1\}$ and indices of the components ranging over $F \cup \{0,2\}$, let the subset $\mathcal{D}'(i; w) \subseteq [0,1) \times [0, \gamma^{-1})^{\frac{1}{2}m+1}$ be given by those vectors $(s_0', \ldots, s_{m-1}')$ for which $v_1', \ldots, v_{m-1}' \in [0,1)$ exist such that $s_0', \ldots, s_{m-1}', v_1', \ldots, v_{m-1}'$ satisfy (1)-(3) in place of $S_0', \ldots, S_{m-1}', V_1', \ldots, V_{m-1}'$ and, additionally, the component s_i' and its predecessor must result in w when inserted into whichever of (5), (6) or (7) describes W_i'. Thus, for example, by (6) the vectors $(s_0', \ldots, s_{m-1}')$ in $\mathcal{D}'(1;1)$ satisfy $s_1' < s_0'(R^2 (\mathrm{mod}\, M))/R$, and by (7) those in $\mathcal{D}'(2i-1;1)$ satisfy $s_{2i-1}' < s_{2i-3}' s_2' \gamma$.*
(ii) For $i \le j \le i+f-1$, $w \in \{0,1\}$ and $t \in \mathcal{T}$, define $\mathcal{D}_f(i,j; w,t) \subseteq [0, \gamma^{-1})^{f+1}$ to be the subset of vectors $(s_{i-1}, \ldots, s_{i+f-1})$ for which there are $v_i, \ldots, v_{i+f-1} \in [0,1)$ such that $s_{i-1}, \ldots, s_{i+f-1}, v_i, \ldots, v_{i+f-1}$ satisfy (4) in place of $S_{i-1}, \ldots, S_{i+f-1}, V_i, \ldots, V_{i+f-1}$ with the assumption that $T(j) = t$ and, additionally, the component s_j and its predecessor must result in w when inserted into the instance of (8) which describes W_j for $T(j) = t$. Thus, for example, the elements of $\mathcal{D}_f(i,j;1,\text{'}M_k\text{'})$ satisfy the constraint $s_j < s_{j-1} s_k' \gamma$.

Remark 1. As already mentioned in Sect. 5, there may also be scenarios of practical interest which imply distributions of S_0' other than the uniform distribution of assertion (x). For signing with constant prefixed padding, for instance, the random variable S_0' can be assumed to have a Dirac (= single-point) distribution due to the definition of S_1'. Generally speaking, the distribution of S_0' influences the conditional density $g(\cdot|\cdot)$ in Lemma 2(i) and hence implicitly the conditional probabilities in Theorem 1 and the optimal decision strategy. Then it is necessary to adjust the derivation of $g(\cdot|\cdot)$ to the concrete distribution of S_0' but otherwise the remaining steps in this and the forthcoming sections pass through identically.

The (conditional) probability of a given er-vector can be described using the sets of Definition 3:

Lemma 2. *(i) Assume the random variable S_0' is equidistributed on $[0,1)$. Then the conditional distribution of the random vector $(S_1', \ldots, S_{m-1}')$ on $[0,1)^{\frac{1}{2}m+1}$ under the condition $W_1' = w_1', \ldots, W_{m-1}' = w_{m-1}'$ has Lebesgue probability density*

$$g(s_1', \ldots, s_{m-1}' \mid w_1', \ldots, w_{m-1}') :=$$

$$\frac{\int_0^1 1_{\bigcap_{i \in F \cup \{2\}} \mathcal{D}'(i; w_i')}(s_0', s_1', \ldots, s_{m-1}')\, ds_0'}{\int_{[0,1) \times [0,\gamma^{-1})^{\frac{1}{2}m+1}} 1_{\bigcap_{i \in F \cup \{2\}} \mathcal{D}'(i; w_i')}(s_0', s_1', \ldots, s_{m-1}')\, ds_0' ds_1' \cdots ds_{m-1}'}. \quad (9)$$

(ii) The distribution of S_{i-1} has a Lebesgue density, say h_{i-1}. Moreover,

$$\text{Prob}(W_i = w_i, \ldots, W_{i+f-1} = w_{i+f-1} \mid W_1' = w_1', \ldots, W_{m-1}' = w_{m-1}') = (10)$$

$$\int_{[0,\gamma^{-1})^{\frac{1}{2}m+f+2}} g(s_1', \ldots, s_{m-1}' \mid w_1', \ldots, w_{m-1}') \cdot h_{i-1}(s_{i-1}) \times$$

$$\times 1_{\bigcap_{u=i}^{i+f-1} \mathcal{D}_f(i,u;w_i,T(u))}(s_{i-1}, \ldots, s_{i+f-1}) \, ds_1' \cdots ds_{m-1}' ds_{i-1} ds_i \cdots ds_{i+f-1}.$$

(iii) If $S_0 = S_r'$ then

$$\text{Prob}(W_1 = 1 \mid W_1' = w_1', \ldots, W_{m-1}' = w_{m-1}') = \tag{11}$$

$$\int_{[0,\gamma^{-1})^{\frac{1}{2}m+1}} g(s_1', \ldots, s_{m-1}' \mid w_1', \ldots, w_{m-1}') \cdot \max\{0, 1 - \gamma^{-1} + s_r'^2 \gamma\} \, ds_1' \cdots ds_{m-1}'.$$

Proof. We first note that $\{(s_0', \ldots, s_{m-1}') \in [0,1) \times [0, \gamma^{-1})^{\frac{1}{2}m+1} \mid W_1' = w_1', \ldots,$ $W_{m-1}' = w_{m-1}'\} = \bigcap_{i \in F \cup \{2\}} \mathcal{D}'(i, w_i')$. For the moment let $\Psi: [0,1)^{m/2+2} \to$ $[0,1) \times [0, \gamma^{-1})^{m/2+1}$ be given by $\Psi(s_0', v_1', \ldots, v_{m-1}') := (s_0', s_1', \ldots, s_{m-1}')$. By this we mean that the coordinates s_i' of image points are determined by instances of the equation (3) of the form $s_i' = \chi\left(s_{i-2}' s_2' \gamma + v_i'\right)$ when $i > 2$ and the similar ones from equations (1) and (2) for $i = 1, 2$. The mapping Ψ is injective (though not surjective) and differentiable almost everywhere with Jacobian 1. As $\Psi(S_0', V_1', \ldots, V_{m-1}') = (S_0', S_1', \ldots, S_{m-1}')$ and since the random variables $S_0', V_1', \ldots, V_{m-1}'$ are independent and equidistributed on $[0,1)$ the transformation theorem (applied to the inverse Ψ^{-1} where it is differentiable) implies that the random vector $(S_0', S_1', \ldots, S_{m-1}')$ has constant density on the image $\Psi([0,1)^{m/2+2})$. The definition of conditional probabilities and computing the marginal density with respect to s_0' proves (9). The first assertion of (ii) follows immediately from Lemma 3 (in the Appendix) with $G = \mathbb{R}$, $\mu = \lambda$, $\nu = \lambda \mid_{[0,1)}$ and τ (depending on $T(i-1)$) denoting the distribution of $S_{i-2}^2 \gamma$ or $S_{i-2} S_j' \gamma$ for a particular index $j \in F$. Equation (10) can be verified in a similar way to (9). For fixed $s_0', \ldots, s_{m-1}'$ we define $\Psi_s: [0, \gamma^{-1}) \times [0,1)^f \to [0, \gamma^{-1})^{f+1}$ by $\Psi(s_{i-1}, v_i, v_{i+f-1}) := (s_{i-1}, \ldots, s_{i+f-1})$. Again, Ψ_s is almost everywhere differentiable with Jacobian 1, and the transformation theorem completes the proof of (ii) as the random variables $S_{i-1}, V_i, \ldots, V_{i+f-1}$ are independent and equidistributed on $[0, \gamma^{-1})$ or $[0,1)$, resp. (Note that $\Psi_s^{-1}(s_{i-1}, *) = (s_{i-1}, *)$.) The first Montgomery multiplication in the computation phase is a squaring. Hence $\text{Prob}(W_1 = 1) = \text{Prob}(S_r'^2 \gamma + V_1 \geq \gamma^{-1})$. As $S_r'^2 \gamma \in [0, \gamma^{-1})$ this proves (iii). $\square$

Remark 2. In the Appendix, Theorem 2 (i) to (iv) and (v) respectively consider the fictional situations that the computation phase consists only of squarings or multiplications by a fixed table entry. The distributions of the $S_1, S_2, \ldots$ then converge respectively to $f \cdot \lambda_{[0,\gamma^{-1})}$ and $f_{(s_j')} \cdot \lambda_{[0,\gamma^{-1})}$ for fixed $s_j' \in [0, \gamma^{-1})$. These were graphed in Figure 1 for $\gamma = 0.525$ and $C^{(j)} = 1.5\gamma R$. In fact, the density $h_{i-1}(\cdot)$ depends on $T(1), \ldots, T(i-1)$ and $s_1', \ldots, s_{m-1}'$ (cf. Sect. 10).

Theorem 1 quantifies the probabilities for the different type vectors of the Montgomery multiplications $i, \ldots, i+f-1$ given the observed extra reductions:

Theorem 1. *Let* $\theta = (\omega_i, \ldots, \omega_{i+f-1}) \in \mathcal{T}^f$. *If* $T(i) = \omega_i, \ldots, T(i+f-1) = \omega_{i+f-1}$ *then let* $p_\theta\left((\boldsymbol{q}_{i,\ldots,i+f-1;k})_{1 \le k \le N} \mid (\boldsymbol{q}'_k)_{1 \le k \le N}\right)$ *denote the conditional probability for the er-vectors* $(\boldsymbol{q}_{i,\ldots,i+f-1;k})_{1 \le k \le N}$ *if* $(\boldsymbol{q}'_k)_{1 \le k \le N}$ *were observed in the initialization phase. Then,*

$$p_\theta\left((\boldsymbol{q}_{i,\ldots,i+f-1;k})_{1 \le k \le N} \mid (\boldsymbol{q}'_k)_{1 \le k \le N}\right) \approx \prod_{k=1}^{N} \int_{[0,\gamma-1)^{\frac{1}{2}m+f+2}} g(s'_1, \ldots, s'_{m-1} \mid q'_{1,k}, \ldots, q'_{m-1,k}) \times$$

$$\times \, h_{i-1}(s_{i-1}) \cdot 1_{\bigcap_{u=i}^{i+f-1} \mathcal{D}_f(i,u;w_u,\omega_u)}(s_{i-1}, \ldots, s_{i+f-1}) \, ds'_1 \ldots ds'_{m-1} ds_{i-1} \ldots ds_{i+f-1} \quad (12)$$

If r *is the left-most block (i.e. digit) of the secret exponent after recoding then*

$$\text{Prob}\left((q_{1,k})_{1 \le k \le N} \mid (\boldsymbol{q}'_k)_{1 \le k \le N}\right) \approx \quad (13)$$

$$\prod_{k=1}^{N} \int_{[0,\gamma-1)^{\frac{1}{2}m+1}} g(s'_1, \ldots, s'_{m-1} \mid q'_{1,k}, \ldots, q'_{m-1,k}) \max\{0, 1-\gamma^{-1}+s'^2_r \gamma\} \, ds'_1 \ldots ds'_{m-1}.$$

Proof. According to our mathematical model we interpret the observed er-vectors $\boldsymbol{q}'_k$ and $\boldsymbol{q}_{i,\ldots,i+f-1;k}$ as realizations of random variables $W'_{1,k}, \ldots, W'_{m-1,k}$ and $W_{i,k}, \ldots, W_{i+f-1,k}$ respectively, where the latter correspond to $T(i) = \omega_i, \ldots,$ $T(i+f-1) = \omega_{i+f-1}$. Theorem 1 is an immediate consequence of Lemma 2 and the mathematical model.   □

7 A Priori Distribution

In Section 9 we determine an optimal decision strategy for simultaneous guessing of the types $T(i), \ldots, T(i+f-1)$ of the ith,$\ldots,(i+f-1)$th Montgomery multiplications. It seems to be reasonable for the attacker to choose the hypothesis θ within the set $\Theta \subseteq \mathcal{T}^f$ of all admissible hypotheses which is the most likely one, given the observed er-vectors. In the sliding window exponentiation scheme a multiplication using a particular table entry is preceded by at least $b = \log_2 m$ squarings. Consequently, for the chosen f,

$$\Theta = \theta_0 \cup \{\theta_{k,j} \mid 1 \le k \le f; 1 \le j \le m-1 \text{ for odd } j\} \qquad \text{if } f \le b+1 \quad (14)$$

where
$\theta_0 := (\text{'}S\text{'}, \ldots, \text{'}S\text{'})$ means $T(i) = \text{'}S\text{'}, \ldots, T(i+f-1) = \text{'}S\text{'}$ and
$\theta_{k,j} := (\text{'}S\text{'}, \ldots, \text{'}S\text{'}, \text{'}M_j\text{'}, \text{'}S\text{'}, \ldots, \text{'}S\text{'})$ means
    $T(i+k-1) = \text{'}M_j\text{'}$ but $T(v) = \text{'}S\text{'}$ for $v \ne i+k-1$.

However, the admissible hypotheses occur with different probabilities. The optimal decision strategy in Section 9 exploits this fact. In the present section we determine a distribution η on Θ which approximates the exact distribution of the admissible hypotheses and which depends on the secret key. We call η the (approximate) *a priori* distribution, and $\eta_{k,j}$ denotes the approximate probability that $(T(i), \ldots, T(i+f-1)) = \theta_{k,j}$ for randomly chosen i.

When re-coding, the secret exponent d is divided into blocks (digits) of length 1 and b. If d is assumed to be random then both block lengths should occur with the same frequency, and the average block length is about $(b+1)/2$. Hence we should expect about $n/(b+1)$ blocks of length b and $n/(b+1)$ blocks of length 1 where n is the bit length of the modulus M. Consequently, about $2n/(b+1)m$ blocks of length b should equal any given odd exponent digit j. Thus we expect this many vectors $(T(i), \ldots, T(i+f-1))$ of type $\theta_{k,j}$ for $1 \leq k \leq f$. As there are about $n + n/(b+1) = (b+2)n/(b+1)$ Montgomery multiplications (including squares) we set

$$\eta_{1,1} := \cdots := \eta_{f,m-1} := \frac{n(b+1)}{(b+1)\frac{1}{2}m(b+2)n} = \frac{1}{\frac{1}{2}m(b+2)} \quad \text{and}$$

$$\eta_0 := 1 - \frac{\frac{1}{2}mf}{(b+2)\frac{1}{2}m} = \frac{b+2-f}{b+2} \quad \text{if } f \leq b+1. \tag{15}$$

8   Error Detection and Correction

It seems to be unhelpful to consider error detection and error correction strategies before the decision strategy itself has been derived. However, the optimal decision strategy considers the different types of possible error. Roughly speaking, it tries to avoid estimation errors but 'favours' those kinds of errors which are easier to detect and correct than others. The following example illuminates the situation.

Example 1. Let $b = 2$ (i.e. $m = 4$), and assume that the secret exponent d is given by $\ldots |0|01|0|01|0| \ldots$. The correct type sequence is then given by
$$\ldots, `S', `S', `S', `M_1', `S', `S', \quad `S', `M_1', `S', \ldots$$
whereas the following a), b) and c) are possible estimation sequences:
a) $\ldots, `S', `S', `S', `M_1', `S', `M_3', `S', `M_1', `S', \ldots$
b) $\ldots, `S', `S', `S', `S', \quad `S', `S', \quad `S', `M_1', `S', \ldots$
c) $\ldots, `S', `S', `S', `M_3', `S', `S', \quad `S', `M_1', `S', \ldots$

Each of the subsequences a), b), and c) contains exactly one false guess. The error in a) ('M_3') is obvious as the number of squarings between two multiplications must be at least $b = 2$. This *type-a error* ('M_j' instead of 'S') is usually easy to detect if its occurrence is isolated, i.e. if there are no further type-a or *type-b errors* ('S' instead of 'M_j') within a neighbourhood of the error. Then one or at most two positions remain for which exactly one guess is false, and we call the type-a error *locally correctable*. A type-a error is not locally correctable if it occurs within a long series of squarings or if bursts of type-a and type-b errors occur. Then we call it a *global* type-a error. The type-b errors and *type-c errors* ('M_i' instead of 'M_j') illustrated in sequences b) and c) are less obvious. If all type-a errors have been corrected the attacker knows the number of type-b errors (since the number of squarings equals the bit length d minus 1) but not their positions. In particular, type-b and type-c errors are global errors. Of course, to correct type-b and type-c errors it is reasonable first to change those guesses where the respective decisions have been "close" and then to check the new exponent estimator (cf. Sect. 10).

9   The Optimal Decision Strategy

The preliminary work has now been done. Here the pieces are assembled to derive an optimal decision strategy for guessing the types of f consecutive Montgomery multiplications simultaneously when $1 \leq f \leq b+1$. Therefore, we interpret the estimation of $T(i), \ldots, T(i+f-1)$ as a statistical decision problem.

Roughly speaking, in a statistical decision problem the statistician (here the attacker) observes a sample $\omega \in \Omega$ (here the observed extra reduction vectors $(\boldsymbol{q}'_k, \boldsymbol{q}_{i,\ldots,i+f-1,k})_{1 \leq k \leq N}$) which he interprets as a realization of a random variable X. The distribution p_θ of X depends on the unknown parameter $\theta \in \Theta$ which has to be guessed (here $\theta = (T(i), \ldots, T(i+f-1))$). The decision strategy clearly depends on the observation ω but also considers the *a priori* distribution η (cf. Sect. 7) which quantifies the likeliness of the possible parameters and the 'damage' caused by the possible guessing errors. The 'damage' is quantified by the loss function $s(\theta, a)$ where $\theta \in \Theta$ denotes the true parameter whereas $a \in \Theta$ stands for a potential guess. In our case the loss function corresponds to the effort which is necessary for the detection, localization and correction of wrong guesses (cf. Sects. 8 and 10). Of course, correct decisions do not cause any loss, i.e. $s(\theta, \theta) = 0$ for all $\theta \in \Theta$. A decision strategy is optimal if its expected loss attains a minimum.

Optimal Decision Strategy. Let the *a priori* distribution η be given by (15). Let $\tau_{opt}((\boldsymbol{q}'_k, \boldsymbol{q}_{i,\ldots,i+f-1,k})_{1 \leq k \leq N}) := a^*$ if the sum

$$\sum_{\theta \in \Theta} s(\theta, a') p_\theta \left((\boldsymbol{q}_{i,\ldots,i+f-1;k} \mid \boldsymbol{q}'_k)_{1 \leq k \leq N} \right) \eta(\theta) \qquad (16)$$

is minimal for $a' = a^*$ (i.e. the attacker picks $a^* \in \Theta$ when he observes the vector $(\boldsymbol{q}'_k, \boldsymbol{q}_{i,\ldots,i+f-1,k})_{1 \leq k \leq N}$). Then τ_{opt} is optimal among all decision strategies which estimate $T(i), \ldots, T(i+f-1)$ simultaneously.

Proof. The proof of the analogous assertion in Section 7 of [12] can be re-applied here almost literally. The conditional probabilities $p_\theta(\cdot|\cdot)$ were computed in Theorem 1.   □

10   Experimental Results

Although theoretical considerations were not significantly more difficult for $f \geq 1$, in this section we fix $f=1$ in order to simplify the calculations (cf. [12], §8). Thus the types of the particular Montgomery multiplications are guessed separately. The restriction on f means it is convenient to attack a sliding window exponentiation scheme, and $m = 4$ (i.e. $b = 2$) was chosen for a simulation. Extra reduction values $q'_{1,k}, \ldots, q'_{m-1,k}$ and $q_{1,k}, q_{2,k}, \ldots$ were obtained from the simulation using pseudo-randomly chosen moduli M and inputs $C_1, \ldots, C_N$.

We first determined the optimal decision strategy given in Section 9. We used the loss function given by $s(`S', `M_j') = 1$ (type-a error), $s(`M_j', `S') = 1.5$ (type-b error) and $s(`M_i', `M_j') = 2.5$ for $i \neq j$ (type-c error). Equation (15) gives the a priori distribution $\eta(`S') = 0.75$ and $\eta(`M_1') = \eta(`M_3') = 0.125$. Next, we computed approximations as follows for the density $h_{i-1(k)}$ for each of the three alternatives $\theta = `S'$, $\theta = `M_1'$ and $\theta = `M_3'$. First, Theorem 2 was applied to obtain the 'pure' limit densities f (cf. Thm. 2(iii)) and $f_{(s)}$ (cf. Thm. 2(v)). The iterates of the densities $f_{up}(x) := 1_{[0,1)}(x)$ and $f_{low}(x) := 1_{[\gamma^{-1}-1, \gamma^{-1})}$ (cf. Thm. 2(iv)) squeeze the respective limit distribution. The convergence is monotonic and exponentially fast (Thm. 2(iv)). If $T(i) = `M_j'$ then at least two squarings had been carried out just before. As the convergence to $f$ is exponentially fast we assumed $h_{i-1(k)} \approx f$ in that case. For the hypothesis $T(i) = `S'$ we set $h_{i-1(k)} := \eta_0 \cdot f + \eta_1 \cdot f_{(s_1)} + \eta_3 \cdot f_{(s_3)}$ with $\eta_1 = \eta_3 = 0.125$ and $\eta_0 = 0.75$, and s_j denoting the ratio of table entry j divided by the modulus M. Then we put the pieces together, determined the conditional probabilities $p_{\cdot S'}(\cdot)$, $p_{\cdot M_1'}(\cdot)$ and $p_{\cdot M_3'}(\cdot)$ using Theorem 1, and so derived the optimal decision strategy.

Table 1. Average number of errors per 100 guesses with $b = 2$, $f = 1$.

M/R	N	type-a	global type-a	type-b	type-c
0.99	350	0.53	0.11	0.29	0.67
0.99	400	0.37	0.07	0.21	0.04
0.85	400	0.74	1.58	0.12	0.06
0.85	450	0.54	0.11	0.62	0.03
0.85	500	0.44	0.08	0.03	0.25
0.70	700	1.24	0.19	0.22	0.35

Applying this optimal decision strategy we obtained guesses $\widetilde{T}(1), \widetilde{T}(2), \ldots$. A large number of simulation runs gave the results in Tables 1 to 3. The "type-a" column in Table 1 covers all errors of type a, namely both the locally correctable and global type-a errors. In a first step the attacker corrects the locally correctable type-a errors. Usually, he knows a reference equation $y^d \equiv x \pmod{M}$, e.g. a signature. Using this he can check whether a guess $\widetilde{d}$ for d is correct. As already observed in Section 8, the number of global errors is relevant for the practical feasibility of the attack. This is the sum of the type-b, type-c and global type-a errors. Table 2 gives the percentage of trials for which the number of such global errors is no more than a given bound. For example, at most one global error occurred in 76% of the trials for the parameter set $M/R \approx 0.85$, $N = 500$, $n = 512$. Clearly, for the sake of efficiency the attacker first tries to change those guesses for which the decision has been 'close'.

Table 2. Number of global errors.

M/R	N	n	0	≤ 1	≤ 2	≤ 3
0.99	350	512	10%	31%	49%	64%
0.99	400	512	16%	46%	62%	78%
0.85	400	512	19%	43%	60%	71%
0.85	450	512	33%	62%	80%	90%
0.85	500	512	46%	76%	90%	97%
0.70	700	512	35%	60%	71%	76%

A successful attack on a 512-bit exponent requires about 680 correct guesses. For this, about $2 \cdot 680 = 1360$ hypotheses have to be rejected. To each rejected hypothesis about the types of a sequence of $f{=}1$ multiplicative operations, we assign the ratio between the expected loss if this hypothesis had been chosen divided by the expected loss for the hypothesis chosen in this decision. The rejected hypotheses are ordered using these ratios, that with the smallest ratio (i.e. the most likely alternative) first. If the estimator $\tilde{d}$ is false the attacker replaces one, two or three guesses respectively by those from the rejected list, beginning with the first, i.e. that with lowest ratio.

Table 3 gives the average rank of the lowest correct hypothesis which has been rejected for a given number of global errors. For instance, if there were three global errors, and the correct guesses for them were ranked 3, 29 and 53 in the list, then the rank of the lowest would be 53. The second row of the table says that if there are exactly three global errors for parameters $M/R \approx 0.99$, $N = 400$, $n = 512$ then 57 is the average rank of the lowest correct hypothesis which was initially rejected. If the lowest rank is ≤ 100 (which would normally be the case for an average of 57), the correction requires at most $\binom{100}{3} = 161700$ evaluations of the reference equation (neglecting the unsuccessful efforts to correct exactly 1 or 2 global errors). This is clearly computationally feasible.

Table 3. Average rank of the last correct hypothesis for 1, 2 or 3 global errors.

M/R	N	n	1	2	3
0.99	350	512	31	66	63
0.99	400	512	30	25	57
0.85	400	512	39	57	55
0.85	450	512	22	37	59
0.85	500	512	24	57	70
0.70	700	512	63	132	271

The conditional probabilities (Section 6), and hence the optimal decision strategy, only depend on the ratio $\gamma = M/R$. In our simulations we assumed that the attacker knows this ratio. However, our attack is also feasible if the attacked device uses the Chinese Remainder Theorem (CRT). Then the attacker uses the extra reductions within the initialization phase and known squarings from the computation stage to estimate the parameter γ. The moduli M for the exponentiations are the prime factors of the RSA modulus $M'=p_1p_2$. For the secret RSA key d', the attacker guesses the exponents actually used, namely $d = d'(\bmod (p_i-1))$ for $i = 1$ or 2. If the guess $\tilde{d}$ for d is correct then $\gcd(C^{d'}(\bmod M') - C^{\tilde{d}}(\bmod M'), M') = p_i$ and, similarly, $\gcd(C - C^{\tilde{d}e}(\bmod M'), M') = p_i$. Since the parameter γ has to be estimated the error probability for a singular decision increases somewhat (cf. [12], first paragraph of Sect.10). However, since CRT involves an exponent of only half the length, the number of wrong guesses per exponentiation will be smaller and so the attack more likely to be successful.

11   Conclusion

A timing attack on RSA implementations has been given further detail in the more complex situation of modular reductions being driven by a register length bound rather than a modulus bound. Graphs of the limiting distributions were drawn illustrating one source of the attack. Another source was the combination of exact conditional probabilities for the modular reductions with statistical decision theory for treating sequences of modular multiplications. This reduces the sample size necessary to deduce the secret RSA key from side channel leakage. The resulting powerful methods reduce the number of errors far enough for their correction to be computationally feasible for keys with standard lengths using data obtained well within the normal lifespan of a key.

References

1. J.-F. Dhem, F. Koeune, P.-A. Leroux, P. Mestré, J.-J. Quisquater & J.-L. Willems, *A practical implementation of the Timing Attack*, Proc. CARDIS 1998, J.-J. Quisquater & B. Schneier (editors), LNCS **1820**, Springer-Verlag, 2000, pp. 175–190.
2. D. E. Knuth, The Art of Computer Programming, vol. 2, *Seminumerical Algorithms*, 2nd Edition, Addison-Wesley, 1981, pp. 441–466.
3. Ç. K. Koç, *Analysis of Sliding Window Techniques for Exponentiation*, Computers and Mathematics with Applications **30**, no. 10, 1995, pp. 17–24.
4. P. Kocher, *Timing attack on implementations of Diffie-Hellman, RSA, DSS, and other systems*, Proc. Crypto 96, N. Koblitz (editor), LNCS **1109**, Springer-Verlag, 1996, pp. 104–113.
5. P. Kocher, J. Jaffe & B. Jun, *Differential Power Analysis*, Advances in Cryptology – Crypto '99, M. Wiener (editor), LNCS **1666**, Springer-Verlag, 1999, pp. 388–397.
6. R. Mayer-Sommer, *Smartly Analyzing the Simplicity and the Power of Simple Power Analysis on Smartcards*, Cryptographic Hardware and Embedded Systems (Proc CHES 2000), C. Paar & Ç. Koç (editors), LNCS **1965**, Springer-Verlag, 2000, pp. 78–92.
7. T. S. Messerges, E. A. Dabbish, R. H. Sloan, *Power Analysis Attacks of Modular Exponentiation in Smartcards*, Cryptographic Hardware and Embedded Systems (Proc CHES 99), C. Paar & Ç. Koç (editors), LNCS **1717**, Springer-Verlag, 1999, pp. 144–157.
8. V. Miller, *Use of Elliptic Curves in Cryptography*, Proc. CRYPTO '85, H. C. Williams (editor), LNCS **218**, Springer-Verlag, 1986, pp. 417–426.

9. P. L. Montgomery, *Modular Multiplication without Trial Division*, Mathematics of Computation **44**, no. 170, 1985, pp. 519–521.
10. R. L. Rivest, A. Shamir & L. Adleman, *A Method for obtaining Digital Signatures and Public-Key Cryptosystems*, Comm. ACM **21**, 1978, pp. 120–126.
11. W. Schindler, *A Timing Attack against RSA with Chinese Remainder Theorem*, Cryptographic Hardware and Embedded Systems (Proc CHES 2000), C. Paar & Ç. Koç (editors), LNCS **1965**, Springer-Verlag, 2000, pp. 109–124.
12. W. Schindler, *A Combined Timing and Power Attack*, Public Key Cryptography (Proc PKC 2002), P. Paillier & D. Naccache (editors), LNCS **2274**, Springer-Verlag, 2002, pp. 263–279.
13. W. Schindler, *Optimized Timing Attacks against Public Key Cryptosystems*, Statistics & Decisions **20**, 2002, pp. 191–210.
14. C. D. Walter, *Montgomery Exponentiation Needs No Final Subtractions*, Electronics Letters **35**, no. 21, October 1999, pp. 1831–1832.
15. C. D. Walter, *Precise Bounds for Montgomery Modular Multiplication and Some Potentially Insecure RSA Moduli*, Proc. CT-RSA 2002, B. Preneel (editor), LNCS **2271**, Springer-Verlag, 2002, pp. 30–39.
16. C. D. Walter & S. Thompson, *Distinguishing Exponent Digits by Observing Modular Subtractions*, Topics in Cryptology − CT-RSA 2001, D. Naccache (editor), LNCS **2020**, Springer-Verlag, 2001, pp. 192–207.

Appendix

Lemma 3. *Let τ and ν denote probability measures on a locally compact abelian group G. If μ is a Haar measure on G and ν has a μ-density then the convolution product $\tau * \nu$ also has a μ-density.*

Proof. For any measurable $B \subseteq G$ we have $\tau * \nu(B) = \int_G \nu(B-x)\,\tau(dx)$. If $\mu(B) = 0$ then $\mu(B-x) = 0$ and hence $\nu(B-x) = 0$. This proves the lemma. $\square$

Theorem 2. *Suppose $\gamma \in (0.5, 1)$, and let F and $\chi \colon [0, \gamma^{-1}+1) \to \mathbb{R}$ be defined as in Section 6. Assume further that $U_0, V_1, V_2, \ldots$ denote independent random variables where U_0 assumes values on $[0, \gamma^{-1})$, while $V_1, V_2, \ldots$ are equidistributed on the unit interval $[0, 1)$. Finally, let $U_{n+1} := \chi(U_n^2\gamma + V_{n+1})$ for all $n \in \mathbb{N}$.*
(i) Regardless of the distribution of U_0, for each $n \geq 1$ the distribution μ_n of U_n has a Lebesgue density f_n. Moreover, $f_n(x) = 1$ for $x \in [\gamma^{-1}-1, 1)$ and $f_n(x) + f_n(x+1) = 1$ for $x \in [0, \gamma^{-1}-1)$.
(ii) For $n \geq 2$ we have

$$f_{n+1}(x) = \int_0^{\sqrt{x\gamma^{-1}}} f_n(u)\,du \qquad \text{for } x \in [0, \gamma^{-1}-1). \tag{17}$$

(iii) Regardless of the distribution of U_0 we have

$$\|f_{n+1} - f_n\|_\infty := \sup_{x \in [0, \gamma^{-1})} |f_{n+1}(x) - f_n(x)| \leq \left(\gamma^{-1}-1\right)^{n-1} \|f_2 - f_1\|_\infty \tag{18}$$

for all $n \geq 2$. *In particular, the sequence* $f_1, f_2, \ldots$ *converges uniformly to a probability density* $f: [0, \gamma^{-1}) \to [0, 1]$ *which does not depend on the distribution of* U_0. *To be precise,* $f(x) = 1$ *for* $x \in [\gamma^{-1}-1, 1)$, $f(x) + f(x+1) = 1$ *for* $x \in [0, \gamma^{-1}-1)$, *and* f *is the unique solution of (17) possessing these properties, i.e. (17) holds with* f *in place of* f_n *and* f_{n+1}.

(iv) *If* $f_1 \leq f$ *on* $[0, \gamma^{-1}-1)$ *then also* $f_n \leq f$ *for all* $n \geq 1$. *If* $f_1 \leq f_2$ *then the sequence* $f_1, f_2, \ldots$ *is monotonically increasing on* $[0, \gamma^{-1}-1)$. *Similarly,* $f_1 \geq f$ *on* $[0, \gamma^{-1}-1)$ *implies* $f_n \geq f$ *for all* $n \geq 1$, *and if* $f_1 \geq f_2$ *then the sequence* $f_1, f_2, \ldots$ *is monotonically decreasing on* $[0, \gamma^{-1}-1)$. *If the distribution of* U_0 *has a Lebesgue density* f_0 *then the assertions of (iv) even hold for* $f_0, f_1, \ldots$ *in place of* $f_1, f_2, \ldots$.

(v) *Let* $U_0' := U_0$ *and* $U_{n+1}' := \chi(U_n' s \gamma + V_{n+1})$ *for all* $n \in \mathbb{N}$ *where* $s \in [0, \gamma^{-1})$ *is fixed. Then all assertions from (i) to (iv) can be transferred almost literally to this case. In particular, for* $x \in [0, \gamma^{-1}-1)$ *we have*

$$f_{n+1}(x) = \begin{cases} \int_0^{x/s\gamma} f_n(u) \, du & \text{if } x < s \\ 1 & \text{otherwise.} \end{cases} \tag{19}$$

The limit distribution $f_{(s)}$ *of* $f_1, f_2, \ldots$ *is the unique solution of (19) with* $f_{(s)}(x) = 1$ *for* $x \in [\gamma^{-1}-1, 1)$ *and such that* $f_{(s)}(x) + f_{(s)}(x+1) = 1$ *for* $x \in [0, \gamma^{-1}-1)$.

Proof. To prove the first assertion of (i) we apply Lemma 3 with $G = \mathbb{R}$ and $\mu = \lambda$ while τ is the distribution of $U_{n-1}^2 \gamma$ and ν the restriction of the Lebesgue measure to $[0, 1)$. If $\gamma^{-1}-1 \leq a < b \leq 1$ we have $U_n \in [a, b)$ iff $U_{n-1}^2 \gamma + V_n \in [a, b) \cup [a+1, b+1)$. As U_{n-1} and V_n are independent, and χ coincides with the reduction modulo 1 on this union and V_n is equidistributed on $[0, 1)$, this proves the second assertion of (i). However, for any $0 \leq a < b \leq \gamma^{-1}-1$, we have $U_n \in [a, b) \cup [a+1, b+1)$ iff $U_{n-1}^2 \gamma + V_n \in [a, b) \cup [a+1, b+1) \cup [a+2, b+2)$. From $\mod 1 = \mod 1 \circ \chi$ we obtain the final assertion of (i) as V_n is equidistributed on $[0, 1)$. For $x \in [0, \gamma^{-1}-1)$ the pre-image $\chi^{-1}(x)$ equals x and hence

$$\text{Prob}(U_{n+1} \leq x) = \int_0^x f_{n+1}(u) \, du = \int_0^{\sqrt{x\gamma^{-1}}} f_n(u) \text{Prob}(u^2 \gamma + V_n \leq x) \, du$$

$$= \int_0^{\sqrt{x\gamma^{-1}}} f_n(u)(x - u^2 \gamma) \, du$$

$$= x \int_0^{\sqrt{x\gamma^{-1}}} f_n(u) \, du - \gamma \int_0^{\sqrt{x\gamma^{-1}}} f_n(u) u^2 \, du.$$

If f_n is continuous at $\sqrt{x\gamma^{-1}}$ differentiating the left- and the right-hand side verifies assertion (ii) for $f_{n+1}(x)$. We denote the density of $U_1^2 \gamma + V_2$ by h_2 for the moment. Applying the convolution formula for densities (to that of $U_1^2 \gamma$ and V_2) yields $|h_2(x+s) - h_2(x)| \leq \text{Prob}(U_1^2 \gamma \in [-s, 0] \cup [1, 1+s])$ for $|s| < 0.5$. As $[-s, 0] \cup [1, 1+s]$ converges to $\{0, 1\}$ the right-hand probability converges to 0 as $s \to 0$, i.e. h_2 is continuous at x. As x was arbitrary this proves the continuity of

h_2, and f_2 has at most two discontinuity points in $(0, \gamma^{-1})$, namely $\gamma^{-1}-1$ and 1. This proves (ii) for $n = 1$ since densities may be defined arbitrarily on zero sets. For arbitrary n, assertion (ii) follows by induction. Clearly,

$$\|f_{n+1} - f_n\|_\infty = \sup_{x \in [0, \gamma^{-1}-1)} |f_{n+1}(x) - f_n(x)|$$

$$= \sup_{x \in [0, \gamma^{-1}-1)} \left| \int_0^{\sqrt{x\gamma^{-1}}} (f_n(u) - f_{n-1}(u)) \, du \right|$$

$$\leq \sup_{y,z \in [0, \gamma^{-1}-1)} \left| \int_y^z (f_n(u) - f_{n-1}(u)) \, du \right| \leq (\gamma^{-1}-1) \|f_n - f_{n-1}\|_\infty.$$

The first inequality follows from the fact that $f_n(x) = f_{n-1}(x) = 1$ on $[\gamma^{-1}-1, 1)$ while $f_n(x) + f_n(x+1) = 1$ for $x \in [0, \gamma^{-1}-1)$. Equation (18) follows by induction. Similarly, one concludes $\|f_n - f_n^*\|_\infty \leq (\gamma^{-1}-1)^{n-2} \|f_2 - f_2^*\|_\infty$ for arbitrary sequences $f_1, f_2, \dots$ and $f_1^*, f_2^*, \dots$. This shows the uniqueness of f and, as $f_1, f_2, \dots$ converges uniformly, f fulfils (17). If, in addition, $f_1(x) \leq f_2(x)$ for all $x \in [0, \gamma^{-1}-1)$ then by induction we obtain

$$f_{n+1}(x) - f_n(x) = \int_0^{\sqrt{x\gamma^{-1}}} (f_n(u) - f_{n-1}(u)) \, du$$

$$= \int_{[0,\sqrt{x\gamma^{-1}}) \cap [0, \gamma^{-1}-1) \cap \{u:u > \sqrt{x\gamma^{-1}-1}\}} (f_n(u) - f_{n-1}(u)) \, du \geq 0.$$

Replacing f_n by f we obtain the first assertion of (iv), and the second part can be shown similarly. Assertion (v) can be verified in same way as (i) to (iv).   □

Predicting the Inversive Generator

Simon R. Blackburn[1], Domingo Gomez-Perez[2], Jaime Gutierrez[2], and
Igor E. Shparlinski[3]

[1] Department of Mathematics, Royal Holloway, University of London
Egham, Surrey, TW20 0EX, UK
s.blackburn@rhul.ac.uk
[2] Faculty of Science, University of Cantabria
E-39071 Santander, Spain
{gomezd,jaime.gutierrez}@unican.es
[3] Department of Computing, Macquarie University,
NSW 2109, Australia
igor@comp.mq.edu.au

Abstract. Let p be a prime and let a and b be integers modulo p. The
inversive congruential generator (ICG) is a sequence (u_n) of pseudorandom numbers defined by the relation $u_{n+1} \equiv au_n^{-1} + b \bmod p$. We show
that if b and sufficiently many of the most significant bits of three consecutive values u_n of the ICG are given, one can recover in polynomial time
the initial value u_0 (even in the case where the coefficient a is unknown)
provided that the initial value u_0 does not lie in a certain small subset
of exceptional values.

1 Introduction

For a prime p, denote by $\mathbb{F}_p$ the field of p elements and always assume that it is
represented by the set $\{0, 1, \dots, p-1\}$. Accordingly, sometimes, where obvious,
we treat elements of $\mathbb{F}_p$ as integer numbers in the above range.

For fixed $a, b \in \mathbb{F}_p^*$, let $\psi_{a,b}$ be the permutation of $\mathbb{F}_p$ defined by

$$\psi_{a,b}(w) = \begin{cases} aw^{-1} + b, & \text{if } w \neq 0, \\ b, & \text{if } w = 0. \end{cases}$$

We refer to the coefficients a and b as the *multiplier* and *shift*, respectively.

We define the *inversive generator* (u_n) of elements of $\mathbb{F}_p$ by the recurrence
relation

$$u_{n+1} = \psi_{a,b}(u_n), \qquad n = 0, 1, \dots, \tag{1}$$

where u_0 is the *initial value*.

This generator has proved to be extremely useful for Quasi-Monte Carlo type
applications, and in particular exhibits very attractive uniformity of distribution

K.G. Paterson (Ed.): Cryptography and Coding 2003, LNCS 2898, pp. 264–275, 2003.

and nonlinearity properties, see [11,12,13,14] for surveys or recent results. This paper concentrates on the cryptographic properties of the inversive generator.

In the cryptographic setting, the initial value u_0 and the constants a and b are assumed to be the secret key, and we want to use the output of the generator as a stream cipher. Of course, if several consecutive values u_n are revealed, it is easy to find u_0, a and b. So in this setting, we output only the most significant bits of each u_n in the hope that this makes the resulting output sequence difficult to predict. In a recent paper [2], we have shown that not too many bits can be output at each stage: the inversive generator is unfortunately polynomial time predictable if sufficiently many bits of its consecutive elements are revealed, so long as a small number of secret keys are excluded. However, most of the results of [2] only hold after excluding a small set of pairs (a, b). If this small set is not excluded, the algorithm for finding the secret information may fail. An optimist might hope that by deliberately choosing the pair (a, b) to lie in this excluded set, one can generate cryptographically stronger sequences. This paper aims to show that this strategy is unlikely to succeed. Namely we introduce some modifications and additions to the method of [2] which allow us to attack the generators no matter how the values of a and b are chosen. We demonstrate our approach in the special case when b is public. Of course, the assumption that b is public reduces the relevance of the problem to cryptography. But we believe that the extra strength of the result we obtain makes this situation of interest in its own right. We also believe this approach can be extended to the case when both a and b are secret.

Assume that the sequence (u_n) is not known but, for some n, approximations w_j of 3 consecutive values u_{n+j}, $j = 0, 1, 2$, are given. We show that if b is public, the values u_{n+j} and a can be recovered from this information in polynomial time if the approximations w_j are sufficiently good and if a certain small set of initial values u_0 are excluded. (The results in [2] exclude a small set of pairs (a, b) in addition to values of u_0, and so in this sense our result here is stronger.)

Throughout the paper the term polynomial time means polynomial in $\log p$. Our results involve another parameter Δ which measures how well the values w_j approximate the terms u_{n+j}. This parameter is assumed to vary independently of p subject to satisfying the inequality $\Delta < p$ (and is not involved in the complexity estimates of our algorithms).

We should emphasise that this paper is concerned with rigorous results (see [2] for a discussion of both rigorous and heuristic methods).

The remainder of the paper is structured as follows.

We start with a short outline of some basic facts about lattices in Section 2.1 and rational functions Section 2.2. In Section 3.1 we formulate our main results and outline the plan of the proof, which is given in Section 3.2. Finally, Section 4 makes some final comments and poses several open questions.

Acknowledgment.   The authors would like to thank Harald Niederreiter for his interest and helpful discussions. This paper was written during visits of I.S. to the University of Cantabria (partially supported by MECD grant SAB2000-

0260) and to Royal Holloway, University of London (supported by an EPSRC Visiting Fellowship). D. G.-P. and J.G. were partially supported by Spanish Ministry of Science grant BFM2001-1294. The support and hospitality of all these organisations are gratefully acknowledged.

2   Lattices and Rational Functions

2.1   Background on Lattices

Here we collect several well-known facts about lattices which form the background to our algorithms.

We review several related results and definitions on lattices which can be found in [3]. For more details and more recent references, we also recommend consulting [1,4,5,8,9,10].

Let $\{\mathbf{b}_1, \dots, \mathbf{b}_s\}$ be a set of linearly independent vectors in $\mathbb{R}^r$. The set

$$\mathcal{L} = \{\mathbf{z} \; : \; \mathbf{z} = c_1 \mathbf{b}_1 + \dots + c_s \mathbf{b}_s, \quad c_1, \dots, c_s \in \mathbb{Z}\}$$

is called an *s-dimensional lattice* with *basis* $\{\mathbf{b}_1, \dots, \mathbf{b}_s\}$. If $s = r$, the lattice L is of *full rank*.

To each lattice $\mathcal{L}$ one can naturally associate its *volume*

$$\mathrm{vol}\,(\mathcal{L}) = \Big(\det\left(\langle \mathbf{b}_i, \mathbf{b}_j \rangle \right)_{i,j=1}^{s} \Big)^{1/2},$$

where $\langle \mathbf{a}, \mathbf{b} \rangle$ denotes the inner product, which does not depend on the choice of the basis $\{\mathbf{b}_1, \dots, \mathbf{b}_s\}$.

For a vector $\mathbf{u}$, let $\|\mathbf{u}\|$ denote its *Euclidean norm*. The famous Minkowski theorem, see Theorem 5.3.6 in Section 5.3 of [3], gives the upper bound

$$\min\,\{\|\mathbf{z}\|\colon \; \mathbf{z} \in \mathcal{L} \setminus \{\mathbf{0}\}\} \leq s^{1/2}\,\mathrm{vol}\,(\mathcal{L})^{1/s} \tag{2}$$

on the shortest nonzero vector in any s-dimensional lattice $\mathcal{L}$ in terms of its volume. In fact $s^{1/2}$ can be replaced by the *Hermite constant* $\gamma_s^{1/2}$, for which we have

$$\frac{1}{2\pi e}s + o(s) \leq \gamma_s \leq \frac{1.744}{2\pi e}s + o(s), \qquad s \to \infty.$$

The Minkowski bound (2) motivates a natural question: how to find the shortest vector in a lattice. The celebrated *LLL algorithm* of Lenstra, Lenstra and Lovász [7] provides a desirable solution in practice, and the problem is known to be solvable in deterministic polynomial time (polynomial in the bit-size of the basis of $\mathcal{L}$) provided that the dimension of $\mathcal{L}$ is fixed (see Kannan [6, Section 3], for example). The lattices in this paper are of fixed dimension. (Note that there are several indications that the shortest vector problem is **NP**-complete when the dimension grows.)

In fact, in this paper we consider only very special lattices. Namely, only lattices which are consisting of integer solutions $\mathbf{x} = (x_0, \ldots, x_{s-1}) \in \mathbb{Z}^s$ of the system of congruences

$$\sum_{i=0}^{s-1} a_{ij} x_i \equiv 0 \bmod q_j, \qquad j = 1, \ldots, m,$$

modulo some integers $q_1, \ldots, q_m$. Typically (although not always) the volume of such a lattice is the product $Q = q_1 \ldots q_m$. Moreover all the aforementioned algorithms, when applied to such a lattice, become polynomial in $\log Q$.

2.2 Zeros of Rational Functions

Our second basic tool is essentially the theorem of Lagrange which asserts that a non-zero polynomial of degree N over any field has no more than N zeros in this field. In fact we apply it to rational functions which require only obvious adjustments.

The rational functions we consider belong to a certain family of functions parametrised by small vectors in a certain lattice, thus the size of the family can be kept under control. Zeros of these rational functions correspond to potentially "bad" initial values of the inversive generator (1). Thus, if all rational functions in this family are not identical to zero modulo p then we have an upper bound on the number of such "bad" initial values. Hence, a crucial part of our approach is to study possible vanishing of functions in the above family and to show that this may happen only for very few values of the coefficients of the generator (1). To establish this property we repeatedly use the fact that non-trivial linear combinations of rational functions with pairwise distinct poles do not vanish identically.

3   Predicting the Inversive Generator with Unknown Multiplier

3.1   Formulation of the Main Result and Plan of Proof

Assume the multiplier a of the inversive generator is unknown, but shift b is given to us. We show that we can recover u_0 and a for all but $O(\Delta^5)$ values of u_0 when given approximations to three consecutive values u_n, u_{n+1}, u_{n+2} produced by the inversive generator, except when u_0 lies in a small set of exceptional values. To simplify the notation, we assume that $n = 0$ from now on.

Theorem. *Let p be a prime number and let Δ be an integer such that $p > \Delta \geq 1$. Let $a, b \in \mathbb{F}_p^*$. There exists a set $\mathcal{U}(\Delta; a, b) \subseteq \mathbb{F}_p$ of cardinality $\#\mathcal{U}(\Delta; a, b) = O(\Delta^5)$ with the following property. Whenever $u_0 \notin \mathcal{U}(\Delta; a, b)$ then, given approximations $|w_j - u_j| \leq \Delta$, $j = 0, 1, 2$ to three consecutive values u_0, u_1, u_2 produced by the inversive generator (1), and given the value of b, one can recover u_0 and a in deterministic polynomial time.*

An outline of the algorithm given in the proof of this Theorem goes as follows. The algorithm is divided into six stages.

Stage 1: We assume that the two exceptional values 0 and $-a/b$ lie in $\mathcal{U}(\Delta; a, b)$. We construct a certain lattice $\mathcal{L}$ (see (5) below) of dimension five; this lattice depends on the approximations w_0, w_1, w_2 and the integer b. We also show that a certain vector $\mathbf{e}$ directly related to missing information about u_0, u_1, u_2 is a very short vector in this lattice. A shortest nonzero vector $\mathbf{f} = (f_0, \ldots, f_4)$ in $\mathcal{L}$ is found; see [6] for the appropriate algorithm.

Stage 2: We show that $\mathbf{f}$ provides some valuable information about $\mathbf{e}$ for all initial values u_0 except for u_0 from a certain exceptional set $\mathcal{V}(\Delta; a, b) \subseteq \mathbb{F}_p$ of cardinality $\#\mathcal{V}(\Delta; a, b) = O(\Delta^5)$ (which is defined as a set of zeros of a certain parametric family of rational functions).

Stage 3: We show that if $f_0 \neq 0$ then recovering $\mathbf{e}$ (and hence the secret information u_0 and a) from $\mathbf{f}$ is straight forward. If $f_0 \neq 0$, the algorithm terminates at this stage.

Stage 4: We show that if $f_0 = 0$ then the vector $\mathbf{f}$ enables us to compute small integers r and s such that $b = r/s \bmod p$. (In fact these integers can be found independently by the continued fraction algorithm.) The algorithm uses this information, together with the integers w_0, w_1, w_2 and b, to compute a second lattice $\mathcal{L}'$ of dimension four. There is a short vector $\mathbf{e}'$ in $\mathcal{L}$, and again this vector is closely related to the secret information u_0 and a.

Stage 5: We show that all short vectors in $\mathcal{L}'$ are parallel to $\mathbf{e}'$ for all initial values u_0 except for u_0 from another exceptional set $\mathcal{V}'(\Delta; a, b) \subseteq \mathbb{F}_p$ of cardinality $\#\mathcal{V}'(\Delta; a, b) = O(\Delta^5)$ (which is also defined as a set of zeros of a certain parametric family of rational function).

Stage 6: We find a shortest nonzero vector $\mathbf{f}'$ in $\mathcal{L}'$ and show that if $u_0 \notin \mathcal{U}(\Delta; a, b)$, where

$$\mathcal{U}(\Delta; a, b) = \{0, -a/b\} \cup \mathcal{V}(\Delta; a, b) \cup \mathcal{V}'(\Delta; a, b)$$

then recovering $\mathbf{e}'$ (and thus finding the secret information) from $\mathbf{f}$ and $\mathbf{f}'$ is now straightforward.

3.2   Proof of the Main Result

The theorem is trivial when $\Delta^5 \geq p$, and so we assume that $\Delta^5 < p$. Let us fix $a, b \in \mathbb{F}_p^*$. We assume that $u_0 \in \mathbb{F}_p$ is chosen so as not to lie in a certain subset $\mathcal{U}(\Delta; a, b)$ of $\mathbb{F}_p^*$. This subset is of cardinality $O(\Delta^5)$, but as its definition is fairly complicated we define it gradually as we move through the proof.

Stage 1: Building the lattice $\mathcal{L}$. We begin by defining a lattice $\mathcal{L}$, and showing how knowing a short vector in $\mathcal{L}$ usually leads to the recovery of the secret information.

We may assume that $u_0 u_1 \not\equiv 0 \bmod p$, for clearly there are at most two values of u_0, namely $u_0 \equiv 0 \pmod p$ and $u_0 \equiv -a/b \pmod p$ for which this does not hold, and we place these two values in $\mathcal{U}(\Delta; a, b)$. From

$$u_1 \equiv a u_0^{-1} + b \bmod p \qquad \text{and} \qquad u_2 \equiv a u_1^{-1} + b \bmod p$$

we derive

$$u_1 u_0 \equiv a + b u_0 \bmod p \qquad \text{and} \qquad u_1 u_2 \equiv a + b u_1 \bmod p. \tag{3}$$

Therefore,

$$u_1(u_2 - u_0) \equiv b(u_1 - u_0) \bmod p. \tag{4}$$

For $j \in \{0, 1, 2\}$, define $\varepsilon_j = u_j - w_j$. We have that $|\varepsilon_j| \leq \Delta$. Now (4) becomes

$$(w_1 + \varepsilon_1)(w_2 - w_0 + \varepsilon_2 - \varepsilon_0) \equiv b(w_1 - w_0) + b(\varepsilon_1 - \varepsilon_0) \bmod p.$$

Writing

$$A \equiv (w_1(w_2 - w_0) - b(w_1 - w_0))\Delta^{-2} \bmod p, \quad B_0 \equiv -(w_1 + b)\Delta^{-1} \bmod p,$$
$$B_1 \equiv (w_2 - w_0 - b)\Delta^{-1} \bmod p, \quad B_2 \equiv w_1 \Delta^{-1} \bmod p \quad \text{and} \quad C \equiv 1 \bmod p,$$

we obtain

$$A\Delta^2 + B_0\Delta\varepsilon_0 + B_1\Delta\varepsilon_1 + B_2\Delta\varepsilon_2 + C\varepsilon_1(\varepsilon_2 - \varepsilon_0) \equiv 0 \bmod p.$$

Therefore the lattice $\mathcal{L}$ consisting of solutions $\mathbf{x} = (x_0, x_1, x_2, x_3, x_4) \in \mathbb{Z}^5$ of the congruences

$$Ax_0 + B_0 x_1 + B_1 x_2 + B_2 x_3 + C x_4 \equiv 0 \bmod p,$$
$$x_0 \equiv 0 \bmod \Delta^2, \tag{5}$$
$$x_1 \equiv x_2 \equiv x_3 \equiv 0 \bmod \Delta,$$

contains a vector

$$\mathbf{e} = (\Delta^2 e_0, \Delta e_1, \Delta e_2, \Delta e_3, e_4) = (\Delta^2, \Delta\varepsilon_0, \Delta\varepsilon_1, \Delta\varepsilon_2, \varepsilon_1(\varepsilon_2 - \varepsilon_0)).$$

We have

$$e_0 = 1, \qquad |e_1|, |e_2|, |e_3| \leq \Delta, \qquad |e_4| \leq 2\Delta^2$$

thus the Euclidean norm $\|\mathbf{e}\|$ of $\mathbf{e}$ satisfies the inequality

$$\|\mathbf{e}\| \leq (\Delta^4 + \Delta^4 + \Delta^4 + \Delta^4 + 4\Delta^4)^{1/2} \leq 3\Delta^2.$$

Let $\mathbf{f} = (\Delta^2 f_0, \Delta f_1, \Delta f_2, \Delta f_3, f_4) \in \mathcal{L}$ be a shortest nonzero vector in $\mathcal{L}$. So $\|\mathbf{f}\| \leq \|\mathbf{e}\| \leq 3\Delta^2$. We have

$$|f_0| \leq \|\mathbf{f}\|\Delta^{-2} \leq 3, \qquad |f_1|, |f_2|, |f_3| \leq \|\mathbf{f}\|\Delta^{-1} \leq 3\Delta, \qquad |f_4| \leq \|\mathbf{f}\| \leq 3\Delta^2.$$

Note that we may compute $\mathbf{f}$ in polynomial time from the information we are given.

Stage 2: Defining the first exceptional set $\mathcal{V}(\Delta; a, b)$. The vector $\mathbf{d}$ defined by $f_0 \mathbf{e} - e_0 \mathbf{f}$ lies in $\mathcal{L}$ and has first component 0. We might hope that $\mathbf{e}$ and $\mathbf{f}$ are always parallel, in which case $\mathbf{d}$ would be the zero vector. Sadly, this is not always the case. So we claim that something weaker is true: namely that $d_2 = 0$ and $d_3 - d_1 = 0$ unless u_0 belongs to the set $\mathcal{V}(\Delta; a, b)$ which we define below. Before we establish this claim, we prove some facts about the vector $\mathbf{d}$.

Using the first congruence in (5), we find that

$$B_0 \Delta d_1 + B_1 \Delta d_2 + B_2 \Delta d_3 + C d_4 \equiv 0 \bmod p \tag{6}$$

where we define

$$d_i = f_0 e_i - e_0 f_i = f_0 e_i - f_i \text{ for } i \in \{0, 1, 2, 3\}. \tag{7}$$

Note that $|d_i| \leq 3|e_i| + |f_i|$ and hence

$$|d_1|, |d_2|, |d_3| \leq 6\Delta \qquad \text{and} \qquad |d_4| \leq 9\Delta^2. \tag{8}$$

Using the definitions of B_0, B_1, B_2 and C, we find that

$$-(w_1 + b)d_1 + d_2(w_2 - w_0 - b) + d_3 w_1 + d_4 \equiv 0 \bmod p,$$

and after the substitutions $w_i = u_i - \varepsilon_i$ we find

$$(d_3 - d_1)u_1 + d_2 u_2 - d_2 u_0 \equiv b(d_2 + d_1) + E \bmod p \tag{9}$$

where

$$E = -d_4 - \varepsilon_0 d_2 + \varepsilon_1(d_3 - d_1) + \varepsilon_2 d_2.$$

The bound (8) implies that $|E| \leq 33\Delta^2$. We now write this equality as a rational function of u_0. Setting

$$\Psi_1(u) = \frac{bu + a}{u} \qquad \text{and} \qquad \Psi_2(u) = \frac{(a + b^2)u + ab}{a + bu},$$

we have that $u_i = \Psi(u_0)$ for $i \in \{1, 2\}$. So (9) becomes

$$(d_3 - d_1)\Psi_1(u_0) + d_2 \Psi_2(u_0) - d_2 u_0 \equiv b(d_2 + d_1) + E \bmod p. \tag{10}$$

Let us consider the rational function

$$\Phi_{\mathbf{d}}(u) = (d_3 - d_1)\Psi_1(u) + d_2 \Psi_2(u) - d_2 u$$

corresponding to the left hand side of (10). Clearly, $\Phi_{\mathbf{d}}(u)$ can be written as the quotient of a polynomial of degree at most 3 and a polynomial of degree at most 2.

We assert that if $d_2 \neq 0$ or $d_3 - d_1 \neq 0$ then $\Phi_{\mathbf{d}}(u)$ is not a constant function. We prove the contrapositive implication. So assume that $\Phi_{\mathbf{d}}(u)$ is constant. Now $\Psi_1(u)$ is not constant, since $a \not\equiv 0 \bmod p$. So $\Psi_1(u)$ has a pole at 0 (and has no other poles). Similarly, $\Psi_2(u)$ is not constant and so has a pole at $-a/b$ (and no other poles). The functions $\Psi_1(u)$ and $\Psi_2(u)$ have poles at distinct places and u has no finite poles at all, so the only way that $\Phi_{\mathbf{d}}(u)$ can be the constant function is if $d_2 \equiv 0 \bmod p$ and $d_3 - d_1 \equiv 0 \bmod p$. But our bounds (8) on the size of $\mathbf{d}$ now imply that $d_2 = 0$ and $d_3 - d_1 = 0$. This establishes our assertion about $\Phi_{\mathbf{d}}(u)$.

Suppose that $d_2 \neq 0$ or $d_3 - d_1 \neq 0$. Since $\Phi_{\mathbf{d}}(u)$ is a nonconstant quotient of two polynomials of degree at most 3, the congruence (10) can be satisfied for at most 3 values of u_0 once d_1, d_2, d_3 and E have been chosen. There are $O(\Delta)$ choices for each of d_1, d_2 and d_3, by (8). There are $O(\Delta^2)$ choices for E since $|E| \leq 33\Delta^2$. Hence there are only $O(\Delta^5)$ values of u_0 that satisfy some congruence of the form (10) where $\mathbf{d}$ and E satisfy the appropriate bounds. We place these $O(\Delta^5)$ values of u_0 in $\mathcal{V}(\Delta; a, b)$, and once this is done we see that the case when $d_2 \neq 0$ or $d_3 - d_1 \neq 0$ cannot occur (for then (10) would imply that $u_0 \in \mathcal{V}(\Delta; a, b)$).

This establishes the claim we made in the first paragraph of Stage 2, so we may assume that $d_2 = 0$ and $d_3 - d_1 = 0$.

Stage 3: Predicting the generator when $f_0 \neq 0$. Suppose that $f_0 \neq 0$. The definition (7) of d_2 shows that $0 = d_2 = f_0 \varepsilon_1 - f_2$. Thus $\varepsilon_1 \equiv f_2 / f_0 \bmod p$ and so we may compute the secret information ε_1. To obtain the remainder of the secret information, we note that the following three congruences hold:

$$a + b(\varepsilon_0 + w_0) \equiv (\varepsilon_0 + w_0)(\varepsilon_1 + w_1) \bmod p,$$
$$a + b(\varepsilon_1 + w_1) \equiv (\varepsilon_1 + w_1)(\varepsilon_2 + w_2) \bmod p, \qquad (11)$$
$$f_0 \varepsilon_0 - f_1 \equiv f_0 \varepsilon_2 - f_2 \bmod p.$$

The first two of these congruences follow from (3), and the second follows from the fact that $d_1 = d_3$ together with the definition (7) of $\mathbf{d}$. But, since ε_1 is now known, the system (11) is linear in the variables a, ε_0 and ε_2. These equations have a unique solution if and only if $bf_0 \not\equiv 0 \bmod p$ (as can be seen by calculating the appropriate 3×3 determinant). Our assumption that $f_0 \neq 0$ together with our bound on $|f_0|$ shows that $f_0 \not\equiv 0 \bmod p$. Since $b \in \mathbb{F}_p^*$, we find that $bf_0 \not\equiv 0 \bmod p$ and so we may solve the system (11) to find ε_0, ε_2 and a. Finally, we compute u_0 from w_0 and ε_0 and so the algorithm terminates successfully in this case. So we are done when $f_0 \neq 0$.

Stage 4: Building the lattice $\mathcal{L}'$. We may now assume that $f_0 = 0$. So $d_i = -f_i$, $i = 1, 2, 3, 4$. We aim to show that b must have a special form.

The fact that $d_2 = 0$ and $d_3 - d_1 = 0$ means that the congruence (9) becomes

$$0 \equiv bd_1 + E \bmod p.$$

Using the definition of E, we find that $bd_1 \equiv d_4 \bmod p$, and so $bf_1 \equiv f_4 \bmod p$. It is easy to see that $f_1 \not\equiv 0 \bmod p$ (for the congruences $f_2 \equiv -d_2 \equiv 0 \bmod p$, $f_3 \equiv -d_3 \equiv -d_1 \equiv f_1 \bmod p$ and $f_4 \equiv bf_1 \bmod p$ would contradict the fact that $\mathbf{f}$ is a nonzero vector). Hence $b \equiv f_4 / s_1 \bmod p$ and so we may write

$$b \equiv r/s \bmod p, \text{ where } r = f_4 / \gcd(f_1, f_4) \text{ and } s = f_1 / \gcd(f_1, f_4).$$

Note that r and s are coprime, $|r| \leq 3\Delta^2$ and $|s| \leq 3\Delta$. Moreover we know r and s since we have computed $\mathbf{f}$. Also note that r and s are determined by b, up to sign. To see this, suppose that r' and s' are coprime integers such that $|r'| \leq 3\Delta^2$, $|s'| \leq 3\Delta$ and $r'/s' \equiv b \equiv r/s \bmod p$. Then $rs' \equiv sr' \bmod p$ and since both rs' and sr' have absolute value at most $9\Delta^3$ we find that $rs' = sr'$. But since $\gcd(r, s) = \gcd(r', s') = 1$ we now find that $r = \sigma r'$ and $s = \sigma s'$ for some element $\sigma \in \{1, -1\}$.

We now consider a new lattice: the lattice $\mathcal{L}'$ consisting of solutions $\mathbf{x} = (x_0, x_1, x_2, x_3) \in \mathbb{Z}^4$ of the congruences

$$A'x_0 + B'x_1 + B'_1 x_2 + C'x_4 \equiv 0 \bmod p,$$
$$x_0 \equiv 0 \bmod \Delta^3, \tag{12}$$
$$x_1 \equiv x_2 \equiv 0 \bmod \Delta^2,$$

where

$$A' \equiv sA\Delta^{-1} \bmod p, \quad B' \equiv sw_1\Delta^{-2} \bmod p,$$
$$B'_1 \equiv s(w_2 - w_0)\Delta^{-2} \bmod p \quad \text{and} \quad C' \equiv 1 \bmod p.$$

It is easy to check that the lattice (12) contains the vector

$$\mathbf{e}' = \left(\Delta^3 e'_0, \Delta^2 e'_1, \Delta^2 e'_2, e'_3\right),$$

where

$$\mathbf{e}' = \left(\Delta^3, \Delta^2(\varepsilon_2 - \varepsilon_0), \Delta^2\varepsilon_1, s\varepsilon_1(\varepsilon_2 - \varepsilon_0) - r(\varepsilon_1 - \varepsilon_0)\right).$$

We have

$$e'_0 = 1, \qquad |e'_1| \leq 2\Delta, \qquad |e'_2| \leq \Delta, \qquad |e'_3| \leq 24\Delta^3$$

thus the Euclidean norm $\|\mathbf{e}'\|$ of $\mathbf{e}'$ satisfies the inequality

$$\|\mathbf{e}\| \leq \left(\Delta^6 + 4\Delta^6 + \Delta^6 + 576\Delta^6\right)^{1/2} \leq 25\Delta^3.$$

Stage 5: Defining the second exceptional set $\mathcal{V}'(\Delta; a, b)$. We now show that all short vectors in $\mathcal{L}'$ are parallel to $\mathbf{e}$ unless u_0 belongs to the set $\mathcal{V}'(\Delta; a, b)$ which we define below.

Assume, for a contradiction, that there is another vector

$$\mathbf{f}' = (\Delta^3 f'_0, \Delta^2 f'_1, \Delta^2 f'_2, f'_3) \in \mathcal{L}'$$

with $\|\mathbf{f}'\| \leq \|\mathbf{e}'\| \leq 25\Delta^3$ which is not parallel to $\mathbf{e}'$. The vector $\mathbf{d}'$ defined by $\mathbf{d}' = f'_0\mathbf{e}' - e'_0\mathbf{f}'$ lies in $\mathcal{L}'$ and has first component 0. Using the first congruence in (12), we find that

$$B'\Delta^2 d'_1 + B'_1\Delta^2 d'_2 + C'd'_3 \equiv 0 \bmod p \tag{13}$$

where for $i \in \{1, 2, 3\}$ we define $d'_i = f'_0 e'_i - e'_0 f'_i = f'_0 e'_i - f'_i$. Note that $|d'_i| \leq 25|e'_i| + |f'_i|$ and hence

$$|d'_1| \leq 75\Delta, \qquad |d'_2| \leq 50\Delta, \qquad |d'_3| \leq 25^2\Delta^3. \tag{14}$$

Using the definitions of B', B_1' and C', we find that

$$sw_1 d_1' + s(w_2 - w_0)d_2' + d_3' \equiv 0 \bmod p,$$

and after the substitutions $w_i = u_i - \varepsilon_i$ we find

$$u_1 s d_1' + s(u_2 - u_0)d_2' \equiv E' \bmod p \tag{15}$$

where

$$E' = -d_3' + s\varepsilon_1 d_1' - s(\varepsilon_0 - \varepsilon_2)d_2'.$$

The bounds (14) imply that $|E'| \leq 25^3 \Delta^3$. We now write this equality as a rational function of u_0. Then (15) becomes

$$s d_1' \Psi_1(u_0) + s d_2' \Psi_2(u_0) - s d_2' u_0 \equiv E' \bmod p. \tag{16}$$

Let us consider the rational function

$$\Phi_{\mathbf{d}'}'(u) = s d_1' \Psi_1(u) + s d_2' \Psi_2(u) - d_2' u$$

corresponding to the left hand side of (16). Clearly, $\Phi_{\mathbf{d}'}'(u)$ can be written as the quotient of a polynomial of degree at most 3 and a polynomial of degree at most 2.

Now, $\Phi_{\mathbf{d}'}'(u)$ is a non-constant rational function of u. Suppose $\Phi_{\mathbf{d}'}'(u)$ is constant. Then (arguing as for $\Phi_{\mathbf{d}}(u)$ above) we must have that $d_1' \equiv d_2' \equiv 0 \bmod p$. But then (13) shows that $d_3' \equiv 0 \bmod p$, and so our bounds (14) on the absolute value of d_1', d_2' and d_3' imply that $d_1' = d_2' = d_3' = 0$. This implies that $\mathbf{d}' = 0$ and so $\mathbf{e}'$ and $\mathbf{f}'$ are parallel. This contradicts our choice of $\mathbf{f}'$, and so we must have that $\Phi_{\mathbf{d}'}'(u)$ is a non-constant rational function of u.

Since $\Phi_{\mathbf{d}'}'(u)$ is of degree at most 3, the congruence (16) can be satisfied for at most 3 values of u_0 once s, d_1', d_2' and E' have been chosen. There are at most 2 choices for s (as s is determined up to sign by b). There are $O(\Delta)$ choices for each of d_1', and d_2', by (14). There are $O(\Delta^3)$ choices for E' since $|E'| \leq 25^3 \Delta^3$. Hence there are only $O(\Delta^5)$ values of u_0 that satisfy some congruence of the form (16) where the d_i' and E' satisfy the appropriate bounds. We place these $O(\Delta^5)$ values of u_0 in $\mathcal{V}'(\Delta; a, b)$, and so we get a contradiction to our assumption that $\mathbf{f}'$ and $\mathbf{e}'$ are not parallel. So all short vectors in $\mathcal{L}'$ are parallel to $\mathbf{e}'$ whenever $u_0 \notin \mathcal{V}'(\Delta; a, b)$.

Stage 6: Predicting the generator for $f_0 = 0$. We apply a deterministic polynomial time algorithm for the shortest vector problem in a finite dimensional lattice to find a shortest nonzero vector $\mathbf{f}'$ in $\mathcal{L}'$, and this vector must be parallel to $\mathbf{e}'$. We recover $\mathbf{e}'$ by using the fact that $\mathbf{e}' = \mathbf{f}'/f_0'$. This gives us ε_1 which is used to calculate u_1. In order to compute u_0 we have to solve the following linear system of congruences in the unknowns ε_0 and ε_2:

$$\begin{aligned} f_0'(\varepsilon_2 - \varepsilon_0) &\equiv f_1' \bmod p, \\ f_0'(s\varepsilon_1(\varepsilon_2 - \varepsilon_0) - r(\varepsilon_1 - \varepsilon_0)) &\equiv f_3' \bmod p, \end{aligned} \tag{17}$$

which has a unique solution. Finally, a can be calculated by using the fact that $a \equiv u_0 u_1 - b u_0 \bmod p$. Defining

$$\mathcal{U}(\Delta; a, b) = \{0, -a/b\} \cup \mathcal{V}(\Delta; a, b) \cup \mathcal{V}'(\Delta; a, b)$$

which finishes the proof. $\square$

4   Remarks and Open Questions

Obviously our result is nontrivial only for $\Delta = O(p^{1/5})$. Thus increasing the size of the admissible values of Δ (even at the cost of considering more consecutive approximations) is of prime importance.

One can presumably obtain a very similar result in the dual case, where a is given but the shift b is unknown.

As we have mentioned several other results about predictability of inversive and other nonlinear generators have recently been obtained in [2]. However, they are somewhat weaker than the present result because each of them excludes a certain small exceptional set of pairs of parameters (a, b). We believe that the approach of this work may help to eliminate this drawback. Certainly this question deserves further study.

We do not know how to predict the inversive (and other generators considered in [2]) in the case when the modulus p is secret as well. We remark that in the case of the linear congruential generator a heuristic approach to this problem has been proposed in [4]. However it is not clear how to extend this (even just heuristically) to the case of nonlinear generators.

References

1. M. Ajtai, R. Kumar and D. Sivakumar, 'A sieve algorithm for the shortest lattice vector problem', *Proc. 33rd ACM Symp. on Theory of Comput. (STOC 2001)*, Association for Computing Machinery, 2001, 601–610.
2. S. R. Blackburn, D. Gomez-Perez, J. Gutierrez and I. E. Shparlinski, 'Predicting nonlinear pseudorandom number generators', *Preprint*, 2003.
3. M. Grötschel, L. Lovász and A. Schrijver, *Geometric algorithms and combinatorial optimization*, Springer-Verlag, Berlin, 1993.
4. A. Joux and J. Stern, 'Lattice reduction: A toolbox for the cryptanalyst', *J. Cryptology*, **11** (1998), 161–185.
5. R. Kannan, 'Algorithmic geometry of numbers', *Annual Review of Comp. Sci.*, **2** (1987), 231–267.
6. R. Kannan, 'Minkowski's convex body theorem and integer programming', *Math. Oper. Res.*, **12** (1987), 415–440.
7. A. K. Lenstra, H. W. Lenstra and L. Lovász, 'Factoring polynomials with rational coefficients', *Mathematische Annalen*, **261** (1982), 515–534.
8. D. Micciancio and S. Goldwasser, *Complexity of lattice problems*, Kluwer Acad. Publ., 2002.

9. P. Q. Nguyen and J. Stern, 'Lattice reduction in cryptology: An update', in: W. Bosma (Ed), *Proc. ANTS-IV, Lect. Notes in Comp. Sci. Vol. 1838*, Springer-Verlag, Berlin, 2000, 85–112.

10. P. Q. Nguyen and J. Stern, 'The two faces of lattices in cryptology', in: J.H. Silverman (Ed), *Cryptography and Lattices Lect. Notes in Comp. Sci. Vol. 2146* , Springer-Verlag, Berlin, 2001, 146–180.

11. H. Niederreiter, 'New developments in uniform pseudorandom number and vector generation', in: H. Niederreiter and P.J. Shiue (Eds), *Monte Carlo and Quasi-Monte Carlo Methods in Scientific Computing, Lect. Notes in Statistics Vol. 106*, Springer-Verlag, Berlin, 1995, 87–120.

12. H. Niederreiter, 'Design and analysis of nonlinear pseudorandom number generators', in G.I. Schueller and P. D. Spanos (Eds) *Monte Carlo Simulation*, A.A. Balkema Publishers, Rotterdam, 2001, 3–9.

13. H. Niederreiter and I. E. Shparlinski, 'Recent advances in the theory of nonlinear pseudorandom number generators', in: K.-T. Fang, F.J. Hickernell and H. Niederreiter (Eds), *Proc. Conf. on Monte Carlo and Quasi-Monte Carlo Methods, 2000*, Springer-Verlag, Berlin, 2002, 86–102.

14. H. Niederreiter and I. E. Shparlinski, 'Dynamical systems generated by rational functions', in: Marc Fossorier, Tom Høholdt and Alain Poli (Eds), *Applied Algebra, Algebraic Algorithms and Error Correcting Codes – AAECC-15, Lect. Notes in Comp. Sci. Vol. 2643*, Springer-Verlag, Berlin, 2003, 6–17.

A Stochastical Model and Its Analysis for a Physical Random Number Generator Presented At CHES 2002

Werner Schindler

Bundesamt für Sicherheit in der Informationstechnik (BSI)
Godesberger Allee 185–189
53175 Bonn, Germany
Werner.Schindler@bsi.bund.de

Abstract. In [3] an elementary attack against a true (physical) random number generator presented at CHES 2002 ([9]) was described under the condition that the attacker knows all (i.e. also the unpublished) details. In the present paper the general design is analyzed and conditions of use are formulated under which the random number generator is unconditionally secure.

Keywords: True (physical) random number generator, entropy, stochastical model, AIS 31.

1 Introduction

Random number generators (*RNGs*) are essential components of many cryptosystems. Weak random number generators may reduce the security of these cryptosystems considerably, e.g. because it is possible to guess randomly generated session keys with high probability. Unlike in stochastic simulations random numbers used for cryptographic applications (e.g. as session keys or signature parameters) usually have to be unpredictable. Unpredictability cannot be ensured merely by means of statistical black box tests. For true (physical) random number generators (*TRNGs*) this requires an extensive analysis of the specific design. We further point out that a TRNG evaluation should also consider the suitability of the online tests ([8]). The interested reader is further referred to ([1], [6]) where an approach for the evaluation of physical random number generators due to CC (Common Criteria [2]), resp. ITSEC (Information Technology Security Evaluation Criteria [5]), is specified.

In [9] an unorthodox design of a true (physical) random number generator is proposed. Two independent ring oscillators clock a linear feedback shift register and a cellular linear automaton, resp. The frequencies of the ring oscillators are not controlled. Upon external request 32 bits of both the linear shift register and the cellular automaton are masked out, permuted, XORed, and the resulting 32 bit vector is output. It is demanded that the linear feedback shift register and the cellular automaton should at least clock 86, resp. 74, times between

K.G. Paterson (Ed.): Cryptography and Coding 2003, LNCS 2898, pp. 276–289, 2003.

two consecutive samplings. In [9] statistical tests have been applied to the RNG output. However, [9] does not contain an analysis of the RNG design.

In [3] an attack against this RNG is described *under the condition* that the attacker knows the positions of the masked bits, the permutations, (at least approximately) the mean cycle lengths of both ring oscillators and the intermediate times between consecutive samplings which are assumed to be close to the specified minimum. In the end, this is not very surprising as the 'source' of randomness is the jitter of the ring oscillators clocking the linear feedback shift register and the cellular automaton. Apparently, it is not possible to increase the overall entropy of the system by 32 bit within less than 100 cycles of the ring oscillators. It is plausible to assume that the designers of the TRNG have been aware of this fact as the positions of the masked bits, the permutations and technical details have been kept secret. Without this knowledge the attack described in [3] is not feasible. Alternatively, a one-way postprocessing could have been chosen to ensure computational security.

However, for true random number generators the situation is more favourable than for pseudorandom number generators as one need not only rely on practical or computational security. In fact, if the entropy per random bit is sufficiently large this even implies unconditional security. In the present case it seems to be intuitively clear that the entropy per random output bit can be increased by increasing the minimum intermediate time between consecutive samplings and maybe decreasing the number of random bits per output. In this paper we will quantify this intuition. Therefore we derive a stochastical model of the TRNG and analyze this model. We point out that in Germany (besides other requirements) this was necessary for an evaluation of the TRNG with regard to ITSEC and CC (cf. [1,6]). We further point out that the obtained results do not help to improve Dichtl's ([3]) or any similar attacks. On the contrary, the results could be used by the system designers to choose parameters for which the entropy per output bit is close to 1.

This paper is organized as follows: In Chapter 2 we describe the RNG from [9] and sketch Dichtl's attack ([3]). In Sect. 3 we outline the general plan, and Sect. 4 provides definitions and a useful lemma which will be needed in Sect. 5 where a stochastical model of the RNG is formulated and analyzed. Lower and upper bounds for the entropy per output random bit are derived. The paper closes with final remarks.

2   Description of the TRNG and Dichtl's Attack

First we describe the TRNG presented in [9]. Then we sketch Dichtl's attack.

Description of the TRNG (I). (cf. [9]) The TRNG consists of a 43-bit shift register (LFSR) and a 37-bit cellular automaton shift register (CASR) which are clocked by two independent ring oscillators. The frequencies of the ring oscillators are not controlled and drift with variations in temperatures and at voltage. The LFSR has the primitive feedback polynomial

$$p_1(x) := x^{43} + x^{41} + x^{20} + x + 1 \tag{1}$$

while the cells $a_0, \ldots, a_{36}$ of the CASR are renewed via

$$a_i(t+1) := a_{i-1(\bmod 37)}(t) \oplus a_{i+1(\bmod 37)}(t) \text{ for } i \neq 28 \qquad (2)$$
$$a_{28}(t+1) := a_{27}(t) \oplus a_{28}(t) \oplus a_{29}(t) \qquad (3)$$

where t denotes number of the clock cycle. The states are not reset after power-up. Upon external request 32 specified cells of the LFSR and CASR are masked out, permutated and XORed. The 32-bit XOR-sum is output. Even when no random numbers are used the TRNG is active. The minimum sampling time is chosen so that the LFSR and the CASR clock at least 86 or 74 times, resp.

Dichtl's attack – Assumptions. (cf. [3]) The attacker knows the positions of the LFSR and CASR which are masked out and both permutations. Further, it is assumed that the attacker knows at least three consecutive 32-bit output blocks and the intervals between the sampling times. (Assumed scenario: The TRNG generates session keys and the attacker is a legitimate receiver of one such key.) The attacker also knows (at least approximately) the mean cycle lengths of both ring oscillators under the current environmental conditions. The intermediate times between consecutive samplings are assumed not to be considerably larger than the prescribed minimum.

Dichtl's attack. (cf. [3]) If the attacker knew the exact number of clocks of the LFSR and CASR between the sampling times of the three output blocks he could formulate a system of 92 linear equations over $GF(2)$ in $43 + 37 = 80$ variables. Its solution gave the states of the LFSR and CASR when the first, resp. the third block has been generated. In 'real life' the attacker does not know the exact number of clocks. Instead he would try numbers in a neighbourhood of the expected values (= intermediate times / average clock length of the respective ring oscillator). Note that the system of linear equations is over-determined. This can be used for a first check: If it has no solution the guessed numbers of cycles must be wrong. If the intermediate times are near the prescribed minimum the attacker has a good chance to guess the searched numbers of clock cycles in a practically feasible number of trials. After having determined the values of the LFSR and CASR the attacker tries to find the preceding and subsequent keys generated by the TRNG in a similar manner. We point out that [3] considers the case where the attacker may only know approximate values for the mean cycle lengths of the ring oscillators.

Remark 1. Under the assumed conditions the described attack should work in practice if the intermediate times between successive samplings are not considerably larger than the prescribed minimum bound. Apparantly, the whole system (i.e. the states of the LFSR and the CASR) cannot collect 32 bits of entropy within such a small number of cycles.

3   How to Quantify the Entropy – Central Ideas

Normally, the physical noise source of a TRNG generates analog signals which are digitized after uniform time intervals, e.g. by a comparator. These values are called *digitized analog signals*, or shortly, *das-random numbers*. In many cases these values are deterministically postprocessed yielding the so-called *internal random numbers* (notation as in [8]). Finally, one is interested in the entropy per internal number. Apart from trivial cases where the postprocessing is 1-1 and does not influence the average entropy per internal random number the impact of the digital postprocessing can only be analyzed after the underlying distribution of the das random numbers has been determined or at least estimated. Although the situation seems to be very different in the present case we follow an analogous approach. Therefore, we use the terminology introduced above to formulate an alternate description of the TRNG. This might seem to be artificial at first sight but turns out to be very convenient for the later analysis. We point out that the LFSR and CASR are both finite state machines with linear transition maps A_1 and A_2, resp. In particular, we will represent A_1 and A_2 as matrices over GF(2) in the following.

Description of the TRNG (II). The m-bit outputs (in [9]: $m = 32$) of the TRNG are the internal random numbers. Assume that the internal random numbers are output at times $s_1, s_1 + s_2, s_1 + s_2 + s_3, \ldots$, i.e. s_j is the intermediate time between sampling the $(j-1)^{th}$ and the jth m-bit block. The das-random numbers $(v_{1;j}, v_{2;j})_{j \in \mathbb{N}}$ stand for the number of clock cycles of the LFSR and CASR within the time interval $(s_1 + \ldots + s_{j-1}, s_1 + \cdots + s_j]$ where the first index denotes the automaton ('1' for the LFSR and '2' for the CASR). At time $s_0 = 0$ the two state machines are initialized by values $b_{1;0}$ and $b_{2;0}$, resp. The states at the times $s_1, s_1 + s_2, \ldots$ are denoted with $b_{1;j}$ and $b_{2;j}$. In particular, $b_{i;j} = A_i^{v_{i;j}} b_{i;j-1}$. The internal random numbers result from the states of the LFSR and CASR by applying an output function $\psi \colon \mathrm{GF}(2)^{43+37} \to \mathrm{GF}(2)^m$.

Roughly speaking, the source of entropy is the variation of the number of clock cycles of both ring oscillators within a time interval. By clocking the LFSR and CASR the entropy is 'transferred' to the states of these automatons, yielding uncertainty about their states, and, transitively uncertainty on the output bits (internal random numbers). To be viewed as secure against any potential attacker the conditional entropy per bit of the internal random numbers with respect to their predecessors should be almost 1. In Sect. 5 we will study these phenomena and quantify a lower and an upper bound for the increase of entropy per internal random number.

4   Definitions and Preparatory Lemmata

The present section provides definitions and preparatory lemmata which will be needed in the next section. Though useful the lemmata are rather technical and may be skipped in the first reading.

Definition 1. *Random variables are denoted with capital letters, realizations of these random variables, i.e. values assumed by these random variables, with the respective small letters. The notation $X \sim \nu$ means that the random variable X is ν-distributed. As usually, iid stands for 'independent and identically distributed'. A sequence of random variables $X_1, X_2, \ldots$ is called* stationary *if for each $r \geq 1$ the distribution of $(X_{t+1}, \ldots, X_{t+r})$ does not depend on the shift parameter t. The sequence $X_1, X_2, \ldots$ is called h-dependent if the vectors $(X_1, \ldots, X_a)$ and $(X_b, \ldots, X_n)$ are independent if $a+h < b$. As usually, the term $N(\mu, \sigma^2)$ denotes a normal distribution with mean μ and variance σ^2. The cumulative distribution function of the standard normal distribution $N(0,1)$ is denoted with Φ.*

We point out that a sequence of random variables $X_1, X_2, \ldots$ is 0-dependent iff it is independent.

Lemma 1. *Let $T_1, T_2, \ldots$ denote a stationary h-dependent sequence of nonnegative random variables with $E(T_j) = \mu > 0$, $\mathrm{Var}(T_j) = \sigma_T^2 > 0$ and $E(|T_j|^3) < \infty$.*
(i) For $s \in [0, \infty)$ let

$$V_{(s)} := \inf \left\{ \tau \in \mathbb{N} \mid \sum_{j=1}^{\tau+1} T_j > s \right\}. \tag{4}$$

For $k \geq 1$ we have

$$\mathrm{Prob}(V_{(s)} = k) = \mathrm{Prob}\,(T_1 + \cdots + T_k \leq s) - \mathrm{Prob}\,(T_1 + \cdots + T_{k+1} \leq s) \ and \tag{5}$$
$$\mathrm{Prob}(V_{(s)} = 0) = 1 - \mathrm{Prob}\,(T_1 \leq s). \tag{6}$$

In particular,
$$\mathrm{Prob}(V_{(s)} = \infty) = 0. \tag{7}$$

(ii) Let

$$\sigma^2 := \sigma_T^2 + 2 \sum_{i=1}^{q} E\left((T_i - \mu)(T_{q+1} - \mu)\right). \tag{8}$$

Then the distributions of the random variables $(\sum_{j=1}^{k} T_j - k\mu)/(\sqrt{k}\sigma)$ tend to the standard normal distribution as k tends to infinity. (Here and in the following we tacitly assume that σ is the positive root of σ^2.) In particular,

$$\mathrm{Prob}\left(\frac{T_1 + \cdots + T_k - k\mu}{\sqrt{k}\sigma} \leq x\right) \xrightarrow{k \to \infty} \Phi(x). \tag{9}$$

for each $x \in \mathbb{R}$.
(iii) Let $s = r\mu$ with $r > 0$. Unless k is very small in good approximation

$$\mathrm{Prob}\left(V_{(r\mu)} = k\right) \approx \Phi\left(\frac{r-k}{\sqrt{k}} \cdot \frac{\mu}{\sigma}\right) - \Phi\left(\frac{r - (k+1)}{\sqrt{k+1}} \cdot \frac{\mu}{\sigma}\right) \ for \ k \geq 1 \tag{10}$$

$$\mathrm{Prob}\left(V_{(r\mu)} = 0\right) \approx 1 - \Phi\left((r-1)\frac{\mu}{\sigma}\right). \tag{11}$$

(iv) If the sequence $T_1, T_2, \ldots$ is iid the condition $E(|T_j|^3) < \infty$ may be dropped. In particular, $\sigma^2 = \sigma_T^2$. If additionally $T_j \sim N(\mu, \sigma^2)$ in (10) and (11) '=' holds instead of '≈'.

Proof. Equation (5) follows immediately from

$$\mathrm{Prob}(V_{(s)} = k) = \mathrm{Prob}(V_{(s)} \geq k) - \mathrm{Prob}(V_{(s)} \geq k+1),$$

and (6) can be verified in a similar manner. As the T_j are non-negative

$$\mathrm{Prob}\left(V_{(s)} = \infty\right) = \mathrm{Prob}\left(\sum_{j=1}^{\infty} T_j \leq s\right) \leq \mathrm{Prob}\left(\sum_{j=1}^{\infty} T_{j(h+1)} \leq s\right)$$
$$\leq \mathrm{Prob}\left(T_{j(h+1)} \geq \mu/2 \text{ for at most } \lfloor 2s/\mu \rfloor \text{ many indices}\right).$$

As the random variables $T_{(h+1)}, T_{2(h+1)}, \ldots$ are iid and $\mathrm{Prob}(T_{j(h+1)} \geq \mu/2) > 0$ the right-hand probability equals zero which proves (7). By assumption $E(|T_j|^3) < \infty$ and $T_1, T_2, \ldots$ is an h-dependent stationary sequence of random variables. The first assertion of (ii) hence follows from Theorem 1 in [4] wheras the second is a consequence of the first. Assertion (iii) follows immediately from (i) and (ii). If the random variables T_j are iid (ii) follows from the 'usual' Central Limit Theorem for iid random variables. This proves the first assertion of (iv), and the second follows from the properties of normal distributions.   □

Definition 2. *Let X denote a random variable assuming values in a countable set Ω (e.g. $\Omega = \mathbb{N}_0$). Extending the definition of entropy for random variables with finite range (cf. [7], Sect. 2.2.1) we define analogously*

$$H(X) := -\sum_{\omega \in \Omega} \mathrm{Prob}(X = \omega) \log_2(\mathrm{Prob}(X = \omega)). \tag{12}$$

Remark 2. We note that $H(X) \in [0, \infty]$ where the value ∞ is possible. The random variable $V_{(s)}$ defined in Lemma 1, i.e. the case we are interested in, has finite entropy for each $s \in (0, \infty)$ (cf. Lemma 2(ii)).

5   Analysis of the RNG

In this section we determine lower and upper bounds for the (average) increase of entropy per internal random bit. (To be precise, entropy is not a property of random numbers but of the underlying random variables.) Clearly, the entropy bound depends on the technical parameters $\mu_1, \mu_2, \sigma_1, \sigma_2$ and the intermediate times $s_1, s_2, \ldots$ between successive samplings. Here μ_i denotes the mean cycle length of the ith ring oscillator while σ_i is defined as in (8) with $\sigma_{T;i}^2$ being the variance of a random cycle length. We first formulate a stochastical model for the RNG.

Stochastical model of the RNG. In the following the das random numbers $(v_{1;j}, v_{2;j})_{j \in \mathbb{N}}$ are viewed as realizations of random variables $(V_{1;j}, V_{2;j})_{j \in \mathbb{N}}$. The random variables $V_{1;1}, V_{2;1}, V_{1;2}, V_{2;2}, V_{1;3}, \ldots$ are assumed to be independent, and $V_{i;j}$ is distributed as $V_{(r_{i;j} \mu_i)}$ in Lemma 1(iii) with $r_{i;j} := s_j / \mu_i$. Analogously, we view the inner states $b_{i;j}$ at times $s_0 = 0, s_1, s_1 + s_2, \ldots$ as realizations of random variables $(B_{1;j}, B_{2;j})_{j \in \mathbb{N}_0}$. In particular, $B_{i;0}$ describes the random initial value of the ith automaton (LFSR or CASR, resp.), and $B_{i;j} = A_i^{V_{i;j}} B_{i;j-1}$ for $(i, j) \in \{1, 2\} \times \mathbb{N}$. For both $i = 1, 2$ the random variables $B_{i;1}, B_{i;2}, \ldots$ are Markovian, i.e. $\text{Prob}(B_{i;j} = b_j \mid B_{i;0} = b_0, \ldots, B_{i;j-1} = b_{j-1})$ $= \text{Prob}(B_{i;j} = b_j \mid B_{i;j-1} = b_{j-1})$ for all $b_0, \ldots, b_j$. If $s_j = s$ for all $j \in \mathbb{N}$ both Markov chains are homogeneous. The Markov chains $(B_{1;j})_{j \in \mathbb{N}_0}$ and $(B_{2;j})_{j \in \mathbb{N}_0}$ are independent. Finally, the internal random numbers $c_1, c_2, \ldots$ are viewed as realizations of random variables $C_1, C_2, \ldots$ with $C_j := \psi(B_{1;j}, B_{2;j})$.

Remark 3. The distributions of the random variables $V_{i;j}, B_{i;j}$ and C_j clearly depend on the intermediate times $s_1, s_2, \ldots$. Consequently, they actually should be denoted by $V_{i;j(s_j)}, B_{i;j(s_j)}$ and $C_{j(s_j)}$, resp. However, for the sake of readability we suppress the term (s_j).

Justification of the stochastical model. (i) The lengths of the particular clock cycles of the ring oscillators may be viewed as realizations of stationary sequences of random variables $(T_{1;j})_{j \in \mathbb{N}}$ and $(T_{2;j})_{j \in \mathbb{N}}$ with finite memory. The sequences $(T_{1;j})_{j \in \mathbb{N}}$ and $(T_{2;j})_{j \in \mathbb{N}}$ are independent as the oscillators are independent, whereas the random variables $T_{1;1}, T_{1;2}, \ldots$ (resp. $T_{2;1}, T_{2;2}, \ldots$) need not be independent.

(ii) At time $s_1 + \ldots + s_{j-1}$ the states of the two automatons attain the values $b_{1;j-1}$ and $b_{2;j-1}$. As the current cycles of the ring oscillators may not have just been completed when c_{j-1} was sampled the first cycle within the next time period $(s_1 + \ldots + s_{j-1}, s_1 + \cdots + s_j]$ is likely to be shorter than the others. Consequently, $V_{i;j-1}$ and $V_{i;j}$ are not independent in a strict sense. To be on the safe side when determining the entropy bounds and since $r_{i;j} \gg 1$ under realistic assumptions we neglect this effect. Moreover, the distribution of $V_{i;j}$ is very robust in the sense that it demands only weak assumptions on the distribution of the sequence $(T_{i;j})_{j \in \mathbb{N}}$ besides its stationarity and the finite memory property (cf. Lemma 1). Consequently, we view both sequences $(V_{1;j})_{j \in \mathbb{N}}$ and $(V_{2;j})_{j \in \mathbb{N}}$ as independent where $V_{i;j}$ is distributed as $V_{(r_{i;j} \mu_i)}$ in Lemma 1(iii) with $\mu = \mu_i$, $\sigma^2 = \sigma_i^2$, and $r_{i;j} = s_j / \mu_i$.

(iii) The Markov property of $(B_{i;j})_{j \in N}$ follows immediately from the fact that the random variables $V_{i;1}, V_{i;2}, \ldots$ are independent.

Definition 3. *To simplify the notation we introduce the abbreviations* $\Omega_1 := GF(2)^{43}$, $\Omega_2 := GF(2)^{37}$ *and* $\Omega_K := GF(2)^m$. *Further,* $\rho_i \colon \Omega_i \to \Omega_K$ *stands for the selection function which masks m bits from Ω_i out and permutes them; i.e. the output function can be written as* $\psi(b_1, b_2) = \rho_1 b_1 \oplus \rho_2 b_2$. *Further,* $\nu_{i;j}$ *denotes the distribution of the random variable* $B_{i;j}$.

Theorem 1. *(i) For $i = 1, 2$ let $r_{i;j} := s_j/\mu_i$. Then*

$$\text{Prob}(V_{i;j} = k) = \begin{cases} \Phi\left(\frac{r_{i;j}-k}{\sqrt{k}} \cdot \frac{\mu_i}{\sigma_i}\right) - \Phi\left(\frac{r_{i;j}-(k+1)}{\sqrt{k+1}} \cdot \frac{\mu_i}{\sigma_i}\right) & \text{for } k \geq 1 \\ 1 - \Phi\left((r_{i;j} - 1)\frac{\mu_i}{\sigma_i}\right) & \text{for } k = 0. \end{cases} \quad (13)$$

(ii) Moreover,

$$H(B_{i;j+1} \mid B_{i;1}, \ldots, B_{i;j}) = H(B_{i;j+1} \mid B_{i;j}) = H(V_{i;j+1} \, (\text{mod}(|\Omega_i| - 1))) \quad (14)$$

$$\frac{1}{\tau}H(C_{j+1}, \ldots, C_{j+\tau} \mid C_1, \ldots, C_j) = \frac{1}{\tau}\sum_{u=1}^{\tau} H(C_{j+u} \mid C_1, \ldots, C_{j+u-1}) \quad (15)$$

$$\geq \frac{1}{\tau}\sum_{u=1}^{\tau} H(C_{j+u} \mid B_{1;j+u-1}, B_{2;j+u-1}).$$

Proof. Assertion (i) follows immediately from Lemma 1 (iii) and the stochastical model formulated above. The Markov property of $B_{i;1}, B_{i;2}, \ldots$ imply the first equation in (14), and the second follows immediately from the fact that $b, A_i b, A_i^2 b, \ldots$ defines a reordering of $\Omega_i \setminus \{0\}$ for each $b \neq 0$. The left-hand equality in (15) follows from elementary computation rules for entropies. Replacing the conditions $C_1, \ldots, C_{j+u-1}$ by $B_{1;1}, \ldots, B_{1;j+u-1}, B_{2;1}, \ldots, B_{2;j+u-1}$ clearly gives '$\geq$'. As $C_{j+u} = \psi(B_{1;j+u}, B_{2;j+u})$ and since the Markov chains $(B_{1;j})_{j\in\mathbb{N}_0}$ and $(B_{2;j})_{j\in\mathbb{N}_0}$ are independent this completes the proof of (15). $\square$

Under realistic assumptions the distribution of $V_{i;j}$ has concentrated nearly all of its mass on a small subset (compared with $|\Omega_i| - 1$), and restricted to this subset the mapping $v_{i;j+1} \mapsto v_{i;j+1}(\text{mod}(|\Omega_i| - 1))$ is injective. Consequently, it is reasonable to assume

$$H(B_{i;n+1} \mid B_{i;n}) = H(V_{i;n+1}) \quad (16)$$

in the following. Of course, our interest lies in the term $H(C_{j+1}, \ldots, C_{j+\tau} \mid C_1, \ldots, C_j)$ which is a measure for the security of the RNG. This belongs to 'real-life' situations (cf. Sect. 2) where the attacker is the legitimate owner of previous session keys or if the RNG is also used to generate openly transmitted challenges or initialization vectors for block ciphers, for instance. Unlike $(B_{i;j})_{j\in\mathbb{N}_0}$ the sequence $(C_j)_{j\in\mathbb{N}}$ is not Markovian which makes a direct computation of the conditional entropy very difficult. The right-hand side of (15) provides a lower bound which is more accessible for concrete computations (cf. Theorem 2).

Remark 4. The term $H(C_{n+1} \mid B_{1;n}, B_{2;n})$ quantifies the entropy of C_{n+1} under the condition that the attacker knows the inner states of both automatons at time $s_1 + \cdots + s_n$, i.e. $b_{1;n}$ and $b_{2;n}$. Note that the latter might be the consequence that the preceding internal random numbers (maybe openly transmitted data) have been generated too quickly (cf. Sect. 2).

Corollary 1. *Assume that* $s = s_1 = s_2 = \cdots$. *For each* $\epsilon > 0$ *there exists a number* N_ϵ *such that*

$$H(C_{n+1} \mid C_1, \ldots, C_n) - \epsilon \leq H(B_{1;n+1}, B_{2;n+1} \mid B_{1;n}, B_{2;n})$$
$$= H(V_{1;n+1}) + H(V_{2;n+1}). \tag{17}$$

for all $n > N_\epsilon$.

Proof. This corollary is an immediate consequence from Lemma 2 (ii) and (iii) with $X_n := (B_{1;n}, B_{2;n})$ and $\phi := \psi$, and (16). □

Remark 5. The right-hand side of (17) gives an upper bound if n is sufficiently large, i.e. if the attacker has observed a number of internal random numbers $c_1, c_2, \ldots, c_n$. We point out that this inequation may not be valid for small indices n as the knowledge of c_1, for instance, gives less information than the knowledge of $(b_{1;1}, b_{2;1})$ (cf. Example 1). Anyway, it is not recommendable to mask out more than $H(V_{1;n+1}) + H(V_{2;n+1})$ many bits.

Finally, we determine the right-hand terms of (15). The definition of conditional entropy yields

$$H(C_{n+1} \mid B_{1;n}, B_{2;n}) = \tag{18}$$
$$\sum_{b_1 \in \Omega_1} \sum_{b_2 \in \Omega_2} \nu_{1;n}(b_1) \nu_{2;n}(b_2) H(C_{n+1} \mid B_{1;n} = b_1, B_{2;n} = b_2).$$

However, for concrete computations (18) is useless as we had to compute 2^{80} conditional entropies each of which consisting of infinitely many summands. Moreover, it is difficult to determine the distributions $\nu_{i;n}$. The following theorem simplifies the necessary computations considerably.

Theorem 2. *For* $i = 1, 2$ *let* $D_i := \{w_i, w_i + 1, \ldots, w_i + d_i - 1\} \subseteq \mathbb{N}$ *with* $md_1 \leq 43$ *and* $md_2 \leq 37$ *such that* $\mathrm{Prob}(V_{i;n} \in D_i) \approx 1$. *Assume further that*

$$\mathrm{rank} \begin{pmatrix} \rho_i \mathbf{1}_i \\ \rho_i A_i \\ \cdots \\ \rho_i A_i^{d_i - 1} \end{pmatrix} = md_i \tag{19}$$

for $i = 1, 2$ *where* $\mathbf{1}_i$ *denotes the identity matrix on* Ω_i. *Then*

$$H(C_{n+1} \mid B_{1;n}, B_{2;n}) \approx \tag{20}$$
$$-2^{-m(d_1 + d_2 - 2)} \sum_{(y_1, \ldots, y_{d_1 - 1}) \in \Omega_K^{d_1 - 1}} \sum_{(y_1', \ldots, y_{d_2 - 1}') \in \Omega_K^{d_2 - 1}} \times$$
$$\times \sum_{z \in \Omega_K} q(z; y_1, \ldots, y_{d_1 - 1}, 0, y_1', \ldots, y_{d_2 - 1}', 0) \times$$

$$\times \log_2(q(z; y_1, \ldots, y_{d_1-1}, 0, y'_1, \ldots, y'_{d_2-1}, 0)).$$

with $q: \Omega_K \times \Omega_K^{d_1} \times \Omega_K^{d_2} \to \mathbb{R}$, $q(z; y_1, \ldots, y'_{d_2}) :=$

$$\sum_{s=1}^{d_1} \sum_{t=1}^{d_2} 1_z(y_s \oplus y'_t) \mathrm{Prob}(V_{1;n+1} = w_1 + s - 1) \mathrm{Prob}(V_{2;n+1} = w_2 + t - 1)$$

Proof. Rewriting the right-hand side of (18) gives the equivalent term

$$- \sum_{b_1 \in \Omega_1} \sum_{b_2 \in \Omega_2} \nu_{1;n}(b_1) \nu_{2;n}(b_2) \sum_{z \in \Omega_K} \widetilde{q}(z; b_1, b_2) \log_2 \widetilde{q}(z; b_1, b_2)$$

$$\approx - \sum_{b_1 \in \Omega_1} \sum_{b_2 \in \Omega_2} \nu_{1;n}(b_1) \nu_{2;n}(b_2) \sum_{z \in \Omega_K} \widetilde{\widetilde{q}}(z; b_1, b_2) \log_2 \widetilde{\widetilde{q}}(z; b_1, b_2)$$

with $\displaystyle \widetilde{q}(z; b_1, b_2) := \sum_{v_1, v_2 \in \mathbb{N}_0} 1_z(\rho_1 A_1^{v_1} b_1 \oplus \rho_2 A_2^{v_2} b_2) \mathrm{Prob}(V_{1;n+1} = v_1) \mathrm{Prob}(V_{2;n+1} = v_2)$

and $\displaystyle \widetilde{\widetilde{q}}(z; b_1, b_2) := \sum_{v_1 \in D_1; v_2 \in D_2} 1_z(\rho_1 A_1^{v_1} b_1 \oplus \rho_2 A_2^{v_2} b_2) \mathrm{Prob}(V_{1;n+1} = v_1) \mathrm{Prob}(V_{2;n+1} = v_2)$

where we made use of the fact that $V_{1;n}$ and $V_{2;n}$ are independent. (We remark that $\widetilde{q}(z; b_1, b_2) = \mathrm{Prob}(C_{n+1} = z \mid B_{1;n} = b_1, B_{2;n} = b_2)$.) The mapping $\Psi: \Omega_i \to \Omega_K^{d_i}$, $\Psi(b_i) := (\rho_i A_i^{w_i}, \ldots, \rho_i A_i^{w_i + d_i - 1})$ is $GF(2)$-linear and surjective by assumption (19) (multiply that matrix by $A_i^{w_1}$ from the right). Reordering the second line and collecting equal terms yields

$$\sum_{y_1, \ldots, y_{d_1} \in \Omega_K \; y'_1, \ldots, y'_{d_2} \in \Omega_K} \nu_1(\Psi_1^{-1}(y_1, \ldots, y_{d_i})) \nu_2(\Psi_2^{-1}(y'_1, \ldots, y'_{d_i})) \quad \times$$

$$\times \sum_{z \in \Omega_K} q(z; y_1, \ldots, y_{d_1}, y'_1, \ldots, y'_{d_2}) \log_2(q(z; y_1, \ldots, y_{d_1}, y'_1, \ldots, y'_{d_2})).$$

By (19) Ψ_i is a surjective linear mapping. For $x \in \Omega_K^{d_i}$ we hence conclude $|\Psi_i^{-1}(x)| = 2^{43 - md_1}$ for $i = 1$, resp. $= 2^{37 - md_2}$ for $i = 2$. It can easily be checked that $q(z; y_1, \ldots, y_{d_1}, y'_1, \ldots, y'_{d_2}) = q(z \oplus y_{d_1} \oplus y'_{d_2}; \overline{\overline{y_1}} \ldots, \overline{\overline{y_{d_1}-1}}, 0, \overline{y'_1}, \ldots, \overline{y'_{d_2}-1}, 0)$ where $\overline{\overline{y}} := y \oplus y_{d_1}$ and $\overline{y} := y \oplus y'_{d_2}$ for the moment. In particular, this induces an equivalence relation on $\Omega_K^{d_1} \times \Omega_K^{d_2}$: Namely, the pairs (y, y') and (y'', y''') are related iff the corresponding components of y and y'' as well as the corresponding components of y' and y''' differ by arbitrary constants y_a and y_b, resp. Each class has exactly one representative (y_0, y'_0) having a zero in the right-most component of both vectors. Note that for any elements belonging to the same equivalence class the sets $\{q(z; y, y') \mid z \in \Omega_K\}$ are equal. Thus the equivalence relation on $\Omega_K^{d_1} \times \Omega_K^{d_2}$ induces an equivalence relation on $\Omega_1 \times \Omega_2$. In fact, $\Omega_1 \times \Omega_2$ can be divided into unions of disjoint pre-images $\Psi_1^{-1}(y) \times \Psi_2^{-1}(y')$, each class containing 2^{2m} such pre-images. For each pair $(b_{1;0}, b_{2;0})$ the mapping

$$((v_1 + \cdots + v_j)(\mathrm{mod}\ 2^{43} - 1), (v_1 + \cdots + v_j)(\mathrm{mod}\ 2^{37} - 1)) \mapsto$$
$$(A_1^{v_1 + \cdots + v_j} b_{1;0}, A_2^{v_1 + \cdots + v_j} b_{2;0})$$

is bijective. It is hence easy to see that the distributions $\nu_{1;n} \otimes \nu_{2;n}$ converge to the equidistribution on $(\Omega_1 \setminus \{0\}) \times (\Omega_2 \setminus \{0\})$ as j tends to infinity. For our purposes, however, it is fully sufficient that $\nu_{1;n} \otimes \nu_{2;n}$ has (nearly) equal mass on all equivalence classes on $\Omega_1 \times \Omega_2$. This convergence is yet much faster. Assuming the latter completes the proof of this theorem (cf. Remark 6 (iii)).   □

Remark 6. (i) For $m = 1$ Assumption (19) is asymptotically fulfilled for each selection function $\rho_i \colon \Omega_i \to \Omega_K = \{0,1\}$ as the LFSR and CASR have no invariant subspaces besides $\{0\}$.

(ii) If the conditions (19), $md_1 \leq 43$ and $md_2 \leq 37$ are dropped in (20) only those d_i-tuples $(y_1, \ldots, y_{d_1})$ and $(y'_1, \ldots, y'_{d_2})$ have to be considered which lie in the image of the linear mappings $\Psi_i \colon \Omega_i \to \Omega_K^{d_i}$.

(iii) The assumption that the product measure $\nu_{1;n} \otimes \nu_{2;n}$ has approximately the mass $2^{-m(d_1-1)} 2^{-m(d_2-1)}$ on each equivalence class (defined in the proof of Theorem 2) is rather mild if $m(d_i - 1)$ is considerably smaller than 43 or 37, resp. It should already be fulfilled for small indices n, i.e. in an early stage after the start of the RNG (cf. the proof of Theorem 2).

We applied Theorem 1 and Theorem 2 to compute the entries in Table 1. For simplicity we assumed $\mu_1 = \mu_2$ and $\sigma_1 = \sigma_2$. The first column contains the intermediate times between two samplings of internal random numbers (expressed as multiples of the mean values $\mu_1 = \mu_2$). In 'real-life' the ratio σ_i/μ_i clearly depends on the used technical components. In Table 1 we exemplarily considered the case $\sigma_i/\mu_i = 0.01$. To obtain similar entropy bounds for $\sigma_i/\mu_i < 0.01$ (resp. for $\sigma_i/\mu_i > 0.01$) the respective values $r_{i;n+1}$ have to be chosen larger (may be chosen smaller). The second column quantifies the gain of entropy of the total system (i.e. the states of the LFSR and CASR) within this time interval (conservative estimator; cf. 'Justification of the stochastical model'). These values provide upper bounds for the conditional entropy of the internal random numbers. To be precise, for large indices n the term $\min\{H(V_{1;n+1}) + H(V_{2;n+1}), m\}$ can be viewed as an upper bound for the conditional entropy $H(C_{n+1} \mid C_1, \ldots, C_n)$ (cf. (17)) which belongs to the scenario that the attacker has observed a large number of internal random numbers $c_1, c_2, \ldots, c_n$. On the other hand the right-most column provides a lower bound for this conditional probability. For $r_{i;n+1} = 20000$ and $m = 2$, for instance, the (conditional) entropy of a 128-bit session key lies in the interval $[64 \cdot 1.910, 64 \cdot 2] = [122.24, 128]$. Dividing the right-most terms in Table 1 by the block size m gives a lower entropy bound per output bit.

Remark 7. (i) For fixed μ_i and σ_i the entropy essentially increases when s_j, resp. $r_{i;n+1}$, increases. Consequently, e.g. the first row gives entropy bounds for $r_{i;n+1} \geq 10000$.

(ii) However, the entropy function is not monotonous in a strict sense: To see this, assume for the moment that $r_{i;n+1}$ is rather small. If $r_{i;n+1} := k$ is an integer then $V_{i;n+1}$ assumes both values k and $k - 1$ with almost equal probability ≈ 0.5, i.e. $H(V_{i;n+1}) \approx 1$. If $r_{i;n+1} = k + 0.5$, however, it is very likely then $V_{i;n+1}$ assumes the value k which means $H(V_{i;n+1}) \approx 0$. The described effect vanishes

Table 1.   Example: $\sigma_i/\mu_i = 0.01$

$r_{i;n+1}$	$H(V_{1;n+1}) + H(V_{2;n+1})$	m	$H(C_{n+1} \mid B_{1;n}, B_{2;n})$
10000	4.209	1	0.943
10000	4.209	2	1.827
10000	4.209	3	2.600
20000	5.152	1	0.971
20000	5.152	2	1.910
60000	6.698	1	0.991

when $r_{i;n+1}$ and hence also $H(V_{i;n+1})$ increase (as in Table 1) since then the mass of the distribution of $V_{i;n+1}$ is spread over an interval of natural numbers. (iii) When the intermediate time $r_{i;n+1}$ increases (for fixed ratio σ_i/μ_i) the cardinality of the sets D_i (cf. Theorem 2) must also increase to fulfil the condition $\text{Prob}(V_{i;n+1} \in D_i) \approx 1$. We point out that the computational effort to determine $H(C_{n+1} \mid B_{1;n}, B_{2;n})$ increases exponentially in d_i and m.
(iv) The results derived in this paper do not help to improve Dichtl's attack or any similar attack as an attacker will intuitively begin guessing values for $(v_{1;n+1}, v_{2;n+1})$ which are close to the mean values $(r_{1;n+1}, r_{2;n+1})$ (cf. Sect. 2, 'Dichtl's attack') anyway. Moreover, even an attacker with little knowledge in stochastics could easily determine an approximate distribution of $V_{i;n+1}$ (cf. Theorem 1) by stochastic simulations. Theorem 2 provides an approximate expression for $H(C_{n+1} \mid B_{1;n}, B_{2;n}) \leq H(C_{n+1} \mid C_1, \ldots, C_n)$ which should be of little profit to perform any attack. The intention of this paper, however, is the opposite: It provides results which enable to quantify parameter sets yielding unconditional security.

6   Conclusions

We have formulated and analyzed a stochastical model for the TRNG presented in [9]. We determined lower and upper entropy bounds per random output bit. Applying these results parameter sets $(r_{i;n+1}, m)$ can be determined for which the entropy per output bit is close to 1 yielding unconditional security. Compared with the parameters proposed in [9] this reduces the output rate of the TRNG noticeably where the concrete numbers clearly depend on the properties of the technical components.

References

1. AIS 31: Functionality Classes and Evaluation Methodology for Physical Random Number Generators. Version 1 (25.09.2001) (mandatory if a German IT security certificate is applied for; English translation).
   www.bsi.bund.de/zertifiz/zert/interpr/ais31e.pdf
2. Common Criteria for Information Technology Security Evaluation, Part 1–3; Version 2.1, August 1999 and ISO 15408:1999.

3. M. Dichtl: How to Predict the Output of a Hardware Random Number Generator. In: C.D. Walter, Ç.K. Koç, C. Paar (eds.): Cryptographic Hardware and Embedded Systems – CHES 2003, Springer, Lecture Notes in Computer Science 2779, Berlin 2003, 181–188. Preversion also in IACR Cryptology ePrint Archive, Report 2003/051.

4. W. Hoeffding, H. Robbins: The Central Limit Theorem for Dependent Random Variables. Duke Math. J. **15** (1948), 773–780.

5. Information Technology Security Evaluation Criteria (ITSEC); Provisional Harmonised Criteria, Version 1.2, June 1991.

6. W. Killmann, W. Schindler: A Proposal for: Functionality Classes and Evaluation Methodology for True (Physical) Random Number Generators. Version 3.1 (25.09.2001), mathematical-technical reference of [1] (English translation); www.bsi.bund.de/zertifiz/zert/interpr/trngk31e.pdf

7. A.J. Menezes, P.C. van Oorschot, S.C. Vanstone: Handbook of Applied Cryptography. CRC Press, Boca Raton (1997).

8. W. Schindler, W. Killmann: Evaluation Criteria for True (Physical) Random Number Generators Used in Cryptographic Applications. In: B.S. Kaliski Jr., Ç.K. Koç, C. Paar (eds.): Cryptographic Hardware and Embedded Systems – CHES 2002, Springer, Lecture Notes in Computer Science 2523, Berlin 2003, 431–449.

9. T. Tkacik: A Hardware Random Number Generator. In: B.S. Kaliski Jr., Ç.K. Koç, C. Paar (eds.): Cryptographic Hardware and Embedded Systems – CHES 2002, Springer, Lecture Notes in Computer Science 2523, Berlin 2003, 450–453.

Appendix

Lemma 2. *(i) Assume that X is a random variable assuming values on a countable set Ω and that its distribution is given by the sequence $p_1, p_2, \ldots$. If there exists a real number $\alpha \in (0,1)$ with $\sum_{j=1}^{\infty} p_i^{1-\alpha} < \infty$ then $H(X) < \infty$.*
(ii) Let the random variable $V_{(r\mu)}$ be defined as in Lemma 1. Then $H(V_{(r\mu)}) < \infty$ for each $r \in (0, \infty)$.
(iii) Let $X_0, X_1, \ldots$ denote a stationary sequence of discrete random variables assuming values on a countable set Ω with $H(X_0) < \infty$. For any mapping $\phi \colon \Omega \to \Omega'$ we have

$$H(X_{n+1} \mid X_0, \ldots, X_n) \searrow h_X \quad \text{and} \quad H(\phi(X_{n+1}) \mid \phi(X_0), \ldots, \phi(X_n)) \searrow h_{\phi, X} \tag{21}$$

for non-negative real numbers h_X and $h_{\phi,X}$ as $n \to \infty$. In particular,

$$0 \leq h_{\phi, X} \leq h_X. \tag{22}$$

Proof. Assertion (i) follows immediately from the fact that $x^\alpha \log_2 x$ converges to zero as x tends to zero for each positive α. That is, $|\log_2 x| < x^{-\alpha}$ if x is sufficiently small. Within the proof of (ii) we set $c := \mu/\sigma$. Elementary arithmetic yield

$$\frac{r-k}{\sqrt{k}} c - \frac{r-(k+1)}{\sqrt{k+1}} c = \frac{c(r + \sqrt{k}\sqrt{k+1})}{\sqrt{k}\sqrt{k+1}(\sqrt{k+1} + \sqrt{k})} < \frac{c}{\sqrt{k}} < 1$$

if $k > k_0$ for a suitable threshold $k_0 \in \mathbb{N}$. If $k > r$ the density of the $N(0,1)$-distribution attains its maximum in $[c(r-(k+1))/\sqrt{k+1}, c(r-k)/\sqrt{k}]$ at the right-hand boundary of the interval. In particular,

$$\frac{1}{\sqrt{2\pi}} e^{-\frac{1}{2}\left(\frac{c(r-k)}{\sqrt{k}}\right)^2} < \frac{1}{\sqrt{2\pi}} e^{-\frac{c^2 k}{8}}$$

for $k > 2r$. With Lemma 1(iii) we obtain $\mathrm{Prob}(V_{(r\mu)} = k)^{0.5} < (2\pi)^{0.25} e^{-c^2 k/16}$ for $k > k_0, 2r$. Applying (i) with $\alpha = 0.5$ completes the proof of (ii) as $\sum_{k=1}^{\infty} e^{-c'k} < \infty$ for each $c' > 0$. As the random variables $X_1, X_2, \ldots$ in (iii) are stationary we conclude

$$H(X_{n+2} \mid X_0, \ldots, X_{n+1}) \leq H(X_{n+2} \mid X_1, \ldots, X_{n+1}) = H(X_{n+1} \mid X_0, \ldots, X_n).$$

As a non-increasing sequence of non-negative real numbers converges to a non-negative limit we obtain

$$\frac{H(X_0, \ldots, X_n)}{n} = \frac{H(X_0)}{n} + \sum_{j=1}^{n} \frac{H(X_j \mid X_0, \ldots, X_{j-1})}{n} \to h_X$$

which proves the left-hand side of (21). With $\phi(X_j)$ instead of X_j we obtain an analogous formula for $h_{\phi, X}$. As $H(X_0, \ldots, X_{n+1}) \geq H(\phi(X_0), \ldots, \phi(X_{n+1}))$ this completes the proof of (22). $\square$

Lemma 2 says that the increase of entropy $H(\phi(X_{n+1}) \mid \phi(X_0), \ldots, \phi(X_n))$ is not larger than $H(X_{n+1} \mid X_0, \ldots, X_n)$ as n tends to infinity. However, the following counterexample shows that $H(\phi(X_{j+1}) \mid \phi(X_0), \ldots, \phi(X_j)) > H(X_{j+1} \mid X_0, \ldots, X_j)$ is possible for a small index j although $H(\phi(X_{j+1})) \leq H(X_{j+1})$. The reason for this phenomenon is that the sequence $\phi(X_0), \ldots, \phi(X_n)$ usually provides less information than $X_0, \ldots, X_n$.

Example 1. Let $\Omega = \{0, 1, 2, 3\}$, $\Omega' = \{0, 1\}$, and $\phi: \Omega \to \Omega'$ maps $\omega \in \Omega$ onto its most significnt bit. Further, the random variable X_0 is equidistributed on Ω, and $X_{n+1} := X_n + 1 (\mathrm{mod}\, 4)$ for $n \geq 0$. Then $H(X_0) = 2$, $H(X_{n+1} \mid X_0, \ldots, X_n) = 0$ for $n \geq 0$ whereas $H(\phi(X_0)) = 1$, $H(\phi(X_1) \mid \phi(X_0)) = 1$, and $H(\phi(X_{n+1}) \mid \phi(X_0), \ldots, \phi(X_n)) = 0$ for $n \geq 1$.

Analysis of Double Block Length Hash Functions

Mitsuhiro Hattori, Shoichi Hirose, and Susumu Yoshida

Graduate School of Informatics, Kyoto University, Kyoto, 606–8501 JAPAN
hattori@hanase.kuee.kyoto-u.ac.jp

Abstract. The security of double block length hash functions and their compression functions is analyzed in this paper. First, the analysis of double block length hash functions by Satoh, Haga, and Kurosawa is investigated. The focus of this investigation is their analysis of the double block length hash functions with the rate 1 whose compression functions consist of a block cipher with the key twice longer than the plaintext/ciphertext. It is shown that there exists a case uncovered by their analysis. Second, the compression functions are analyzed with which secure double block length hash functions may be constructed. The analysis shows that these compression functions are at most as secure as the compression functions of single block length hash functions.

1 Introduction

A hash function is a mapping from the set of all binary sequences to the set of binary sequences of some fixed length. It is one of the most important primitives in cryptography[1]. A hash function dedicated to cryptography is called a cryptographic hash function. Cryptographic hash functions are classified into unkeyed hash functions and keyed hash functions. In this paper, the unkeyed hash functions are discussed and they are simply called hash functions.

A hash function usually consists of a compression function. A compression function is the function $f : \{0,1\}^a \times \{0,1\}^b \to \{0,1\}^a$. There are two major methods for construction of a compression function, namely, from scratch and based on a block cipher. The topic of this paper is the latter method. The main motivation of this construction is the minimization of design and implementation effort, which is supported by the expectation that secure hash functions can be constructed from secure block ciphers.

Hash functions based on block ciphers are classified into two categories: single block length hash functions and double block length hash functions. A single block length hash function is a hash function the length of whose output is equal to that of the block cipher. The length of the output of a double block length hash function is twice larger than that of the block cipher. The length of the output of a widely used block cipher is 64 or 128. Thus, single block length hash functions are no longer secure.

The compression functions of double block length hash functions are classified by the number of encryptions and the key length of the block cipher. The double block length hash functions with the compression functions with two encryptions

K.G. Paterson (Ed.): Cryptography and Coding 2003, LNCS 2898, pp. 290–302, 2003.

of an (m, m) block cipher were analyzed in [2,3], where an (m, κ) block cipher is the one with the length of the plaintext/ciphertext m and the length of the key κ. Satoh, Haga, and Kurosawa [4] analyzed the double block length hash functions with the compression functions with one encryption of an (m, m) or $(m, 2m)$ block cipher. They also analyzed the double block length hash functions with the compression functions with two encryptions of an $(m, 2m)$ block cipher. They stated that no effective attacks were found for the double block length hash functions with the compression functions, with two encryptions of an $(m, 2m)$ block cipher, satisfying the property called "exceptional" defined by them.

In this paper, first, the analysis of double block length hash functions by Satoh, Haga, and Kurosawa is investigated. The focus of the investigation is their analysis of the double block length hash functions with the rate 1 whose compression functions consist of two encryptions of an $(m, 2m)$ block cipher. This investigation shows that there exists a case uncovered by their analysis. This result implies that there exist double block length hash functions whose compression functions do not satisfy the property "exceptional" and on which no effective attacks are found.

Second, for the double block length hash functions on which no effective attacks are known, their compression functions are analyzed. It is shown that all of these compression functions are at most as secure as those of single block length hash functions.

The paper is organized as follows. Some definitions are introduced and mathematical facts are described in Section 2. Block-cipher-based hash functions are defined in Section 3. The analysis by Satoh et.al. is investigated in Section 4. In Section 5, the analysis of compression functions is described. Finally Section 6 concludes this paper with future work.

2   Preliminaries

$\mathbb{N}$ denotes the set of natural numbers. $\oplus$ denotes the bitwise exclusive OR. $a\|b$ denotes the concatenation of $a \in \{0,1\}^i$ and $b \in \{0,1\}^j$, where $a\|b \in \{0,1\}^{i+j}$.

2.1   Block Ciphers

A block cipher is a keyed function which maps an m-bit plaintext block to an m-bit ciphertext block. Let $\kappa, m \in \mathbb{N}$. An (m, κ) block cipher is a mapping $E : \{0,1\}^\kappa \times \{0,1\}^m \to \{0,1\}^m$. For each $k \in \{0,1\}^\kappa$, the function $E_k(\cdot) = E(k, \cdot)$ is a one-to-one mapping from $\{0,1\}^m$ to $\{0,1\}^m$. $\{0,1\}^\kappa$ and $\{0,1\}^m$ in the domain $\{0,1\}^\kappa \times \{0,1\}^m$ and $\{0,1\}^m$ in the range are called the key space, the plaintext space, and the ciphertext space, respectively. m is called the block length and κ is called the key length.

2.2   Hash Functions

A hash function is a mapping from the set of all binary sequences to the set of binary sequences of some fixed length. A hash function is denoted by $h : \{0,1\}^* \to \{0,1\}^a$, where $\{0,1\}^* = \bigcup_{i>0}\{0,1\}^i$.

A hash function $h : \{0,1\}^* \to \{0,1\}^a$ usually consists of a compression function $f : \{0,1\}^a \times \{0,1\}^b \to \{0,1\}^a$ and an initial value $IV \in \{0,1\}^a$. h is computed by the iterated application of f to the given input. Thus, h is called an iterated hash function. The output of the hash function h for an input $M \in \{0,1\}^*$, $h(M)$, is calculated as follows. M is called a message.

Step 1. The message M is divided into the blocks of the equal length b. If the length of M is not a multiple of b, M is padded using an unambiguous padding rule. Let $M_1, M_2, \ldots, M_n$ be the blocks from the (padded) message M, where $M_i \in \{0,1\}^b$ for $i = 1, 2, \ldots, n$.

Step 2. $H_i = f(H_{i-1}, M_i)$ is calculated for $i = 1, 2, \ldots, n$, where $H_i \in \{0,1\}^a$ and $H_0 = IV$. H_n is the output of h for the message M, that is, $H_n = h(M)$. If the initial value should be specified, the equation is described as $H_n = h(H_0, M)$.

2.3   Properties Required for Hash Functions

For a hash function h, there exist many pairs $(M, \hat{M})$ such that $h(M) = h(\hat{M})$ and $M \neq \hat{M}$. For cryptographic use, the hash function h must satisfy the following properties.

Preimage resistance. Given a hash value H, it is computationally infeasible to find a message M such that $h(M) = H$.

Second preimage resistance. Given a message M, it is computationally infeasible to find a message $\hat{M}$ such that $h(M) = h(\hat{M})$ and $M \neq \hat{M}$.

Collision resistance. It is computationally infeasible to find a pair of messages, M and $\hat{M}$, such that $h(M) = h(\hat{M})$ and $M \neq \hat{M}$.

The relationships among the properties are [1]:

- If a hash function satisfies the second preimage resistance, then it also satisfies the preimage resistance, and
- If a hash function satisfies the collision resistance, then it also satisfies the second preimage resistance.

Therefore, it is the easiest to satisfy preimage resistance, and it is the most difficult to satisfy collision resistance.

2.4   Attacks on Hash Functions

The following attacks [3] are against the properties listed in Section 2.3.

The preimage attack. Given an initial value H_0 and a hash value H, find a message M such that $H = h(H_0, M)$.

The second preimage attack. Given an initial value H_0 and a message M, find a message $\hat{M}$ such that $h(H_0, M) = h(H_0, \hat{M})$ and $M \neq \hat{M}$.

The free-start preimage attack. Given a hash value H, find an initial value H_0 and a message M such that $h(H_0, M) = H$.

The free-start second preimage attack. Given an initial value H_0 and a message M, find an initial value $\hat{H}_0$ and a message $\hat{M}$ such that $h(H_0, M) = h(\hat{H}_0, \hat{M})$ and $(H_0, M) \neq (\hat{H}_0, \hat{M})$.

The collision attack. Given an initial value H_0, find two messages M, $\hat{M}$ such that $h(H_0, M) = h(H_0, \hat{M})$ and $M \neq \hat{M}$.

The semi-free-start collision attack. Find an initial value H_0 and two messages M, $\hat{M}$ such that $h(H_0, M) = h(H_0, \hat{M})$ and $M \neq \hat{M}$.

The free-start collision attack. Find two initial values H_0, $\hat{H}_0$ and two messages M, $\hat{M}$ such that $h(H_0, M) = h(\hat{H}_0, \hat{M})$ and $(H_0, M) \neq (\hat{H}_0, \hat{M})$.

The following two propositions [5] are often used to estimate the amount of computation of the attacks.

Proposition 1. *Suppose that a sample of size r is drawn from a set of N elements with replacement. If $r, N \to \infty$, then the probability that a given element is drawn converges to*

$$1 - \exp\left(-\frac{r}{N}\right) . \tag{1}$$

Proposition 2 (Birthday Paradox). *Suppose that a sample of size r is drawn from a set of N elements with replacement. If $r, N \to \infty$ and r is $O(\sqrt{N})$, then the probability that there is at least one coincidence is converges to*

$$1 - \exp\left(-\frac{r^2}{2N}\right) . \tag{2}$$

3   Hash Functions Based on Block Ciphers

3.1   Compression Function Construction

There are two major methods for constructing compression functions: construction based on block ciphers and construction from scratch. The topic of this paper is the former construction.

3.2   Single Block Length Hash Functions and Double Block Length Hash Functions

Let $h : \{0,1\}^* \to \{0,1\}^a$ be an iterated hash function and $E : \{0,1\}^\kappa \times \{0,1\}^m \to \{0,1\}^m$ be a block cipher used in the compression function of h. If $a = m$, then h is called a single block length hash function. If $a = 2m$, then h is called a double block length hash function..

Let σ be the number of the encryptions of the block cipher used in the compression function. Let $b = |M_i|$. Then, the rate is defined as $b/(\sigma \cdot m)$ and is used as a speed index.

3.3 Hash Functions Considered in This Paper

We consider double block length hash functions with the rate 1, whose compression functions are composed of two encryptions of an $(m, 2m)$ block cipher.

Let $M_i = (M_i^1, M_i^2) \in \{0,1\}^{2m}$ be a message block, where $M_i^1, M_i^2 \in \{0,1\}^m$. The compression function $H_i = f(H_{i-1}, M_i)$ is defined by the two functions f^1, f^2 such as

$$\begin{cases} H_i^1 = f^1(H_{i-1}^1, H_{i-1}^2, M_i^1, M_i^2) \\ H_i^2 = f^2(H_{i-1}^1, H_{i-1}^2, M_i^1, M_i^2) \end{cases} \tag{3}$$

where $H_j = (H_j^1, H_j^2)$ and $H_j^1, H_j^2 \in \{0,1\}^m$, for $j = i-1, i$. Each of f^1 and f^2 contains one encryption of the $(m, 2m)$ block cipher E. H_i^1 and H_i^2 are represented by

$$\begin{cases} H_i^1 = E_{A\|B}(C) \oplus D \\ H_i^2 = E_{W\|X}(Y) \oplus Z \end{cases} \tag{4}$$

where $A, B, C, D, W, X, Y, Z \in \{0,1\}^m$. A, B, C, D, W, X, Y and Z are represented by linear combinations of $H_{i-1}^1, H_{i-1}^2, M_i^1$ and M_i^2 as follows:

$$\begin{pmatrix} A \\ B \\ C \\ D \end{pmatrix} = L_1 \begin{pmatrix} H_{i-1}^1 \\ H_{i-1}^2 \\ M_i^1 \\ M_i^2 \end{pmatrix} \tag{5}$$

$$\begin{pmatrix} W \\ X \\ Y \\ Z \end{pmatrix} = L_2 \begin{pmatrix} H_{i-1}^1 \\ H_{i-1}^2 \\ M_i^1 \\ M_i^2 \end{pmatrix} , \tag{6}$$

where L_1 and L_2 are 4×4 binary matrices.

Let $\boldsymbol{a}$, $\boldsymbol{b}$, $\boldsymbol{c}$ and $\boldsymbol{d}$ denote row vectors of L_1 and let $\boldsymbol{w}$, $\boldsymbol{x}$, $\boldsymbol{y}$ and $\boldsymbol{z}$ denote row vectors of L_2.

3.4 The Black-Box Model

The black-box model [6] is used in our analysis. In this model, a block cipher is assumed to be random, that is, $E_k : \{0,1\}^m \to \{0,1\}^m$ is a random permutation for each $k \in \{0,1\}^\kappa$. Two oracles, E and E^{-1}, are available. E simply returns $E_k(x)$ on an input (k, x). E^{-1}, on an input (k, y), returns x such that $E_k(x) = y$.

4 A Comment on the Analysis by Satoh, Haga, and Kurosawa

Satoh, Haga, and Kurosawa [4] have analyzed the security of the double block length hash functions defined in Section 3.3. In this section, their analysis is reconsidered. It is shown that there exists a case uncovered by their analysis.

4.1   Analysis by Satoh et.al.

Satoh, Haga, and Kurosawa [4] presented effective attacks against hash functions whose compression functions does not satisfy the property called "exceptional" defined in their paper. This property is defined as follows.

Definition 1. *Let L be a 4×4 binary matrix. Let L_r be the 4×2 submatrix of L, where L_r consists of the right half elements of L. Let L_r^3 be the 3×2 submatrix of L_r such that the third row of L_r is deleted. Let L_r^4 be the 3×2 submatrix of L_r such that the fourth row of L_r is deleted. L is called exceptional if $\mathrm{Rank}(L) = 4$ and $\mathrm{Rank}(L_r^3) = \mathrm{Rank}(L_r^4) = 2$.*

The following claim is Theorem 16 in their paper.

Claim 1. *For the double block length hash functions of the rate 1, whose round function has the form of (4), suppose that at least one of L_1 and L_2 is not exceptional. Then, there exist second preimage and preimage attacks with about 4×2^m complexity. Furthermore, there exists a collision attack with about $3 \times 2^{m/2}$ complexity.*

Throughout this paper, as in the above claim, the complexity of an attack is the required number of encryptions and decryptions of the block cipher. This is the number of the queries to the oracles.

4.2   A Comment

In this section, it is shown that the attacks by Satoh et.al. do not work on some hash functions as is expected though their compression functions do not satisfy "exceptional".

Let N_2 be the 2×2 submatrix of L_r, where N_2 consists of the upper half elements of L_r. Satoh et.al. presented their proof of Claim 1 for two cases: (i) $\mathrm{Rank}(L) = 3$ and $\mathrm{Rank}(N_2) = 2$ and (ii) $\mathrm{Rank}(L) = 4$. The first case is investigated in the remaining part. Their proof proceeds as follows.

Since $\mathrm{Rank}(N_2) = 2$, one can find (by elementary row operations) $\alpha, \beta = 0, 1$ such that

$$L' = \begin{pmatrix} a \\ b \\ c \\ d \oplus \alpha a \oplus \beta b \end{pmatrix} = \begin{pmatrix} N_1 & N_2 \\ N_3' & N_4' \end{pmatrix}, \tag{7}$$

where

$$N_4' = \begin{pmatrix} * & * \\ 0 & 0 \end{pmatrix}. \tag{8}$$

Let

$$
\begin{pmatrix} A \\ B \\ C \\ D' \end{pmatrix} = L' \begin{pmatrix} H_{n-1}^1 \\ H_{n-1}^2 \\ M_n^1 \\ M_n^2 \end{pmatrix}. \tag{9}
$$

Then, $D' = 0$, H_{n-1}^1, H_{n-1}^2 or $H_{n-1}^1 \oplus H_{n-1}^2$.

Subsequently, they stated in their proofs that $c = \lambda_1 a \oplus \lambda_2 b$ when $D' \neq 0$. However, in general, there may be a case that $c = \lambda_1 a \oplus \lambda_2 b \oplus d$ even if $D' \neq 0$. Furthermore, in this case, their attack for the case that $\text{Rank}(L) = 3$, $\text{Rank}(N_2) = 2$, and $D' \neq 0$ cannot be applied.

In their attack, the adversary chooses random triples (A, B, C) such that $C = \lambda_1 A \oplus \lambda_2 B$ and computes $D = E_{A\|B}(C) \oplus H_n^1$. Then the adversary computes $D' = D \oplus \alpha A \oplus \beta B$. However, if $c = \lambda_1 a \oplus \lambda_2 b \oplus d$, C is calculated by A, B, and D. Therefore, the adversary cannot compute D by $E_{A\|B}(C) \oplus H_n^1$.

5   Collision-Resistance of Compression Functions

From the results by Satoh, Haga, and Kurosawa and the discussion in the last section, no effective attacks are found on the double block length hash functions defined in Section 3.3 with compression functions satisfying

(i) "exceptional", or
(ii) $\text{Rank}(L_1) = \text{Rank}(L_2) = 3$,
   $c \oplus d = \lambda_1 a \oplus \lambda_2 b$ for some $\lambda_1, \lambda_2 \in \{0, 1\}$,
   $y \oplus z = \lambda_3 w \oplus \lambda_4 x$ for some $\lambda_3, \lambda_4 \in \{0, 1\}$,
   and $\text{Rank}(N_{1,2}) = \text{Rank}(N_{2,2}) = 2$, where $N_{i,2}$ is the upper right 2×2 submatrix of L_i for $i = 1, 2$.

For collision resistance, Merkle [7] and Damgård [8] independently showed that, from any algorithm for the collision attack on a hash function, an algorithm for the free-start collision attack on its compression function is constructed. The complexity of the latter algorithm is almost equal to that of the former one. Thus, if there exists no effective free-start collision attack on a compression function, then there exists no effective collision attack on the hash function composed of the compression function.

In this section, however, it is shown that there exist effective free-start collision attacks with complexity $O(2^{m/2})$ on all compression functions satisfying $\text{Rank}(L_1) \geq 3$ and $\text{Rank}(L_2) \geq 3$. Thus, it is impossible to prove collision resistance of double block length hash functions composed of the compression functions satisfying the condition (i) or (ii) above based on the result of Merkle and Damgård.

Effective free-start (second) preimage attacks are also presented in this section.

Theorem 1. *Let f be any compression function represented by the equation (4) such that $\text{Rank}(L_1) \geq 3$ and $\text{Rank}(L_2) \geq 3$. Then, there exist a free-start second preimage attack and a free-start collision attack on f with complexities about 2×2^m and $2 \times 2^{m/2}$, respectively.*

This theorem is proved for the following two cases:

(i) $\text{Rank}(L_1) = 4$ or $\text{Rank}(L_2) = 4$,
(ii) $\text{Rank}(L_1) = \text{Rank}(L_2) = 3$.

The following lemma is for the former case.

Lemma 1. *Suppose that $\text{Rank}(L_1) \geq 3$ and $\text{Rank}(L_2) \geq 3$. If $\text{Rank}(L_1) = 4$ or $\text{Rank}(L_2) = 4$, then there exist a free-start second preimage attack and a free-start collision attack with complexities about 2×2^m and $2 \times 2^{m/2}$, respectively.*

Proof. Without loss of generality, suppose that $\text{Rank}(L_1) = 4$.
Since $\text{Rank}(L_1) = 4$, from (5),

$$
\begin{pmatrix} H_{i-1}^1 \\ H_{i-1}^2 \\ M_i^1 \\ M_i^2 \end{pmatrix} = L_1^{-1} \begin{pmatrix} A \\ B \\ C \\ D \end{pmatrix}.
\tag{10}
$$

The free-start second preimage attack. The adversary *Adv* proceeds as follows.

Step 0. *Adv* computes the output (H_i^1, H_i^2) for the given input $(H_{i-1}^1, H_{i-1}^2, M_i^1, M_i^2)$.
Step 1. *Adv* chooses 2^m random triples $(\tilde{A}, \tilde{B}, \tilde{C})$ and computes $\tilde{D} = E_{\tilde{A} \| \tilde{B}}(\tilde{C}) \oplus H_i^1$. Since the block cipher is assumed to be random, $\tilde{D}$ is also random.
Step 2. For each 4-tuples $(\tilde{A}, \tilde{B}, \tilde{C}, \tilde{D})$, *Adv* computes $(\tilde{H}_{i-1}^1, \tilde{H}_{i-1}^2, \tilde{M}_i^1, \tilde{M}_i^2)$ with (10). Since $(\tilde{A}, \tilde{B}, \tilde{C}, \tilde{D})$ is random, $(\tilde{H}_{i-1}^1, \tilde{H}_{i-1}^2, \tilde{M}_i^1, \tilde{M}_i^2)$ is also random.
Step 3. For each $(\tilde{H}_{i-1}^1, \tilde{H}_{i-1}^2, \tilde{M}_i^1, \tilde{M}_i^2)$, *Adv* computes $(\tilde{W}, \tilde{X}, \tilde{Y}, \tilde{Z})$ with (6) and computes $\tilde{H}_i^2 = E_{\tilde{W} \| \tilde{X}}(\tilde{Y}) \oplus \tilde{Z}$.

Since $\text{Rank}(L_2) \geq 3$, $(\boldsymbol{w}, \boldsymbol{x}) \neq (\boldsymbol{0}, \boldsymbol{0})$. Therefore, at least one of $\tilde{W}$ and $\tilde{X}$ is expressed by a linear combination of $\tilde{H}_{i-1}^1$, $\tilde{H}_{i-1}^2$, $\tilde{M}_i^1$ and $\tilde{M}_i^2$. Since $(\tilde{H}_{i-1}^1, \tilde{H}_{i-1}^2, \tilde{M}_i^1, \tilde{M}_i^2)$ is random, $E_{\tilde{W} \| \tilde{X}}(\tilde{Y})$ is random, and $\tilde{H}_i^2$ is also random. Thus, according to Proposition 1, *Adv* can find $\tilde{H}_i^2$ such that $H_i^2 = \tilde{H}_i^2$ with probability about 0.63. The total complexity is about 2×2^m.

The free-start collision attack. *Adv* chooses arbitrary H_i^1. Then it chooses $2^{m/2}$ random triples $(\tilde{A}, \tilde{B}, \tilde{C})$ and computes $\tilde{H}_i^2$ in the same way as in the Steps 1–3 above. According to Proposition 2, *Adv* can find a collision of f^2 with probability about 0.39. The total complexity is about $2 \times 2^{m/2}$. □

The next lemma is for the case that $\text{Rank}(L_1) = \text{Rank}(L_2) = 3$.

Lemma 2. *Suppose that* $\mathrm{Rank}(L_1) = \mathrm{Rank}(L_2) = 3$. *Then, there exist a free-start second preimage attack and a free-start collision attack with complexities about* 2×2^m *and* $2 \times 2^{m/2}$, *respectively.*

This lemma is lead from the following two lemmas.

Lemma 3. *Suppose that* $\mathrm{Rank}(L_1) = \mathrm{Rank}(L_2) = 3$. *If* $c \oplus d$ *is not represented by any linear combination of* a *and* b, *or* $y \oplus z$ *is not represented by any linear combination of* w *and* x, *then there exist a free-start second preimage attack and a free-start collision attack with complexities about* 2×2^m *and* $2 \times 2^{m/2}$, *respectively.*

Proof. Without loss of generality, suppose that $c \oplus d$ is not represented by any linear combination of a and b.

Since $\mathrm{Rank}(L_1) = 3$, the following three cases should be considered:

(a) a and b are linearly dependent,
(b) a and b are linearly independent and $c = \lambda_1 a \oplus \lambda_2 b$ for some $\lambda_1, \lambda_2 \in \{0, 1\}$,
(c) a and b are linearly independent and $d = \lambda_1 a \oplus \lambda_2 b$ for some $\lambda_1, \lambda_2 \in \{0, 1\}$.

Case (a)

In this case, either a or b is 0, or $a = b$. Thus the key $A\|B$ is $A\|0$, $0\|B$, or $A\|A$. Suppose that the key is $A\|A$. For the other two cases, the proofs are almost same. From the equation (5),

$$\begin{pmatrix} A \\ C \\ D \end{pmatrix} = \hat{L}_1 \begin{pmatrix} H_{i-1}^1 \\ H_{i-1}^2 \\ M_i^1 \\ M_i^2 \end{pmatrix} , \qquad (11)$$

where $\hat{L}_1$ is the 3×4 submatrix of L_1 such that the second row is deleted. Let

$$\hat{L}_1 = \begin{pmatrix} h^1 & h^2 & m^1 & m^2 \end{pmatrix} , \qquad (12)$$

where h^1, h^2, m^1, and m^2 are column vectors. Since $\mathrm{Rank}(\hat{L}_1) = 3$, one column vector of $\hat{L}_1$ is expressed by a linear combination of the other column vectors. Without loss of generality, suppose that h^1 is expressed by a linear combination of the other vectors. Then,

$$\begin{pmatrix} A \\ C \\ D \end{pmatrix} = \tilde{L}_1 \begin{pmatrix} H_{i-1}^2 \\ M_i^1 \\ M_i^2 \end{pmatrix} \oplus h^1 H_{i-1}^1 , \qquad (13)$$

where $\tilde{L}_1 = \begin{pmatrix} h^2 & m^1 & m^2 \end{pmatrix}$. Since $\mathrm{Rank}(\tilde{L}_1) = 3$,

$$\begin{pmatrix} H_{i-1}^2 \\ M_i^1 \\ M_i^2 \end{pmatrix} = \tilde{L}_1^{-1} \left(\begin{pmatrix} A \\ C \\ D \end{pmatrix} \oplus h^1 H_{i-1}^1 \right) . \qquad (14)$$

The free-start second preimage attack.

Step 0. *Adv* computes the output (H_i^1, H_i^2) for the given input $(H_{i-1}^1, H_{i-1}^2, M_i^1, M_i^2)$.

Step 1. *Adv* chooses 2^m random 2-tuples $(\tilde{A}, \tilde{C})$ and computes $\tilde{D} = E_{\tilde{A}\|\tilde{A}}(\tilde{C}) \oplus H_i^1$.

Step 2. For each $(\tilde{A}, \tilde{C}, \tilde{D})$, *Adv* chooses a random $\tilde{H}_{i-1}^1$ and computes $(\tilde{H}_{i-1}^2, \tilde{M}_i^1, \tilde{M}_i^2)$ from (14).

Step 3. For each $(\tilde{H}_{i-1}^1, \tilde{H}_{i-1}^2, \tilde{M}_i^1, \tilde{M}_i^2)$, *Adv* computes $(\tilde{W}, \tilde{X}, \tilde{Y}, \tilde{Z})$ with (6) and computes $\tilde{H}_i^2 = E_{\tilde{W}\|\tilde{X}}(\tilde{Y}) \oplus \tilde{Z}$.

In the above process, Step 1 can be executed since $\boldsymbol{a}, \boldsymbol{c}, \boldsymbol{d}$ of L_1 are linearly independent. For Step 3, at least one of $\tilde{W}$ and $\tilde{X}$ is expressed by a linear combination of $\tilde{H}_{i-1}^1, \tilde{H}_{i-1}^2, \tilde{M}_i^1$ and $\tilde{M}_i^2$ since $\mathrm{Rank}(L_2) = 3$. Since $(\tilde{H}_{i-1}^1, \tilde{H}_{i-1}^2, \tilde{M}_i^1, \tilde{M}_i^2)$ is random, $E_{\tilde{W}\|\tilde{X}}(\tilde{Y})$ is random, and $\tilde{H}_i^2$ is also random. Thus, according to Proposition 1, *Adv* can find $\tilde{H}_i^2$ such that $H_i^2 = \tilde{H}_i^2$ with probability about 0.63. The total complexity is about 2×2^m.

The free-start collision attack. *Adv* chooses arbitrary H_i^1. *Adv* chooses $2^{m/2}$ random 2-tuples $(\tilde{A}, \tilde{C})$ and computes $\tilde{D} = E_{\tilde{A}\|\tilde{A}}(\tilde{C}) \oplus H_i^1$. After that, it computes $\tilde{H}_i^2$ in the same way as in the Steps 2–3 above. According to Proposition 2, *Adv* can find a collision of f^2 with probability about 0.39. The total complexity is about $2 \times 2^{m/2}$.

Case (b)

The free-start second preimage attack.

Step 0. *Adv* computes the output (H_i^1, H_i^2) for the given input $(H_{i-1}^1, H_{i-1}^2, M_i^1, M_i^2)$.

Step 1. *Adv* chooses 2^m random triples $(\tilde{A}, \tilde{B}, \tilde{C})$ such that $\tilde{C} = \lambda_1 \tilde{A} \oplus \lambda_2 \tilde{B}$ and computes $\tilde{D} = E_{\tilde{A}\|\tilde{B}}(\tilde{C}) \oplus H_i^1$.

Step 2. For each $(\tilde{A}, \tilde{B}, \tilde{C}, \tilde{D})$, *Adv* chooses a random 4-tuple $(\tilde{H}_{i-1}^1, \tilde{H}_{i-1}^2, \tilde{M}_i^1, \tilde{M}_i^2)$ which satisfies (5).

Step 3. For each $(\tilde{H}_{i-1}^1, \tilde{H}_{i-1}^2, \tilde{M}_i^1, \tilde{M}_i^2)$, *Adv* computes $(\tilde{W}, \tilde{X}, \tilde{Y}, \tilde{Z})$ with (6) and computes $\tilde{H}_i^2 = E_{\tilde{W}\|\tilde{X}}(\tilde{Y}) \oplus \tilde{Z}$.

In the above process, Step 1 can be executed since $\boldsymbol{a}, \boldsymbol{b}, \boldsymbol{d}$ of L_1 are linearly independent. For Step 3, since $\mathrm{Rank}(L_2) = 3$, at least one of $\tilde{W}$ and $\tilde{X}$ is expressed by a linear combination of $\tilde{H}_{i-1}^1, \tilde{H}_{i-1}^2, \tilde{M}_i^1$ and $\tilde{M}_i^2$. Since $(\tilde{H}_{i-1}^1, \tilde{H}_{i-1}^2, \tilde{M}_i^1, \tilde{M}_i^2)$ is random, $E_{\tilde{W}\|\tilde{X}}(\tilde{Y})$ is random, and $\tilde{H}_i^2$ is also random. Thus, according to Proposition 1, *Adv* can find $\tilde{H}_i^2$ such that $H_i^2 = \tilde{H}_i^2$ with probability about 0.63. The total complexity is about 2×2^m.

The free-start collision attack. Adv chooses arbitrary H_i^1. Then it chooses $2^{m/2}$ random triples $(\tilde{A}, \tilde{B}, \tilde{C})$ such that $\tilde{C} = \lambda_1 \tilde{A} \oplus \lambda_2 \tilde{B}$ and computes $\tilde{H}_i^2$ in the same way as in the Steps 1–3 above. According to Proposition 2, Adv can find a collision of f^2 with probability about 0.39. The total complexity is about $2 \times 2^{m/2}$.

Case (c)

The proof is omitted because the attacks in this case are almost similar to those in Case (b). □

Lemma 4. *Suppose that* $\mathrm{Rank}(L_1) = \mathrm{Rank}(L_2) = 3$. *If* $c \oplus d = \lambda_1 a \oplus \lambda_2 b$ *and* $y \oplus z = \lambda_3 w \oplus \lambda_4 x$ *for some* $\lambda_1, \lambda_2, \lambda_3, \lambda_4 \in \{0, 1\}$, *then there exist a free-start second preimage attack and a free-start collision attack with complexities about* 2^m *and* $2^{m/2}$, *respectively.*

Proof. Suppose that both w and x are represented by linear combinations of a, b, c, d. Then, y is represented by a linear combination of a, b, c, d iff z is represented by a linear combination of a, b, c, d. This is because $y \oplus z = \lambda_3 w \oplus \lambda_4 x$ for some $\lambda_3, \lambda_4 \in \{0, 1\}$. Thus, to prove the lemma, the following three cases are considered.

(a1) All of w, x, y, z are represented by linear combinations of a, b, c, d.
(a2) Both w and x are represented by linear combinations of a, b, c, d and neither y nor z are represented by linear combinations of a, b, c, d.
(b) At least one of w and x is not represented by any linear combination of a, b, c, d.

Let $\mathcal{V}(A, B, C, D)$ be the set of 4-tuples $(H_{i-1}^1, H_{i-1}^2, M_i^1, M_i^2)$ such that

$$\begin{pmatrix} A \\ B \\ C \\ D \end{pmatrix} = L_1 \begin{pmatrix} H_{i-1}^1 \\ H_{i-1}^2 \\ M_i^1 \\ M_i^2 \end{pmatrix}. \tag{15}$$

Since $\mathrm{Rank}(L_1) = 3$, the number of the elements of $\mathcal{V}(A, B, C, D)$ is 2^m. In the case (a1), (W, X, Y, Z) is constant for every element in $\mathcal{V}(A, B, C, D)$. In the case (a2), both Y and Z takes 2^m different values each of which corresponds to an element in $\mathcal{V}(A, B, C, D)$. In the case (b), $W \| X$ takes 2^m different values each of which corresponds to an element in $\mathcal{V}(A, B, C, D)$.

The free-start second preimage attack.

Step 0. Adv computes the output (H_i^1, H_i^2) for the given input $(H_{i-1}^1, H_{i-1}^2, M_i^1, M_i^2)$.

Step 1. *Adv* repeatedly chooses a 4-tuple $(\tilde{H}_{i-1}^1, \tilde{H}_{i-1}^2, \tilde{M}_i^1, \tilde{M}_i^2)$ in $\mathcal{V}(A, B, C, D)$ randomly, where

$$\begin{pmatrix} A \\ B \\ C \\ D \end{pmatrix} = L_1 \begin{pmatrix} H_{i-1}^1 \\ H_{i-1}^2 \\ M_i^1 \\ M_i^2 \end{pmatrix}, \tag{16}$$

and computes $(\tilde{W}, \tilde{X}, \tilde{Y}, \tilde{Z})$ by (6) and $\tilde{H}_i^2 = E_{\tilde{W} \| \tilde{X}}(\tilde{Y}) \oplus \tilde{Z}$ until $H_i^2 = \tilde{H}_i^2$.

This procedure succeeds

- with probability about 1 and with complexity 3 for the case (a1),
- with probability about 0.63 and with complexity about 2^m for the case (a2) and the case (b).

The free-start collision attack.

Step 0. *Adv* chooses arbitrary $(H_{i-1}^1, H_{i-1}^2, M_i^1, M_i^2)$ and computes H_i^1.
Step 1. *Adv* repeatedly chooses a 4-tuple $(\tilde{H}_{i-1}^1, \tilde{H}_{i-1}^2, \tilde{M}_i^1, \tilde{M}_i^2)$ in $\mathcal{V}(A, B, C, D)$ randomly, where

$$\begin{pmatrix} A \\ B \\ C \\ D \end{pmatrix} = L_1 \begin{pmatrix} H_{i-1}^1 \\ H_{i-1}^2 \\ M_i^1 \\ M_i^2 \end{pmatrix}, \tag{17}$$

and computes $(\tilde{W}, \tilde{X}, \tilde{Y}, \tilde{Z})$ by (6) and $\tilde{H}_i^2 = E_{\tilde{W} \| \tilde{X}}(\tilde{Y}) \oplus \tilde{Z}$ until a collision of f^2 is found.
This procedure succeeds
- with probability about 1 and with complexity 3 for the case (a1),
- with probability about 0.39 and with complexity about $2^{m/2}$ for the case (a2) and the case (b).

$\square$

Note 1. If the compression function satisfies the conditions in Lemma 1 or 3, then there exists a free-start preimage attack with almost the same probability and complexity as the second preimage attack. This attack is obtained simply by skipping Step 0.

If the compression function satisfies the conditions in Lemma 4 and those of the case (a2) or (b), then there exists a free-start preimage attack with probability 0.63^2 and with complexity 2×2^m. This attack is obtained from the free-start second preimage attack in the proof of Lemma 4 by replacing Step 0 with the following Step 0'.

Step 0'. *Adv* repeatedly chooses $(H_{i-1}^1, H_{i-1}^2, M_i^1, M_i^2)$ at random until the output corresponding to it is equal to the first half of the given output (H_i^1, H_i^2).

This step succeeds with probability about 0.63 and with complexity about 2^m.

6   Conclusion

The security of double block length hash functions and their compression functions has been analyzed. First, the analysis by Satoh, Haga, and Kurosawa of double block length hash functions has been investigated and it has been shown that there is a case uncovered by their analysis. Then, some effective attacks have been presented on the compression functions which may produce secure double block length hash functions.

Future work includes the analysis of the double block length hash functions whose security remains unclear.

Acknowledgements. The authors would like to thank the anonymous referees for their valuable comments.

References

1. A. Menezes, P. van Oorschot, and S. Vanstone, *"Handbook of Applied Cryptography,"* CRC Press, 1996.
2. L. Knudsen and X. Lai, "New attacks on all double block length hash functions of hash rate 1, including the parallel-DM," *EUROCRYPT'94, Lecture Notes in Computer Science,* vol. 950, pp.410–418, 1995.
3. L. Knudsen, X. Lai, and B. Preneel, "Attacks on fast double block length hash functions," *Journal of Cryptology,* vol. 11, no. 1, pp.59–72, 1998.
4. T. Satoh, M. Haga, and K. Kurosawa, "Towards secure and fast hash functions," *IEICE Transactions of Fundamentals,* vol. 82-A, no. 1, pp.55–62, 1999.
5. M. Girault, R. Cohen, and M. Campana, "A generalized birthday attack," *EURO-CRYPT'88, Lecture Notes in Computer Science,* vol. 330, pp.129–156, 1988
6. J. Black, P. Rogaway, and T. Shrimpton, "Black-box analysis of the block-cipher-based hash-function constructions from PGV," *CRYPTO'02, Lecture Notes in Computer Science,* vol. 2442, pp.320–335, 2002.
7. R. Merkle, "One way hash functions and DES," *CRYPTO'89, Lecture Notes in Computer Science,* vol. 435, pp.428–446, 1990.
8. I. Damgård, "A design principle for hash functions," *CRYPTO'89, Lecture Notes in Computer Science,* vol. 435, pp.416–427, 1990.

Cryptography in Wireless Standards

(Invited Paper)

Valtteri Niemi

Nokia Research Center
P.O. Box 407 FIN-00045 Nokia Group Finland
Valtteri.Niemi@nokia.com

We survey the use of cryptographic mechanisms in the context of some of the most important wireless technologies. On the area of cellular systems, we first take a look at security solutions in the GSM technology, the dominant global cellular standard. We show how the security model and security mechanisms were extended and enhanced in the successor of the GSM system, i.e. in the Universal Mobile Telecommunications System (UMTS), specified in a global initiative, the 3rd Generation Partnership Project (3GPP). In the area of short-range wireless technologies we examine Wireless LAN security, standardized by IEEE, and Bluetooth security, specified by an industry consortium called the Bluetooth SIG (Special Interest Group). The radio frequencies used by WLAN and Bluetooth are close to each other, even partially overlapping. The most notable difference between the two systems is the range of the area of communication. Typical use case of WLAN is access to Internet through a WLAN base station from distances up to several hundred meters while typical Bluetooth use case is communication between two devices, e.g. a mobile phone and a headset, where distance between the devices is in the order of ten meters.

Cryptography provides the most efficient tools for design of security mechanisms in digital communication technologies. This is true in particular in the case of wireless communications where no inherent physical security can be assumed. Cryptographic algorithms provide a solid basis for wireless security but it is also a nontrivial task to define how the algorithms are used as a part of the communication system architecture. For instance, if message content and message origin identity is authenticated in a certain layer of the communication (e.g. network layer), this does not guarantee anything about the content of the same message (e.g. application level identities) in a higher layer. Key management is another example of a tricky issue.

In the GSM system the most essential cryptographic algorithms are used for the following purposes. The A3 algorithm is used in a challenge-response protocol for user authentication, and in the same context the A8 algorithm is used for key generation. This generated key is subsequently used in the radio interface physical layer encryption algorithm A5. The GSM security architecture is modular in the sense that each algorithm may be replaced by another one (maybe more modern) without affecting the rest of the security system. Therefore, the symbols A3, A8 and A5 refer rather to the input-output structure of the algorithm, the internal structures are not fixed. For the radio interface encryption, three different stream ciphers A5/1, A5/2 and A5/3 have been standardized

K.G. Paterson (Ed.): Cryptography and Coding 2003, LNCS 2898, pp. 303–305, 2003.

so far. The situation is even more fragmented for A3 and A8. This is because these algorithms need not be standardized; the algorithms are executed in two places, in the SIM card of the user terminal, and in the Authentication Centre of the user's home network. Hence, each network operator may use proprietary algorithms.

In the packet switched domain of the GSM system, i.e. in GPRS (General Packet Radio Service), the A5 encryption on radio layer one is replaced by GEA encryption placed on layer three of the radio network. Here the changes in the cryptographic algorithm input-output structure are small but the change of layer has much more substantial effect: the protection is extended further in the network, i.e. from the terminal all the way to the core network. So far, three different stream ciphers GEA1, GEA2 and GEA3 have been standardized.

The GSM security architecture is a result of a trade-off between cost and security. The most vulnerable parts of the system have been protected while others are left without cryptographic protection because of performance or implementation cost reasons. As already hinted above, in circuit-switched part of GSM the encryption protects only the radio interface between the terminal and the base station. Also, the terminal does not execute any explicit authentication of the network, leaving the terminal vulnerable against certain types of active attacks.

The UMTS system builds on top of GSM: the core network part is a straightforward evolution from GSM while the radio network part is a result of a more revolutionary development. The major shift is from TDMA (Time Division Multiple Access) technology to WCDMA (Wideband Code Division Multiple Access) technology. The evolution vs. revolution aspect is also visible in the security features. The UMTS authentication and key agreement mechanisms involve the core network part, and the mechanisms were created by extending GSM-type challenge-response user authentication protocol with a sequence number based network authentication mechanism. In the radio network, encryption is provided by f8 algorithm on the radio layer two. As a completely new feature, integrity protection algorithm f9 is applied to signaling messages on the radio layer three. Here again, the symbols f8 and f9 refer to the input-output structure of the algorithms, the internal structure is not fixed. Currently, one version of both algorithms has been standardized, both based on the KASUMI block cipher. The 3GPP has begun the process of adding another pair of algorithms to the set of specifications.

The IEEE 802.11 group specified the original security mechanisms for Wireless LAN under the name Wired Equivalent Privacy (WEP). As the name indicates, the goal was the same as in GSM: to provide security level comparable to that of wired networks. Unfortunately, the original design of WEP contains several weaknesses. For example, the RC4 cipher is used with short initialization values, key management is weak in many implementations and the system lacks integrity protection and replay protection. The IEEE 802.11 is in process of creating a completely new set of security mechanisms (with new cryptoalgorithms as well). The industry consortium Wi-Fi Alliance has already endorsed an intermediate set of mechanisms but the complete IEEE 802.11i specification set is scheduled for completion during the year 2004.

The Bluetooth system contains authentication and key agreement between two peer devices. The cryptoalgorithm SAFER++ was chosen for this purpose. Here the authentication algorithm has to be standardized because it is run in terminal devices. The keys used in authentication are agreed first in a specific pairing procedure. For confidentiality, the basic Bluetooth specification set contains a stream cipher tailor-made for this purpose.

On the Correctness of Security Proofs for the 3GPP Confidentiality and Integrity Algorithms

Tetsu Iwata and Kaoru Kurosawa

Department of Computer and Information Sciences,
Ibaraki University
4–12–1 Nakanarusawa, Hitachi, Ibaraki 316-8511, Japan
{iwata,kurosawa}@cis.ibaraki.ac.jp

Abstract. $f8$ and $f9$ are standardized by 3GPP to provide confidentiality and integrity, respectively. It was claimed that $f8$ and $f9'$ are secure if the underlying block cipher is a PseudoRandom Permutation (PRP), where $f9'$ is a slightly modified version of $f9$. In this paper, however, we disprove both claims by showing a counterexample. We first construct a PRP F with the following property: There is a constant $\mathtt{Cst}$ such that for any key K, $F_K(\cdot) = F_{K \oplus \mathtt{Cst}}^{-1}(\cdot)$. We then show that $f8$ and $f9'$ are completely insecure if F is used as the underlying block cipher. Therefore, PRP assumption does *not* necessarily imply the security of $f8$ and $f9'$, and it is *impossible* to prove their security under PRP assumption. It should be stressed that these results do not imply the original $f8$ and $f9$ (with KASUMI as the underlying block cipher) are insecure, or broken. They simply undermine their provable security.

1 Introduction

$f8$ and $f9$ are standardized by 3GPP to provide confidentiality and integrity, respectively, where 3GPP is the body standardizing the next generation of mobile telephony. See the home page of ETSI (the European Telecommunications Standards Institute), www.etsi.org. Both $f8$ and $f9$ are modes of operations based on the block cipher KASUMI.

- KASUMI is a 64-bit block cipher whose key length is 128 bits [2],
- $f8$ is a symmetric encryption scheme which is a variant of the Output Feedback (OFB) mode, with 64-bit feedback [1], and
- $f9$ is a message authentication code (MAC) which a variant of the CBC MAC [1].

Many of the block cipher modes of operations are provably secure assuming that the underlying block cipher is a PseudoRandom Permutation (PRP assumption). For example, we have: CTR mode [3] and CBC encryption mode [3] for symmetric encryption, PMAC [7] and OMAC [11] for message authentication, and IAPM [9], OCB mode [16], EAX mode [6] and CWC mode [14] for authenticated encryption.

Therefore, it is tempting to prove the security of $f8$ and $f9$. Now suppose that the underlying block cipher used in $f8$ and $f9$ is a PRP. Then:

K.G. Paterson (Ed.): Cryptography and Coding 2003, LNCS 2898, pp. 306–318, 2003.

- At Asiacrypt 2001, Kang, Shin, Hong and Yi claimed that $f8$ is a secure symmetric encryption scheme in the sense of left-or-right indistinguishability [12, Theorem 4], and
- At FSE 2003, Hong, Kang, Preneel and Ryu claimed that $f9'$ is a secure MAC [10, Theorem 2], where $f9'$ is a slightly modified version of $f9$.

In this paper, however, we disprove both claims by showing a counterexample. We first construct a block cipher F with the following properties:

- F is a PRP, and
- there is a constant Cst such that for any key K, $F_K(\cdot) = F_{K \oplus \mathtt{Cst}}^{-1}(\cdot)$.

We then show that $f8$ and $f9'$ are completely insecure if F is used as the underlying block cipher. Therefore, PRP assumption does *not* necessarily imply the security of $f8$ and $f9'$, and it is *impossible* to prove their security under PRP assumption.

It should be stressed that these results do not imply the original $f8$ and $f9$ (with KASUMI as the underlying block cipher) are insecure, or broken. They simply undermine their provable security.

Related works. Initial security evaluation of KASUMI, $f8$ and $f9$ can be found in [8]. Knudsen and Mitchell analyzed the security of $f9'$ against forgery and key recovery attacks [13].

2   Preliminaries

2.1   Block Ciphers, Symmetric Encryption Schemes, and MACs

Block cipher, E. A block cipher E is a function $E : \{0,1\}^k \times \{0,1\}^n \to \{0,1\}^n$, where, for each $K \in \{0,1\}^k$, $E(K, \cdot)$ is a permutation over $\{0,1\}^n$. We write $E_K(\cdot)$ for $E(K, \cdot)$. k is called the key length and n is called the block length. For KASUMI, $k = 128$ and $n = 64$, and for the AES, $k = 128, 192, 256$ and $n = 128$.

Symmetric encryption scheme, $\mathcal{SE}$. A symmetric encryption scheme $\mathcal{SE} = (\mathcal{E}, \mathcal{D})$ consists of two algorithms. An encryption algorithm $\mathcal{E}$ takes a key $K \in \{0,1\}^k$ and a plaintext $M \in \{0,1\}^*$ to return a ciphertext C. We write $C = \mathcal{E}_K(M)$. It uses and then updates a state that is maintained across invocations (in this paper, we only consider a stateful encryption scheme. See [3] for a randomized encryption scheme). The decryption algorithm $\mathcal{D}$ is deterministic and stateless. It takes the key K and a string C to return either the corresponding plaintext M or the symbol $\perp$. We require $\mathcal{D}_K(\mathcal{E}_K(M)) = M$ for all $M \in \{0,1\}^*$.

MAC algorithm, MAC. A MAC algorithm is a function MAC $: \{0,1\}^k \times \{0,1\}^* \to \{0,1\}^l$. It takes a key $K \in \{0,1\}^k$ and a message $M \in \{0,1\}^*$ to return a l-bit tag $T \in \{0,1\}^l$. We write $\mathrm{MAC}_K(\cdot)$ for $\mathrm{MAC}(K, \cdot)$.

2.2   Security Definitions

Our definitions follow from those given in [15] for PRP and SPRP, [3] for left-or-right indistinguishability, and [4] for the security of MACs.

Security of block ciphers (PRP) [15]. Let $\mathrm{Perm}(n)$ denote the set of all permutations on $\{0,1\}^n$. We say that P is a random permutation if P is randomly chosen from $\mathrm{Perm}(n)$.

The security of a block cipher $E : \{0,1\}^k \times \{0,1\}^n \to \{0,1\}^n$ as a pseudorandom permutation (PRP) is quantified as $\mathsf{Adv}_E^{\mathsf{prp}}(\mathcal{A})$, the advantage of an adversary $\mathcal{A}$ that tries to distinguish $E_K(\cdot)$ (with a randomly chosen key K) from a random permutation $P(\cdot)$. Let $\mathcal{A}^{E_K(\cdot)}$ denote $\mathcal{A}$ with an oracle which, in response to a query X, returns $E_K(X)$, and let $\mathcal{A}^{P(\cdot)}$ denote $\mathcal{A}$ with an oracle which, in response to a query X, returns $P(X)$. After making queries, $\mathcal{A}$ outputs a bit. Then the advantage is defined as

$$\mathsf{Adv}_E^{\mathsf{prp}}(\mathcal{A}) \stackrel{\mathrm{def}}{=} \left| \Pr(\mathcal{A}^{E_K(\cdot)} = 1) - \Pr(\mathcal{A}^{P(\cdot)} = 1) \right| ,$$

where:

- in $\Pr(\mathcal{A}^{E_K(\cdot)} = 1)$, the probability is taken over the random choice of $K \stackrel{R}{\leftarrow} \{0,1\}^k$ and $\mathcal{A}$'s coin toss (if any), and
- in $\Pr(\mathcal{A}^{P(\cdot)} = 1)$, the probability is taken over the random choice of $P \stackrel{R}{\leftarrow} \mathrm{Perm}(n)$ and $\mathcal{A}$'s coin toss (if any).

We say that E is a PRP if $\mathsf{Adv}_E^{\mathsf{prp}}(\mathcal{A})$ is sufficiently small for any $\mathcal{A}$.

Security of block ciphers (SPRP) [15]. Similarly, the security of a block cipher $E : \{0,1\}^k \times \{0,1\}^n \to \{0,1\}^n$ as a super-pseudorandom permutation (SPRP) is quantified as $\mathsf{Adv}_E^{\mathsf{sprp}}(\mathcal{A})$, the advantage of an adversary $\mathcal{A}$ that tries to distinguish $E_K(\cdot), E_K^{-1}(\cdot)$ (with a randomly chosen key K) from a random permutation $P(\cdot), P^{-1}(\cdot)$. Let $\mathcal{A}^{E_K(\cdot),E_K^{-1}(\cdot)}$ denote $\mathcal{A}$ with oracles which, in response to a query X, returns $E_K(X)$, and in response to a query Y, returns $E_K^{-1}(Y)$. Similarly, let $\mathcal{A}^{P(\cdot),P^{-1}(\cdot)}$ denote $\mathcal{A}$ with oracles which, in response to a query X, returns $P(X)$, and in response to a query Y, returns $P^{-1}(Y)$. After making queries, $\mathcal{A}$ outputs a bit. Then the advantage is defined as

$$\mathsf{Adv}_E^{\mathsf{sprp}}(\mathcal{A}) \stackrel{\mathrm{def}}{=} \left| \Pr(\mathcal{A}^{E_K(\cdot),E_K^{-1}(\cdot)} = 1) - \Pr(\mathcal{A}^{P(\cdot),P^{-1}(\cdot)} = 1) \right|$$

where

- in $\Pr(\mathcal{A}^{E_K(\cdot),E_K^{-1}(\cdot)} = 1)$, the probability is taken over the random choice of $K \stackrel{R}{\leftarrow} \{0,1\}^k$ and $\mathcal{A}$'s coin toss (if any), and
- in $\Pr(\mathcal{A}^{P(\cdot),P^{-1}(\cdot)} = 1)$, the probability is taken over the random choice of $P \stackrel{R}{\leftarrow} \mathrm{Perm}(n)$ and $\mathcal{A}$'s coin toss (if any).

We say that E is a SPRP if $\mathsf{Adv}_E^{\mathsf{sprp}}(\mathcal{A})$ is sufficiently small for any $\mathcal{A}$.

Security of symmetric encryption schemes (left-or-right indistinguishability, lor)
[3]. Let $\mathcal{SE} = (\mathcal{E}, \mathcal{D})$ be a (stateful) symmetric encryption scheme. The adversary
is allowed queries of the form (M_0, M_1), where M_0 and M_1 are equal-length
messages. We define the left-or-right oracle $\mathcal{E}_K(\mathcal{LR}(\cdot, \cdot, b))$, where $b \in \{0, 1\}$. It
takes input (M_0, M_1) and if $b = 0$ then it computes $C = \mathcal{E}_K(M_0)$ and returns C,
else it computes $C = \mathcal{E}_K(M_1)$ and returns C. After making queries, $\mathcal{A}$ outputs
a bit. Then we define the advantage as

$$\mathsf{Adv}^{\mathsf{lor}}_{\mathcal{SE}}(\mathcal{A}) \stackrel{\text{def}}{=} \left| \Pr\left(\mathcal{A}^{\mathcal{E}_K(\mathcal{LR}(\cdot,\cdot,0))} = 1\right) - \Pr\left(\mathcal{A}^{\mathcal{E}_K(\mathcal{LR}(\cdot,\cdot,1))} = 1\right) \right| ,$$

where the probabilities are taken over the random choice of $K \stackrel{R}{\leftarrow} \{0,1\}^k$ and
$\mathcal{A}$'s coin toss (if any).

We say that $\mathcal{SE}$ is secure in the sense of left-or-right indistinguishability if
$\mathsf{Adv}^{\mathsf{lor}}_{\mathcal{SE}}(\mathcal{A})$ is sufficiently small for any $\mathcal{A}$.

Security of MACs [4]. Let MAC : $\{0,1\}^k \times \{0,1\}^* \rightarrow \{0,1\}^l$ be a MAC algorithm. Let $\mathcal{A}^{\mathrm{MAC}_K(\cdot)}$ denote $\mathcal{A}$ with an oracle which, in response to a query
$M \in \{0,1\}^*$, returns $\mathrm{MAC}_K(M) \in \{0,1\}^l$. We say that an adversary $\mathcal{A}^{\mathrm{MAC}_K(\cdot)}$
forges if $\mathcal{A}$ outputs (M, T), where $T = \mathrm{MAC}_K(M)$ and $\mathcal{A}$ never queried M to
its oracle $\mathrm{MAC}_K(\cdot)$. Then we define the advantage as

$$\mathsf{Adv}^{\mathsf{mac}}_{\mathrm{MAC}}(\mathcal{A}) \stackrel{\text{def}}{=} \Pr(\mathcal{A}^{\mathrm{MAC}_K(\cdot)} \text{ forges}) ,$$

where the probability is taken over the random choice of $K \stackrel{R}{\leftarrow} \{0,1\}^k$ and $\mathcal{A}$'s
coin toss (if any).

We say that a MAC algorithm is secure if $\mathsf{Adv}^{\mathsf{mac}}_{\mathrm{MAC}}(\mathcal{A})$ is sufficiently small for
any $\mathcal{A}$.

2.3 3GPP Confidentiality Algorithm *f*8 [1]

*f*8 is a stateful symmetric encryption scheme standardized by 3GPP. It uses a
block cipher $E : \{0,1\}^k \times \{0,1\}^n \rightarrow \{0,1\}^n$ as the underlying primitive. Let
*f*8[E] be *f*8, where E is used as the underlying block cipher. It takes a k-
bit key K, a message $M \in \{0,1\}^*$ and returns a ciphertext $C = f8[E_K](M)$,
which is the same length as M. It also maintains an n-bit state **state**, which
is, essentially, a counter. It is important to note that the rightmost bit of **state**
is always 0. (In the original specification, **state** is a 64-bit string of the form
state $=$ COUNT$\|$BEARER$\|$DIRECTION$\|$0 . . . 0, where COUNT is a 32-bit
counter, BEARER is a 5-bit bearer identity, DIRECTION is a 1-bit direction,
followed by 26 zeros.) It also uses a k-bit constant KM $=$ 0x55...55, called the key
modifier.

Let $M \in \{0,1\}^*$ be a message, and $m = \lceil |M|/n \rceil$ be a block length of M,
where $|M|$ denotes the bit length of M. Then the encryption proceeds as follows:

1. Let $KS[0] = 0^n$.
2. Let $A = E_{K \oplus \mathrm{KM}}(\textbf{state})$.

3. For $i = 1$ to m do:
     Let $I[i] = A \oplus \text{num2str}_n(i-1) \oplus KS[i-1]$
     Let $KS[i] = E_K(I[i])$
4. Let $C = M \oplus$ the leftmost $|M|$ bits of $KS[1]\| \cdots \|KS[m]$
5. Return C

In the above description, $\text{num2str}_n(i-1)$ denotes n-bit binary representation of $i-1$. We emphasize that state is an n-bit counter whose rightmost bit is 0. See Fig. 1. In Fig. 1, $M[i]$ denotes the i-th n-bit block of M and $C[i]$ denotes the i-th n-bit block of C. If $|M[m]| < n$ then the right $n - |M[m]|$ bits of $KS[m]$ is ignored. Decryption is done in an obvious way.

At Asiacrypt 2001, Kang, Shin, Hong and Yi claimed that $f8[E]$ is a secure symmetric encryption scheme in the sense of left-or-right indistinguishability, if E is a PRP. More precisely:

Claim (Kang, Shin, Hong and Yi [12, Theorem 4]). Suppose that, for any adversary $\mathcal{B}$ that makes at most q queries, $\text{Adv}_E^{\text{prp}}(\mathcal{B}) \leq \epsilon$. Then for any adversary $\mathcal{A}$ that makes at most q queries, and the queries have at most μ bits in total,

$$\text{Adv}_{f8[E]}^{\text{lor}}(\mathcal{A}) \leq 2\epsilon + \frac{(\mu/n)(\mu/n+1)}{2^{n-1}} \ .$$

2.4   3GPP Integrity Algorithm $f9$

Specification of $f9$ [1]. $f9$ is a MAC algorithm that uses a block cipher $E : \{0,1\}^k \times \{0,1\}^n \to \{0,1\}^n$ as the underlying primitive. Similarly to $f8$, let $f9[E]$ be $f9$, where E is used as the underlying block cipher. It takes a k-bit key K, a message $M \in \{0,1\}^*$ and returns a l-bit tag $T = f9[E_K](M)$. It also uses a k-bit constant $\text{KM} = \text{0xAA...AA}$, called the key modifier.

Let $M = M[1]\| \cdots \|M[m]$ be a message, where each $M[i]$ ($1 \leq i \leq m - 1$) is n bits. The last block $M[m]$ may have fewer than n bits. We define a function $\text{pad}_n(M)$ works as follows: first, it concatenates COUNT, FRESH, M and DIRECTION, and then appends a single "1" bit, followed by between 0 and $n - 1$ "0" bits so that the total length is a multiple of n. More precisely,

$$\text{pad}_n(M) = \text{COUNT}\|\text{FRESH}\|M\|\text{DIRECTION}\|1\|0^{n-2-(|M| \bmod n)} \ ,$$

where the length of COUNT$\|$FRESH is n bits, COUNT is a counter, FRESH is a random string, and DIRECTION is a one-bit direction.

Then we generate a tag T as follows:

1. Break $\text{pad}_n(M)$ into n-bit blocks $PS[1]\| \cdots \|PS[m]$
2. Let $A = 0^n$ and $B = 0^n$.
3. For $i = 1$ to m do:
     Let $A = E_K(A \oplus PS[i])$
     Let $B = B \oplus A$
4. Let $B = E_{K \oplus \text{KM}}(B)$
5. Let $T = $ the leftmost l bits of B
6. Return T

See Fig. 2.

Specification of f9' [13,10]. We define a slightly modified version of $f9$, called $f9'$. In $f9'$, the function $\text{pad}_n(M)$ is modified as follows: it simply appends a single "1" bit, followed by between 0 and $n-1$ "0" bits so that the total length is a multiple of n bits. More precisely,

$$\text{pad}_n(M) = M\|1\|0^{n-1-(|M| \bmod n)} .$$

Thus, we simply ignore COUNT, FRESH, and DIRECTION. Or, equivalently, we consider COUNT, FRESH, and DIRECTION as a part of the message. The rest of the algorithm is the same as $f9$. See Fig. 3.

At FSE 2003, Hong, Kang, Preneel and Ryu claimed that $f9'[E]$ is a secure MAC if E is a PRP. More precisely:

Claim (Hong, Kang, Preneel and Ryu [10, Theorem 2]). Suppose that, for any adversary $\mathcal{B}$ that makes at most σ queries, $\text{Adv}_E^{\text{prp}}(\mathcal{B}) \leq \epsilon$. Then for any adversary $\mathcal{A}$ that makes at most q queries, and the queries have at most σ blocks in total (one block counts as n bits),

$$\text{Adv}_{f9'[E]}^{\text{mac}}(\mathcal{A}) \leq \epsilon + \frac{\sigma^2 + q^2}{2^n} + \frac{1}{2^l} .$$

Also, Knudsen and Mitchell analyzed $f9'$ against forgery and key recovery attacks [13].

3   Construction of a PRP, F

In this section, we construct a PRP $F : \{0,1\}^k \times \{0,1\}^n \rightarrow \{0,1\}^n$, with the following property: Let Cst be some k-bit constant. For any $K \in \{0,1\}^k$,

$$F_K(\cdot) = F_{K \oplus \text{Cst}}^{-1}(\cdot) .$$

Let $E : \{0,1\}^{k-1} \times \{0,1\}^n \rightarrow \{0,1\}^n$ be a block cipher. It uses a $(k-1)$-bit key K' to encrypt an n-bit plaintext X into an n-bit ciphertext $Y = E_{K'}(X)$, where $E_{K'}(X) \overset{\text{def}}{=} E(K', X)$. For each $K' \in \{0,1\}^{k-1}$, $E_{K'}(\cdot)$ is a permutation over $\{0,1\}^n$.

Now we construct a new block cipher $F : \{0,1\}^k \times \{0,1\}^n \rightarrow \{0,1\}^n$ from E as in Fig. 4. The inputs to the algorithm are a block cipher E and some non-zero k-bit constant Cst. The output is a new block cipher F.

- For a k-bit string Cst $= (\text{Cst}_0, \text{Cst}_1, \ldots, \text{Cst}_{k-1})$, nzi(Cst) denotes the smallest index of non-zero element. That is, nzi(Cst) $= j$ such that $\text{Cst}_0 = \cdots = \text{Cst}_{j-1} = 0$ and $\text{Cst}_j = 1$. For example, if $k = 4$ and Cst $= \text{0xA} = 1010$, then nzi(Cst) $= 0$, and if Cst $= \text{0x5} = 0101$, then nzi(Cst) $= 1$.
- num2str$_{k-1}(i)$ is a $(k-1)$-bit binary representation of i. For example, if $k = 4$ then num2str$_{k-1}(0) = (0,0,0)$ and num2str$_{k-1}(6) = (1,1,0)$.

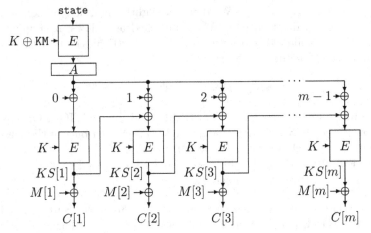

Fig. 1. Illustration of $f8$.

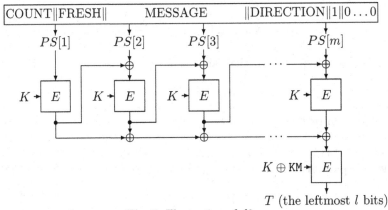

Fig. 2. Illustration of $f9$.

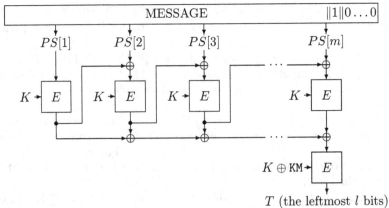

Fig. 3. Illustration of $f9'$.

```
Construction of F from E and Cst
j ← nzi(Cst);
for i = 0 to 2^(k−1) − 1 do {
        K' ← num2str_(k−1)(i);
        K_1 ← first_(0..j−1)(K')‖0‖last_(j..k−2)(K');
        K_2 ← K_1 ⊕ Cst;
        F_(K_1) ← E_(K');
        F_(K_2) ← E_(K')^(−1); }
```

Fig. 4. Construction of F from E and Cst.

- For a $(k-1)$-bit string $K' = (K'_0, \ldots, K'_{k-2})$ and an integer $0 \le j \le k-1$, $\text{first}_{0..j-1}(K')$ denotes the first j bits of K'. That is, $\text{first}_{0..j-1}(K') = (K'_0, \ldots, K'_{j-1})$. For example, if $j = 2$ and $K' = (1,1,0)$ then we have $\text{first}_{0..j-1}(K') = (1,1)$, and if $j = 1$ and $K' = (0,1,0)$ then we have $\text{first}_{0..j-1}(K') = (0)$. If $j = 0$, then $\text{first}_{0..j-1}(K')$ is an empty string.
- Similarly, for a $(k-1)$-bit string $K' = (K'_0, \ldots, K'_{k-2})$ and an integer $0 \le j \le k-1$, $\text{last}_{j..k-2}(K')$ denotes the last $(k-1) - j$ bits of K'. That is, $\text{last}_{j..k-2}(K') = (K'_j, \ldots, K'_{k-2})$. For example, if $j = 2$ and $K' = (1,1,0)$ then $\text{last}_{j..k-2}(K') = (0)$, and if $j = 1$ and $K' = (0,1,0)$ then $\text{last}_{j..k-2}(K') = (1,0)$. If $j = k-1$, then $\text{last}_{j..k-2}(K')$ is an empty string.
- $a\|b$ denotes the concatenation of a and b. For example, if $a = 1$ and $b = (1,0,1)$ then $a\|b = (1,1,0,1)$.

Observe that F_K is well defined for all $K \in \{0,1\}^k$. Indeed, "for loop" in the third line contains 2^{k-1} iterations, and for each loop, two Fs are assigned. Let $K'^{(i)}$, $K_1^{(i)}$ and $K_2^{(i)}$ denote K', K_1 and K_2 in the i-th iteration. Then we see that for any distinct i and i',

- $K_1^{(i)} \ne K_1^{(i')}$ and $K_2^{(i)} \ne K_2^{(i')}$ (since $K'^{(i)} \ne K'^{(i')}$), and
- $K_1^{(i)} \ne K_2^{(i')}$ and $K_2^{(i)} \ne K_1^{(i')}$ (since they differ in the j-th bit).

That is, $K_1^{(i)}$ and $K_2^{(i)}$ in the i-th iteration will not be assigned in the i'-th iteration.

Also observe that we have, for any $K \in \{0,1\}^k$, $F_K(\cdot) = F_{K \oplus \text{Cst}}^{-1}(\cdot)$.

We show two small examples. First, let $k = 4$, Cst = 0xA = 1010 and

$$E = \{E_{000}, E_{001}, E_{010}, E_{011}, E_{100}, E_{101}, E_{110}, E_{111}\},$$

where each $E_{K'}$ is a permutation over $\{0,1\}^n$. In this case, $j = 0$, and for $K' = (K'_0, K'_1, K'_2)$, $K_1 = (0, K'_0, K'_1, K'_2)$, and $K_2 = (1, K'_0, K'_1 \oplus 1, K'_2)$. Then we obtain

$$F = \{F_{0000}, F_{0001}, F_{0010}, F_{0011}, F_{0100}, F_{0101}, F_{0110}, F_{0111},$$
$$F_{1000}, F_{1001}, F_{1010}, F_{1011}, F_{1100}, F_{1101}, F_{1110}, F_{1111}\}$$

Algorithm $\mathcal{A}^{\mathcal{O}, \mathcal{O}^{-1}}$

bit $\xleftarrow{R} \{0, 1\}$;
if bit $= 0$ then $\mathcal{O}' \leftarrow \mathcal{O}$;
           else $\mathcal{O}' \leftarrow \mathcal{O}^{-1}$;
when $\mathcal{B}$ asks its r-th query X_r:
    return $\mathcal{O}'(X_r)$;
when $\mathcal{B}$ halts and output b:
    output b;

Fig. 5. Construction of $\mathcal{A}$.

where
$$\begin{cases} F_{0000} = E_{000}, F_{0001} = E_{001}, F_{0010} = E_{010}, F_{0011} = E_{011}, \\ F_{0100} = E_{100}, F_{0101} = E_{101}, F_{0110} = E_{110}, F_{0111} = E_{111}, \\ F_{1000} = E_{010}^{-1}, F_{1001} = E_{011}^{-1}, F_{1010} = E_{000}^{-1}, F_{1011} = E_{001}^{-1}, \\ F_{1100} = E_{110}^{-1}, F_{1101} = E_{111}^{-1}, F_{1110} = E_{100}^{-1}, F_{1111} = E_{101}^{-1}. \end{cases}$$

Next, let $k = 4$, and $\mathtt{Cst} = \mathtt{0x5} = \mathtt{0101}$. In this case, $j = 1$, and for $K' = (K_0', K_1', K_2')$, $K_1 = (K_0', 0, K_1', K_2')$, and $K_2 = (K_0', 1, K_1', K_2' \oplus 1)$. Then we obtain
$$\begin{cases} F_{0000} = E_{000}, F_{0001} = E_{001}, F_{0010} = E_{010}, F_{0011} = E_{011}, \\ F_{0100} = E_{001}^{-1}, F_{0101} = E_{000}^{-1}, F_{0110} = E_{011}^{-1}, F_{0111} = E_{010}^{-1}, \\ F_{1000} = E_{100}, F_{1001} = E_{101}, F_{1010} = E_{110}, F_{1011} = E_{111}, \\ F_{1100} = E_{101}^{-1}, F_{1101} = E_{100}^{-1}, F_{1110} = E_{111}^{-1}, F_{1111} = E_{110}^{-1}. \end{cases}$$

We note that F can be computed efficiently if E can be computed efficiently. Suppose that we are given a k-bit key K and a plaintext X, and we want to compute $F_K(X)$. Now, let $j \leftarrow \mathtt{nzi}(\mathtt{Cst})$, and check if the j-th bit of K is zero. If it is, let $K' \leftarrow \mathtt{first}_{0..j-1}(K) \| \mathtt{last}_{j+1..k-1}(K)$ and return $E_{K'}(X)$. Otherwise let $K' \leftarrow \mathtt{first}_{0..j-1}(K \oplus \mathtt{Cst}) \| \mathtt{last}_{j+1..k-1}(K \oplus \mathtt{Cst})$ and return $E_{K'}^{-1}(X)$.

We now show that if E is a SPRP, then F is a PRP. More precisely, we have the following theorem.

Theorem 1. *If* $\mathtt{Adv}_E^{\mathsf{sprp}}(\mathcal{A}) \leq \epsilon$ *for any adversary* $\mathcal{A}$ *that makes at most* q *queries, then* $\mathtt{Adv}_F^{\mathsf{prp}}(\mathcal{B}) \leq \epsilon$ *for any adversary* $\mathcal{B}$ *that makes at most* q *queries.*

Proof. We prove through a contradiction argument. Suppose that there exists an adversary $\mathcal{B}$ such that $\mathtt{Adv}_F^{\mathsf{prp}}(\mathcal{B}) > \epsilon$ where $\mathcal{B}$ asks at most q queries. By using $\mathcal{B}$, we construct an adversary $\mathcal{A}$ such that $\mathtt{Adv}_E^{\mathsf{sprp}}(\mathcal{A}) > \epsilon$ where $\mathcal{A}$ asks at most q queries.

The construction is given in Fig. 5. $\mathcal{A}$ first randomly chooses a bit $\mathtt{bit}$, and let $\mathcal{O}' \leftarrow \mathcal{O}$ if $\mathtt{bit} = 0$, and $\mathcal{O}' \leftarrow \mathcal{O}^{-1}$ otherwise. Then $\mathcal{A}$ simply uses $\mathcal{O}'$ to answer $\mathcal{B}$'s queries. Finally $\mathcal{A}$ outputs b which is the output of $\mathcal{B}$.

First, suppose that $(\mathcal{O}, \mathcal{O}^{-1}) = (P, P^{-1})$. Then $\mathcal{A}$ gives $\mathcal{B}$ a perfect simulation of a random permutation (regardless of the value of $\mathtt{bit}$). Therefore, we have

$$\Pr(P \xleftarrow{R} \mathrm{Perm}(n) : \mathcal{B}^{P(\cdot)} = 1) = \Pr(P \xleftarrow{R} \mathrm{Perm}(n) : \mathcal{A}^{P(\cdot), P^{-1}(\cdot)} = 1) \ .$$

Next, suppose that $(\mathcal{O}, \mathcal{O}^{-1}) = (E, E^{-1})$. Then $\mathcal{A}$ gives $\mathcal{B}$ a perfect simulation of F, since from the $\mathcal{B}$'s point of view, each

$$E_{0,\ldots,0}, \ldots, E_{1,\ldots,1}, E_{0,\ldots,0}^{-1}, \ldots, E_{1,\ldots,1}^{-1}$$

is chosen with probability $(1/2) \times (1/2^{k-1}) = 1/2^k$, which is a precise simulation of F. Therefore, we have

$$\Pr(K \overset{R}{\leftarrow} \{0,1\}^k : \mathcal{B}^{F_K(\cdot)} = 1)$$
$$= \Pr(K \overset{R}{\leftarrow} \{0,1\}^{k-1}, \texttt{bit} \overset{R}{\leftarrow} \{0,1\} : \mathcal{A}^{E_K(\cdot), E_K^{-1}(\cdot)} = 1) \ .$$

$\square$

Note that F is a secure block cipher: it is a PRP. On the other hand, for any $K \in \{0,1\}^k$, we have $F_K(\cdot) = F_{K \oplus \texttt{Cst}}^{-1}(\cdot)$. That is, F is a completely insecure block cipher against, so called, the related key attacks [5].

4   PRP Does NOT Imply the Security of $f8$

Let $F : \{0,1\}^k \times \{0,1\}^n \to \{0,1\}^n$ be a PRP, with the following property: For any $K \in \{0,1\}^k$,

$$F_K(\cdot) = F_{K \oplus \texttt{KM}}^{-1}(\cdot) \ ,$$

where $\texttt{KM} = \texttt{0x55}\ldots\texttt{55}$ (k bits).

Let $f8[F]$ denote $f8$, where F is used as the underlying block cipher.

We now show that $f8[F]$ is a completely insecure encryption scheme. We have the following theorem.

Theorem 2. $f8[F]$ *is not secure in the sense of left-or-right indistinguishability. There exists an adversary $\mathcal{A}$ that makes 1 query and* $\text{Adv}_{f8[F]}^{\text{lor}}(\mathcal{A}) = 1$.

Proof. The adversary $\mathcal{A}$ has an oracle $f8[F_K](\mathcal{LR}(\cdot, \cdot, b))$ which randomly chooses K, and in response to a query (M_0, M_1), returns $f8[F_K](M_b)$ for $b \in \{0,1\}$. The adversary is given in Fig. 6. In Fig. 6, $\texttt{lsb}(C)$ denotes the least significant bit of C (the rightmost bit).

Algorithm $\mathcal{A}^{f8[F_K](\mathcal{LR}(\cdot,\cdot,b))}$
let $M_0 \leftarrow 0^n$;
let $M_1 \leftarrow 1^n$;
$C \leftarrow f8[F_K](\mathcal{LR}(M_0, M_1, b))$;
if $\texttt{lsb}(C) = 0$ then $b' \leftarrow 0$;
                    else $b' \leftarrow 1$;
output b';

Fig. 6. Construction of $\mathcal{A}$.

We see that the first block of key stream $SK[1]$ is given by

$$SK[1] = F_K(F_{K \oplus \text{KM}}(\text{state})) = F_K(F_K^{-1}(\text{state})) = \text{state} .$$

Since the right most bit of state is 0, we see that the rightmost bit of $0^n \oplus \text{state}$ is always 0 and the rightmost bit of $1^n \oplus \text{state}$ is always 1. Therefore the adversary given in Fig. 6 succeeds with probability 1. $\square$

5   PRP Does NOT Imply the Security of $f9'$

Let $F : \{0,1\}^k \times \{0,1\}^n \to \{0,1\}^n$ be a PRP, with the following property: For any $K \in \{0,1\}^k$,

$$F_K(\cdot) = F_{K \oplus \text{KM}}^{-1}(\cdot) ,$$

where $\text{KM} = \text{0xAA} \ldots \text{AA}$ (k bits).

Let $f9'[F]$ denote $f9'$, where F is used as the underlying block cipher.

We now show that $f9'[F]$ is a completely insecure MAC. We have the following theorem.

Theorem 3. $f9'[F]$ *is not a secure MAC. There exists an adversary* $\mathcal{A}$ *without making any query and* $\text{Adv}_{f9'[F]}^{\text{mac}}(\mathcal{A}) = 1$.

Proof. The adversary $\mathcal{A}$ has an oracle $f9'[F_K](\cdot)$ which randomly chooses K, and in response to a query $M \in \{0,1\}^*$, returns $f9'[F_K](M)$ (but it will not use this oracle). The adversary is given in Fig. 7.

Algorithm $\mathcal{A}^{f9'[F_K](\cdot)}$
let M **be any message such that** $1 \le |M| < n$;
let T **be the left** l **bits of** $M\|1\|0^{n-1-(|M| \bmod n)}$;
output (M, T);

Fig. 7. Construction of $\mathcal{A}$.

We see that for any message such that $1 \le |M| < n$,

$$\begin{aligned}
T &= \text{the leftmost } l \text{ bits of } F_{K \oplus \text{KM}}(F_K(M\|1\|0^{n-1-(|M| \bmod n)})) \\
&= \text{the leftmost } l \text{ bits of } F_K^{-1}(F_K(M\|1\|0^{n-1-(|M| \bmod n)})) \\
&= \text{the leftmost } l \text{ bits of } M\|1\|0^{n-1-(|M| \bmod n)}.
\end{aligned}$$

Therefore, it is easy to verify that the adversary given in Fig. 7 succeeds with probability 1. $\square$

6   Discussions

About our F. It seems that our F is a useless block cipher to use in any practical applications. Still, it is a PRP. It may be interesting that such a weak PRP exists. This is because the definition of a PRP treats only *one* randomly chosen key K, and has nothing to do with the associated *second* key $K \oplus \mathtt{Cst}$. In this sense, one may consider that assuming the block cipher being a PRP is a weak assumption.

Flaws in the previous proofs in [12,10]. For both $f8$ and $f9'$, $E_K(\cdot)$ and $E_{K \oplus \mathtt{Cst}}(\cdot)$ are treated as two independent random permutations. The fact is that, if we only assume that E is a PRP, there is a case where $E_K(\cdot)$ and $E_{K \oplus \mathtt{Cst}}(\cdot)$ are not independent, or strongly correlated (as our F satisfies $F_K(\cdot) = F_{K \oplus \mathtt{Cst}}^{-1}(\cdot)$).

What our results mean. Our results do not show that the original $f8$ and $f9$ (with KASUMI as the underlying block cipher) are insecure, or broken, since KASUMI does not have a property of our F. Also, we note that $f9$ does not allow one block padded message while $f9'$ does. Still, it should not be misunderstood that our results have nothing to do with the security of $f8$ and $f9$. One interpretation of our results is:

- $f8$ is *less* secure than, for example, CTR mode, since CTR mode achieves left-or-right indistinguishability under PRP assumption, while it is *impossible* for $f8$ to achieve this security notion.
- $f9'$ is *less* secure than, for example, OMAC, since OMAC is a secure MAC under PRP assumption, while it is *impossible* for $f9'$ to achieve this security notion.

7   Conclusions

In this paper, we showed that it is impossible to prove the security of $f8$ and $f9'$ under PRP assumption. The security claims made on $f8$ and $f9'$ in [12,10] are both flawed, and we do not know any rigorous security proofs for them. It might be possible to prove their security in the ideal cipher model, or assuming that the underlying block cipher achieves the definition of related key attacks resistance proposed by Bellare and Kohno [5].

References

1. 3GPP TS 35.201 v 3.1.1. Specification of the 3GPP confidentiality and integrity algorithms, Document 1: $f8$ and $f9$ specification. Available at
   http://www.3gpp.org/tb/other/algorithms.htm.
2. 3GPP TS 35.202 v 3.1.1. Specification of the 3GPP confidentiality and integrity algorithms, Document 2: KASUMI specification. Available at
   http://www.3gpp.org/tb/other/algorithms.htm.

3. M. Bellare, A. Desai, E. Jokipii, and P. Rogaway. A concrete security treatment of symmetric encryption. *Proceedings of the 38th Annual Symposium on Foundations of Computer Science, FOCS '97*, pp. 394–405, IEEE, 1997.

4. M. Bellare, J. Kilian, and P. Rogaway. The security of the cipher block chaining message authentication code. *JCSS*, vol. 61, no. 3, 2000. Earlier version in *Advances in Cryptology – CRYPTO '94, LNCS 839*, pp. 341–358, Springer-Verlag, 1994.

5. M. Bellare, and T. Kohno. A theoretical treatment of related-key attacks: RKA-PRPs, RKA-PRFs, and applications. *Advances in Cryptology – EUROCRYPT 2003, LNCS 2656*, pp. 491–506, Springer-Verlag, 2003.

6. M. Bellare, P. Rogaway, and D. Wagner. A conventional authenticated-encryption mode. See Cryptology ePrint Archive, Report 2003/069, http://eprint.iacr.org/.

7. J. Black and P. Rogaway. A block-cipher mode of operation for parallelizable message authentication. *Advances in Cryptology – EUROCRYPT 2002, LNCS 2332*, pp. 384–397, Springer-Verlag, 2002.

8. Evaluation report (version 2.0). Specification of the 3GPP confidentiality and integrity algorithms, Report on the evaluation of 3GPP confidentiality and integrity algorithms. Available at http://www.3gpp.org/tb/other/algorithms.htm.

9. C.S. Jutla. Encryption modes with almost free message integrity. *Advances in Cryptology – EUROCRYPT 2001, LNCS 2045*, pp. 529–544, Springer-Verlag, 2001.

10. D. Hong, J-S. Kang, B. Preneel and H. Ryu. A concrete security analysis for 3GPP-MAC. Pre-proceedings of *Fast Software Encryption, FSE 2003*, pp. 163–178, 2003. To appear in *LNCS*, Springer-Verlag.

11. T. Iwata and K. Kurosawa. OMAC: One-Key CBC MAC. Pre-proceedings of *Fast Software Encryption, FSE 2003*, pp. 137–162, 2003. To appear in *LNCS*, Springer-Verlag. See http://crypt.cis.ibaraki.ac.jp/.

12. J-S. Kang, S-U. Shin, D. Hong and O. Yi. Provable security of KASUMI and 3GPP encryption mode f8. *Advances in Cryptology – ASIACRYPT 2001, LNCS 2248*, pp. 255–271, Springer-Verlag, 2001.

13. L.R. Knudsen and C.J. Mitchell. Analysis of 3gpp-MAC and two-key 3gpp-MAC. *Discrete Applied Mathematics*, vol. 128, no. 1, pp. 181–191, 2003.

14. T. Kohno, J. Viega, and D. Whiting. The CWC authenticated encryption (associated data) mode. See Cryptology ePrint Archive, Report 2003/106, http://eprint.iacr.org/.

15. M. Luby and C. Rackoff. How to construct pseudorandom permutations from pseudorandom functions. *SIAM J. Comput.*, vol. 17, no. 2, pp. 373–386, April 1988.

16. P. Rogaway, M. Bellare, J. Black, and T. Krovetz. OCB: a block-cipher mode of operation for efficient authenticated encryption. *Proceedings of ACM Conference on Computer and Communications Security, ACM CCS 2001*, ACM, 2001.

A General Attack Model on Hash-Based Client Puzzles

Geraint Price

Information Security Group,
Mathematics Department,
Royal Holloway University of London,
Egham, Surrey, TW20 0EX,
United Kingdom.
geraint.price@rhul.ac.uk

Abstract. In this paper, we present a general attack model against hash-based client puzzles. Our attack is generic in that it works against many published protocols. We introduce a new protocol and subsequently attack our new construction as well. We conclude by drawing two requirements of client puzzle protocols that would overcome our attack.

1 Introduction

With Internet based Denial of Service (DoS) attacks on the increase, there has been much research within the security community to find ways of preventing these attacks. One of the research avenues explored in recent years is broadly classed as *client puzzle* protocols [JB99,DS01,ANL00]. In these protocols, clients wishing to use a resource on a server have to solve a puzzle before the server is willing to commit the resource to the client.

The nature of these puzzles is such that they are easy to generate and verify, but relatively expensive to solve. The most common means of providing these puzzles is to use hash functions.

In this paper, we review some published protocols, in doing so we point out a flaw in the client puzzle model. We then present a general attack model against hash-based client puzzles which allows an adversary to solve puzzles cheaply and thus mount an attack on a server.

Our attack model is powerful in that it allows clients to solve many instances of client puzzles in parallel. The adversary does not need to compute any of the puzzles herself, drawing on the resources of benign users. This can greatly reduce the computational resources an adversary needs to have at her disposal to mount such an attack.

As well as pointing out these flaws in the published protocols, we provide the design principles required to defeat our attacks.

The remainder of this paper is organised as follows. Section 2 provides a more detailed overview of the work done on client puzzles to date and introduces our generic attack. Sections 3.1, 3.2 and 3.3 provide attacks against published

K.G. Paterson (Ed.): Cryptography and Coding 2003, LNCS 2898, pp. 319–331, 2003.
© Springer-Verlag Berlin Heidelberg 2003

protocols. Section 4 outlines a new protocol which adds client puzzle capability to an SSL-style handshake [FKK96]. Although we are also able to attack our new protocol, our discussion in Section 4 leads us to identify the design principles required to overcome our attack model. We draw our conclusions in Section 5.

2   Client Puzzles: an Overview

In this section we provide a more detailed overview of client puzzles, outlining their basic form and discussing protocols in the literature. We also describe our general method of attacking client puzzle protocols.

The basic outline of a *client puzzle* [JB99] is as follows. A client requesting a resource from a server is asked to solve a puzzle before the server is willing to commit the resource to the client. The rationale is that a legitimate client is willing to accept a slight delay to gain the resource. Conversely, an attacker who wishes to mount a Denial of Service (DoS) attack needs to request multiple instances of the same resource concurrently. The accumulative delay from having to solve these puzzles in parallel subsequently makes the cost of the attack prohibitive.

The nature of the puzzle should be such that a server can generate a puzzle very efficiently. The puzzle itself should be moderately hard for the client to solve and then easy for the server to verify.

Dwork and Noar [DN92] were the first to introduce the concept of requiring a user to compute a moderately hard function to gain access to a resource. They provide *junk e-mail* as the setting for their work, where a client wishing to send an e-mail to a recipient would need to calculate the solution to a puzzle before the e-mail would be delivered. While this might impose a delay of a few seconds for someone sending an e-mail, it is assumed that a legitimate sender would tolerate a short delay. The scheme aims to defeat people who send mass unsolicited e-mails, for which the accumulated cost would be prohibitive. The functions they use in their scheme are *broken signature functions*. They note the lack of work in the literature on the moderately-hard functions required for their puzzle schemes.

Juels and Brainard [JB99] were the first to introduce the term *client puzzle*, although their work is clearly influenced by Dwork and Noar. Juels and Brainard's work defends against connection depletion attacks, specifically the TCP SYN flooding attack. The TCP SYN attack makes use of the commitment of resources at the server end (in this case a "half-open" connection). The attacker requests a connection, but does not complete the three-way protocol, which leaves the connection in a committed, but unusable state. Jules and Brainard use a hash function as the building block for their puzzle, which has been adopted in most of the client puzzle work since. The server computes a hash $h(s)$ of an n-bit secret s. The client is then provided with $h(s)$, along with an m-bit section of the preimage. The task for the client is to provide the server with remaining $n - m$ bits of the preimage. The assumption here is that the fastest means of computing a preimage is a brute-force attack. The length of $n - m$ can be modified in order to make the puzzle harder or easier.

Client puzzles were then used by Aura et al. [ANL00] for protecting against DoS attacks on authentication protocols. Their work generalises the concept to protect a two-way authentication protocol as a precursor to securing a resource. In a similar manner to the work of Juels and Brainard, they use hash functions as the core of their puzzle.

Dean and Stubblefield [DS01] provide a modified version of the TLS protocol as a means of protecting it against DoS attacks.

More recently, Wang and Reiter [WR03] modified the concept to one of a *puzzle auction*. In a puzzle auction, the clients requesting a resource solve a puzzle in a similar manner to other client puzzles. In other forms of client puzzles [JB99, ANL00] the server sets the difficulty of the puzzle according to the perceived threat to the service. In Wang and Reiter's puzzle auction, it is up to the client to decide the level of difficulty of the puzzle they are going to solve. The server then provides the next free resource to the client who solved the most difficult puzzle. The rationale behind the auction is that a legitimate client is likely to be willing to spend more effort computing a solution than an attacker. They back this claim up by quoting a recent survey of Distributed Denial of Service (DDoS) attack code [GW00], which is used to infect multiple hosts on the Internet to provide the attacker with *zombie* computers. The study found that the code run on zombie machines was likely to restrict its processing activity so as not to alert the legitimate user that their machine has been infected.

While most of the work carried out on client puzzles makes use of hash functions, Abadi et al. [ABMW03] note that they may not provide the most appropriate means of setting puzzles. A brute-force search of a hash function is invariably a CPU-bound computation. They note there is a large disparity, in terms of processing power, of machines connected to the Internet. They propose the use of memory-bound functions as an alternative, where there is much less disparity in non-cache memory access times between high and low-end machines. While we agree with their statements, we concentrate our work on hash-based puzzles for the remainder of this paper.

2.1 A Generic Weakness in Puzzle Protocols

We now look at at what is a generic weakness in puzzle-based protocols. Specifically, those that employ hash functions as a basis for their puzzles.

Our attack is based on the observation that, if not managed correctly, an adversary can recruit unwitting clients to solve the server puzzles on her behalf. Because of the similarity to the man-in-the-middle attack – where an adversary can forge or pass messages between legitimate participants to fool them into thinking they are communicating with each other – we name our attack the *Man-in-the-Riddle* attack.

The basic vulnerability in these protocols comes from the weakness in the binding between a particular instance of the puzzle and the legitimate participants of the protocol.

We describe our attack model here, and it is shown in figure 1. Mallory (M) generates a request for a resource from server (S). At the same time, Mallory advertises her own service on the Internet. When S provides Mallory with a puzzle (P) to solve, she passes this puzzle on to a legitimate client (C) [1] of hers to solve on her behalf. Mallory then uses the solution (P_{sol}) to convince S to provide her with the resource. S thinks that Mallory has committed resources in solving the puzzle, and as long as Mallory provides C with the resource it requested, then C is unlikely to be aware that he has been used by Mallory as a puzzle solving oracle.

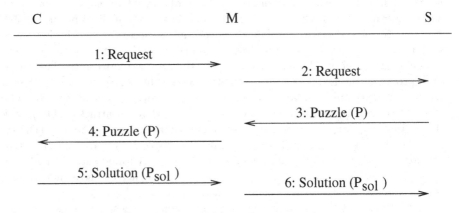

Fig. 1. An overview of our attack model

The ability of this attack to succeed depends on Mallory's ability to convince C that S's puzzles are her own. Many of the protocols in the literature include mechanisms to bind the identity of the communicating parties into the protocol run as a means of defending against such attacks. As we see from our work in Section 3 we are able to circumvent these mechanisms, thus removing the ability of the client puzzle protocol to protect the server. This is compounded by Mallory's ability to maintain a potentially large number of independent protocols running in parallel.

Because of the very nature of DoS attacks, for our attack to be a mitigate success, we require Mallory to be able to recruit unwitting clients at a sufficiently high rate. We believe that, with the large number of music sharing and pornography sites on the Internet, an attacker should find it relatively easy to maintain a cover site with a sufficient number of clients. The ability to launch attacks in this manner was demonstrated recently in the case of the *Sobig.F* virus, which was thought to have been spread initially via pornographic web sites [USA03].

---

[1] In the remainder of this paper, for simplicity, we will use C to represent a single instance of a benign client. In reality, Mallory will be looking to use many such clients in parallel to solve puzzles for her.

If the attack requires a large number of solved puzzles, then it would also be straightforward for Mallory to replicate her site a small number of times to multiply the number of available clients.

Another assumption we make in outlining this attack is that managing connections is far less resource intensive for Mallory than solving the puzzles herself. We believe this is trivially true, otherwise the puzzle would not represent a challenge and hence the deterrent to an attacker in the first place.

The service being attacked might be able to cope with the influx in the strict sense (i.e. that Mallory does not cause enough traffic to crash the server). However, it could be Mallory's goal to procure a significant portion of S's resources at a limited cost to herself. This could ultimately starve S of legitimate clients, which could be important if the server is a commercial site such as www.amazon.com, where attracting legitimate customers to pay for products is important.

3   Attacks against Published Protocols

In this section we will briefly overview how our generic attack can be adapted for specific client puzzle protocols in the literature. We keep our discussion about each attack relatively short for the sake of brevity, outlining the basic protocol along with how our attack is realised in each scenario.

3.1   Aura et al.

Aura et al. [ANL00] propose a protocol which uses classical authentication mechanisms in combination with client puzzles to defeat DoS attacks against a server.

Our attack on the protocol is demonstrated in figure 2. If Mallory was a legitimate client, the original protocol would be represented by messages 2, 3, 6 and 7.

Our attack works as follows: Mallory receives a legitimate request from C (message 1) and she immediately forwards this to S (msg 2) claiming it as her own. S sends back a signed response [2] (msg 3) which is the same message that is sent to all clients requesting a service. In the message is a nonce N_s that ensures all responses to puzzles are recent. The nonce N_s is refreshed periodically, but generating a new one and signing it does not place a high computational load on S. k is the difficulty parameter of the puzzle, and T_s is an optional time stamp. M forwards the content to C (msg 4). For the attack to work, M needs to be able to claim to be S. Her ability to do this will depend on the particular implementation of the protocol. If S's identity is tightly bound to the communication protocol, then this is difficult to achieve, because the certificate accompanying the signature should bind to S's identity within the communication mechanism (e.g. IP address or DNS name). In their paper, Aura et al. mention that the signature on the message 3 can be left out if we are

---

[2] The notation $S_x(\ldots)$ is used to indicate a message signed by X.

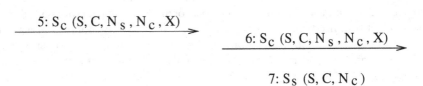

Fig. 2. An attack on the Aura et al. authentication protocol

not concerned about DoS attacks on C. Unfortunately this also leaves S much more open to abuse by M, as M now finds it easier to masquerade as S. This is because she does not have to generate anything claiming to be from S initially.

Upon receiving message 4, C solves the puzzle. This is done by using a brute-force search on the following hash: $h(C, N_s, N_c, X) = 0_1 0_2 \ldots 0_k Y$, where k is the security parameter in message 3 and X is the solution to the puzzle. C then signs the solution and sends it to M (msg 5), which is then forwarded to S (msg 6). S will then verify the validity of the message, commit to the resource and respond with a confirmation (msg 7). In a similar manner to message 4, this portion of the attack requires M to be able to masquerade as C. Again, if the binding between C's identity and the underlying communication protocol is not tight, this should be relatively easy for M to achieve. We envisage that clients are unlikely to have names that bind themselves to a particular IP address in the first place. For example, this would certainly be true for roaming solutions[3] that require the user to download their certificate if they wish to use it on a host that is not their own. Even strong bindings are of little use if the checks are not carried out effectively, as can be seen from some poor SSL implementations [Ben02].

As a final note in this section, even if the identities of the parties at the communication level are tightly bound to the signatures, our stronger variant of the attack we deploy against the Wang and Reiter protocol in Section 3.3 can be adapted to work here.

---

[3] Such as the Arcot soft certificate www.arcot.com.

3.2 Dean and Stubblefield

Dean and Stubblefield [DS01] provide an implementation of TLS which includes
client puzzles to stop a DoS attack against the server running the secure protocol.
Theirs is the easiest published protocol to attack because there is no binding
between the puzzle and the protocol in which it is running.

The attack on the Dean and Stubblefield protocol is shown in figure 3. If
M was a legitimate client, then their original protocol would be represented by
messages 2, 3, 4, 8 and 9. Their puzzle also uses a hash construction, although
the way their puzzle is constructed differs from that of Aura et al. In this case
the puzzle is constructed from the triple $(n, x', h(x))$, where x' is x with its n
lowest bits set to 0. They assume that $h(x)$ is a preimage resistant hash function,
and the solution to the puzzle is the full value of x. Because the hash function is
preimage resistant, then the best way for a client to generate x, is to try values
in the domain bounded by 0 and x' until a match is found.

The entire block is random to prevent an attacker from effectively precom-
puting all possible puzzle values.

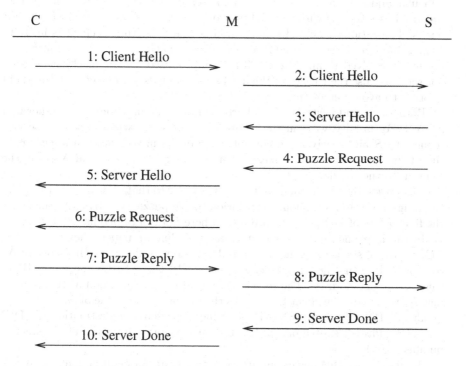

Fig. 3. An attack on the Dean and Stubblefield TLS protocol

This attack is simpler in its form than that of the attack on Aura et al. This is because S does not sign the puzzle. This allows M to forward exactly the same puzzle to C with impunity.

In their paper, Dean and Stubblefield state that they do not concern themselves with attackers that are capable of launching multiple attacks in parallel. Although in their original estimation this might not be an issue, we believe that our attack changes the means by which such an statement can be made. If their assumption is to ignore the threat of an attacker with potentially huge computing power at her disposal, our attack circumvents this by surreptitiously harnessing the computing power of multiple benign clients. This means that while Mallory is an attacker which they believe their protocol is secure against – she in computationally bounded – our attack allows her to harness the power of a much stronger adversary.

3.3   Wang and Reiter

Wang and Reiter [WR03] propose a modification of the client puzzle which they call a *puzzle auction*. In their protocol a client solves a puzzle of a particular difficulty and submits the solution to the server. If the server is under attack, then it choses the *bid* with the highest work function as the next client which it serves. The rationale behind this method is that a legitimate client is likely to commit more resources to solving a puzzle than an attacker. This is because most machines involved in an attack will be compromised *zombie* machines who do not want to raise their usage of the machines resources above a certain threshold in order to avoid detection.

Both Wang and Reiter's original protocol (a) and our attack (b) are shown in figure 4. For brevity, we omit the original SYN message, where C sends a service request to S and receives the current value of N_s in response. In a legitimate run of the protocol r_c is a request, with N_c being C's nonce, and X being the solution to the puzzle.

The puzzle that Wang and Reiter use is similar to that of Aura et al [ANL00]. In Wang and Reiter's version, the solution to the puzzle is a string X, such that the first n bits of $h(N_s, N_c, X)$ are zero, where $h()$ is a public hash function.

In the implemented example in Wang and Reiter's paper, they modify the TCP protocol stack to include the puzzle auction. In doing so, they specify N_s to be a hash of the the triple (secret, timer, SIP), where SIP is the source IP, in this case S's IP address. The secret is changed periodically, and at this point a new N_s is issued. The client then solves the puzzle for the value of N_c that is the quadruple (DIP, SP, DP, ISN). These values represent the destination IP (DIP – C's IP address), source port (SP), destination port (DP) and initial sequence number (ISN).

At first glance this makes our attack impractical. As we discuss in our attack on Aura et al. in Section 3.1, we recommend that each puzzle be bound to the communication protocol to avoid spoofing. Wang and Reiter achieve this through hashing the IP addresses and port numbers into their puzzle. This would seem to make it impossible for Mallory to pass off S's puzzle as her own. Unfortunately

C   M   S

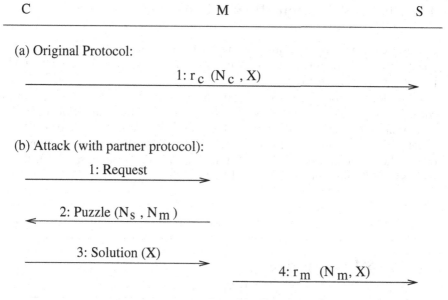

Fig. 4. An attack on the Wang and Reiter protocol

that is only true if Mallory runs an exact copy of the Wang and Reiter's protocol with C. We now describe a stronger variant of our general attack model in we define a *partner protocol* [4] which is run in conjunction with Wang and Reiter's protocol. This is shown in figure 4(b).

Our partner protocol is also a client puzzle protocol, which is of a similar design to Dean and Stubblefield's protocol. In our case, C is presented with a partial preimage and asked to find the remaining portion of the preimage. In our case, we forward the concatenated tuple (N_s, N_m), claiming it to be our partial preimage, and ask C to produce X as the solution to the puzzle.

In a fully mounted attack, Mallory would only need to send the initial SYN request every time N_s was changed by the server. She could then generate the (N_s, N_m) tuples to hand out to each new C that came along during the lifetime of each N_s.

Another benefit our attack has over the assumptions Wang and Reiter make in their paper is that the clients solving our puzzles are not zombies. Their auction works on the principle that a benign client will work harder than a zombie client. Because Mallory manages to convince multiple legitimate clients to solve what it thinks is a legitimate puzzle, it is willing to expend as much resource as any other legitimate client. This negates one of the driving factors behind their auction. In some respects this makes our attack much easier to mount, given that Mallory does not need to infect a potentially large number of zombie computers before the attack can take place.

---

[4] We note that this concept is similar to Kelsey et al's *chosen protocol attack* [KSW97].

4   A New SSL Client Puzzle Protocol

In this section we present – and ultimately break – our own variant of the client puzzle protocol. In doing so we discover what is required for a client puzzle protocol to remain secure.

Our new protocol, shown in figure 5(a), is based on an SSL-type protocol [FKK96], where a client C wants to set up a secure connection with S. It uses a similar construct to Aura et al's protocol, asking C to find a value X which satisfies the property: $h(IP_s, IP_c, N_s, N_c, X) = 0_1 0_2 \ldots 0_k Y$ for any value of Y. The other inputs to the hash function are S's IP address (IP_s), C's IP address (IP_c) and a nonce generated by C (N_c).

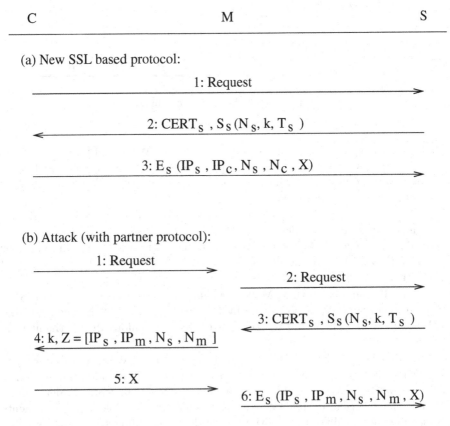

Fig. 5. A new SSL based client puzzle protocol

S signs the broadcast puzzle $S_s(N_s, k, T_s)$. If S's certificate binds its name to the underlying communication mechanism, this should make it more difficult for Mallory to claim S's puzzles as her own.

Once C has solved the puzzle, he encrypts the solution X, binding it to his own IP address and to the IP address of the server. This should also protect against Mallory masquerading as C to S.

Unlike Aura et al.'s protocol, we do not get C to sign the puzzle solution. They use this as a means of achieving mutual authentication. Given that client puzzles are aimed at use in a widely distributed environment such as the Internet, the lack of client-side SSL certificates would mean that there is little use for this part of the protocol.

Unfortunately, as we can see from figure 5(b), our new protocol can also be attacked. The attack is similar to our attack against Wang and Reiter's protocol, where we use a partner protocol of Mallory's choosing. In the attack, M receives a legitimately signed puzzle from S, she then provides C with a puzzle of the form (k, Z). She claims that Z is her own nonce N_s, in a similar manner to Dean and Stubblefield's protocol. She then asks C to solve $h(Z, X) = 0_1 0_2 \ldots 0_k Y$ for any value of Y.

Our new protocol uses a similar binding to the communication medium that Wang and Reiter use. Unfortunately, as we can see from the attack, this seems to provide little defence in the face of an adversary that can run a parallel protocol of her design.

A variant of our attack here could be used against Aura et al.'s protocol, even if the variant of their protocol is used which signs the server's nonce and there is a tight binding between the identities and the communication addresses.

It would appear that the only way to resist our partner protocol variant of the attack is through the following combination of design decisions:

- The server nonce N_s should be tightly bound to S's identity using a signature. The certificate that authenticates this signature should be tightly bound to the communication medium (e.g. IP address).
- C should have the freedom to chose a nonce N_c that influences the known portion of the preimage of the hash function. This would prevent Mallory from forcing the entirety of the known portion of the preimage upon C. This would prevent the scenario above, where in the attack of figure 5(b), where Mallory claims the entirety of the preimage as her nonce Z.
- All client puzzles would have to conform to the above two principles. This is the most difficult condition to achieve. As long as Mallory is free to use client puzzle protocols of her own design, she should be able to convince a sufficient number of benign Cs to solve puzzles on her behalf. If the protocol does not use the previous two principles, then Mallory can easily pass off S's puzzles as her own.

5   Conclusions and Future Work

In this paper we have presented a general attack model against hash-based client puzzles. We demonstrate our attack against three published protocols. We design our own variant of a client puzzle protocol and demonstrate that our new protocol is also vulnerable to the attack.

330   G. Price

Our attack is powerful because it allows an adversary to solve large numbers
of client puzzles in parallel without expending the required computational resources herself. The puzzles are solved through surreptitiously using the resources
of benign clients. The main requirement of our attack is that the adversary maintain a sufficiently attractive service of her own to hand off the puzzles at the
required rate. Given the prevalence of music sharing and pornography sites on
the Internet, we believe that this requirement is a relatively easy one to fulfil.

We conclude, from our discussion in Section 4, that the only way to defeat our
attack model completely is to standardise various elements of all client puzzle
protocols. The two requirements are as follows:

- The server should sign its component of the puzzle.
- The client should have freedom to chose a portion of the puzzle prior to
  generating the solution.

The combination of these two factors would make it impossible for the adversary to force a legitimate client to solve puzzles from a separate protocol
run.

There would appear to be a great deal of similarity between client puzzles
and the use of nonces for freshness in authentication protocols. If Alice sends out
a nonce which she then receives back in a protocol message – without additional
mechanisms protecting the use of the nonce – all she can convince herself of
is that *someone* has recently seen the nonce. In a similar manner here, if S
generates a puzzle and receives a reply, without additional constraints on the
form and use of the puzzle, all S can be sure of is that *someone* has calculated
the solution.

Although we present our attack against hash-based client puzzles, it would
be interesting to see if we could expand our attack to emerging memory-bound
puzzles [ABMW03].

Another element of our work that requires further attention are the potential conditions under which puzzle based schemes might be secure. We present
what we believe to be the two main requirements in Section 4. A more rigorous
demonstration of whether these requirements were sufficient and necessary will
follow.

Acknowledgements. We would like to thank Kenneth Paterson for his comments in helping clarify some of the points presented in this paper.

References

_

re.

[ABMW03] Martín Abadi, Mike Burrows, Mark Manasse, and Ted Wobber. Moderately hard, memory-bound functions. In *Proceedings of the 10th Annual
Network and Distributed System Security Symposium*, 2003.
[ANL00] Tuomas Aura, Pekka Nikander, and Jussipekka Leiwo. DOS-resistant authentication with client puzzles. In *Proceedings of the 8th International
Workshop on Security Protocols*, volume 2133 of *Lecture Notes in Computer Science*, pages 170–177, April 2000.

[Ben02]   Mike Benham. Internet Exproler SSL Vulnerability.
            http://www.securiteam.com/windowsntfocus/5JP0E0081M.html,
            August 2002.

[DN92]   C. Dwork and M. Naor. Pricing via processing or combatting junk mail.
            In E. Birckell, editor, *Proceedings of Advances in Cryptology – CRYPTO
            92*, volume 1328 of *Lecture Notes in Computer Science*, pages 139–147,
            1992.

[DS01]   Drew Dean and Adam Stubblefield. Using client puzzles to protect TLS.
            In *Proceedings of 10th Annual USENIX Security Symposium*, 2001.

[FKK96]   A Freier, P. Karlton, and P. Kocher. The SSL protocol - version
            3.0. Available at http://home.netscape.com/eng/ssl3/ssl-toc.html,
            March 1996.

[GW00]   X. Geng and A. Whinston. Defeating distributed denial of service attacks.
            IEEE IT Professional, 2(4):36–41, July—August 2000.

[JB99]   Ari Juels and John Brainard. Client puzzles: A cryptographic counter-
            measure against connection depletion attacks. In S. Kent, editor, *Proceed-
            ings of the 1999 Network and Distributed System Security Symposium*,
            pages 151–165, 1999.

[KSW97]   J. Kelsey, B. Schneier, and D. Wagner. Protocol interactions and the
            chosen protocol attack. In B. Christianson, B. Crispo, M. Lomas, and
            M. Roe, editors, *Security Protocols 1997*, volume 1361 of *Lecture Notes
            in Computer Science*, pages 91–104, April 1997.

[USA03]   Sobig.F worm believed to start at Web porn site. Available at
            http://www.usatoday.com/tech/news/computersecurity/2003-08-22
            -sobig-start_x.htm, August 2003.

[WR03]   XiaoFeng Wang and Michael K. Reiter. Defending against denial-of-
            service attacks with puzzle auctions. In *Proceedings of the 2003 IEEE
            Symposium on Security and Privacy*, 2003.

Tripartite Authenticated Key Agreement Protocols from Pairings

S.S. Al-Riyami and K.G. Paterson

Information Security Group,
Royal Holloway, University of London, Egham, Surrey, TW20 0EX
{s.al-riyami,kenny.paterson}@rhul.ac.uk

Abstract. Joux's protocol [29] is a one round, tripartite key agreement protocol that is more bandwidth-efficient than any previous three-party key agreement protocol. But it is insecure, suffering from a simple man-in-the-middle attack. This paper shows how to make Joux's protocol secure, presenting several tripartite, authenticated key agreement protocols that still require only one round of communication and no signature computations. A pass-optimal authenticated and key confirmed tripartite protocol that generalises the station-to-station protocol is also presented. The security properties of the new protocols are studied using provable security methods and heuristic approaches. Applications for the protocols are also discussed.

Areas: Secure protocols; key agreement; authentication; pairings.

1 Introduction

Asymmetric key agreement protocols are multi-party protocols in which entities exchange public information allowing them to create a common secret key that is known only to those entities and which cannot be predetermined by any party. This secret key, commonly called a *session key*, can then be used to create a confidential or integrity-protected communications channel amongst the entities. Beginning with the famous Diffie-Hellman protocol [19], a huge number of two-party key agreement protocols have been proposed (see [10] and [38, Chapter 12.6] for surveys). This reflects the fundamental nature of key exchange as a cryptographic primitive.

The situation where three or more parties share a secret key is often called *conference keying*. The three-party (or tripartite) case is of most practical importance not only because it is the most common size for electronic conferences but because it can be used to provide a range of services for two parties communicating. For example, a third party can be added to chair, or referee a conversation for ad hoc auditing, data recovery or escrow purposes.

One of the most exciting developments in recent years in the area of key agreement is Joux's tripartite key agreement protocol [29]. This protocol makes

K.G. Paterson (Ed.): Cryptography and Coding 2003, LNCS 2898, pp. 332–359, 2003.

use of *pairings* on elliptic curves and requires each entity to transmit only a single broadcast message containing a short-term key. This should be contrasted with the obvious extension of the Diffie-Hellman protocol to three parties, which requires two broadcasts per entity. However, just like the raw Diffie-Hellman, protocol, Joux's protocol is unauthenticated and suffers from man-in-the-middle attacks. See Section 3 for a description of Joux's protocol and an overview of pairings on elliptic curves.

1.1   Our Contribution

In this paper we show how to transform Joux's protocol into a secure tripartite protocol that still requires only a single broadcast per entity. In fact, we present four different one round, tripartite, authenticated key agreement (TAK) protocols in Section 4. These protocols all have the same protocol messages, but have different methods for calculating session keys from those messages. Our protocols are specifically designed to avoid the use of expensive signature computations. It is clear how to augment Joux's protocol with signatures on the short-term keys so as to prevent man-in-the-middle attacks.

Note that we do *not* make the naive claim that the possibility of generating four different session keys from one protocol run means that our protocol is four times as efficient as other protocols – there are far better ways of deriving multiple session keys than this! Rather we present four protocols and analyze their different security properties. Our one round protocols build on Joux's protocol and draw on ideas from the Unified Model [4], MTI [36] and MQV [32,33] protocols.

In Section 5, we consider proofs of security for our protocols. Our proofs use an adaptation of the Bellare-Rogaway model [7] to the public key setting. Our model is similar to that given by Blake-Wilson, Johnson and Menezes in [9] (though our model extends to the three-party situation, and our extension is different to that presented in the symmetric-key setting in [8]). Our proofs show that the first of the TAK protocols is secure and has perfect forward secrecy provided that the Bilinear Diffie-Hellman Problem (BDHP) is hard. This computational problem forms the basis for the security of the identity-based encryption scheme of [11] – see Section 3 for more details on the BDHP. We do not provide a proof of security for our other protocols. But we note that the MQV protocol, on which our fourth TAK protocol is modelled, has never been proven secure in any model, though it has been widely adopted for standardisation [2, 3,27,28]. In view of the incomplete analysis currently available through provable approaches, we choose to supplement our proofs of security with ad hoc analysis in Section 6. This allows us to consider attacks on our protocols not included in the security model. In Section 7, we look at the scenario where one of the three parties is off-line. The protocol we give for this situation can be applied to key escrow with an off-line escrow agent.

In our penultimate section, Section 8, we examine pairing-based authenticated key agreement with key confirmation in the non-broadcast setting. The main point we make in that section is that a properly authenticated and key confirmed protocol based on pairings can be no more efficient (in terms of protocol passes) than the obvious extension of the station-to-station protocol [20] to three parties. Thus the apparent efficiency of Joux's protocol is lost when it is made secure and when an appropriate measure of efficiency is used.

The final section, Section 9, contains some conclusions and ideas for future work.

1.2   Related Work and Version History

As we have already noted, our work builds on that of Joux [29], and our one round protocols draw upon the Unified Model [4], MTI [36] and MQV [32,33] protocols. The tripartite AKC non-broadcast protocol presented builds on the STS [20] protocol. Our security model is inspired by Blake-Wilson, Johnson and Menezes' extension [9] of the Bellare-Rogaway model [7]. More recent work on models for secure protocols (in particular key agreement protocols) can be found in [6,14,16,17,44]. Work pertinent to the efficient implementation of our protocols can be found in [5,21,22]. Other work, reflecting the recent explosion of interest in using pairings to construct cryptographic schemes, can be found in [11,12,18, 24,25,31,40,42,45].

This version of our paper includes some major improvements and changes to our original work [1]. In particular one of the protocols of [1], namely TAK-1 is proved to have perfect forward secrecy and another, TAK-2, is no longer explored in any detail because of a fatal man-in-the-middle attack that was discoverd by Shim [43].

2   Protocol Goals and Attributes

Here we discuss the various desirable attributes and goals that one may wish a key agreement protocol to possess.

2.1   Extensional Security Goals

An *extensional* goal [13,41] for a protocol is defined to be a design goal that is independent of the protocol details. Here we list two desirable and widely-agreed extensional goals for key agreement protocols. Further discussion can be found in [38, Chapter 12]. The first goal we try to achieve is *implicit key authentication*. This goal, if met, assures an entity that only the intended other entities *can* compute a particular key. This level of authentication results in what is known as an *authenticated key agreement* (AK) protocol. If each entity is also assured that the intended other entities actually *have* computed the key, then the resulting protocol is called an *authenticated key agreement with confirmation* (AKC) protocol. Another goal is to provide a *good key*. This goal states that the key is selected uniformly at random from the key space, so that no adversary has an information-theoretic advantage when mounting a guessing strategy to determine the key.

2.2 Security Attributes

A number of desirable security attributes have been identified for key agreement protocols [9,10,32,33] and we borrow our definitions from these sources. Depending on the application scenario, these attributes can be vital in excluding realistic attacks.

Known session key security.   A protocol is *known session key secure* if it still achieves its goal in the face of an adversary who has learned some previous session keys.

(Perfect) forward secrecy.   A protocol enjoys *forward secrecy* if, when the long-term private keys of one or more entities are compromised, the secrecy of previous session keys is not affected. *Perfect* forward secrecy refers to the scenario when the long term private keys of all the participating entities are compromised.

No key-compromise impersonation.   Suppose A's long-term private key is disclosed. Then of course an adversary can impersonate A in any protocol in which A is identified by this key. We say that a protocol resists *key-compromise impersonation* when this loss does not enable an adversary to impersonate *other* entities as well and obtain the secret key.

No unknown key-share.   In an *unknown key-share attack*, an adversary convinces a group of entities that they share a key with the adversary, whereas in fact the key is shared between the group and another party. This situation can be exploited in a number of ways by the adversary when the key is subsequently used to provide encryption or integrity [30].

No key control.   It should not be possible for any of the participants (or an adversary) to force the session key to a preselected value or predict the value of the session key. (See [39] for a discussion of how protocol participants can partially force the values of keys to particular values and how to prevent this using commitments).

2.3 Further Attributes

Two important computational attributes are low *computation overhead* and the ability to perform *precomputation*. It may be desirable that one or more entities (perhaps with limited computational environments) should perform less computation than the others.

It is an advantage when a protocol has low *communication overhead*, which means that only a small amount of data is transmitted. Designing protocols with a minimal number of *passes* and *rounds* is always desirable. The number of passes is the total number of messages exchanged in the protocol. A *round* consists of all the messages that can be sent and received in parallel within one time unit. A *broadcast* message is a message that is sent to every party in a protocol. These notions of communication complexity are each more or less appropriate depending on the particular network architecture that is in use. For example, wireless systems operate exclusively in broadcast mode, so that every packet is simultaneously available to all nodes. Then the number of rounds and broadcasts is a more natural way of measuring a protocol's communication complexity. On the other hand, the Internet Protocol running over a public network like the internet usually makes use of point-to-point communications, and then the number of passes is the right measure. As we shall see with Joux's protocol, communication advantages that a protocol apparently possesses can disappear when one either considers a different network architecture or more stringent security requirements.

If the messages transmitted in a protocol are *independent* of each other, in the sense that they can be sent and received in any order, then the protocol is *message independent*. A protocol is *role symmetric* when messages transmitted and computations performed by all the entities have the same structure. In the context of public key cryptography, *short-term public keys* (or *values*) are generally only used once to establish a session, and are sometimes called *ephemeral keys*. On the other hand, *long-term public keys* are static keys used primarily to authenticate the protocol participants.

2.4 Attacks

An attack occurs when the intended goals of a protocol are not met or the desired security attributes do not hold. A *passive* attack occurs when an adversary can prevent the protocol from accomplishing its goals by simply observing the protocol runs. Conversely, an *active* attack is one in which the adversary may delete, inject, alter or redirect messages, by interleaving multiple instantiations of the same protocol and the like. The formal model of security that we introduce in Section 5 is powerful enough to capture many kinds of active attack.

3   Key Agreement via Pairings on Elliptic Curves

Here, we briefly review some background on pairings on elliptic curves, and then present Joux's protocol [29].

3.1   Pairings

We use the same notation as in [12]. We let $\mathbb{G}_1$ be an additive group of prime order q and $\mathbb{G}_2$ be a multiplicative group of the same order q. We assume the existence of an efficiently computable bilinear map $\hat{e}$ from $\mathbb{G}_1 \times \mathbb{G}_1$ to $\mathbb{G}_2$. Typically, $\mathbb{G}_1$ will be a subgroup of the group of points on an elliptic curve over a finite field, $\mathbb{G}_2$ will be a subgroup of the multiplicative group of a related finite field and the map $\hat{e}$ will be derived from either the Weil or Tate pairing on the elliptic curve. We also assume that an element $P \in \mathbb{G}_1$ satisfying $\hat{e}(P, P) \neq 1_{\mathbb{G}_2}$ is known. By $\hat{e}$ being bilinear, we mean that for $Q, W, Z \in \mathbb{G}_1$, both

$$\hat{e}(Q, W + Z) = \hat{e}(Q, W) \cdot \hat{e}(Q, Z) \quad \text{and} \quad \hat{e}(Q + W, Z) = \hat{e}(Q, Z) \cdot \hat{e}(W, Z)$$

hold.

When $a \in \mathbb{Z}_q$ and $Q \in \mathbb{G}_1$, we write aQ for Q added to itself a times, also called scalar multiplication of Q by a. As a consequence of bilinearity, we have that, for any $Q, W \in \mathbb{G}_1$ and $a, b \in \mathbb{Z}_q$:

$$\hat{e}(aQ, bW) = \hat{e}(Q, W)^{ab} = \hat{e}(abQ, W) = \dots$$

a fact that will be used repeatedly in the sequel without comment.

We refer to [5,12,11,21,22] for a more comprehensive description of how these groups, pairings and other parameters should be selected in practice for efficiency and security. We simply assume in what follows that suitable groups $\mathbb{G}_1$ and $\mathbb{G}_2$, a map $\hat{e}$ and an element $P \in \mathbb{G}_1$ have been chosen, and that elements of $\mathbb{G}_1$ and $\mathbb{G}_2$ can be represented by bit strings of the appropriate lengths. We note that the computations that need to be carried out by entities in our protocols will always involve pairing computations, and that the complexity of these will generally dominate any other calculations. However, with recent advances in efficient implementation of pairings, [5,22], the complexity of a pairing computation is now of a similar order to that of elliptic curve point multiplication.

3.2   Joux's Protocol

In [29], Joux introduced a very simple and elegant one-round protocol in which the secret session key for three parties could be created using just one broadcast per entity. In Joux's protocol, $a, b, c \in \mathbb{Z}_q^*$ are selected uniformly at random by A, B and C respectively. The ordering of protocol messages is unimportant and any of the three entities can initiate the protocol. In all the protocol messages "Sends to" is denoted by "$\rightarrow$".

Protocol messages:
$A \rightarrow B, C$: aP (1)
$B \rightarrow A, C$: bP (2)
$C \rightarrow A, B$: cP (3)
Protocol 1 (Joux's one round protocol [1]).

---

[1] The reader will notice that our version of Joux's protocol is simpler than the original. It uses a modified pairing such that $\hat{e}(P, P) \neq 1$; this allows us to avoid sending two points per participant.

Protocol description: Once the communication is over, A computes $K_A = \hat{e}(bP, cP)^a$, B computes $K_B = \hat{e}(aP, cP)^b$ and C computes $K_C = \hat{e}(aP, bP)^c$. By bilinearity of $\hat{e}$, these are all equal to $K_{ABC} = \hat{e}(P, P)^{abc}$. This can serve as the secret key shared by A, B and C.

Although not explicitly mentioned in [29], the success of this protocol in achieving its aim of agreeing a good key for the three entities in the face of passive adversaries is related to the hardness of the Bilinear Diffie-Hellman problem (BDHP) for the pair of groups $\mathbb{G}_1$, $\mathbb{G}_2$. This problem, formalised in [12], can be stated as:

Given P, aP, bP, cP with a, b and c chosen uniformly at random from $\mathbb{Z}_q^*$, compute $\hat{e}(P, P)^{abc}$.

More properly, the session key should be derived by applying a suitable key derivation function to the quantity $\hat{e}(P, P)^{abc}$. For otherwise, an attacker might be able to get partial information about session keys even if the BDHP is hard. This motivates study of hard bits for the BDHP [23].

It is known that the BDHP is no harder than the computational Diffie-Hellman problems in either $\mathbb{G}_1$ or $\mathbb{G}_2$. The reverse relationship is not known. Typically then, one chooses parameters so that $\mathbb{G}_1$, a subgroup of the group of points on an elliptic curve, has around 2^{160} elements and so that $\mathbb{G}_2$ is a subgroup of $\mathbb{F}_r$ where r has roughly 1024 bits. See [12,21] for details.

Unfortunately just like the unauthenticated two-party Diffie-Hellman protocol, Joux's protocol is subject to a simple man-in-the-middle attack. The attack is easily realised, and allows an adversary E to masquerade as any entity to any other entity in the network. We leave the construction of such an attack as a straightforward exercise.

4   One Round Tripartite Authenticated Key Agreement Protocols

The advantage of Joux's tripartite protocol over any previous tripartite key agreement protocol is that a session key can be established in just one round. The disadvantage is that this key is not authenticated, and this allows a man-in-the-middle attack.

In this section, we develop protocols which also need just one round, but which provide a key which is implicitly authenticated to all entities. Our AK protocols are generalisations of the standardised [2,3,27] Unified Model protocol [4], the MTI family of protocols [36] and the MQV protocol [32,33] to the setting of pairings. In fact we present a single protocol with four different methods for deriving a session key.

In order to provide session key authentication, some form of authenticated long-term private/public key pairs are needed. As with the other protocols, a certification authority (CA) is used in the initial set-up stage to provide certificates which bind users' identities to long-term keys. The certificate for entity A will be of the form: $\text{Cert}_A = (\mathcal{I}_A \| \mu_A \| P \| \mathcal{S}_{CA}(\mathcal{I}_A \| \mu_A \| P))$. Here $\mathcal{I}_A$ denotes the identity string of A, $\|$ denotes the concatenation of data items, and $\mathcal{S}_{CA}$ denotes the CA's signature. Entity A's long-term public key is $\mu_A = xP$, where $x \in \mathbb{Z}_q^*$ is the long-term private key of A. Element P is a public value and is included in order to specify which element is used to construct μ_A and the short-term public values. Similarly Cert_B and Cert_C are the certificates for entities B and C, with $\mu_B = yP$ and $\mu_C = zP$ as their long-term public keys. Certificates could contain further information, such as validity periods.

As usual, in the protocol below, short-term keys $a, b, c \in \mathbb{Z}_q^*$ are selected uniformly at random by A, B and C respectively.

Protocol messages:

$$A \rightarrow B, C: \quad aP \| \text{Cert}_A \quad (1)$$
$$B \rightarrow A, C: \quad bP \| \text{Cert}_B \quad (2)$$
$$C \rightarrow A, B: \quad cP \| \text{Cert}_C \quad (3)$$

Protocol 2 (Tripartite Authenticated Key Agreement (TAK) protocol).

Protocol description: An entity A broadcasting to B and C, sends his fresh short-term public value aP along with a certificate Cert_A containing his long-term public key. Corresponding values and certificates are broadcast by B and C to A, C and A, B respectively. Notice that the protocol messages are just the same as in Joux's protocol, except for the addition of certificates. Each party verifies the authenticity of the two certificates he receives. If any check fails, the protocol should be aborted. When no check fails, one of four possible session keys described below should be computed. Below, H denotes a suitable hash function.

TAK key generation:

1. **Type 1 (TAK-1):**
   The keys computed by the entities are:
   $$K_A = H(\hat{e}(bP, cP)^a \| \hat{e}(yP, zP)^x),$$
   $$K_B = H(\hat{e}(aP, cP)^b \| \hat{e}(xP, zP)^y),$$
   $$K_C = H(\hat{e}(aP, bP)^c \| \hat{e}(xP, yP)^z).$$
   By bilinearity, all parties now share the session key
   $$K_{ABC} = H(\hat{e}(P, P)^{abc} \| \hat{e}(P, P)^{xyz}).$$

2. **Type 2 (TAK-2):**
   The keys computed by the entities are:
   $$K_A = \hat{e}(bP, zP)^a \cdot \hat{e}(yP, cP)^a \cdot \hat{e}(bP, cP)^x,$$
   $$K_B = \hat{e}(aP, zP)^b \cdot \hat{e}(xP, cP)^b \cdot \hat{e}(aP, cP)^y,$$
   $$K_C = \hat{e}(aP, yP)^c \cdot \hat{e}(xP, bP)^c \cdot \hat{e}(aP, bP)^z.$$
   The session key is $K_{ABC} = \hat{e}(P, P)^{(ab)z + (ac)y + (bc)x}$.

3. **Type 3 (TAK-3):**
   The keys computed by the entities are:
   $$K_A = \hat{e}(yP, cP)^x \cdot \hat{e}(bP, zP)^x \cdot \hat{e}(yP, zP)^a,$$
   $$K_B = \hat{e}(aP, zP)^y \cdot \hat{e}(xP, cP)^y \cdot \hat{e}(xP, zP)^b,$$
   $$K_C = \hat{e}(aP, yP)^z \cdot \hat{e}(xP, bP)^z \cdot \hat{e}(xP, yP)^c.$$
   The session key is $K_{ABC} = \hat{e}(P, P)^{(xy)c+(xz)b+(yz)a}$.

4. **Type 4 (TAK-4):**
   The keys computed by the entities are:
   $$K_A = \hat{e}(bP + H(bP\|yP)yP, cP + H(cP\|zP)zP)^{a+H(aP\|xP)x},$$
   $$K_B = \hat{e}(aP + H(aP\|xP)xP, cP + H(cP\|zP)zP)^{b+H(bP\|yP)y},$$
   $$K_C = \hat{e}(aP + H(aP\|xP)xP, bP + H(bP\|yP)yP)^{c+H(cP\|zP)z}.$$
   The session key is
   $$K_{ABC} = \hat{e}(P, P)^{(a+H(aP\|xP)x)(b+H(bP\|yP)y)(c+H(cP\|zP)z)}.$$

Notes:

- Joux's protocol [29] and our protocols are all vulnerable 'small subgroup' attacks and variants of them observed by Lim and Lee [34]. To protect against this, a verification algorithm should be applied by each entity to ensure that their received elements are actually in $\mathbb{G}_1$. This is fairly cheap to do when $\mathbb{G}_1$ is an elliptic curve group.
- In all four cases, key generation is role symmetric and each entity uses knowledge of both short-term and long-term keys to produce a unique shared secret key. No party has control over the resulting session key K_{ABC} and if any one of a, b, c is chosen uniformly at random, then K_{ABC} is a random element of $\mathbb{G}_2$. Of course delays in receipt of messages from the last entity should not be tolerated by the other entities, because after the last entity sees all the other participants' messages he is capable of fixing a small number of bits in the final shared secret key. See [39] for more details.
- Since all four keys are created after transmitting the same protocol messages, the communication overhead of each protocol version is identical. However, TAK-2 and TAK-3 key generation require more computation compared to TAK-1: in the former, each entity must make three pairing calculations, with the latter, just two. Better still, TAK-4 requires only a single pairing computation per entity. Protocols TAK-1 and TAK-3 can exploit pre-computation if entities know in advance with whom they will be sharing a key. In TAK-1, all entities can pre-compute the term $\hat{e}(P, P)^{xyz}$. With TAK-3, A can pre-compute $\hat{e}(P, P)^{ayz}$, with similar pre-computations for B and C. However, these terms cannot be re-used because fresh a, b and c should be used in each new protocol session.

Rationale for the TAK keys' algebraic forms:

- Protocol TAK-1 is analogous to the Unified Model protocol, whilst protocols TAK-2 and TAK-3 have their roots in the MTI/A0 protocol. The

Unified Model protcol is similar to the MTI protocols but utilises a hash function H with concatenation to combine various components, instead of a multiplication. The MTI-like variant of TAK-1, in which the agreed key is $K_{ABC} = \hat{e}(P, P)^{abc+xyz}$, suffers from a severe form of key-compromise impersonation attack. This attack does not require the adversary to learn a long-term private key. Rather, the adversary only needs to obtain a session key and one of the short-term secret values used in a protocol run to mount the attack. It would be prudent to derive session (and MAC keys for key confirmation if desired) by applying a hash function to each K_{ABC}. This would prevent problems arising from the possible existence of relatively easy bits in the BDHP . Using a hash function in TAK-1 has the additional benefit that it allows a security proof to be given (assuming H is modelled by a random oracle) – see Section 5.

- Protocol TAK-4 is modelled on the MQV protocol but avoids that protocol's unknown key-share weakness [30] by using a hash function H to combine long-term and short-term private keys. Here H's output is assumed to be onto $\mathbb{Z}_q^*$. Note that the protocol resulting from omission of this hash function produces the key $K_{ABC} = \hat{e}(P, P)^{(a+x)(b+y)(c+z)}$. However, this version of the protocol suffers from an unknown key-share weakness similar to that presented for the MQV protocol in [30], wherein the attacker *does* know the private key corresponding to his registered public key. As a consequence, this attack cannot be prevented by requiring the adversary to provide a proof of possession for her private key as part of the registration process. See Section 6.3 for further discussion of unknown key-share attacks.

- Other MTI-like protocols can be produced if A, B and C broadcast the ordered pairs $ayP\|azP$, $bxP\|bzP$ and $cxP\|cyP$ respectively (along with the appropriate certificates). This protocol can be used to produce the MTI/C0-like shared secret key $K_{ABC} = \hat{e}(P, P)^{abc}$, which for example A calculates by $K_{ABC} = \hat{e}(bxP, cxP)^{ax^{-2}}$. It can also be used to produce the MTI/B0-like shared secret key $K_{ABC} = \hat{e}(P, P)^{ab+bc+ca}$, which A can calculate by $K_{ABC} = \hat{e}(bxP, cxP)^{x^{-2}} \cdot \hat{e}(P, bxP)^{ax^{-1}} \cdot \hat{e}(P, cxP)^{ax^{-1}}$. Although these protocols produce a key with forward secrecy, we do not consider them further here because they require significantly higher bandwidth and do not offer a security advantage over our other protocols. For example, both protocols suffer from key compromise impersonation attacks and the MTI/C0 analogue is also vulnerable to known session key attacks.

Our TAK protocols include long-term private keys in the computation of each K_{ABC} in order to prevent man-in-the-middle attacks. Shim [43], however, has shown that simply involving the long-term keys is not sufficient to prevent a man-in-the-middle attack on TAK-2. Due to this severe vulnerability in the TAK-2 protocol, its security will not be discussed further. The TAK-2 protocol should be avoided and merely remains in this section of the paper for completeness and to provide a contrast with our other protocols. For the remaining protocols, other forms of active attack can still occur. We consider such attacks on a case-by-case basis in Section 6, after considering proofs of security.

5 Security Proofs

In this section, we introduce a security model for TAK protocols, and consider in detail the security of protocol TAK-1 in this model.

Our model is similar to those introduced in [7] and [9], with some simplifications that are possible because of the one round nature of our TAK protocols. In particular, we avoid the use of matching conversations (and session IDs introduced in later work [6]). Rather than fully describing our model, we highlight the differences to previous work.

Let k be a security parameter. As usual, we assume a set $\mathcal{I} = \{1, 2, \ldots, T_1(k)\}$ of protocol participants, where $T_1(k)$ is a polynomial bound in k on the number of participants. We will also use $A, B, C, \ldots$ to refer to protocol participants, while E is reserved for our adversary (who is not a participant). Each participant A is modelled by an oracle Π_A^s, which the adversary E can query at will. Here, s is a session number, that determines which random tape Π_A^s will use. In previous work, oracles were of the form $\Pi_{i,j}^s$ and modelled messages sent from participant i to participant j in session s. We remove this dependence on receiving parties; in all our protocols, all messages are broadcast.

Oracles exist in one of several possible states Accept, Reject, or *. In our protocols, an oracle accepts only after receipt of two properly formulated messages (containing different certificates to its own) and the transmission of two messages, not necessarily in that order (and with the received messages possibly originating from the adversary and not the oracles identified in the certificates in those messages). When an oracle accepts, we assume it accepts holding a session key K that is k bits in length. We also assume there is a key generation process $\mathcal{G}$ which produces a description of groups $\mathbb{G}_1$ and $\mathbb{G}_2$ and the map $\hat{e}$, assigns random tapes and oracles as necessary, distributes long-term private keys to participants, and prepares certificated long-term public keys. (Thus our model assumes a perfect certification process and does not capture attacks based on registration weaknesses like those described in Section 6.3). As usual, the benign adversary is defined to be one that simply passes messages to and fro between participants.

Adversary E is assumed to have complete control over the network, and we allow E to make three kinds of query to an oracle Π_A^s: Send, Reveal and Corrupt. These have the usual meanings, as per [9]: Send allows the adversary to send a message of her choice to Π_A^s and to record the response, or to induce Π_A^s to initiate a protocol run with participants of E's choosing. Reveal reveals the session key (if any) held by Π_A^s, while Corrupt produces the long-term private key of the oracle. Notice that when making a Send query, E does not need to specify the intended recipients of the oracle's reply. This is because, in our protocols, the oracle's replies are independent of these recipients anyway. In fact E can relay these messages to any party of her choosing.

In our TAK protocols, each party sends the same message to two other participants and receives two messages from those participants. In our model, we say that three oracles Π_A^s, Π_B^t and Π_C^u have *participated in a matched session* if they have received messages exclusively generated by one another (via the adversary). In other words, the two messages that Π_A^s has received are those generated by Π_B^t and Π_C^u, and these are the two oracles and sessions which received the single message from Π_A^s, and likewise for the other two participants.

In addition to the above queries, we allow E to make one further **Test** query of one of the oracles Π_A^s at any point during her attack. This oracle must be *fresh*: it must have accepted, not have been subject to a **Reveal** query, be uncorrupted, not have participated in a matched session with any revealed oracle, and not have received messages containing the certificate of a corrupted oracle. The reply to this **Test** query is either the session key K held by the oracle, or a random k-bit string, the choice depending on a fair coin toss. The adversary's advantage, denoted $advantage^E(k)$, is the probability that E can distinguish K from the random string. Notice that we remove the unnatural restriction in earlier models [7,9] that this **Test** query be the adversary's last interaction with the model.

As usual, a function $\epsilon(k)$ is negligible, if for every $c > 0$ there exists a $k_c > 0$ such that for all $k > k_c$, $\epsilon(k) > k^{-c}$.

We say that a protocol is a secure TAK protocol if:

1. In the presence of the benign adversary, and when oracles participate in a matched session, all the oracles always accept holding the same session key, which is distributed randomly and uniformly on $\{0,1\}^k$.
2. If uncorrupted oracles Π_A^s, Π_B^t and Π_C^u participate in a matched session, then all three oracles accept and hold the same session key, which is again uniformly distributed on $\{0,1\}^k$.
3. $advantage^E(k)$ is negligible.

The first condition is that a secure TAK protocol does indeed distribute a key of the correct form. The second condition ensures that this remains true even if *all* other oracles are corrupted. The last condition says that no adversary can obtain any information about the session key held by a fresh oracle.

With the description of our security model in hand, we now state:

Theorem 1. *Protocol TAK-1 is a secure TAK protocol, assuming that the adversary makes no **Reveal** queries, that the Bilinear Diffie-Hellman problem (for the pair of groups $\mathbb{G}_1$ and $\mathbb{G}_2$) is hard and provided that H is a random oracle.*

We prove this theorem in Appendix A.

We can also formally model the forward secrecy properties of TAK protocols. The model is largely as before, but now the interaction with E is slightly different. In addition to the usual **Send**, **Reveal** and **Corrupt** queries, we allow E to make one **Test** query of an oracle of her choice, say Π_A^s. We assume that Π_A^s has accepted and has participated in a matched session with oracles Π_B^t and

Π_C^u. Thus Π_A^s will have calculated a session key in common with Π_B^t, Π_C^u. We further assume that none of these three oracles has been the subject of a Reveal query. However, any or all of the oracles Π_A, Π_B, Π_C may be corrupted at any point in E's attack. In response to E's query, E is given the long-term private keys for Π_A^s, Π_B^t and Π_C^u (these oracles are now effectively corrupted). Adversary E is also given either the session key K held by Π_A^s, or a random k-bit string, the choice depending on a fair coin toss. The adversary's advantage, denoted $advantage^{E,fs}(k)$, is the probability that E can distinguish K from the random string. Again, the Test query need not be the adversary's last interaction with the model.

We say that a TAK protocol has perfect forward secrecy if, in the above game, $advantage^{E,fs}(k)$ is negligible.

Theorem 2. *Protocol TAK-1 has perfect forward secrecy, assuming that the adversary makes no* Reveal *queries, that the Bilinear Diffie-Hellman problem (for the pair of groups $\mathbb{G}_1$ and $\mathbb{G}_2$) is hard and provided that H is a random oracle.*

The proof of this theorem is given in Appendix A.

We comment on the significance of Theorems 1 and 2. We emphasise that our proofs of security do not allow the adversary to make Reveal queries of oracles. This means that our proof does not capture known session-key attacks. In fact protocol TAK-1 is vulnerable to a simple attack of this type. We describe the attack in full in Appendix B. The attack only works on TAK-1 because of the symmetry of the short-term components, and attacks of this type do not appear to apply to TAK-2, TAK-3 or TAK-4. The attack is analogous to known session key attacks well-understood for other protocols (see the comments following [9, Theorem 11] for an example).

In Section 8.1, we consider a confirmed version of Joux's protocol. This is obtained in the usual way, by adding three confirmation messages (which are piggy-backed on other messages in a non-broadcast environment), one for each protocol participant. The confirmation messages use encryptions and signatures; these could be replaced with MACs using keys derived from the short term values exchanged during the protocol run. We expect that the techniques used to prove the security of Protocol 2 of [9] could be adapted to prove that the described confirmed version of TAK-1 is indeed a secure AKC protocol.

.   Finally, the techniques used to prove Theorems 1 and 2 do not appear to extend to our other protocols. Nor has any security proof yet appeared for the MQV protocol (on which TAK-4 is based), although the MQV protocol is widely believed to be secure and has been standardised [2,3,27,28]. In any case, as we shall see in the next section, current security models do not handle all the reasonable attack types and so need to be augmented by *ad hoc* analysis.

6   Heuristic Security Analysis of TAK Protocols

We present a variety of attacks on the three TAK protocols that are not captured by the security models of the previous section. These are mostly inspired by earlier attacks on the two-party MTI protocols. Following this analysis, we summarise the security attributes of our TAK protocols in Table 1. In this section, we use E_A to indicate that the adversary E is impersonating A in sending or receiving messages intended for or originating from A. Similarly, $E_{B,C}$ denotes an adversary impersonating both B and C.

6.1   Key-Compromise Impersonation Attack on TAK-1

We present a key-compromise impersonation attack that occurs when E can obtain the long-term private key of one of the entities. Suppose E has obtained x, A's long-term private key. Then E can calculate $\hat{e}(P,P)^{xyz}$ using x and public data in B and C's certificates. Knowledge of this value now allows E to impersonate any entity engaging in a TAK-1 protocol run with A (and not just A).

These kinds of attacks do not appear to apply to TAK-3 or TAK-4 because of the combinations of long-term and short-term key components in computing K_{ABC} used in those protocols. However, Shim [43] presented 'partial key compromise impersonation attacks' on TAK-3 and TAK-4. These attacks use A's long-term key to impersonate C to B. However, A does not compute the same session key as B and C do, and the adversary does not learn A's session key. So this attack is only meaningful in the scenario where A does not then use its session key to communicate with B or C (but B and C use their version of the key to communicate with one another). In the attack, the adversary has A's long-term key and replaces A's short-term key in the protocol run. It is therefore not surprising that B will compute a key that the adversary can also compute. Indeed this attack is tantamount to a simple impersonation attack on A by the adversary.

In fact, no implicitly authenticated conference key agreement protocol, which broadcasts only certificates and short-term values, can prevent an adversary from mounting a 'partial key compromise impersonation attack' of the type described by Shim [43]. This is because preventing such an adversary is equivalent to preventing an adversary from impersonating an entity with knowledge of that entity's long-term and short-term keys. This is of course impossible. Given that not all the entities compute a common shared key in a partial key compromise attack, key confirmation (using MACs and the shared key, say) is sufficient to prevent this form of attack.

6.2   Forward Secrecy Weakness in TAK-3

As we saw in Theorem 2, TAK-1 has perfect forward secrecy. Protocol TAK-4 also appears to have this property because the key K_{ABC} agreed also includes the component $\hat{e}(P, P)^{abc}$. However, it is straightforward to see that if an adversary obtains two long-term private keys in TAK-3 then she has the ability to obtain old session keys (assuming she keeps a record of the public values aP, bP, cP). Thus TAK-3 does not enjoy forward secrecy. The protocol can be made into a perfectly forward secret protocol, at extra computational cost, by using the key $K_{ABC} \cdot \hat{e}(P, P)^{abc}$ in place of the key K_{ABC}.

6.3   Unknown Key-Share Attacks

A basic source substitution attack applies to all our TAK protocols. It utilises a potential registration weakness for public keys to create fraudulent certificates [37].

Adversary E registers A's public key μ_A as her own, creating $\text{Cert}_E = (\mathcal{I}_E \| \mu_A \| P \| \mathcal{S}_{\text{CA}}(\mathcal{I}_E \| \mu_A \| P))$. Then she intercepts A's message $aP \| \text{Cert}_A$ and replaces Cert_A with her own certificate Cert_E. Note that E registered the A's long-term public key xP as her own without knowing the value of x. Therefore she cannot learn the key K_{ABC}. However, B and C are fooled into thinking they have agreed a key with E, when in fact they have agreed a key with A. They will interpret any subsequent encrypted messages emanating from A as coming from E. This basic attack could be eliminated if the CA does not allow two entities to register the same long-term public key. However, this solution may not scale well to large or distributed systems. A better solution is discussed after highlighting another type of source substitution attack.

A second source substitution attack can be applied to TAK-3. Even if the CA does the previous check, the adversary can still obtain a Cert_E from the CA which contains a component μ_E which is a multiple of μ_A. The adversary can then alter the short-term keys in subsequent protocol messages by appropriate multiples. As with the last attack the adversary does not create the shared key. Rather, the attack gives her the ability to fool two participants B, C into believing messages came from her rather then from the third (honest) participant A. This attack is described in Appendix C.

The adversary in our attacks *does not* know her long term private key. Therefore, all these source substitution attacks are easily prevented if the CA insists that each registering party provides a proof of possession of his private key when registering a public key. This can be achieved using a variety of methods. For example, one might use zero-knowledge techniques or regard the pair (x, xP) as an ECDSA signature key pair and have the registering party sign a message of the CA's choice using this key pair. As an alternative, identities can be included in the key derivation function to prevent unknown key-share attacks.

6.4   Insider and Other Attacks

Certain new kinds of insider attack must be considered when we move to tripartite protocols. For example, an insider A might be able to fool B into believing that they have participated in a protocol run with C, when in fact C has not been active. An active A can do this easily in our TAK protocols, simply by choosing C's value cP and injecting it into the network along with C's certificate and stopping B's message from reaching C. This kind of attack can have serious consequences for tripartite protocols: for example, if C acts as an on-line escrow agent, then B believes a shared key has been properly escrowed with C when in fact it has not. On the other hand, when C acts as an off-line escrow agent, as in the protocol we describe in Section 7, this insider attack is only significant if C is providing an auditing function. This lack of auditability is actually beneficial if protocol participants want to have a deniability property, as in protocols like IKE, IKEv2 and JFK [26].

Attacks of this type can be prevented in a number of ways. Adding a confirmation phase, as we do in Section 8, prevents them. An alternative requires a complete protocol re-design, but maintains a one round broadcast protocol: simply use the long-term keys to sign short-term values (rather than combining them with short-term keys as in our TAK protocols) and agree the key $\hat{e}(P, P)^{abc}$. This approach requires each participant to maintain a log of all short-term values used, or to use synchronized clocks and time-stamps, to prevent an attacker simply replaying an old message. This requirement along with the need to create and verify signatures for each protocol run makes this solution somewhat unwieldy.

We note that Shim [43] has also found a rather theoretical 'known-key conspiracy attack' on TAK-2. This appears to be what we have called a known session key attack, but it requires two adversaries to conspire and to reveal three session keys. A similar attack can also be found on TAK-3, but is easily prevented if TAK-3 is augmented with a key derivation function.

In addition to the above attacks, we note that TAK-3 is vulnerable to a triangle attack of the type introduced by Burmester [15]. We do not include it here, as it is somewhat theoretical in nature.

6.5   Security Summary

Table 1 compares the security attributes that we believe our protocols TAK-1, TAK-2, TAK-3 and TAK-4 to possess. We have also included a comparison with the 'raw' Joux protocol.

Based on this table and the analysis in Section 5, we recommend the use of protocol TAK-4 or protocol TAK-3 along with pre-computation (in the event that the use of a hash function is not desirable). If perfect forward secrecy is also required, then TAK-3 can be modified as described in Section 6.2. Protocol TAK-4 has the additional benefit of being the most computationally efficient of all our protocols. Of course, robust certification is needed for all of our TAK protocols in order to avoid unknown key-share attacks.

Table 1. Comparison of security goals and attributes for one round tripartite key agreement protocols.

	Joux	TAK-1	TAK-2	TAK-3	TAK-4
Implicit key authentication	No	Yes	No	Yes	Yes
Known session key secure	No	No[i]	No	Yes[ii]	Yes
Perfect forward secrecy	n/a	Yes	No	No[iii]	Yes
Key-compromise impersonation secure	n/a	No	No	Yes[iv]	Yes[iv]
Unknown key-share secure	No	Yes[v]	Yes[vi]	Yes[vi]	Yes[v]

(i)   See Appendix B.

(ii) Only when a key derivation function is used; see Section 6.4.

(iii) No forward secrecy when two long-term private keys are compromised, but still has forward secrecy if only one is compromised.

(iv) Note, however, that Shim [43] has found a 'partial key compromise impersonation attack' on this scheme.

(v)   If the CA checks that public keys are only registered once, and if inconvenient use (vi).

(vi) If the CA verifies that each user is in possession of the long-term private key corresponding to his public key.

7   Tripartite Protocols with One Offline Party

As we mentioned in the introduction, there is an application of tripartite key exchange protocols to the two-party case when one of the parties acts as an escrow agent. It may be more convenient that this agent be off-line, meaning that he receives messages but is not required to send any messages. In this section, we adapt our earlier protocols to this situation. The protocol below is a modified version of TAK-4. We assume that C is the off-line party and that C's certificate Cert_C is pre-distributed, or is readily available to A and B.

Protocol messages:

$A \rightarrow B, C: \quad aP\|\text{Cert}_A \qquad (1)$

$B \rightarrow A, C: \quad bP\|\text{Cert}_B \qquad (2)$

Protocol 6 (Off-line TAK protocol).

Protocol description: The protocol is as in TAK-4, but without the participation of C. Entities A and B use C's long-term public key zP in place of his short-term public value cP when calculating the session key. Thus, the session key computations carried out by the entities is

$K_{ABC} = \hat{e}(P, P)^{(a+H(aP\|xP)x)(b+H(bP\|yP)y)(z)}$. Note that C can compute the key when required.

This protocol is resistant to all the previous attacks except the simple source substitution attack which is easily thwarted via robust registration procedures. It also has forward secrecy, obviously except when private key z is compromised.

Here z can be viewed as an independent master key, which can be regularly updated along with the corresponding certificate.

8 Non-Broadcast, Tripartite AKC Protocols

Up to this point, we have considered protocols that are efficient in the broadcast setting: they have all required the transmission of one broadcast message per participant. As we mentioned in the introduction, the number of broadcasts is not always the most relevant measure of a protocol's use of communications bandwidth. A good example is the basic broadcast Joux protocol, which offers neither authentication nor confirmation of keys and requires six passes in a non-broadcast network. In this section we introduce a pairing-based tripartite key agreement protocol that also requires six passes, but that offers both key confirmation and key authentication, i.e. is an AKC protocol. We show that *any* such protocol requires at least six passes. We then compare our protocol to a tripartite version of the station-to-station protocol [20].

8.1 A Six Pass Pairing-Based AKC Protocol

Our notation in describing our pairing-based tripartite, authenticated key agreement with key confirmation (TAKC) protocol is largely as before. Additionally, $\mathcal{S}_A(\sigma)$ denotes A's signature on the string σ. We assume now that the CA's certificate Cert_A contains A's signature verification key. Also $E_K(\sigma)$ denotes encryption of string σ using a symmetric algorithm and key K, and χ denotes the string $aP\|bP\|cP$.

Protocol messages:

$$A \to B: \quad aP\|\text{Cert}_A \tag{1}$$
$$B \to C: \quad aP\|\text{Cert}_A\|bP\|\text{Cert}_B \tag{2}$$
$$C \to A: \quad bP\|\text{Cert}_B\|cP\|\text{Cert}_C\|E_{K_{ABC}}(\mathcal{S}_C(\mathcal{I}_A\|\mathcal{I}_B\|\chi)) \tag{3}$$
$$A \to B: \quad cP\|\text{Cert}_C\|E_{K_{ABC}}(\mathcal{S}_C(\mathcal{I}_A\|\mathcal{I}_B\|\chi))\|E_{K_{ABC}}(\mathcal{S}_A(\mathcal{I}_B\|\mathcal{I}_C\|\chi)) \tag{4}$$
$$B \to C: \quad E_{K_{ABC}}(\mathcal{S}_A(\mathcal{I}_B\|\mathcal{I}_C\|\chi))\|E_{K_{ABC}}(\mathcal{S}_B(\mathcal{I}_A\|\mathcal{I}_C\|\chi)) \tag{5}$$
$$B \to A: \quad E_{K_{ABC}}(\mathcal{S}_B(\mathcal{I}_A\|\mathcal{I}_C\|\chi)) \tag{6}$$

Protocol 3 (TAKC protocol from pairings).

Protocol description: Entity A initiates the protocol execution with message (1). After receiving message (2), entity C is able to calculate the session key $K_{ABC} = \hat{e}(P, P)^{abc}$. The same session key is calculated after receiving messages (3) and (4), for A and B respectively. Messages (3) and onwards contain signatures on the short-term values and identities in the particular protocol run. This provides key authenticity. These signatures are transmitted in encrypted form using the session key K_{ABC} and this provides key confirmation. The confirmations from C to B, A to C and B to A are piggy-backed and forwarded by the intermediate party in messages (3), (4) and (5) respectively. More properly, encryptions should use a key derived from K_{ABC} rather than K_{ABC} itself. The symmetric encryptions can be replaced by appending MACs to the signatures with the usual safeguards.

If the expected recipients' identities were not included in the signatures this protocol would be vulnerable to an extension of an attack due to Lowe [35]. This attack exploits an authentication error and allows a limited form of unknown key-share attack. To perform it, we assume adversary D has control of the network. The attack is as follows.

1. D_C intercepts message (2), then D forwards (2) replacing Cert_B with Cert_D to C as if it originated from D. Thus, C assumes he is sharing a key with A and D.
2. D_A intercepts message (3) en route to A. Now D_C forwards this message replacing Cert_D with Cert_B to A.
3. Entity A receives (3) and continues with the protocol, sending message (4).
4. D blocks messages (5) and (6), so C and A assume an incomplete protocol run has occurred and terminate the protocol. However, on receipt of message (4), entity B already thinks he has completed a successful protocol run with A and C, whilst C might not even know B exists.

As usual in an unknown key-share attack, D cannot compute the shared key. The attack is limited because A and C end up with an aborted protocol run (rather than believing they have shared a key with B). The attack is defeated in our protocol because the inclusion of identities in signatures causes the protocol to terminate after message (3), when A realises that an illegal run has occurred.

We claim that no tripartite AKC protocol can use fewer than six passes. This can be reasoned as follows. Each of the three entities must receive two short-term keys to construct K_{ABC}, so a total of six short-term values must be received. But the first pass can contain only one short-term key (the one known by the sender in that pass), while subsequent passes can contain two. Thus a minimum of four passes are needed to distribute all the short-term values to all of the parties. So only after at least four passes is the last party (entity B in our protocol) capable of creating the key. This last party needs another two passes to provide key confirmation to the other two entities. So at least six passes are needed in total. Notice that this argument holds whether the network supports broadcasts or not.

8.2 A Six Pass Diffie-Hellman Based AKC Protocol

The station-to-station (STS) protocol is a three pass, two-party AKC protocol designed by Diffie, van Oorschot and Wiener [20] to defeat man-in-the-middle attacks. Here we extend the protocol to three parties and six passes, a pass-optimal protocol by the argument above.

An appropriate prime p and generator g mod p are selected. In Protocol 4 below, $a, b, c \in \mathbb{Z}_p^*$ are randomly generated short-term values and χ denotes the concatenation $g^a \| g^b \| g^c$. As before, $E_{K_{ABC}}(\cdot)$ donates symmetric encryption under session key K_{ABC}, and as before $\mathcal{S}_A(\cdot)$ denotes A's signature. Again, we assume that authentic versions of signature keys are available to the three participants. We have omitted modulo p operations for simplicity of presentation.

Sequence of protocol messages:

$$
\begin{align}
A \to B: &\quad g^a \| \mathrm{Cert}_A \tag{1} \\
B \to C: &\quad g^a \| \mathrm{Cert}_A \| g^b \| \mathrm{Cert}_B \| g^{ab} \tag{2} \\
C \to A: &\quad g^b \| \mathrm{Cert}_B \| g^c \| \mathrm{Cert}_C \| g^{bc} \| E_{K_{ABC}}(\mathcal{S}_C(\mathcal{I}_A \| \mathcal{I}_B \| \chi)) \tag{3} \\
A \to B: &\quad g^c \| \mathrm{Cert}_C \| g^{ac} \| E_{K_{ABC}}(\mathcal{S}_C(\mathcal{I}_A \| \mathcal{I}_B \| \chi)) \| E_{K_{ABC}}(\mathcal{S}_A(\mathcal{I}_B \| \mathcal{I}_C \| \chi)) \tag{4} \\
B \to C: &\quad E_{K_{ABC}}(\mathcal{S}_A(\mathcal{I}_B \| \mathcal{I}_C \| \chi)) \| E_{K_{ABC}}(\mathcal{S}_B(\mathcal{I}_A \| \mathcal{I}_C \| \chi)) \tag{5} \\
B \to A: &\quad E_{K_{ABC}}(\mathcal{S}_B(\mathcal{I}_A \| \mathcal{I}_C \| \chi)) \tag{6}
\end{align}
$$

Protocol 4 (TAKC protocol generalising STS protocol).

Protocol description: The protocol is similar to protocol 3 in operation, with an additional extra computation done before steps (2)(3) and (4). The shared session key is $K_{ABC} = g^{abc}$ mod p.

8.3   Analysis of AKC Protocols

Two immediate conclusions from our analysis can be drawn. Firstly, we have given a pairing-based, tripartite AKC protocol using just the same number of passes as are needed in Joux's protocol (but with the penalty of introducing message dependencies). Secondly, this AKC version of Joux's protocol is no more efficient in terms of passes than a 3-party version of the STS protocol! Thus when one considers confirmed protocols in a non-broadcast environment, the apparent advantage that Joux's protocol enjoys disappears. Of course, there is a two round broadcast version of the TAKC protocol (requiring 6 broadcasts and 12 passes). Both of our six pass AKC protocols can be done in 5 rounds in a broadcast environment.

9   Conclusions

We have taken Joux's one round tripartite key agreement protocol and used it to construct one round TAK protocols. We have considered security proofs and heuristic security analysis of our protocols, as well as an off-line version of our protocols. We have preserved the innate communications efficiency of Joux's protocol while enhancing its security functionality. We also considered tripartite variants of the STS protocol, suited to non-broadcast networks, showing that in this case, pairing-based protocols can offer no communication advantage over more traditional Diffie-Hellman style protocols.

Future work should consider the security of our protocols in more robust models, capturing a larger set of realistic attacks. Constructing multi-party key agreement protocols using our TAK protocols as a primitive might result in bandwidth-efficient protocols. Finally, it would be interesting to see if the methods of [14] could be emulated in the setting of pairings to produce TAK protocols secure in the standard model.

Acknowledgement. We would like to thank Chris Mitchell, Steven Galbraith and the anonymous referees for their comments on earlier versions of this paper.

References

1. S.S. Al-Riyami and K.G. Paterson. Authenticated three party key agreement protocols from pairings. Cryptology ePrint Archive, Report 2002/035, 2002. http://eprint.iacr.org/.
2. American National Standards Institute – ANSI X9.42. Public key cryptography for the financial services industry: Agreement of symmetric keys using discrete logarithm cryptography, 2001.
3. American National Standards Institute – ANSI X9.63. Public key cryptography for the financial services industry: Key agreement and key transport using elliptic curve cryptography, 2001.
4. R. Ankney, D. Johnson, and M. Matyas. The Unified Model – contribution to X9F1, October 1995.
5. P.S.L.M. Barreto, H.Y. Kim, B. Lynn, and M. Scott. Efficient algorithms for pairing-based cryptosystems. In *Advances in Cryptology – CRYPTO 2002*, volume 2442 of *LNCS*, pages 354–368. Springer-Verlag, 2002.
6. M. Bellare, D. Pointcheval, and P. Rogaway. Authenticated key exchange secure against dictionary attacks. In B. Preneel, editor, *Advances in Cryptology – EUROCRYPT 2000*, volume 1807 of *LNCS*, pages 139–155. Springer-Verlag, 2000.
7. M. Bellare and P. Rogaway. Entity authentication and key distribution. In D.R. Stinson, editor, *Advances in Cryptology – CRYPTO'93*, volume 773 of *LNCS*, pages 232–249. Springer-Verlag, 1994.
8. M. Bellare and P. Rogaway. Provably secure session key distribution: The three party case. In *Proceedings of the Twenty-Seventh Annual ACM Symposium on Theory of Computing STOC*, pages 57–66. ACM, 1995.
9. S. Blake-Wilson, D. Johnson, and A. Menezes. Key agreement protocols and their security analysis. In *Proceedings of the sixth IMA International Conference on Cryptography and Coding*, volume 1355 of *LNCS*, pages 30–45. Springer-Verlag, 1997.
10. S. Blake-Wilson and A. Menezes. Authenticated Diffie-Hellman key agreement protocols. In S. Tavares and H. Meijer, editors, *5th Annual Workshop on Selected Areas in Cryptography (SAC'98)*, volume 1556 of *LNCS*, pages 339–361. Springer-Verlag, 1998.
11. D. Boneh and M. Franklin. Identity-based encryption from the Weil pairing. In J. Kilian, editor, *Advances in Cryptology – CRYPTO 2001*, volume 2139 of *LNCS*, pages 213–229. Springer-Verlag, 2001.
12. D. Boneh and M. Franklin. Identity-based encryption from the Weil pairing. *SIAM J. Computing*, 32(3):586–615, 2003. http://www.crypto.stanford.edu/~dabo/abstracts/ibe.html, full version of [11].

13. C. Boyd. Towards extensional goals in authentication protocols. In *Proceedings of the 1997 DIMACS Workshop on Design and Formal Verification of Security Protocols*, 1997. http://www.citeseer.nj.nec.com/boyd97towards.html/.

14. E. Bresson, O. Chevassut, and D. Pointcheval. Dynamic group Diffie-Hellman key exchange under standard assumptions. In L.R. Knudsen, editor, *Advances in Cryptology – EUROCRYPT 2002*, volume 2332 of *LNCS*, pages 321–336. Springer-Verlag, 2002.

15. M. Burmester. On the risk of opening distributed keys. In Y. Desmedt, editor, *Advances in Cryptology – CRYPTO'94*, volume 839 of *LNCS*, pages 308–317. Springer-Verlag, 1994.

16. R. Canetti and H. Krawczyk. Analysis of key-exchange protocols and their use for building secure channels. In B. Pfitzmann, editor, *Advances in Cryptology – EUROCRYPT 2001*, volume 2045 of *LNCS*, pages 453–474. Springer-Verlag, 2001.

17. R. Canetti and H. Krawczyk. Universally composable notions of key exchange and secure channels. In L.R. Knudsen, editor, *Advances in Cryptology – EUROCRYPT 2002*, volume 2332 of *LNCS*, pages 337–351. Springer-Verlag, 2002.

18. J.C. Cha and J.H. Cheon. An identity-based signature from gap Diffie-Hellman groups. In Y. Desmedt, editor, *Public Key Cryptography – PKC 2003*, volume 2567 of *LNCS*, pages 18–30. Springer-Verlag, 2002.

19. W. Diffie and M. Hellman. New directions in cryptography. *IEEE Transactions on Information Theory*, IT-22(6):644–654, 1976.

20. W. Diffie, P.C. van Oorschot, and M. Wiener. Authentication and authenticated key exchanges. *Designs, Codes and Cryptography*, 2:107–125, 1992.

21. S.D. Galbraith. Supersingular curves in cryptography. In C. Boyd, editor, *Proceedings of AsiaCrypt 2001*, volume 2248 of *LNCS*, pages 495–513. Springer-Verlag, 2001.

22. S.D. Galbraith, K. Harrison, and D. Soldera. Implementing the Tate pairing. In *Algorithmic Number Theory 5th International Symposium, ANTS-V*, volume 2369 of *LNCS*, pages 324–337. Springer-Verlag, 2002.

23. S.D. Galbraith, H.J. Hopkins, and I.E. Shparlinski. Secure Bilinear Diffie-Hellman bits. Cryptology ePrint Archive, Report 2002/155, 2002. http://eprint.iacr.org/.

24. C. Gentry and A. Silverberg. Heirarchical ID-based cryptography. In Y. Zheng, editor, *Advances in Cryptology – ASIACRYPT 2002*, volume 2501 of *LNCS*, pages 548–566. Springer-Verlag, 2002.

25. F. Hess. Efficient identity based signature schemes based on pairings. In K. Nyberg and H. Heys, editors, *Selected Areas in Cryptography 9th Annual International Workshop, SAC 2002*, volume 2595 of *LNCS*, pages 310–324. Springer-Verlag, 2003.

26. P. Hoffman. Features of proposed successors to IKE. Internet Draft, draft-ietf-ipsec-soi-features-01.txt, 2002.

27. IEEE P1363. Standard specifications for public key cryptography, 2000. http://grouper.ieee.org/groups/1363/index.html.

28. ISO/IEC 15946-3. Information technology – security techniques – cryptographic techniques based on elliptic curves – part 3: Key establishment, awaiting publication.

29. A. Joux. A one round protocol for tripartite Diffie-Hellman. In W. Bosma, editor, *Proceedings of Algorithmic Number Theory Symposium – ANTS IV*, volume 1838 of *LNCS*, pages 385–394. Springer-Verlag, 2000.

30. B. Kaliski, Jr. An unknown key-share attack on the MQV key agreement protocol. *ACM Trans. on Information and Systems Security*, 4(3):275–288, 2001.

31. M. Kim and K. Kim. A new identification scheme based on the bi-linear Diffie-Hellman problem. In *Proc. ACISP 2002*, volume 2384 of *LNCS*, pages 362–378. Springer-Verlag, 2002.

32. L. Law, A. Menezes, M. Qu, J. Solinas, and S.A. Vanstone. An efficient protocol for authenticated key agreement. Technical Report CORR 98-05, Department of C & O, University of Waterloo, 1998. To appear in Designs, Codes and Cryptography.

33. L. Law, A. Menezes, M. Qu, J. Solinas, and S.A. Vanstone. An efficient protocol for authenticated key agreement. *Designs, Codes and Cryptography*, 28(2):119–134, 2003.

34. C. H. Lim and P. J. Lee. A key recovery attack on discrete log-based schemes using a prime order subgroup. In *Advances in Cryptology – CRYPTO'97*, volume 1294 of *LNCS*, pages 249–263. Springer-Verlag, 1997.

35. G. Lowe. Some new attacks upon security protocols. In *PCSFW: Proceedings of The 9th Computer Security Foundations Workshop*, pages 162–169. IEEE Computer Society Press, 1996.

36. T. Matsumoto, Y. Takashima, and H. Imai. On seeking smart public-key-distribution systems. *Trans. IECE of Japan*, E69:99–106, 1986.

37. A. Menezes, M. Qu, and S. Vanstone. Some new key agreement protocols providing mutual implicit authentications. *2nd Workshop on Selected Areas in Cryptography (SAC'95)*, pages 22–32, May 1995.

38. A. Menezes, P.C. van Oorschot, and S. Vanstone. *Handbook of Applied Cryptography*. CRC Press, Boca Raton, 1997.

39. C. Mitchell, M. Ward, and P. Wilson. Key control in key agreement protocols. *Electronics Letters*, 34:980–981, 1998.

40. K.G. Paterson. ID-based signatures from pairings on elliptic curves. *Electronics Letters*, 38(18):1025–1026, 2002.

41. A. Roscoe. Intensional specifications of security protocols. In *Proceedings 9th IEEE Computer Security Foundations Workshop*, pages 28–38. IEEE Computer Society Press, 1996.

42. R. Sakai, K. Ohgishi, and M. Kasahara. Cryptosystems based on pairing. In *The 2000 Symposium on Cryptography and Information Security*, Okinawa, Japan, January 2000.

43. K. Shim. Cryptanalysis of Al-Riyami-Paterson's authenticated three party key agreement protocols. Cryptology ePrint Archive, Report 2003/122, 2003. http://eprint.iacr.org/.

44. V. Shoup. On formal models for secure key exchange. IBM Technical Report RZ 3120, 1999. http://shoup.net/papers.

45. N.P. Smart. An identity based authenticated key agreement protocol based on the Weil pairing. *Electronics Letters*, 38(13):630–632, 2002.

Appendix A: Proof of Theorems

Proof of Theorem 1

We provide a proof of Theorem 1.

Conditions 1 and 2. Given the assumption that H is a random oracle, these conditions follow directly from the protocol description.

Condition 3. Suppose that $advantage^E(k) = n(k)$ is non-negligible. We show how to construct from E an algorithm F which solves the BDHP with non-negligible probability. We describe F's operation. F's input is a description of the groups $\mathbb{G}_1$, $\mathbb{G}_2$ and the map $\hat{e}$, a non-identity element $P \in \mathbb{G}_1$, and a triple of elements $x_A P, x_B P, x_C P \in \mathbb{G}_1$ with x_A, x_B, x_C chosen randomly from $\mathbb{Z}_q^*$. F's task is to compute and output the value $g^{x_A x_B x_C}$ where $g = \hat{e}(P, P)$.

F operates as follows. F chooses a triple $A, B, C \in \mathcal{I}$ uniformly at random. F simulates the running of the key generation algorithm $\mathcal{G}$, choosing all participants' long-term private keys randomly itself, and computing the corresponding long-term public keys and certificates, but with the exception of Π_A, Π_B and Π_C's keys. As public values for Π_A, Π_B and Π_C, F chooses the values $x_A P$, $x_B P$, $x_C P$ respectively. Then F starts adversary E.

In E's attack, F will simulate all the oracles Π_i, $i \in \mathcal{I}$. So F must answer all the oracle queries that E makes. F answers E's queries as follows. F simply answers E's distinct H queries at random, maintaining a table of queries and responses as he proceeds. Note that we do not allow our adversary to make Reveal queries, so F does not need to answer any queries of this type. F answers any Corrupt queries by revealing the long-term private key, except for Corrupt queries on Π_A, Π_B or Π_C. In the event of such queries, F aborts. F replies to Send queries in the usual way, with correctly formulated responses $a_{i,s} P \| Cert_{\Pi_i}$ for all oracles Π_i^s, where $a_{i,s} \in_R \mathbb{Z}_q^*$.

Finally, we consider how F responds to the Test query on oracle Π_i. F generates a random bit $b \in \{0, 1\}$. If $b = 0$, F should respond with the key held by Π_i, while if $b = 1$, F should respond with a random k-bit value. Now F is capable of answering the Test query correctly except when $b = 0$ and the tested oracle is an instance of Π_A, Π_B or Π_C. In this last case, F's response should be of the form $H(Q \| g^{x_A x_B x_C})$ where $Q \in \mathbb{G}_2$, involving the invocation of the random oracle. This use of H should be consistent with previous and future uses, but of course F does not know $g^{x_A x_B x_C}$, so cannot properly simulate the oracle in this case. Instead, F responds with a random k-bit value. This potentially introduces an imperfection into F's simulation, but we will argue below that this has no effect on success probabilities.

The final stage is as follows. Let $T_2(k)$ denote a polynomial bound on the number of H queries answered by F in the course of E's attack. F picks a value ℓ uniformly at random from $\{1, \ldots, T_2(k)\}$. Now F parses the ℓ-th H query into the form $Q \| W$ where $Q, W \in \mathbb{G}_2$. If this is not possible, F aborts. If it is, then F outputs W as its guess for the value $g^{x_A x_B x_C}$ and stops.

Now we must evaluate the probability that F's output is correct. Notice that E's view of F's simulation of the oracles is indistinguishable from E's view in a real attack provided that F is not forced to abort when asked a Corrupt query and that E does not detect that F's simulation of the random oracle was deficient when responding to the Test query – more on these situations later. Now E picks some accepted oracle $\Pi_{i_1}^s$ for its Test query in a real attack. Suppose that $\Pi_{i_1}^s$ has received two messages containing certificates of oracles Π_{i_2}

and Π_{i_3}. The session key held by oracle $\Pi_{i_1}^s$ will be of the form $H(Q\|g^{x_{i_1}x_{i_2}x_{i_3}})$ where $Q \in \mathbb{G}_2$ and x_{i_j} is the long-term private key of Π_{i_j}, $1 \leq j \leq 3$. Since by definition, the oracles Π_{i_j} are uncorrupted, and E does not ask any Reveal queries, if E is to succeed in distinguishing this session key from a random string with non-negligible probability $n(k)$, then E must have queried H on an input of the form $Q\|g^{x_{i_1}x_{i_2}x_{i_3}}$ at some point in its attack with some non-negligible probability $n'(k)$. The probability that this event occurs in F's simulation of the oracles is therefore also $n'(k)$. Since F outputs a random query from the list of $T_2(k)$ queries, has randomly distributed public keys $x_A P$, $x_B P$, $x_C P$ amongst the $T_1(k)$ participants, and is only deemed successful if he does not abort and his output is of the form $g^{x_A x_B x_C}$, we see that F is successful with probability better than:

$$\frac{n'(k)}{T_1(k)^3 T_2(k).}$$

However, this is still non-negligible in k.

We claim that our argument is not affected by the imperfection introduced into F's simulation when E asks a Corrupt query that F cannot answer: to be successful, E must ask a Test query of an uncorrupted but accepted oracle which has received messages containing certificates of two further uncorrupted oracles. This means that for E to be successful, at least three distinct, uncorrupted oracles must remain. So F, having made a random choice of where to place public keys $x_A P, x_B P, x_C P$, has at least a $\frac{1}{T_1(k)^3}$ chance of not facing an unanswerable Corrupt query whenever E is successful. This factor is already taken into account in our analysis.

One problem remains: what effect on E's behaviour does F's deviation in giving a random response to the "difficult" Test query have? In particular, what effect does it have on success probabilities? E's behaviour can only differ from that in a true attack run if E detects any inconsistency in F's simulation of the random oracle. In turn, this can only happen if at some point in the attack, E queries H on an input of the form $Q\|g^{x_A x_B x_C}$. For otherwise, no inconsistency arises. At this point, E's behaviour becomes undefined. (In this situation, E might guess that F's response to the Test query is a random key ($b = 1$) rather than the "correct" key ($b = 0$). But we must also consider the possibility that E simply might not terminate.) So we assume that F simply aborts his simulation whenever E's attack lasts longer than some polynomial bound $T_3(k)$ on the length of a normal attack. Notice that H has been queried on an input of the form $Q\|g^{x_A x_B x_C}$ at some point in F's simulation, and that up until this point, E's view is indistinguishable from that in a real attack. Thus the number of H queries made by E will still be bounded by $T_2(k)$ up to this point, and an input of the required type will occur amongst these. So F's usual guessing strategy will be successful with probability $1/T_2(k)$ even when E's behaviour is affected F's inability to correctly respond to the Test query. Since this is the same success probability for guessing in the situation where everything proceeds normally, it is now easy to see that F's overall success probability is still at least

$$\frac{n'(k)}{T_1(k)^3 T_2(k)}.$$

Proof of Theorem 2

Suppose that $advantage^{E,f^s}(k) = n(k)$ is non-negligible. We show how to construct from E an algorithm F which solves the BDHP with non-negligible probability. We describe F's operation. F's input is a description of the groups $\mathbb{G}_1$, $\mathbb{G}_2$ and the map $\hat{e}$, a non-identity element $P \in \mathbb{G}_1$, and a triple of elements $x_A P, x_B P, x_C P \in \mathbb{G}_1$ with x_A, x_B, x_C chosen randomly from $\mathbb{Z}_q^*$. F's task is to compute and output the value $g^{x_A x_B x_C}$ where $g = \hat{e}(P, P)$.

F operates as follows. F simulates the running of the key generation algorithm $\mathcal{G}$, choosing all participants' long-term private keys randomly itself, and computing the corresponding long-term public keys and certificates.

F also chooses a triple $A, B, C \in \mathcal{I}$ uniformly at random, and three positive integers s, t, u that are all bounded above by the number $T_3(k)$ of different sessions that E enters into across all the oracles. Then F starts adversary E.

F must answer all the oracle queries that E makes. F answers E's queries as follows. F simply answers E's distinct H queries at random, maintaining a table of queries and responses as he proceeds. Note that we do not allow our adversary to make **Reveal** queries, so F does not need to answer any queries of this type. F answers any **Corrupt** queries by revealing the long-term private key that it holds. F replies to **Send** queries in the usual way, with correctly formulated responses $a_{i,r} P \| Cert_{\Pi_i}$ for all oracles Π_i^r, where $a_{i,r} \in_R \mathbb{Z}_q^*$, except when queried for responses for oracles Π_A^s, Π_B^t and Π_C^u. In these special cases, F responds with $x_A P \| Cert_{\Pi_A}$, $x_B P \| Cert_{\Pi_B}$ and $x_C P \| Cert_{\Pi_C}$, respectively.

Finally, we consider how F responds to the **Test** query on oracle Π_i. F generates a random bit $b \in \{0, 1\}$. If $b = 0$, F should respond with the key held by Π_i, while if $b = 1$, F should respond with a random k-bit value. Now F is capable of answering the **Test** query correctly except in one special case: this is when $b = 0$, when the tested oracle is Π_A^s, Π_B^t or Π_C^u and when the tested oracle has participated in a matched session which comprises exactly these three oracles. In this last case, F's response should be of the form $H(g^{x_A x_B x_C} \| W)$ where $W \in \mathbb{G}_2$, but F cannot properly simulate the oracle in this case. Instead, F responds with a random k-bit value. This potentially introduces an imperfection into F's simulation, but this has no effect on success probabilities; this can be argued just as in the proof of Theorem 1.

Let $T_2(k)$ denote a polynomial bound on the number of H queries answered by F in the course of E's attack. F picks a value ℓ uniformly at random from $\{1, \ldots, T_2(k)\}$. Now F parses the ℓ-th H query into the form $Q \| W$ where $Q, W \in$

$\mathbb{G}_2$. If this is not possible, F aborts. If it is, then F outputs Q as its guess for the value $g^{x_A x_B x_C}$ and stops. Notice that E's view of F's simulation of the oracles is indistinguishable from E's view in a real attack, provided that E does not detect that F's simulation of the random oracle was deficient when responding to the Test query. Now E picks some accepted oracle $\Pi_{i_1}^r$ for its Test query in a real attack. Suppose that $\Pi_{i_1}^r$ has received two messages containing the short-term values of oracles Π_{i_2} and Π_{i_3}. The session key held by oracle $\Pi_{i_1}^r$ will be of the form $H(g^{x_{i_1} x_{i_2} x_{i_3}} \| W)$ where $W \in \mathbb{G}_2$ and x_{i_j} is the short-term private key of Π_{i_j}, $1 \le j \le 3$. Since E does not ask any Reveal queries, if E is to succeed in distinguishing this session key from a random string with non-negligible probability $n(k)$, then E must have queried H on an input of the form $g^{x_{i_1} x_{i_2} x_{i_3}} \| W$ at some point in its attack with some non-negligible probability $n'(k)$. The probability that this event occurs in F's simulation is therefore also $n'(k)$. Recall that F outputs a random query from the list of $T_2(k)$ queries, has randomly distributed values $x_A P$, $x_B P$, $x_C P$ as short-term keys amongst the $T_3(k)$ sessions, and is only deemed successful if his output is of the form $g^{x_A x_B x_C}$. Recall too that E only attacks oracles that have participated in matched sessions. Combining these facts, we see that F is successful with probability better than:

$$\frac{n'(k)}{T_3(k)^3 T_2(k)}.$$

However, this is still non-negligible in k.

Appendix B: Known Session Key Attack on TAK-1

We present a known session key attack on TAK-1 that makes use of session interleaving and message reflection. In the attack, E interleaves three sessions and reflects messages originating from A back to A in the different protocol runs. The result is that the session keys agreed in the three runs are identical, so E, upon revealing one of them, gets keys for two subsequent sessions as well. In what follows, E_A indicates that the E is impersonating A in sending or receiving messages intended for or originating from A. Similarly, $E_{B,C}$ denotes an adversary impersonating both B and C.

A is convinced to initiate three sessions with E:

   Session $\alpha : A \to E_{B,C} :$ $aP \| \text{Cert}_A$ (1^α)
   Session $\beta : A \to E_{B,C} :$ $a'P \| \text{Cert}_A$ (1^β)
   Session $\gamma : A \to E_{B,C} :$ $a''P \| \text{Cert}_A$ (1^γ)

E reflects and replays pretending to be B and C, to complete session α:

   $E_B \to A :$ $a'P \| \text{Cert}_B$ (2^α)
   $E_C \to A :$ $a''P \| \text{Cert}_C$ (3^α)

Similarly the second session is completed by $E_{B,C}$ sending $a''P\|\text{Cert}_B$ (2^β) and $aP\|\text{Cert}_C$ (3^β) to A. In the third parallel session she sends $aP\|\text{Cert}_B$ (2^γ) and $a'P\|\text{Cert}_C$ (3^γ) to A.

E now obtains the first session key $H(\hat{e}(P,P)^{aa'a''}\|\hat{e}(P,P)^{xyz})$. She then knows the keys for the next two sessions, as these are identical to this first session key.

Appendix C: Source Substitution Attack on TAK-3

We now present in detail the second source substitution attack on TAK-3.

1. A sends $aP\|\text{Cert}_A$ to $E_{B,C}$.
2. E computes $\mu_E = \delta x P$ and registers μ_E as part of her Cert_E.
3. E initiates a run of protocol TAK-3 by sending $aP\|\text{Cert}_E$ to B, C.
4. B sends $bP\|\text{Cert}_B$ to E, C; C sends $cP\|\text{Cert}_C$ to E, B.
5. B and C (following the protocol) compute
   $$K_{EBC} = K_{AE_{B,C}} = \hat{e}(P,P)^{(\delta xy)c+(\delta xz)b+(yz)a}.$$
6. E_B sends $\delta bP\|\text{Cert}_B$ to A.
7. E_C sends $\delta cP\|\text{Cert}_C$ to A.
8. A (following the protocol) computes a key
   $$K_{AE_{B,C}} = K_{AE_{B,C}} = \hat{e}(P,P)^{(\delta xy)c+(\delta xz)b+(yz)a} = K_{EBC}.$$
9. Now E, forwarding A's messages encrypted under key $K_{EBC} = K_{AE_{B,C}}$ to B and C, and fools them into believing that A's messages come from her.

This attack does not seem to apply to TAK-1 or TAK-4 because of the way in which long-term private key components are separated from the short-term components in K_{ABC} in TAK-1 and due to the use of a hash function in TAK-4.

Remote User Authentication Using Public Information*

Chris J. Mitchell

Mobile VCE Research Group, Information Security Group
Royal Holloway, University of London
Egham, Surrey TW20 0EX, UK
C.Mitchell@rhul.ac.uk

Abstract. A method for remote user authentication is proposed that requires only public information to be stored at the verifying host. Like the S/KEY scheme, the new technique uses only symmetric cryptography and is resistant to eavesdropping, but, unlike S/KEY, it is resistant to host impersonation attacks. The avoidance of asymmetric cryptographic techniques makes the scheme appropriate for low cost user authentication devices.

1   Introduction

Authentication of remote users is a problem commonly encountered in distributed computing environments. It is a problem addressed by the S/KEY user authentication system, details of which have been published as an Internet RFC, [2]. A complete software implementation of S/KEY has also been made publicly available (see [2]).

The S/KEY scheme, which is closely based on a scheme devised by Lamport, [6], has been designed to provide users with 'one-time passwords', which can be used to control user access to remote hosts. Of course, as with any such system, after the user authentication process is complete, i.e. after the one-time password has been sent across the network, no protection is offered against subversion of the link by third parties. This fact is pointed out in [2]. Indeed it is stated there that the S/KEY scheme 'does not protect a network eavesdropper from gaining access to private information, and does not provide protection against "inside" jobs or against active attacks where the potential intruder is able to intercept and modify the packet stream'.

It is further claimed that S/KEY 'is not vulnerable to eavesdropping/replay attacks'. Unfortunately, as has been pointed out by a number of authors, (see, for

* The work reported in this paper has formed part of the Software Based Systems area of the Core 2 Research Programme of the Virtual Centre of Excellence in Mobile & Personal Communications, Mobile VCE, **www.mobilevce.com**, whose funding support, including that of the EPSRC, is gratefully acknowledged. More detailed technical reports on this research are available to Industrial Members of Mobile VCE.

K.G. Paterson (Ed.): Cryptography and Coding 2003, LNCS 2898, pp. 360–369, 2003.
© Springer-Verlag Berlin Heidelberg 2003

example, [8] or Note 10.7 of [7]) depending on the definition of 'replay attack', the S/KEY scheme can fail to provide this property. Specifically, the S/KEY scheme is subject to a 'host impersonation' attack, where the false host can obtain information from the remote user which can be used to impersonate the remote user to the genuine host at some later occasion.

One major advantage of S/KEY over other user authentication schemes is that it only requires the verifying host to store public information about the remote user. Knowledge of this public information is not in itself sufficient to enable a third party to masquerade as the remote user, although if the remote user's secret key is poorly chosen then the public information would enable a brute force search to be performed. The purpose of this paper is to propose an alternative user authentication scheme which retains this major advantage of S/KEY, but which resists host impersonation attacks. It also has the practical implementation advantage that it uses only symmetric cryptography, an important issue if the remote user's secret key is stored in a device (e.g. a cheap smart card) with limited computational power.

2   The New Scheme

We suppose that the host H and remote user U have an initial secure session during which the U supplies H with trusted public information. We therefore divide our description into two phases, 'set up' and 'use'.

Prior to these phases, two system parameters t and r are selected, where t and r are positive integers satisfying $r < t$. The choice of these values affects the security of the scheme (see Section 3 below). A method for computing MACs (Message Authentication Codes) must also be agreed; this could be HMAC or a block cipher based CBC-MAC — see, for example, ISO/IEC 9797, [4,5]. Whatever method is chosen must be resistant to both key recovery and forgery attacks. In fact, we require resistance to a slightly generalised version of key recovery. Key recovery attacks normally involve an attacker using a number of (message,MAC) pairs to find the key used to generate the MACs. Here, we do not wish the attacker to be able to find *any* key which maps a given message to a given MAC, regardless of whether or not it was the actual key used. By choosing the algorithm and algorithm parameters carefully, it should be possible to achieve the desired attack resistance.

For the purposes of the discussion below we write $M_K(X)$ for the MAC computed on data X using secret key K.

2.1   Set up Phase

The remote user U chooses a set of t secret keys for the MAC algorithm, denotes by $K_1, K_2, \ldots, K_t$. U then chooses a random data string X and computes

$$V_i = M_{K_i}(X)$$

for every i $(1 \leq i \leq t)$. U then:

- passes the values $V_1, V_2, \ldots, V_t$ and X to H, and
- securely stores $K_1, K_2, \ldots, K_t$ and X.

H securely stores $V_1, V_2, \ldots, V_t$ and X as the public verification information for U. The integrity of this information must be preserved, but secrecy is not required.

2.2   Use of the Scheme

We now suppose that U wishes to authenticate him/herself to host H. The process operates in a series of steps, as follows. Note that if, at any point, one of the checks made by the receiver of a message fails, then, unless otherwise stated, the entire protocol run is deemed to have failed. This assumption applies throughout this paper.

1. H first sends X to U.
2. U first checks the correctness of X, in case of loss of synchronisation between U and H. If X is incorrect then, in certain circumstances, U may check the previous value of X to see if synchronisation can be restored (this possibility is discussed further in Section 3). U then chooses a new set of t secret keys: $K_1', K_2', \ldots, K_t'$ and selects a new random value X'. U also computes two sequences of t values:

$$V_i' = M_{K_i'}(X'), \ W_i' = M_{K_i}(V_1'\|V_2'\|\cdots\|V_t'), \ (1 \leq i \leq t)$$

   where here, as throughout, $\|$ denotes concatenation of data items.
   U now sends $(W_1', W_2', \ldots, W_t')$ to H.
3. H then chooses a random r-subset of $\{1, 2, \ldots, t\}$, say $\{i_1, i_2, \ldots, i_r\}$, and sends this subset to U.
4. U now sends the r secret keys $K_{i_1}, K_{i_2}, \ldots, K_{i_r}$ to H, as well as the set of t values $V_1', V_2', \ldots, V_t'$ and the value X'. U now replaces the stored values $X, K_1, K_2, \ldots, K_t$ with $X', K_1', K_2', \ldots, K_t'$. (In certain cases, U may retain the 'old' values, as discussed below).
5. H now verifies the r MAC values $V_{i_1}, V_{i_2}, \ldots, V_{i_r}$ using the set of r keys supplied by U and the stored value of X. If *all* these values are correct, then H also verifies the r MAC values $W_{i_1}', W_{i_2}', \ldots, W_{i_r}'$ using the set of r keys supplied by U and the values $V_1', V_2', \ldots, V_t'$ supplied by U previously. If all these MACs are also correct, then H accepts U as valid, and replaces $X, V_1, V_2, \ldots, V_t$ with $X', V_1', V_2', \ldots, V_t'$.

3   Discussion and Analysis

We now consider practical and security aspects of the scheme. Observe that elements of the scheme are very similar to the one time signature (OTS) scheme of Rabin [9] (see also section 11.6 of [7]). Indeed, one way of looking at the scheme above is to regard the set up phase as key generation for an OTS scheme, as a result of which the host is equipped with the public key. One iteration of the protocol consists of the user generating a new OTS key pair, signing the new public key with the existing OTS private key, and the host then performing a verification process. On completion the host has a copy of the new OTS public key, ready to start the process again.

The scheme also has features in common with the 'Guy Fawkes' protocol of Anderson et al. [1]. However, the scheme nevertheless has properties distinct from previously specified protocols.

3.1   Choices for t and r

We first consider how t and r should be chosen. To avoid certain attacks, we wish to choose these values so that the probability of a third party successfully guessing the subset $\{i_1, i_2, \ldots, i_r\}$ in advance is negligible. That is we wish to arrange things so that $1/\binom{t}{r}$ is negligible.

Given that we wish to minimise t (to minimise the storage and bandwidth requirements) then this probability is minimised by choosing $r = \lfloor t/2 \rfloor$, since $\binom{t}{\lfloor t/2 \rfloor} \geq \binom{t}{i}$ for all i. Also, since $\sum_{i=0}^{t} \binom{t}{i} = 2^t$, we immediately have that $\binom{t}{\lfloor t/2 \rfloor} > 2^t/(t+1)$ if $t > 1$.

Hence, if we wish to guarantee that the probability of successfully guessing the subset is at most 10^{-9} say, then choosing $t \geq 35$ will suffice.

3.2   Host Impersonation Attacks

Suppose a third party, E say, wishes to impersonate H to U with a view to learning enough to impersonate U to H at some subsequent time. In step 3 of the protocol, E can only choose a random r-subset of $\{1, 2, \ldots, t\}$, and E will then learn a set of r of the secret keys. However, at a later stage, when E impersonates U to H, E will only be successful if he/she knows all the keys in the subset chosen by H. The odds of this will be acceptably small as long as t and r are chosen appropriately (see the discussion in Section 3.1).

3.3   Man in the Middle Attacks

Of course, as with any authentication protocol, it will always be possible for a third party E to simply sit between U and H in the communications channel, and relay messages between the two. This only becomes a genuine threat if E is able to change some part of the messages, and/or to re-order them in some way. We now look at the various messages in turn, to see if this is possible.

- In step 2, E could change some or all of the MAC values W_i'. However, given that at this stage E will not know any of the keys K_i, the probability that the modified values will be correct is negligibly small (since we assume that forgery attacks are not feasible).

- In step 3, E could change the subset, but then the set of keys returned in step 4 will not be correct.
- In step 4, E could modify some or all of the secret keys K_{i_j} and/or some or all of the MAC values V_i'. Successfully changing the values V_i' would require knowledge of the keys K_i', but none of these have yet been divulged by U. Changing the secret keys K_{i_j} is prevented by the fact that H can check them using the values V_{i_j}. (Changing these verification MACs would have required knowledge of the previous set of keys, and changing these previous keys would have required changing the previous verification MACs, and so on).

3.4   Denial of Service Attacks

There is, of course, a simple and effective 'denial of service' attack against the described protocol. A third party E can simply engage in the protocol with U by impersonating H. When U tries to authenticate him/herself to the genuine H, U will have a different value of X to that sent by H in the first step of the protocol.

There are two main ways in which this can be dealt with. Firstly, U could simply abandon the attempt to authenticate to H, and arrange for the system to be re-initialised. Secondly, U could retain 'old' values of X (along with the associated set of keys) and use them to complete the authentication protocol. However, such a process has very serious dangers, depending on the choices of t and r.

With r set to $\lfloor t/2 \rfloor$, even doing the process twice would completely destroy the system security. A malicious third party E could impersonate H to U twice, using two disjoint r-subsets of $\{1, 2, \ldots, t\}$. This would mean that E would obtain all of the keys $K_1, K_2, \ldots, K_t$ (or all but one of them if t is odd). As a result, E would be able to impersonate U to H any number of times.

Hence if we are to allow the same key set to be used more than once then r and t need to be chosen appropriately. Also, U needs to implement a counter to limit the number of times any particular key set is used for the authentication process. The limit for this counter will be determined by the choices for t and r. This issue is discussed further in the next section.

3.5   Resynchronisation

As we have just seen, one way of limiting the impact of denial of service attacks by malicious third parties impersonating the host, is to allow a key set to be used more than once. This may also be necessary if the authentication process between user and host fails part way through, e.g. because of a communications failure.

If a key set is to be used up to a maximum of c times (this being enforced by the counter held by U) then it should be the case then any party with knowledge of c different random r-subsets of the set of t keys $K_1, K_2, \ldots, K_t$ should have a negligible probability of knowing all the members of another randomly chosen r-subset of keys.

To compute the necessary probabilities we make some simplifying assumptions (pessimistic from the point of view of the legitimate system users). We suppose that, by bad luck or by host impersonation, all the c different r-subsets are mutually disjoint. Thus we require that the probability that a randomly chosen r-subset of $\{1, 2, \ldots, t\}$ does not contain any element from a specified subset of size $t - cr$ shall be small. The following result, the proof of which is elementary, is therefore relevant.

Lemma 1. *Suppose c, r and t are positive integers satisfying $r(c + 1) < t$. If S is a subset of $\{1, 2, \ldots, t\}$ of size cr, then the probability that R, a randomly chosen r-subset of $\{1, 2, \ldots, t\}$, is a subset of S is equal to*

$$\frac{\binom{cr}{r}}{\binom{t}{r}}.$$

We thus require that c, r and t should be chosen so that

$$\frac{cr(cr - 1) \ldots (cr - r + 1)}{t(t - 1) \cdots (t - r + 1)},$$

which is bounded above by $(cr/t)^r$, is small. As an example we can put $r = 32$ and $t = 64c$, and we are then guaranteed that the probability of a successful attack is less than 2^{-32}.

4   A Streamlined Version of the Protocol

In the protocol presented above, steps 1 and 2 can be merged with steps 3 and 4 respectively, to give a two-pass protocol. This is at the cost of increasing long-term storage requirements. The modified protocol operates as follows.

4.1   Set up

The remote user U chooses two sets of t secret keys for the MAC algorithm, the *current set*, denoted by $K_1, K_2, \ldots, K_t$, and the *pending set*, denoted by $K_1', K_2', \ldots, K_t'$. U chooses two random data strings used as *key set indicators*, denoted by X and X' (for the current and pending key sets).

U now computes verification MACs for both the current and pending key sets as

$$V_i = M_{K_i}(X) \quad \text{and} \quad V_i' = M_{K_i'}(X')$$

for every i ($1 \leq i \leq t$). U also computes a further set of t MACs

$$W_i' = M_{K_i}(V_1' || V_2' || \cdots || V_t'), \quad (1 \leq i \leq t).$$

U then:

- passes the two sets of verification values and the corresponding key set indicators $(V_1, V_2, \ldots, V_t, X)$ and $(V_1', V_2', \ldots, V_t', X')$ to H,
- passes the t MACs $(W_1', W_2', \ldots, W_t')$ to H, and
- securely stores the two key sets with their respective indicators, i.e. stores $(K_1, K_2, \ldots, K_t, X)$ and $(K_1', K_2', \ldots, K_t', X')$.

H securely stores the information received from U. The integrity of this information must be preserved, but secrecy is not required.

4.2 Use of the Scheme

Suppose that U wishes to authenticate him/herself to host H. The process operates as follows.

1. H chooses a random r-subset of $\{1, 2, \ldots, t\}$, say $\{i_1, i_2, \ldots, i_r\}$, and sends this subset to U along with the current key set indicator X.

2. U first checks the correctness of X, in case of loss of synchronisation between U and H. If X is incorrect then, in certain circumstances, U may check the previous value of X to see if synchronisation can be restored (as discussed in Section 3).

   U then chooses a new set of t secret keys: $K_1'', K_2'', \ldots, K_t''$ and selects a new random key set indicator X''. U also computes two sequences of t values:

$$V_i'' = M_{K_i''}(X'), \quad W_i'' = M_{K_i'}(V_1'' || V_2'' || \cdots || V_t''), \quad (1 \le i \le t).$$

   U now sends X'', $(V_1'', V_2'', \ldots, V_t'')$ and $(W_1'', W_2'', \ldots, W_t'')$ to H. U also sends the r secret keys $K_{i_1}, K_{i_2}, \ldots, K_{i_r}$ to H. U now replaces the stored values of

   - $X, K_1, K_2, \ldots, K_t$ with $X', K_1', K_2', \ldots, K_t'$, and
   - $X', K_1', K_2', \ldots, K_t'$ with $X'', K_1'', K_2'', \ldots, K_t''$.

3. H now verifies the r MAC values $V_{i_1}, V_{i_2}, \ldots, V_{i_r}$ using the set of r keys supplied by U and the stored value of X. If *all* these values are correct, then H also verifies the r MAC values $W_{i_1}', W_{i_2}', \ldots, W_{i_r}'$ using the set of r keys supplied by U and the stored values $V_1', V_2', \ldots, V_t'$. If all these MACs are also correct, then H accepts U as valid, and replaces:

   - $X, V_1, V_2, \ldots, V_t$ with $X', V_1', V_2', \ldots, V_t'$,
   - $X', V_1', V_2', \ldots, V_t'$ with $X'', V_1'', V_2'', \ldots, V_t''$, and
   - $W_1', W_2', \ldots, W_t'$ with $W_1'', W_2'', \ldots, W_t''$.

5   Implementation Issues

We now consider certain practical implementation issues for the protocol.

5.1 Complexity

We start by considering the storage, computation and communications complexity of the scheme as described in Section 2.

- *Storage:* the requirements for the host are to store t MACs and a random value; the requirements for the user are to store t keys. During execution of the protocol, the remote user and host must both store a further $2t$ MACs, and the user must also temporarily store an additional t keys. Note that, for the streamlined scheme of Section 4, the long term storage requirements for host and user increase to $3t$ MACs and $2t$ secret keys respectively. Note also that, if the user retains 'old' key sets for resynchronisation purposes, then this will be at a cost of t keys, a random value and a usage counter for each retained old set.
- *Computation:* the host verifies $2r$ MACs and chooses one random r-subset of $\{1, 2, \ldots, t\}$, and the user generates $2t$ MACs.
- *Communications:* the user sends the host a total of $2t$ MACs, one random value and r secret keys, and the host sends the user one r-subset of $\{1, 2, \ldots, t\}$.

To see what this might mean in practice, suppose we wish to use the basic scheme (of Section 2) in such a way that a particular key set can be used up to $c = 3$ times, and that the user retains one 'old' key set for resynchronisation purposes. We thus choose $r = 32$ and $t = 196$. Suppose that the MAC in use is HMAC based on SHA-1 (see [5] and [3]) with MACs and keys of 160 bits each; suppose also that X contains 128 bits. Then the user must store $2t$ keys, two random values and a counter (which we ignore since it will take negligible space) — this amounts to $392 \times 20 + 32$ bytes, i.e. just under 8 kbytes, with an additional 12 kbytes of short term storage needed during protocol execution). The host must store approximately 4 kbytes of MACs, with an additional 8 kbytes of short term storage needed during protocol execution. During use of the protocol the user will need to compute 392 MACs and the host will need to compute 64 MACs. The total data to be exchanged between host and user during the protocol amounts to around 8.5 kbytes.

Note that the values X do not need to be random or unpredictable - U could generate the values using a modest-sized counter (e.g. of 4 bytes). This would save a small amount of communications and storage costs. It is also tempting to try and reduce the MAC lengths. However, significant length reductions are not possible since, as noted in Section 2, we do not wish the attacker to be able to find *any* key which maps a given message to a given MAC, regardless of whether or not it was the actual key used. This implies that MACs need to be of length close to that assumed immediately above, although a reasonable saving achievable by reducing MACs to, say, 10 bytes is not infeasible. Whilst such modifications will not significantly reduce user storage requirements, the communications requirements are almost halved, as are the host storage requirements.

5.2   A Security Improvement

In cases where $c > 1$, i.e. where key sets may be used more than once, a malicious entity impersonating the host is free to choose the subsets $\{1, 2, \ldots, r\}$ disjointly so as to learn the maximum number of secret keys. This maximises the (small) probability this malicious entity will have of impersonating the user to the genuine host (see Section 3.5). To avoid this, i.e. to increase the difficulty of launching a host impersonation attack, we can modify the protocol so that neither the remote user nor the host chooses the r-subset $\{i_1, i_2, \ldots, i_r\}$ of $\{1, 2, \ldots, t\}$. This can be achieved by prefixing the protocol of Section 4 with two additional messages, and also making use of an appropriate one-way, collision-resistant hash-function (see, for example, [3]). Note that these two additional messages can be merged with the first two messages of the basic protocol of Section 2.

The revised protocol from Section 4.2 starts as follows:

1. H chooses a random value r_H, of length comparable to the key length in use, computes $h(r_H)$ (where h is a pre-agreed hash-function), and sends $h(r_H)$ to U.
2. U chooses a random value r_U, of length the same as r_H, and sends it to H.
3. H computes $h(r_H \| r_U)$ and uses this hash-code to seed a pseudo-random number generator (PRNG) of appropriate characteristics to generate an r-subset $\{i_1, i_2, \ldots, i_r\}$ of $\{1, 2, \ldots, t\}$ — this PRNG could, for example, be based on h. H now sends r_H and X to U.
4. U first checks r_H using the value $h(r_H)$ sent previously. U now recomputes the r-subset $\{i_1, i_2, \ldots, i_r\}$, and continues as in step 2 of the scheme from Section 4.2.

Note that, by sending $h(r_H)$ in the first step, H commits to the random value r_H without revealing it. This prevents either party learning the other party's random value before choosing their own. This, in turn, prevents either party choosing even a small part of the r-subset. Note also that, although this scheme lengthens the protocol, it also slightly reduces the communications complexity, since the r-subset no longer needs to be transferred.

6   Summary and Conclusions

A novel unilateral authentication protocol has been presented, which uses symmetric cryptography and only requires public information to be stored at the verifying host. The computational and storage requirements are non-trivial, but may still be potentially attractive to designers of low-cost remote user authentication devices who wish to avoid the complexity of implementing digital signatures.

References

1. R.J. Anderson, F. Bergadano, B. Crispo, J.-H. Lee, C. Manifavas, and R.M. Needham. A new family of authentication protocols. *ACM Operating Systems Review*, **32**(4):9–20, 1998.

2. N. Haller. *The S/KEY one-time password system*. Bellcore, February 1995. Internet RFC 1760.

3. International Organization for Standardization, Genève, Switzerland. *ISO/IEC 10118-3, Information technology — Security techniques — Hash-functions — Part 3: Dedicated hash-functions*, 1998.

4. International Organization for Standardization, Genève, Switzerland. *ISO/IEC 9797-1, Information technology — Security techniques — Message Authentication Codes (MACs) — Part 1: Mechanisms using a block cipher*, 1999.

5. International Organization for Standardization, Genève, Switzerland. *ISO/IEC 9797-2, Information technology — Security techniques — Message Authentication Codes (MACs) — Part 2: Mechanisms using a hash-function*, 2000.

6. L. Lamport. Password authentication with insecure communication. *Communications of the ACM*, **24**:770–772, 1981.

7. A.J. Menezes, P.C. van Oorschot, and S.A. Vanstone. *Handbook of Applied Cryptography*. CRC Press, Boca Raton, 1997.

8. C.J. Mitchell and L. Chen. Comments on the S/KEY user authentication scheme. *ACM Operating Systems Review*, **30**(4):12–16, October 1996.

9. M.O. Rabin. Digitalized signatures. In R. DeMillo, D. Dobkin, A. Jones, and R. Lipton, editors, *Foundations of Secure Computation*, pages 155–168. Academic Press, 1978.

Mental Poker Revisited

Adam Barnett and Nigel P. Smart

Department of Computer Science,
University of Bristol,
Merchant Venturers Building,
Woodland Road,
Bristol, BS8 1UB,
United Kingdom.
{ab9660,nigel}@cs.bris.ac.uk

Abstract. We discuss how to implement a secure card game without the need for a trusted dealer, a problem often denoted "Mental Poker" in the literature. Our solution requires a broadcast channel between all players and the number of bits needed to represent each card is independent of the number of players. Traditional solutions to "Mental Poker" require a linear relation between the number of players and the number of bits required to represent each card.

1 Introduction

The notion of playing card games "via telephone", or "Mental Poker" as it is sometimes called, has been historically important in the development of cryptography. Mental Poker protocols enable a group of mutually mistrusting players to play cards electronically without the need for trusted dealers. Such a set of protocols should allow the normal card operations (e.g. shuffling, dealing, hiding of card information from players) to be conducted in a way which allows cheaters to be detected.

The original scheme for this problem was developed by Shamir, Rivest and Adleman [15] not long after the publication of the RSA algorithm itself. However, this was soon noticed to be insecure since the naive RSA function leaks information about individual bits. This observation led Goldwasser and Micali [10,11] to develop the notion of semantic security and the need for probabilistic public key encryption. Thus Mental Poker led to the current subject of provable security and the current definitions of what it means for a public key encryption scheme to be secure.

After the work of RSA and Goldwasser and Micali there has been other work on how to conduct two-player card games over a public network with no trusted centre, see [1,6,7,9,12,17]. A full protocol suite for an arbitrary card game with an arbitrary number of players and with no trusted centre is described by Schindelhauer [16]. However, all of these schemes make use of the Goldwasser–Micali probabilistic encryption scheme [11] based on the quadratic residuosity problem, but the Goldwasser-Micali system is very inefficient since it requires a full RSA-style block to encrypt a single bit of information.

K.G. Paterson (Ed.): Cryptography and Coding 2003, LNCS 2898, pp. 370–383, 2003.

As a result the above schemes require

$$l \times \lceil \log_2 M \rceil \times k$$

bits to represent each card where

- l is the number of players in the game.
- M is the number of different cards, usually $M = 52$.
- k is the number of bits in a secure RSA modulus.

Our protocols will give message sizes which are independent of the number of players. In addition our protocol is essentially independent of the number of different cards in the game, at least for any game which one could imagine.

We give two instantiations of our scheme, one based on discrete logarithms and one based on a factoring assumption (actually Paillier's system [13]). We therefore require only k bits to represent each card, where using current security recommendations,

- $k = 322$ if a 160-bit elliptic curve is chosen for the discrete logarithm based variant, and point compression is used.
- $k = 2048$ if a discrete logarithm based variant is used in a finite field of order $\approx 2^{1024}$.
- $k = 2048$ if the Paillier variant is used with an RSA modulus of 1024 bits.

However unlike the system described in [16] for our factoring based protocol we assume that players may not join an existing game, however they may leave. For both discrete logarithm based versions players may both leave and join a game as it proceeds.

The paper is structured as follows. Firstly we define some notions related to proofs of knowledge which we shall require. Then we introduce the notion of a Verifiable l-out-of-l Threshold Masking Function, or VTMF, which is the basic cryptographic operation which we will apply repeatedly to our cards. We then go on to show how the protocols to implement the card game are implemented via a VTMF. Then we give descriptions of two VTMF's, one based on discrete logarithms and one based on Paillier's assumption [13].

We end this introduction by noting that in some sense Mental Poker makes no sense in a model which allows collusion. For example one could use the following protocol suite to create an electronic Bridge game. However, two partners, say East and West, could contact each other by some other means, i.e. not using the broadcast channel, and exchange information which would enable them to have an advantage over North and South. No Mental Poker suite of protocols could detect such cheating. However, the protocols which follow ensure that the colluding parties obtain no more information than if they had colluded in a real game.

2   Proofs of Knowledge

In the following sections we will make reference to various properties of proofs of knowledge. Much of this introductory section summarizes the work of Cramer, Damgård and Schoenmakers [5].

Consider a binary relation $R = \{(x, w)\}$ for which membership can be tested in polynomial time. For any x we say that $w(x)$ is the set of witnesses such that $(x, w) \in R$. A proof of knowledge protocol $\mathcal{P}$ is a two party protocol between a prover P and a verifier V. The prover and verifier are given a common input x and the prover has a private input w. The prover's aim is to convince the verifier that $w \in w(x)$ without revealing what w actually is.

Following Cramer et. al. we restrict to the following subset of such protocols. We assume that the protocol is a three round public coin protocol in that the protocol is an ordered triple

$$m_1, c, m_2$$

where m_1 is called the commitment and comes from the prover, c is a random challenge chosen by the verifier and m_2 is the provers final response.

We assume the following three properties:

1. $\mathcal{P}$ is complete.
   If $w \in w(x)$ then the verifier will accept with probability one.
2. $\mathcal{P}$ has the "special soundness" property.
   For any prover $\mathcal{P}$ given two conversations between P and V with message flows

   $$(m_1, c, m_2) \text{ and } (m_1, c', m_2')$$

   with $c \neq c'$ then an element of $w(x)$ can be computed in polynomial time.
3. $\mathcal{P}$ is honest verifier zero-knowledge.
   There is a simulation $\mathcal{S}$ of $\mathcal{P}$ that on input x produces triples which are indistinguishable from genuine triples between an honest prover and an honest verifier.

To fix ideas consider the Schnorr identification scheme which identifies users on proof of knowledge of a discrete logarithm $h = g^x$. We have

$$m_1 = g^k, m_2 = k + x \cdot c$$

for some random k chosen by P and some random c, chosen by V after P has published m_1. The verifier checks that

$$m_1 = g^{m_2} \cdot h^{-c}.$$

This satisfies all the properties above. Indeed it is the "special soundness" property which allows Pointcheval and Stern [14] to show that the associated signature scheme derived via the Fiat–Shamir paradigm is secure.

Cramer et. al. use the above three round honest verifier proofs of knowledge with the special soundness property to create proofs of knowledge of elements for arbitrary access structures. For example if P the discrete logarithm x_i of one of $h_1 = g^{x_1}, h_2 = g^{x_2}, h_3 = g^{x_3}$. Using the protocols of [5], P can convince V that they know one of the discrete logarithms without revealing which one.

We shall use such protocols to allow players to show that an encryption of a card is an encryption of a card from a given set, or that an encrypted card comes from the set of hearts, or that a player has no hearts left in their hand, etc. etc.

3   Verifiable l-out-of-l Threshold Masking Functions

For our later protocols we will require a set of cryptographic protocols, which we call a Verifiable l-out-of-l Threshold Masking Function, or VTMF for short. In a later section we shall give two examples of such a function, one based on a discrete logarithm assumption and one based on a factoring assumption.

In keeping with the notation in the rest of the paper we shall assume there are l players and there are M values which are to be encrypted (or masked). A VTMF is a set of protocols, to be described in detail below, which produces a semantically secure encryption function (under passive adversaries) from a space $\mathcal{M}$ to a space $\mathcal{C}$. We shall assume that there is a natural encoding

$$\{1, \ldots, M\} \longrightarrow \mathcal{M}$$

which allows us to refer to messages and card values as the same thing. In addition there is a nonce-space $\mathcal{R}$ which is sufficiently large and from which nonces are chosen uniformly at random.

A VTMF consists of the following four protocols:

3.1   Key Generation Protocol

This is a multi-party protocol between the l parties which generates a single public key h. Each player will also generate a secret x_i and a public commitment to this share, denoted h_i. The shares x_i are shares of the unknown private key, x, corresponding to h.

3.2   Verifiable Masking Protocol

A masking function is an encryption function

$$\mathcal{E}_h : \begin{cases} \mathcal{M} \times \mathcal{R} \longrightarrow \mathcal{C} \\ (m, r) \longmapsto \mathcal{E}_h(m; r), \end{cases}$$

and an associated decryption function

$$\mathcal{D}_h : \begin{cases} \mathcal{C} \longrightarrow \mathcal{M} \\ \mathcal{E}_h(m; r) \longmapsto m, \end{cases}$$

with respect to the public key h, we abuse notation and equate knowledge of the private key x with knowledge of the function $\mathcal{D}_h$. In addition there is a honest-verifier zero-knowledge protocol to allow the masking player to verify to any other player that the given ciphertext $\mathcal{E}_h(m; r)$ is an encryption of the message m. Since we will use this protocol in a non-interactive manner using the Fiat-Shamir transform the fact it is only honest-verifier will be no problem, as we will achieve security in the random oracle model. We insist that the encryption function is semantically secure under passive attacks. However, we cannot achieve semantic security under active attacks since we also require the ability to re-mask a message, as we shall now explain.

3.3   Verifiable Re-masking Protocol

Let $\mathcal{C}_m$ denote the set of all possible encryptions of a message m under the masking function above. Given an element c of $\mathcal{C}_m$ there is a function which re-encrypts c to give another representative in $\mathcal{C}_m$, and which can be applied without knowledge of the underlying plaintext m and only knowledge of the public key h. Hence, this function can be applied either by the player who originally masked the card or by any other player. We shall denote this function by $\mathcal{E}'$,

$$\mathcal{E}'_h : \begin{cases} \mathcal{C}_m \times \mathcal{R} \longrightarrow \mathcal{C}_m \\ (c, r) \longmapsto \mathcal{E}'_h(c; r). \end{cases}$$

Again we also insist that there is a honest-verifier zero-knowledge protocol to allow the player conducting the re-masking to verify to any other player that the given ciphertext $\mathcal{E}'_h(c; r)$ is an encryption of the message m, without either player needing to know the underlying plaintext message m.

We insist that if $r \in \mathcal{R}$ is chosen uniformly at random then $\mathcal{E}'_h(c; r)$ is also uniformly distributed over $\mathcal{C}_m$. In addition we also insist that if a user knows r_1 and r_2 such that

$$c_1 = \mathcal{E}'_h(c; r_1) \text{ and } c_2 = \mathcal{E}'_h(c_1, r_2)$$

then they can compute r such that

$$c_2 = \mathcal{E}'_h(c; r).$$

This last property is needed so that certain proofs of knowledge can be executed.

3.4   Verifiable Decryption Protocol

Given a ciphertext $c \in \mathcal{C}$ this is a protocol in which each player generates a value $d_i = \mathcal{D}(c, x_i)$ and an honest-verifier zero-knowledge proof that the value d_i is consistent with both c and the original commitment h_i.

We assume there is a public function

$$\mathcal{D}'(c, d_1, \ldots, d_l) = m$$

which decrypts the ciphertext c given the values $d_1, \ldots, d_l$. This function should be an l-out-of-l threshold scheme in that no subset of $\{d_1, \ldots, d_l\}$ should be able to determine any partial information about m.

4   The Card Protocols

For the purpose of this paper, we assume an authentic broadcast channel between all parties. This can be achieved easily in practice using digital signatures and a passive bulletin board to hold the player communications. There are two types of operations; operations on a single card and operations on a deck (or set) of cards.

4.1   Operations on a Card

We assume that the players have executed the key set up phase for an l-out-of-l VTMF as above, where l is the number of players. Following the approach of [16], we describe the card operations needed in most games. Each card will be represented by an element c of the space $\mathcal{C}$. The value of the card is the unique $m \in \mathcal{M}$ such that $c = \mathcal{E}_h(m; r)$ for some value $r \in \mathcal{R}$. A card is called *open* if $c = \mathcal{E}_h(m; r)$, and r is a publically known value.

Creation of an open card. The creation of an open card requires only the input of one player. To create a card with type m, the player simply creates card value $c = \mathcal{E}_h(m; 1)$. This format can be read and understood by everyone, so verification of the precise card value is trivial.

Masking a card. Masking a card is the application of a cryptographic function that hides the value or type of the card. The homomorphic property of the encryption scheme allows an already masked card to be re-masked. Also, a zero knowledge proof exists that allows the verifier to show that the masked card is the mask of the original card. Thus masking is achieved by use of the verifiable re-masking function $\mathcal{E}'_h$ described above.

Creation of a private card. To create a private card, Alice privately creates a card m and then masks it to give $c = \mathcal{E}_h(m; r)$ which she then broadcasts to all other players. The purpose of broadcasting the masked card is to commit to the generated card and to prevent Alice from generating a wild card. The proof of knowledge here has to be slightly different to that described above since we need to show that the card is a mask of a valid card, i.e. m is a genuine card value. This proof is accomplished by the protocol of Cramer et. al. [5] in which one can show that c is the masking of an element of the required subset $\mathcal{M}$.

Creation of a random covered card. Unlike the scheme proposed by Schindelhauer [16], there is no trivial way of generating a random covered card unless the number of cards is equal to the number of plaintexts of the underlying encryption scheme. In our instantiations below the size of the underlying plaintext space is exponential and so this approach is not going to work.

There are two possible solutions to this problem:

- One option is to enable the entire underlying plaintext space $\mathcal{P}$ to be considered as card values. Hence, one would require a cryptographic hash function

$$H : \mathcal{P} \longrightarrow \mathcal{M}$$

  which maps plaintexts to possible card values. This assumes that every ciphertext corresponds to a valid encryption, which in our instantiations is a valid assumption.
- The second solution is to create a deck containing all the possible cards that could be randomly generated. Each player then shuffles and masks the deck and a card is chosen at random. This is a more costly method, but distributes the probability of selection correctly while giving no player a chance to influence the generation of the card. Operations on a desk, such as shuffling, are considered later.

Opening a card. To open a card, each player executes the verifiable decryption protocol. In the case of publicly opening a card, this information is broadcast to everyone. In the case of privately opening a card one player keeps their decryption information secret. Thus enabling them to read the card. The associated proofs in the verifiable decryption protocol are used to ensure that no player can sabotage the game by providing incorrect decryptions.

There are other possible card opening situations such as when a card needs to be opened by two players, N and S, but no others. Due to the broadcast nature of our network we need to solve this using a two stage process: One player, N say, first re-masks the card and provides a proof of the re-masking. The group minus N and S, then contribute their shares of the decryption of both cards. Player N contributes a decryption of the card to be opened by S, and S contributes a decryption of the card to be opened by N. Similar situations can be dealt with by adapting the above ideas.

4.2   Operations on a Deck

A deck D is modelled as a stack of cards of size n. There is no reason for there to be only one deck in the game, and each players hand could be considered a deck in itself. At the start of the game a player may create the deck. The notation for masking a card $\mathcal{E}_h(m; r)$ is often abused to also mean masking the entire deck in the form $\mathcal{E}_h(D; R)$, where $R = \{r_1 \ldots r_n\}$. In this case, each card is masked individually using the corresponding r from R.

Creating the deck. This operation is equivalent to the creation of several open cards which are then stacked and mask-shuffled. In order to ensure that the player who created the deck does not follow where the cards are, each player must mask-shuffle the deck before play can begin. Any player who does not mask-shuffle the deck could have all of their cards discovered by the collusion of his opponents.

Mask-shuffling the deck. We define a composite operation called mask-shuffle which both shuffles the deck and applies a mask to each individual card in the deck. To mask-shuffle the deck, the player must apply a permutation and then mask each card. If parts of the deck have not yet been masked (either due to the deck just being created or the player adding a card to the deck), the player knows what card has been masked to what value, i.e. it is *privately masked*. Therefore, it is often the case that the deck must be re-masked by all other players to ensure it is *publicly masked*.

Also, the mask-shuffle operation includes a verification stage to ensure the mask-shuffle has been performed correctly, i.e. the resulting deck is semantically the same as the initial deck. This is done using an honest verifier zero-knowledge proof, as in Figure 1.

For the purposes of this proof, it is necessary to keep track of all the permutations and Rs used during the initial masking of D to form D'. However, once the mask-shuffle has been verified as correct they may be discarded. Note that the chance of cheating has been reduced to $\frac{1}{2^s}$, so s should be suitably large to satisfy all parties.

Splitting the deck. Splitting a deck means that a player moves x cards from the top of the deck to the bottom, maintaining the order of the removed cards. This is a similar operation to shuffling the deck, in that a permutation can be used to represent the split. However, the proof needs to be extended in order to show that a split has occurred and *not* a regular shuffle. The proof is similar to the honest verifier zero knowledge proof used for shuffling the deck, on noticing that a split followed by a split is simply another split. Figure 2 shows the protocol in more detail.

Drawing a card from the deck. This operation is equivalent to privately opening a masked card. In this case, the player broadcasts to all players which card he is going to draw. The card is then privately opened once all players have agreed to share enough information to do so. In some instances a player might be restricted to drawing only from the top of the stack. As each player has a copy of the stack this would be possible. Note that if a player has just placed a card on the stack, the player who takes this card will be letting the other player know what card is in his hand. This is acceptable as it would be true in a real game of cards. However, the player who draws this card should re-mask his hand if he ever intends to discard a card. This prevents the other player knowing when he has discarded the card given to him. Discarding cards is considered later.

Discarding a card from hand. A players hand is made up of a series of masked cards which they know the decryption of. However, there are some times when an opponent knows what card you are holding at a certain point (for example, passing cards in Hearts). In this case, the player must always mask-shuffle his hand whenever he intends to discard a card from it. For example, if a player wants to insert a card from his hand into the deck, he must first *discard* this card and then insert it into the deck. This operation ensures that no information about a players hand is revealed.

Alice	Bob
A permutation σ is applied to the deck D, and each element is masked to create D' such that $$D' = \mathcal{E}_h(\sigma(D); R).$$ Alice publishes the mask-shuffled deck D'	
	Bob chooses a security parameter s and sends this to Alice.
Alice generates a set of permutations $\{\sigma_1', \ldots, \sigma_s'\}$ and sends D_i'' to Bob where $$D_i'' = \mathcal{E}_h(\sigma_i'(D'); R_i).$$	
	Bob chooses a random subset of $X \subset \{D_1'', \ldots, D_s''\}$ and sends this to Alice.
For all $D_i'' \in X$, Alice publishes $$\sigma_i', R_i$$ otherwise she publishes $$\sigma \circ \sigma_i', R_i'$$ where R_i' are the masking values used to mask D_i'' from D. These later values should be easily computable from R and R_i	Bob verifies that Alice has supplied the correct masking values and permutations to go from D' to D_i'' if $D_i'' \in X$ or from D to D_i'' if $D_i'' \notin X$.

Fig. 1. Mask-Shuffling the Deck

Rule control. The use of these set tests allow for rule control that are not available in real card games. For example, in whilst style games such as Hearts and Bridge, a player must "Follow Suit" whenever possible. A cheating player could choose to not follow suit for his own advantage. Although the player would discovered to be cheating at the end of the game, it would be preferable to discover the player to be cheating earlier.

By using the protocol of Cramer, Damgård and Schoenmakers [5] a player can always prove when playing a card that it comes from a given set. Hence, one can make sure that players follow rules, such as following suit, as the game progresses.

Alice	Bob
A split S is applied to the deck D, and each element is masked to create D' such that $$D' = \mathcal{E}_h(S(D); R).$$ Alice generates a further set of masked splits $\{S'_1, \ldots, S'_s\}$ and sends D''_i to Bob where $$D''_i = \mathcal{E}_h(S'_i(D'); R_i).$$	Bob chooses a security parameter s and sends this to Alice.
	Bob chooses a random subset of $X \subset \{D''_1, \ldots, D''_s\}$ and sends this to Alice.
For all $D''_i \in X$, Alice publishes $$S'_i, R_i$$ otherwise she publishes $$S \circ S'_i, R'_i$$ where R'_i are the masking values used to mask D''_i from D. These later values should be easily computable from R and R_i	Bob verifies that Alice has supplied the correct masking values and permutations to go from D' to D''_i if $D''_i \in X$ or from D to D''_i if $D''_i \notin X$.

Fig. 2. Splitting the Deck

Leaving the game. Although new players are unable to enter the game, players are able to leave the game whenever they wish. In order to do so, any cards that the player has in their hand must either be returned to the deck or opened and discarded (depending on the rules of the game). Then, each remaining player mask-shuffles the deck and their hands. Once this has been done, the departing player reveals their secret key and verifies it to be correct. This allows all the remaining cards to still be decrypted despite a player not being present. Note that it is not generally possible for the player to re-enter the game once they have left, however for our discrete logarithm based VTMF below we shall show that this is possible.

5   Instantiations of VTMF's

We now present two examples of a VTMF suitable for use in our card protocols. The first is based on the ElGamal encryption scheme whilst the second is based on Paillier's system [13]. Before presenting the instantiations we require the following sub-protocols, both of which are honest-verifier zero-knowledge and possess the "special soundness" property of [5].

Proof of Knowledge of Equality of Discrete Logarithms
The following protocol, due to Chaum and Pedersen [4], provides a proof, that if the verifier is given $x = g^\alpha$ and $y = h^\beta$ then the prover knows α and that $\alpha = \beta$, where g and h have order q.

- The prover sends the commitment $(a, b) = (g^\omega, h^\omega)$ to the verifier, for some random value $\omega \in \mathbb{Z}_q$.
- The verifier sends back a random challenge $c \in \mathbb{Z}_q$.
- The prover responds with $r = \omega + \alpha c \pmod{q}$.
- The verifier accepts the proof if and only if he verifies that $g^r = ax^c$ and $h^r = by^c$.

We shall denote this protocol by $CP(x, y, g, h; \alpha)$.

Proof of nth residuosity modulo n^2
The following protocol, due to Damgård and Jurik [8], provides a proof that a value $u \in \mathbb{Z}_{n^2}$ is a perfect n^{th} power and that the prover knows an nth root v.

- The prover sends the commitment $a = r^n \pmod{n^2}$ for some random value $r \in \mathbb{Z}_{n^2}$.
- The verifier sends back a random challenge $c \in \mathbb{Z}_n$.
- The prover responds with $z = r \cdot v^c \pmod{n^2}$.
- The verifier accepts the proof if and only if $z^n = a \cdot u^c \pmod{n^2}$.

We shall denote this protocol by $DJ(n, u; v)$.

When we apply these protocols the random challenge c will be created from hashing the commitment and the public input values, thus making the protocol non-interactive. In such a situation we also denote the transcript of the proofs by $CP(x, y, g, h; \alpha)$ and $DJ(n, u; v)$.

5.1   Discrete Logarithm Based VTMF

The parties first agree on a finite abelian group G in which the Decision Diffie–Hellman assumption is hard. The users agree on a generate $g \in G$ of order q and set

$$\mathcal{M} = G, \ \mathcal{R} = \mathbb{Z}_q \text{ and } \mathcal{C} = G \times G.$$

Key Generation Protocol

Each player generates a random private key $x_i \in \mathbb{Z}_q$ and publishes $h_i = g^{x_i}$. The public key h is formed from

$$h = \prod_{i=1}^{l} h_i.$$

Verifiable Masking Protocol

The verifiable masking protocol is given by the ElGamal encryption operation

$$(m, r) \longrightarrow (c_1 = g^r, c_2 = h^r \cdot m).$$

The value (c_1, c_2) is published and is accompanied by the proof

$$CP(c_1, c_2/m, g, h; r).$$

That this encryption function is semantically secure under the Decision Diffie-Hellman assumption is an undergraduate exercise.

Verifiable Re-masking Protocol

Given a ciphertext (c_1, c_2) this is re-masked by computing

$$((c_1, c_2), r)) \longrightarrow (c_1' = c_1 \cdot g^r, c_2' = c_2 \cdot h_r).$$

The value (c_1', c_2') is now published and accompanied by the proof

$$CP(c_1'/c_1, c_2'/c_2, g, h; r).$$

Verifiable Decryption Protocol

Given (c_1, c_2), user i publishes $d_i = c_1^{x_i}$ along with a proof $CP(d_i, h_i, c_1, g; x_i)$. Given these values any player can decrypt (c_1, c_2) by computing

$$m = c_2 / \prod_{i=1}^{l} d_i.$$

Joining The Game

Using this VTMF a player may join an existing card game by generating a private key x_{l+1} and public commitment h_{l+1} as above. The public key h for the game is then replaced by

$$h' = h \cdot h_{l+1}.$$

Then each card $c = (c_1, c_2)$ is masked by the new player, under the new public key, by setting

$$c' = (c_1', c_2') = (c_1, c_1^{x_{l+1}} \cdot c_2)$$

along with a proof $CP(h_{l+1}, c_2'/c_2, g, c_1; x_{l+1})$.

5.2 Factoring Based VTMF

Our factoring based assumption will make use of the Paillier probabilistic encryption function [13], which is known to be semantically secure against passive attacks under the n-th residuosity assumption.

Key Generation Protocol
All parties execute a protocol such as that of Boneh and Franklin, see [2,3], to generate a shared RSA modulus $n = p \cdot q$, where each player only knows the value of (p_i, q_i) where $p = \sum_{i=1}^{l} p_i$ and $q = \sum_{i=1}^{l} q_i$. The value n is published and the users generate a share of $\phi = (p-1)(q-1)$ by setting

$$x_i = \begin{cases} n - (p_1 + q_1) - 1 & \text{If } i = 1, \\ -(p_i + q_i) & \text{If } i \geq 2. \end{cases}$$

Note that $\phi = \sum_{i=1}^{l} x_i$. The uses then commit to the value x_i by publishing $h_i = g^{x_i} \pmod{n^2}$, where $g = 1 + n$. We then set publically

$$h = \prod_{i=1}^{m} h_i - 1 \pmod{n^2} = g^{\phi} - 1 \pmod{n^2}.$$

Verifiable Masking Protocol
The verifiable masking protocol is given by Paillier's encryption operation

$$(m, r) \longrightarrow c = g^m r^n \pmod{n^2}.$$

The value c is published and is accompanied by the proof $DJ(n, c/g^m; r)$.

Verifiable Re-masking Protocol
Given a ciphertext c this is re-masked by computing

$$(c, r) \longrightarrow c' = r^n c \pmod{n^2}.$$

The value c' is now published and accompanied by the proof $DJ(n, c'/c; r)$.

Verifiable Decryption Protocol
Given c, user i publishes the value $d_i = c^{x_i} \pmod{n^2}$ along with a proof $CP(d_i, h_i, c, g; x_i)$. Given these values any player can decrypt c by computing

$$\frac{1}{h}\left(\left(\prod_{i=1}^{l} d_i\right) - 1 \pmod{n^2}\right) \pmod{n} = \frac{c^{\phi} - 1 \pmod{n^2}}{g^{\phi} - 1 \pmod{n^2}} \pmod{n}$$

$$= m.$$

6 Conclusion

We have introduced the concept of a VTMF and shown how this can be used to implement a secure multi-party card game with arbitrary number of players and no trusted centre. The number of bits to represent each card in our system is significantly smaller than in previous schemes. We have then gone on to show how a VTMF can be implemented either under a discrete logarithm type assumption or under a factoring based assumption.

The authors would like to thank the referee's for very useful comments which improved the readability of the paper. The authors would also like to thank F. Vercauteren for useful conversations whilst the work in this paper was carried out.

References

1. I. Banary and Z. Füredi. Mental poker with three or more players. *Information and Control*, **59**, 84–93, 1983.
2. D. Boneh and M. Franklin. Efficient generation of shared RSA keys. *Advances in Cryptology – CRYPTO '97*, Springer-Verlag LNCS 1233, 425–439, 1997.
3. D. Boneh and M. Franklin. Efficient generation of shared RSA keys. **J. ACM**, **48**, 702–722, 2001.
4. D. Chaum and T.P. Pedersen. Wallet databases with observers. *Advances in Cryptology – CRYPTO '92*, Springer-Verlag LNCS 740, 89–105, 1993.
5. R. Cramer, I. Damgård and B. Schoenmakers. Proofs of partial knowledge. *Advances in Cryptology – CRYPTO '94*, Springer-Verlag LNCS 839, 174–187, 1994.
6. C. Crépeau. A secure poker protocol that minimises the effect of player coalitions. *Advances in Cryptology – CRYPTO '85*, Springer-Verlag LNCS 218, 73–86, 1986.
7. C. Crépeau. A zero-knowledge poker protocol that achieves confidentiality of the player's strategy, or how to achieve an electronic poker face. *Advances in Cryptology – CRYPTO '86*, Springer-Verlag LNCS 263, 239–247, 1987.
8. I. Damgård and M. Jurik. A generalisation, a simplification and some applications of Paillier's probabilistic public-key system. *Public Key Cryptography – PKC 2001*, Springer-Verlag LNCS 1992, 119–136, 2001.
9. S. Fortune and M. Merrit. Poker protocols. *Advances in Cryptology – CRYPTO '84*, Springer-Verlag LNCS 196, 454–464, 1985.
10. S. Goldwasser and S. Micali. Probabilistic encryption and how to play mental poker keeping secret all partial information. *STOC '82*, 365–377, 1982.
11. S. Goldwasser and S. Micali. Probabilistic encryption. *Journal of Computer and Systems Sciences*, **28**, 270–299, 1984.
12. O. Goldreich, S. Micali and A. Widgerson. How to play any mental game or a completeness theorem for protocols with honest majority. *STOC '87*, 218–229, 1987.
13. P. Paillier. Public key cryptosystems based on composite residue classes. *Advances in Cryptology – EuroCrypt '99*, Springer-Verlag LNCS 1592, 223–238, 1999.
14. D. Pointcheval and J. Stern. Security arguments for digital signatures and blind signatures. *Journal of Cryptology*, **13**, 361–396, 2000.
15. A. Shamir, R. Rivest and L. Adleman. Mental Poker. *MIT Technical Report*, 1978.
16. C. Schindelhauer. A toolbox for mental card games. *Technical Report*, Uni Lubeck, 1998.
17. M. Yung. Subscription to a public key, the secret blocking and the multi-party poker game. *Advances in Cryptology – CRYPTO '84*, Springer-Verlag LNCS 196, 439–453, 1985.

Author Index

Al-Riyami, Sattam S.   332

Barnett, Adam   370
Blackburn, Simon R.   264
Borissov, Yuri   82
Braeken, An   82

Davenport, James H.   207
Dent, Alexander W.   133

Fagoonee, Lina   24
Farrell, Patrick Guy   1

Geißler, Katharina   223
Gomez-Perez, Domingo   264
Granger, Robert   190
Gutierrez, Jaime   264

Hattori, Mitsuhiro   290
Helleseth, Tor   52, 67
Hirose, Shoichi   290
Holt, Andrew J.   207
Honary, Bahram   24
Horadam, K.J.   115

Imai, Hideki   35
Iwata, Tetsu   306

Johansson, Thomas   66

Kiltz, Eike   152
Kurosawa, Kaoru   306

Lee, Pil Joong   167

Malone-Lee, John   152
Mathiassen, John Erik   67
Mitchell, Chris J.   360
Molland, Håvard   67

Nascimento, Anderson C.A.   35
Niederreiter, Harald   183
Niemi, Valtteri   303
Nikova, Svetla   82

Pasalic, Enes   93
Paterson, Kenneth G.   332
Petrides, George   234
Preneel, Bart   82
Price, Geraint   319

Rosnes, Eirik   4

Schaathun, Hans Georg   52
Schindler, Werner   245, 276
Shparlinski, Igor E.   183, 264
Smart, Nigel P.   370

Walter, Colin D.   245
Walton, Richard   125
Winter, Andreas   35

Yoshida, Susumu   290
Ytrehus, Øyvind   4
Yum, Dae Hyun   167

Lecture Notes in Computer Science

For information about Vols. 1–2821
please contact your bookseller or Springer-Verlag

Vol. 2822: N. Bianchi-Berthouze (Ed.), Databases in Networked Information Systems. Proceedings, 2003. X, 271 pages. 2003.

Vol. 2823: A. Omondi, S. Sedukhin (Eds.), Advances in Computer Systems Architecture. Proceedings, 2003. XIII, 409 pages. 2003.

Vol. 2824: Z. Bellahsène, A.B. Chaudhri, E. Rahm, M. Rys, R. Unland (Eds.), Database and XML Technologies. Proceedings, 2003. X, 283 pages. 2003.

Vol. 2825: W. Kuhn, M. Worboys, S. Timpf (Eds.), Spatial Information Theory. Proceedings, 2003. XI, 399 pages. 2003.

Vol. 2826: A. Krall (Ed.), Software and Compilers for Embedded Systems. Proceedings, 2003. XI, 403 pages. 2003.

Vol. 2827: A. Albrecht, K. Steinhöfel (Eds.), Stochastic Algorithms: Foundations and Applications. Proceedings, 2003. VIII, 167 pages. 2003.

Vol. 2828: A. Lioy, D. Mazzocchi (Eds.), Communications and Multimedia Security. Proceedings, 2003. VIII, 265 pages. 2003.

Vol. 2829: A. Cappelli, F. Turini (Eds.), AI*IA 2003: Advances in Artificial Intelligence. Proceedings, 2003. XIV, 552 pages. 2003. (Subseries LNAI).

Vol. 2830: F. Pfenning, Y. Smaragdakis (Eds.), Generative Programming and Component Engineering. Proceedings, 2003. IX, 397 pages. 2003.

Vol. 2831: M. Schillo, M. Klusch, J. Müller, H. Tianfield (Eds.), Multiagent System Technologies. Proceedings, 2003. X, 229 pages. 2003. (Subseries LNAI).

Vol. 2832: G. Di Battista, U. Zwick (Eds.), Algorithms – ESA 2003. Proceedings, 2003. XIV, 790 pages. 2003.

Vol. 2833: F. Rossi (Ed.), Principles and Practice of Constraint Programming – CP 2003. Proceedings, 2003. XIX, 1005 pages. 2003.

Vol. 2834: X. Zhou, S. Jähnichen, M. Xu, J. Cao (Eds.), Advanced Parallel Processing Technologies. Proceedings, 2003. XIV, 679 pages. 2003.

Vol. 2835: T. Horváth, A. Yamamoto (Eds.), Inductive Logic Programming. Proceedings, 2003. X, 401 pages. 2003. (Subseries LNAI).

Vol. 2836: S. Qing, D. Gollmann, J. Zhou (Eds.), Information and Communications Security. Proceedings, 2003. XI, 416 pages. 2003.

Vol. 2837: N. Lavrač, D. Gamberger, H. Blockeel, L. Todorovski (Eds.), Machine Learning: ECML 2003. Proceedings, 2003. XVI, 504 pages. 2003. (Subseries LNAI).

Vol. 2838: N. Lavrač, D. Gamberger, L. Todorovski, H. Blockeel (Eds.), Knowledge Discovery in Databases: PKDD 2003. Proceedings, 2003. XVI, 508 pages. 2003. (Subseries LNAI).

Vol. 2839: A. Marshall, N. Agoulmine (Eds.), Management of Multimedia Networks and Services. Proceedings, 2003. XIV, 532 pages. 2003.

Vol. 2840: J. Dongarra, D. Laforenza, S. Orlando (Eds.), Recent Advances in Parallel Virtual Machine and Message Passing Interface. Proceedings, 2003. XVIII, 693 pages. 2003.

Vol. 2841: C. Blundo, C. Laneve (Eds.), Theoretical Computer Science. Proceedings, 2003. XI, 397 pages. 2003.

Vol. 2842: R. Gavaldà, K.P. Jantke, E. Takimoto (Eds.), Algorithmic Learning Theory. Proceedings, 2003. XI, 313 pages. 2003. (Subseries LNAI).

Vol. 2843: G. Grieser, Y. Tanaka, A. Yamamoto (Eds.), Discovery Science. Proceedings, 2003. XII, 504 pages. 2003. (Subseries LNAI).

Vol. 2844: J.A. Jorge, N.J. Nunes, J.F. e Cunha (Eds.), Interactive Systems. Proceedings, 2003. XIII, 429 pages. 2003.

Vol. 2846: J. Zhou, M. Yung, Y. Han (Eds.), Applied Cryptography and Network Security. Proceedings, 2003. XI, 436 pages. 2003.

Vol. 2847: R. de Lemos, T.S. Weber, J.B. Camargo Jr. (Eds.), Dependable Computing. Proceedings, 2003. XIV, 371 pages. 2003.

Vol. 2848: F.E. Fich (Ed.), Distributed Computing. Proceedings, 2003. X, 367 pages. 2003.

Vol. 2849: N. García, J.M. Martínez, L. Salgado (Eds.), Visual Content Processing and Representation. Proceedings, 2003. XII, 352 pages. 2003.

Vol. 2850: M.Y. Vardi, A. Voronkov (Eds.), Logic for Programming, Artificial Intelligence, and Reasoning. Proceedings, 2003. XIII, 437 pages. 2003. (Subseries LNAI)

Vol. 2851: C. Boyd, W. Mao (Eds.), Information Security. Proceedings, 2003. XI, 443 pages. 2003.

Vol. 2852: F.S. de Boer, M.M. Bonsangue, S. Graf, W.-P. de Roever (Eds.), Formal Methods for Components and Objects. Proceedings, 2003. VIII, 509 pages. 2003.

Vol. 2853: M. Jeckle, L.-J. Zhang (Eds.), Web Services – ICWS-Europe 2003. Proceedings, 2003. VIII, 227 pages. 2003.

Vol. 2854: J. Hoffmann, Utilizing Problem Structure in Planning. XIII, 251 pages. 2003. (Subseries LNAI)

Vol. 2855: R. Alur, I. Lee (Eds.), Embedded Software. Proceedings, 2003. X, 373 pages. 2003.

Vol. 2856: M. Smirnov, E. Biersack, C. Blondia, O. Bonaventure, O. Casals, G. Karlsson, George Pavlou, B. Quoitin, J. Roberts, I. Stavrakakis, B. Stiller, P. Trimintzios, P. Van Mieghem (Eds.), Quality of Future Internet Services. IX, 293 pages. 2003.

Vol. 2857: M.A. Nascimento, E.S. de Moura, A.L. Oliveira (Eds.), String Processing and Information Retrieval. Proceedings, 2003. XI, 379 pages. 2003.

Vol. 2858: A. Veidenbaum, K. Joe, H. Amano, H. Aiso (Eds.), High Performance Computing. Proceedings, 2003. XV, 566 pages. 2003.

Vol. 2859: B. Apolloni, M. Marinaro, R. Tagliaferri (Eds.), Neural Nets. Proceedings, 2003. X, 376 pages. 2003.

Vol. 2860: D. Geist, E. Tronci (Eds.), Correct Hardware Design and Verification Methods. Proceedings, 2003. XII, 426 pages. 2003.

Vol. 2861: C. Bliek, C. Jermann, A. Neumaier (Eds.), Global Optimization and Constraint Satisfaction. Proceedings, 2002. XII, 239 pages. 2003.

Vol. 2862: D. Feitelson, L. Rudolph, U. Schwiegelshohn (Eds.), Job Scheduling Strategies for Parallel Processing. Proceedings, 2003. VII, 269 pages. 2003.

Vol. 2863: P. Stevens, J. Whittle, G. Booch (Eds.), «UML» 2003 – The Unified Modeling Language. Proceedings, 2003. XIV, 415 pages. 2003.

Vol. 2864: A.K. Dey, A. Schmidt, J.F. McCarthy (Eds.), UbiComp 2003: Ubiquitous Computing. Proceedings, 2003. XVII, 368 pages. 2003.

Vol. 2865: S. Pierre, M. Barbeau, E. Kranakis (Eds.), Ad-Hoc, Mobile, and Wireless Networks. Proceedings, 2003. X, 293 pages. 2003.

Vol. 2867: M. Brunner, A. Keller (Eds.), Self-Managing Distributed Systems. Proceedings, 2003. XIII, 274 pages. 2003.

Vol. 2868: P. Perner, R. Brause, H.-G. Holzhütter (Eds.), Medical Data Analysis. Proceedings, 2003. VIII, 127 pages. 2003.

Vol. 2869: A. Yazici, C. Şener (Eds.), Computer and Information Sciences – ISCIS 2003. Proceedings, 2003. XIX, 1110 pages. 2003.

Vol. 2870: D. Fensel, K. Sycara, J. Mylopoulos (Eds.), The Semantic Web - ISWC 2003. Proceedings, 2003. XV, 931 pages. 2003.

Vol. 2871: N. Zhong, Z.W. Raś, S. Tsumoto, E. Suzuki (Eds.), Foundations of Intelligent Systems. Proceedings, 2003. XV, 697 pages. 2003. (Subseries LNAI)

Vol. 2873: J. Lawry, J. Shanahan, A. Ralescu (Eds.), Modelling with Words. XIII, 229 pages. 2003. (Subseries LNAI)

Vol. 2874: C. Priami (Ed.), Global Computing. Proceedings, 2003. XIX, 255 pages. 2003.

Vol. 2875: E. Aarts, R. Collier, E. van Loenen, B. de Ruyter (Eds.), Ambient Intelligence. Proceedings, 2003. XI, 432 pages. 2003.

Vol. 2876: M. Schroeder, G. Wagner (Eds.), Rules and Rule Markup Languages for the Semantic Web. Proceedings, 2003. VII, 173 pages. 2003.

Vol. 2877: T. Böhme, G. Heyer, H. Unger (Eds.), Innovative Internet Community Systems. Proceedings, 2003. VIII, 263 pages. 2003.

Vol. 2878: R.E. Ellis, T.M. Peters (Eds.), Medical Image Computing and Computer-Assisted Intervention - MICCAI 2003. Part I. Proceedings, 2003. XXXIII, 819 pages. 2003.

Vol. 2879: R.E. Ellis, T.M. Peters (Eds.), Medical Image Computing and Computer-Assisted Intervention - MICCAI 2003. Part II. Proceedings, 2003. XXXIV, 1003 pages. 2003.

Vol. 2880: H.L. Bodlaender (Ed.), Graph-Theoretic Concepts in Computer Science. Proceedings, 2003. XI, 386 pages. 2003.

Vol. 2881: E. Horlait, T. Magedanz, R.H. Glitho (Eds.), Mobile Agents for Telecommunication Applications. Proceedings, 2003. IX, 297 pages. 2003.

Vol. 2883: J. Schaeffer, M. Müller, Y. Björnsson (Eds.), Computers and Games. Proceedings, 2002. XI, 431 pages. 2003.

Vol. 2884: E. Najm, U. Nestmann, P. Stevens (Eds.), Formal Methods for Open Object-Based Distributed Systems. Proceedings, 2003. X, 293 pages. 2003.

Vol. 2885: J.S. Dong, J. Woodcock (Eds.), Formal Methods and Software Engineering. Proceedings, 2003. XI, 683 pages. 2003.

Vol. 2886: I. Nyström, G. Sanniti di Baja, S. Svensson (Eds.), Discrete Geometry for Computer Imagery. Proceedings, 2003. XII, 556 pages. 2003.

Vol. 2887: T. Johansson (Ed.), Fast Software Encryption. Proceedings, 2003. IX, 397 pages. 2003.

Vol. 2888: R. Meersman, Zahir Tari, D.C. Schmidt et al. (Eds.), On The Move to Meaningful Internet Systems 2003: CoopIS, DOA, and ODBASE. Proceedings, 2003. XXI, 1546 pages. 2003.

Vol. 2889: Robert Meersman, Zahir Tari et al. (Eds.), On The Move to Meaningful Internet Systems 2003: OTM 2003 Workshops. Proceedings, 2003. XXI, 1096 pages. 2003.

Vol. 2891: J. Lee, M. Barley (Eds.), Intelligent Agents and Multi-Agent Systems. Proceedings, 2003. X, 215 pages. 2003. (Subseries LNAI)

Vol. 2893: J.-B. Stefani, I. Demeure, D. Hagimont (Eds.), Distributed Applications and Interoperable Systems. Proceedings, 2003. XIII, 311 pages. 2003.

Vol. 2894: C.S. Laih (Ed.), Advances in Cryptology - ASIACRYPT 2003. Proceedings, 2003. XIII, 543 pages. 2003.

Vol. 2895: A. Ohori (Ed.), Programming Languages and Systems. Proceedings, 2003. XIII, 427 pages. 2003.

Vol. 2897: O. Balet, G. Subsol, P. Torguet (Eds.), Virtual Storytelling. Proceedings, 2003. XI, 240 pages. 2003.

Vol. 2898: K.G. Paterson (Ed.), Cryptography and Coding. Proceedings, 2003. IX, 385 pages. 2003.

Vol. 2899: G. Ventre, R. Canonico (Eds.), Interactive Multimedia on Next Generation Networks. Proceedings, 2003. XIV, 420 pages. 2003.

Vol. 2901: F. Bry, N. Henze, J. Maluszyński (Eds.), Principles and Practice of Semantic Web Reasoning. Proceedings, 2003. X, 209 pages. 2003.

Vol. 2902: F. Moura Pires, S. Abreu (Eds.), Progress in Artificial Intelligence. Proceedings, 2003. XV, 504 pages. 2003. (Subseries LNAI).

Vol. 2903: T.D. Gedeon, L.C.C. Fung (Eds.), AI 2003: Advances in Artificial Intelligence. Proceedings, 2003. XVI, 1075 pages. 2003. (Subseries LNAI).

Vol. 2904: T. Johansson, S. Maitra (Eds.), Progress in Cryptology – INDOCRYPT 2003. Proceedings, 2003. XI, 431 pages. 2003.

Vol. 2905: A. Sanfeliu, J. Ruiz-Shulcloper (Eds.), Progress in Pattern Recognition, Speech and Image Analysis. Proceedings, 2003. XVII, 693 pages. 2003.

Vol. 2916: C. Palamidessi (Ed.), Logic Programming. Proceedings, 2003. XII, 520 pages. 2003.